Rauert

Rouart

Handbook of Environmental Isotope Geochemistry

VOLUME 1

The Terrestrial Environment, A

HANBOOK OF ENVIRONMENTAL ISOTOPE GEOCHEMISTRY

P. FRITZ and J.Ch. FONTES (Editors)

1. THE TERRESTRIAL ENVIRONMENT, A
2. THE TERRESTRIAL ENVIRONMENT, B
3. THE MARINE ENVIRONMENT, A
4. THE MARINE ENVIRONMENT, B
5. THE HIGH TEMPERATURE ENVIRONMENT

Handbook of Environmental Isotope Geochemistry

Edited by

P. FRITZ

Department of Earth Science, University of Waterloo,
Waterloo, Ontario, Canada

and

J.Ch. FONTES

Laboratoire d'Hydrologie et de Géochimie Isotopique, Université
Paris-Sud, Orsay, France

VOLUME 1

The Terrestrial Environment, A

ELSEVIER SCIENTIFIC PUBLISHING COMPANY
Amsterdam — Oxford — New York   1980

ELSEVIER SCIENTIFIC PUBLISHING COMPANY
335 Jan van Galenstraat
P.O. Box 211, 1000 AE Amsterdam, The Netherlands

Distributors for the United States and Canada:

ELSEVIER/NORTH-HOLLAND INC.
52, Vanderbilt Avenue
New York, N.Y. 10017

Library of Congress Cataloging in Publication Data
Main entry under title:

Handbook of environmental isotope geochemistry.

    Includes bibliographies and indexes.
    CONTENTS: v. 1A. The terrestrial environment.
    1. Isotope geology. 2. Environmental
chemistry. 3. Geochemistry. I. Fritz, Peter,
1937-   II. Fontes, Jean-Charles, 1936-
QE501.4.N9H36   551.9   79-21332
ISBN 0-444-41781-8

ISBN 0-444-41780-X (Vol. 1)
ISBN 0-444-41781-8 (Series)

© Elsevier Scientific Publishing Company, 1980
All rights reserved. No part of this publication may be reproduced, stored in a retrieval system or transmitted in any form or by any means, electronic, mechanical, photocopying, recording or otherwise, without the prior written permission of the publisher, Elsevier Scientific Publishing Company, P.O. Box 330, 1000 AH Amsterdam, The Netherlands

Printed in The Netherlands

PREFACE

Geological Sciences experienced during the past decades very profound changes during which its traditionally descriptive tools were increasingly complemented by rigorous analytical techniques. Chemistry and physics had only a marginal input but play now a crucial role in most geological investigations. This change is also typified by the acceptance of environmental isotope techniques. The theoretical and analytical aspects of this tool were developed by physicists and chemists but applications are mainly in the geologic environment. Naturally occurring stable and radioactive isotopes act thereby as tracers for origin and time.

Early research in isotope geology focussed on dating of terrestrial materials and paleoclimatic studies. The techniques used and results obtained have been widely publicized and are now an integral part of most curricula at universities. The practical importance of these studies was limited, however, and only with the expansion of geochemistry, hydrogeology, and environmental geology within the geosciences did isotope techniques reach the important position they hold today.

Environmental geochemistry has thus as its base the knowledge of classical geology but embraces chemistry and physics as well as many aspects of the life sciences. The field is extremely large and, with the exception of some introductory texts, no comprehensive treatment of the entire subject has yet been attempted. Therefore we decided to present a collection of papers which focus on environmental isotopes in the terrestrial and marine environments. The chapters collected in this series present environmental isotope techniques as complementary and/or basic tools in studies dealing with the dynamics of the geosphere and its interaction with the hydrosphere and biologic systems. This is done without covering in a uniform manner all aspects of isotope geochemistry which may range from the genesis and biogeochemistry of a forest soil to igneous petrology. There will be weaknesses and there will be overlap. However, we have attempted to subdivide the volumes into major subject areas and asked the authors to write each chapter as a virtually self-contained unit. Thus, where the integrity of a chapter would have been compromised by relying too much on data presented elsewhere overlap was considered necessary.

Progress in environmental isotope geochemistry has been remarkable in some areas, notably in isotope hydrology, and slower in others. Thus, some chapters in this series will reflect a well established state of the art whereas others have more the nature of a progress report. However, all should summarize what is known, provide extensive reference lists, and address both the readers relatively new in the field but interested in applications and interpretations as well as scientists which need basic data for reference and comparison. We thus anticipate to find among the readers of this series geologists, hydrologists, geochemists, soil scientists, as well as those who are concerned with our environment, and to whom these isotope techniques could offer new approaches to solve and understand environmental problems.

This collection of papers will be presented in five volumes and will appear under the title of "Handbook of Environmental Isotope Geochemistry". This first volume covers important domains of the continental environment. Isotope hydrology and aqueous geochemistry are emphasized but a first overview on carbon, sulphur and nitrogen isotopes in terrestrial systems is also given. The second volume will deal with additional topics related to the terrestrial environment, whereas Volume 3 and 4 are devoted to the modern and ancient marine environment respectively. Volume 5 will focus on higher-temperature processes and treat topics of interest to economic and hardrock geologists.

To be accessible to a wide readership the collection of manuscripts from different authors under one title demands a certain selectivity, a hierarchy of presentation and documentation as well as completeness and clarity of presentation. The authors who are specialists in their own field and are used to present their work in lectures and public discussions had a difficult task. From our point of view, however, the objectives were met. It remains now for the readers to judge the value of this series but we gratefully acknowledge the efforts made by the authors, as without their collaboration the presentation of these volumes would not have been possible.

J.Ch. Fontes
P. Fritz

Paris and Waterloo, Ont.
October, 1979

LIST OF CONTRIBUTORS

B. BUCHARDT	University of Copenhagen, Copenhagen, Denmark
P. DEINES	Department of Geosciences, Pennsylvania State University, University Park, Pennsylvania, U.S.A.
J.Ch. FONTES	Laboratoire d'Hydrologie et de Géochimie Isotopique, Université Paris-Sud, Orsay, France
P. FRITZ	Department of Earth Science, University of Waterloo, Waterloo, Ontario, Canada
J.R. GAT	Department of Isotope Research, Weizmann Institute of Science, Rehovot, Israel
J.R. HULSTON	Institute of Nuclear Sciences, Lower Hutt, New Zealand
H.R. KROUSE	Department of Physics, University of Calgary, Calgary, Alberta, Canada
R. LÉTOLLE	Laboratoire de Géologie Dynamique, Université Pierre et Marie Curie, Paris, France
W.G. MOOK	Physics Laboratory, University of Groningen, Groningen, The Netherlands
H. MOSER	Institut für Radiohydrometrie, Gesellschaft für Strahlen und Umweltforschung, München, Federal Republic of Germany
J.K. OSMOND	Department of Geology, Florida State University, Tallahassee, Florida, U.S.A.
F.J. PEARSON, Jr.	U.S. Geological Survey, Reston, Virginia, U.S.A.
C.T. RIGHTMIRE	U.S. Geological Survey, Reston, Virginia, U.S.A.
S.M. SAVIN	Department of Earth Sciences, Case Western Reserve University, Cleveland, Ohio, U.S.A.
W. STICHLER	Institut für Radiohydrometrie, Gesellschaft für Strahlen und Umweltforschung, München, Federal Republic of Germany
A.H. TRUESDELL	U.S. Geological Survey, Menlo Park, California, U.S.A.

CONTENTS

Preface V
List of Contributors VII

INTRODUCTION 1
        P. Fritz and J.Ch. Fontes
Definitions 4
Standards 11
References 17

CHAPTER 1. THE ISOTOPES OF HYDROGEN AND OXYGEN IN PRECIPITATION 21
        J.R. Gat
Introduction 21
Tritium in atmospheric waters 22
Stable isotope distribution in atmospheric waters: data 29
Models of the isotope fractionation during evaporation and condensation of water in the atmosphere 33
In-storm variation of isotopic composition, cloud models and hailstone studies ... 37
Stable isotope distribution in atmospheric waters: the global model 40
References 44

CHAPTER 2. CARBON-14 IN HYDROGEOLOGICAL STUDIES 49
        W.G. Mook
Introduction 49
The abundance of ^{14}C 49
The ^{14}C age determination 52
^{14}C dating in groundwater 53
Summary 70
References 71

CHAPTER 3. ENVIRONMENTAL ISOTOPES IN GROUNDWATER HYDROLOGY 75
        J.Ch. Fontes
Introduction 75
Basic principles 76
Groundwater recharge 82
Relations between surface- and groundwaters 98
Mechanism and components of the run-off 102
Leakage between aquifers 108
Isotope hydrology of fractured rocks 113

Mechanism of salinization 121
Groundwater dating 125
Conclusions 132
References 134

CHAPTER 4. ENVIRONMENTAL ISOTOPES IN ICE AND SNOW 141
H. Moser and W. Stichler

Introduction 141
Isotope content of a snow cover in accretion 142
Isotope distribution during the reduction of a temperate snow cover 148
Isotope variations in the transition of snow to glacier ice 154
Snow and ice isotope hydrology 163
Historical glaciology 169
References 174

CHAPTER 5. ISOTOPIC EVIDENCE ON ENVIRONMENTS OF GEOTHERMAL SYSTEMS 179
A.H. Truesdell and J.R. Hulston

Introduction 179
Isotope hydrology of geothermal systems 180
Geothermometry 195
Isotopic dating of geothermal waters 207
Origin of chemical constituents 211
Solid phase studies 214
Summary 216
Appendix — Methods of collection and analysis 217
References 219

CHAPTER 6. SULPHUR AND OXYGEN ISOTOPES IN AQUEOUS SULPHUR COMPOUNDS 227
F.J. Pearson, Jr. and C.T. Rightmire

Introduction 227
Isotope fractionation 227
Geochemistry and isotope distribution of aqueous sulphur compounds 234
Field studies of groundwater systems 241
Summary 254
References 255

CHAPTER 7. URANIUM DISEQUILIBRIUM IN HYDROLOGIC STUDIES 259
J.K. Osmond

Introduction 259
Isotopic fractionation of ^{234}U 261
Mixing studies: continental waters 266
Mixing and uranium balances: marine waters 270
Aquifer interactions 272
Age dating 276
Summary 277
Appendix — Analytical techniques for ^{234}U and ^{238}U analysis 278
References 279

CHAPTER 8. OXYGEN AND HYDROGEN ISOTOPE EFFECTS IN LOW-TEMPERATURE MINERAL-WATER INTERACTIONS 283
S.M. Savin

Introduction 283
Isotopic fractionations between minerals and water 284
Isotope effects during weathering and soil formation 294
Isotopic studies of marine sedimentation, halmyrolysis, authigenesis and early diagenesis 298
Evaporite formation 308
Later diagenetic processes 310
Serpentinization of ultramafic rocks 317
Effect of mineral-water interaction on the isotopic composition of pore water 319
References 321

CHAPTER 9. THE ISOTOPIC COMPOSITION OF REDUCED ORGANIC CARBON 329
P. Deines

Introduction 329
Photosynthesis and the carbon isotopic composition of plants 329
The carbon isotopic composition of organic matter in sediments 344
The carbon isotopic composition of fossil fuels 360
The carbon isotopic composition of atmospheric compounds 384
References 393

CHAPTER 10. NITROGEN-15 IN THE NATURAL ENVIRONMENT 407
R. Létolle

Introduction 407
^{15}N in nature 410
Isotope fractionations 413
^{15}N in organic matter and soils 417
^{15}N in nitrates 420
^{15}N in the hydrosphere 424
References 429

CHAPTER 11. SULPHUR ISOTOPES IN OUR ENVIRONMENT 435
H.R. Krouse

Introduction 435
Terrestrial sulphur isotope abundances and cycling of mobile sulphur compounds .. 435
Elucidation of sources, mixing, and dispersion of sulphur compounds 440
Sulphur isotope fractionation during transformations of atmospheric and aqueous sulphur compounds 453
Sulphur isotope fractionation in the pedosphere 461
Sulphur isotopes elucidate uptake of industrial sulphur compounds by vegetation .. 463
Evaluation of anthropogenic and natural sources of sulphur compounds 465
Summary 466
References 467

CHAPTER 12. ENVIRONMENTAL ISOTOPES AS ENVIRONMENTAL AND CLIMATOLOGICAL INDICATORS 473
B. Buchardt and P. Fritz

Introduction 473
The carbon and oxygen isotope composition of freshwater shells 474

Freshwater lakes and sediments 484
Deuterium in organic matter as paleoclimatic indicators 495
References 500

References Index 505
Subject Index 532

INTRODUCTION

P. FRITZ and J.Ch. FONTES

The 92 naturally occurring elements compromise more than 1000 isotopes. Most occur in terrestrial compounds in trace amounts only but some are sufficiently abundant to be determined quantitatively through routine analyses. The distribution of the different isotopes of the same element between reacting chemical compounds or coexisting phases is not uniform because of the slightly different geochemical and physical behaviour. Responsible are differences in mass and energy contents; the larger these are between isotopes of a given element the more significant will be the isotope fractionation. This effect is more pronounced for light elements although some heavy elements exhibit large enough isotope effects to be of geochemical interest, as for example the uranium isotopes ^{234}U and ^{238}U (J.K. Osmond, this volume, Chapter 7).

Most environmental isotope studies have focussed on the light elements and their isotopes: hydrogen (^{1}H, ^{2}H, ^{3}H), carbon (^{12}C, ^{13}C, ^{14}C), nitrogen (^{14}N, ^{15}N), oxygen (^{16}O, ^{18}O), and sulphur (^{32}S, ^{34}S). These are the most important elements in biologic systems and also participate in most geochemical reactions. Furthermore, they occur in relative great abundance and their contents in different terrestrial compounds can be determined with analytical uncertainties much smaller than natural variations. Therefore, the presentations in this text concentrate on these elements. Table 1 summarizes their average abundances in natural compounds. Also listed are some of the less important and less abundant isotopes of these elements such as ^{17}O, ^{33}S and ^{36}S. The latter may have significance in very specific studies but are usually not included in geochemical or environmental investigations where large numbers of routine analyses are required and where they would yield little additional information.

Table 1 includes also the naturally occurring isotopes of strontium and uranium which in recent years acquired some importance in hydrogeologic and sedimentologic investigations.

Analytical techniques are discussed in many publications and are not repeated here. However, for readers not familiar with environmental isotope techniques a number of key references for the different analytical procedures are given in Table 2.

TABLE 1

Average terrestrial abundance of the isotopes of major elements used in environmental studies

Element	Isotopes	Average terrestrial abundance (%)	Comments
Hydrogen	^{1}H	99.984	
	^{2}H	0.015	
	^{3}H	10^{-14} to 10^{-16}	radioactive, $t_{1/2}$ = 12.35 years
Carbon	^{12}C	98.89	
	^{13}C	1.11	
	^{14}C	$\sim 10^{-10}$	radioactive, $t_{1/2}$ = 5730 years
Oxygen	^{16}O	99.76	
	(^{17}O	0.037) *	
	^{18}O	0.1	
Nitrogen	^{14}N	99.34	
	^{15}N	0.366	
Sulphur	^{32}S	95.02	
	(^{33}S	0.75) *	
	^{34}S	4.21	
	(^{36}S	0.02) *	
Strontium	(^{84}Sr	0.56) *	
	^{86}Sr	9.86	
	^{87}Sr	~ 7.02	
	(^{88}Sr	82.56) *	
Uranium	^{234}U	~ 0.0056	radioactive, $t_{1/2}$ = 2.47 × 10^5 years
	(^{235}U	0.7205) *	radioactive, $t_{1/2}$ = 7.13 × 10^8 years
	^{238}U	99.274	radioactive, $t_{1/2}$ = 4.51 × 10^9 years

* These isotopes are presently not used in environmental studies.

In geochemical systems quantitative considerations are only possible if chemical equilibria are established between reactants and products or if the reaction kinetics are fully understood. Chemical equilibrium is a necessary condition for isotopic equilibrium but does not suffice for it because isotope exchange reactions can proceed without any apparent change in the distribution of the chemical species in a system. If at chemical equilibrium the isotope composition of one component is modified then a new isotopic equilibrium will be established without change in mass or species distribution. For example, the chemical equilibrium between CO_2 (gas) and water is attained more rapidly than oxygen isotope equilibrium between the two.

Each isotope reaction obeys the Law of Mass Action and can be described by a temperature-dependent equilibrium constant. This constant is called the

TABLE 2

References to preparation techniques for stable isotope analyses

Deuterium in water:	Godfrey, 1962; Friedman and Hardcastle, 1970
in organic matter:	Schiegl and Vogel, 1970; Epstein et al., 1976
^{18}O in water:	Epstein and Mayeda, 1953; Sofer and Gat, 1972
in carbonates:	McCrea, 1950; Sharma and Clayton, 1965
in silicates:	Taylor and Epstein, 1962; Clayton and Mayeda, 1963
in sulphates:	Longinelli and Craig, 1967; Sakai and Krouse, 1971
in phosphates:	Tudge, 1960; Longinelli and Nuti, 1965
^{13}C in carbonates:	McCrea, 1950
in organic matter:	Craig, 1953; Hoefs and Schidlowski, 1967
^{15}N in aqueous nitrogen species and in organic compounds:	Kreitler, 1975
Sulphur in aqueous, gaseous and solid sulphur-bearing compounds:	Rafter, 1957; Puchelt and Kullerud, 1970; Thode et al., 1971; Holt, 1975

isotope fractionation factor K, or, more commonly, α. The equilibrium constant of a chemical reaction is then directly related to the isotope fractionation effect of an associated isotope exchange reaction. Mass-balance calculations, the prediction of isotopic changes during geochemical reactions or the recognition of possible reaction paths are thus theoretically possible.

These principles find an important application in the use of "geochemical" correction factors for the determination of "absolute" ^{14}C-ages of ground waters (see W.G. Mook, this volume, Chapter 2, and J.Ch. Fontes, this volume, Chapter 3). It is assumed that chemical and carbon isotope equilibria exist between gaseous, aqueous and solid phases. Model calculations can then be used to describe the geochemical evolution of a groundwater and different carbon sources which contribute to the total inorganic carbon in a water can be identified. This, in turn, permits ^{14}C mass-balance calculations and correction of measured ^{14}C ages.

Applications in hydrothermal systems using sulphur and carbon isotopes have been discussed in great detail by Ohmoto (1972) with special emphasis on the origin of ore-forming fluids and the geochemistry and temperature of ore mineral deposition. Similarly geothermometry based on isotopic differences between syngenetic minerals or coexisting gases in geothermal systems (see A.H. Truesdell and J.R. Hulston, this volume, Chapter 5) are possible, provided chemical and isotopic equilibria are established.

In near-surface environments, however, many processes are unidirectional and neither chemical nor isotopic equilibrium exists between reactants and products. For example, open surface evaporation is generally occurring at less than 100% vapour saturation, i.e. out of equilibrium. Similarly, many biogeochemical reactions are kinetic processes and usually cannot be analyzed quantitatively. However, on the basis of isotope analyses qualitative investigations are often possible. Reaction paths and environmental conditions may be identifiable where the kinetic isotope effects which cause a partitioning of isotopes between different chemical compounds are fairly well known because they are more or less reproducible under given environmental conditions. Such systems can be treated like thermodynamic systems using kinetic partitioning coefficients for isotope reactions. This applies to biologic processes such as sulphate reduction by bacteria (see H.R. Krouse, this volume, Chapter 11) as well as purely inorganic reactions for which the knowledge of the "kinetic isotope effects" permits mass-balance calculations. The determination of water loss from evaporation water bodies can be quoted as one example.

Throughout this handbook relevant isotope fractionation effects are discussed within the section concerned; only a basic introduction is presented in the following paragraphs.

DEFINITIONS

Table 1 shows that, for example, the average abundance of ^{18}O in the terrestrial materials is close to 2000 ppm. This figure varies somewhat from compound to compound because of the isotope effects mentioned above, and the total range of variations for ^{18}O is about 10% relative, i.e. from about 1900 to 2100 ppm. However, it is difficult to determine accurately the absolute ^{18}O abundances in every compound through routine analyses although this is possible with very refined techniques. Fortunately for most geochemical purposes it suffices to know the relative abundances with respect to a standard value. These relative isotope concentrations can be determined rather easily and with great accuracy through differential isotope ratio measurements using double-collecting mass spectrometers. This relative difference is called δ-value" and defined as:

$$\delta x = \frac{R_x - R_{std}}{R_{std}} \tag{1}$$

where R_x represents isotope ratios of a sample ($^2H/^1H$, $^{13}C/^{12}C$, $^{18}O/^{16}O$, $^{34}S/^{32}S$, $^{15}N/^{14}N$, $^{14}C/^{12}C$, etc.) and R_{std} is the corresponding ratio in a standard (see below). The δ-value is generally expressed in parts per thousand (permil, ‰) and written as:

$$\delta x = \left(\frac{R_x}{R_{std}} - 1\right) \times 10^3 \tag{2}$$

A sample with a $\delta^{18}O = +10‰$ is thus enriched in ^{18}O by 10‰ (or 1%)

relative to the standard, $\delta^{18}O = -5‰$ signifies that the sample has 5‰ less ^{18}O than the standard. If the ^{18}O contents of the standard are known (i.e. R_{std}) then the ^{18}O contents in the sample can be determined from these measurements.

Isotope fractionation effects in reaction such as $A \rightleftharpoons B$ are also expressed on the basis of isotope ratios, whereby:

$$\alpha_{A-B} = R_A/R_B \tag{3}$$

Introducing the δ definition (equation 2) one obtains:

$$\alpha_{A-B} = \frac{1000 + \delta_A}{1000 + \delta_B} \tag{4}$$

If the isotopes are randomly distributed in the compounds A and B then under isotopic equilibrium conditions the fractionation factor (α) is related to the equilibrium constant (K) by:

$$\alpha = K^{1/n} \tag{5}$$

where n is the number of atoms exchanged and $K \equiv \alpha$ in a monoatomic reaction. Thus, for simplicity isotope exchange reactions are usually written such that only one atom is exchanged. For example the carbonate (calcite) water ^{18}O exchange would be presented as:

$$\tfrac{1}{3} C^{16}O_3^{2-} + H_2{}^{18}O \rightleftharpoons \tfrac{1}{3} C^{18}O_3^{2-} + H_2{}^{16}O$$

and:

$$K = \frac{(C^{18}O_3^{2-})^{1/3}}{(C^{16}O_3^{2-})^{1/3}} \times \frac{(H_2{}^{16}O)}{(H_2{}^{18}O)} = \frac{(^{18}O/^{16}O)_{CO_3^{2-}}}{(^{18}O/^{16}O)_{H_2O}} = \alpha$$

Most equilibrium isotope fractionation factors are close to unity. For example the above value of 20°C will be about 1.030 with the carbonate being about 30‰ richer in ^{18}O than the water. In order to express this enrichment in permil more conveniently the enrichment * factor ϵ has been introduced with:

$$\epsilon_{A-B} = (\alpha_{A-B} - 1) \times 1000 \tag{6}$$

---

* Because this enrichment is algebraic, ϵ can be positive or negative and thus should better be called isotope separation factor.

It is also useful to note that:

$$10^3(\ln 1.00x) \simeq x$$

i.e. for small fractionation factors the following approximations can easily be deduced from equations (4) and (6) assuming $\delta_B \ll 1000$:

$$\delta_A - \delta_B \simeq \epsilon_{A-B} \simeq 1000 \ln \alpha_{A-B}$$

The introduction of $1000 \ln \alpha_{A-B}$ is insofar of significance as it has been found that α is correlated to the absolute temperature (T) by a relationship of the form

$$1000 \ln \alpha_{A-B} = a + b \cdot T^{-1} + c \cdot T^{-2}$$

where a, b, and c are system specific constants. Within small ranges at low temperatures $1000 \ln \alpha_{A-B}$ becomes practically proportional to T^{-1}.

In a recent U.S. Geological Survey publication Friedman and O'Neil (1977) compiled graphically all known stable isotope fractionation factors of geochemical interest. The reader is referred to this publication for detailed information but Figs. 1 and 2 are given as examples of the temperature dependence of these fractionation factors.

Fig. 1 shows the deuterium (^{2}H) and ^{18}O fractionation in the system vapour-water-ice. In the temperature range shown, liquid and ice are enriched in the heavy isotopes if compared to the vapour phase and, as expected, this enrichment increases with decreasing temperature, i.e. the fractionation factor α becomes larger.

The ^{13}C fractionation effects between calcite and CO_2-gas (Fig. 2) show a somewhat different behaviour: below about 180°C the calcite is enriched in ^{13}C but at that temperature a cross-over occurs and above it the CO_2-gas is richer in ^{13}C than the solid calcite. It is interesting to note that the effects remain rather constant between about 300 and 700°C. In most other systems the α-values decrease more or less continuously until they approach unity.

Important in many physicochemical and isotopic systems is the continuous removal of reaction products: one could name as examples the partial condensation (precipitation) from a vapour reservoir or the incorporation of reduced sulphur species in precipitates (e.g. FeS_2) or as gas (H_2S) resulting from bacterial degradation of aqueous sulphate. If these products cannot re-equilibrate with the reservoir of reactants, the isotopic composition of both reservoir and products will change as a function of residual material left and of the isotope fractionation factors. Such processes can be compared to a "Rayleigh distillation" and thus obey Rayleigh-type equa-

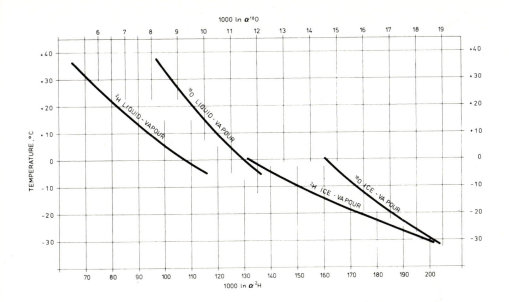

Fig. 1. Deuterium and ^{18}O isotope fractinations between water and vapour and ice and vapour. The ^{18}O values have been calibrated using for the water-vapour fractionation the relationship:

$$1000 \ln \alpha^{18}O = 2.644 - 3.206 \times 10^3 \, T^{-1} + 1.534 \times 10^3 \, T^{-2}$$

(Bottinga and Craig, 1969). The ^{18}O fractionation between ice and vapour has been determined by the same equation and multiplying α-values by the fractionation factor between ice and water at 0°C ($\alpha^{18}O_{ice-water} = 1.0030$) obtained by O'Neil (1968) according to the approximation proposed by Gonfiantini (1971). The deuterium fractionation between vapour and liquid water is described by:

$$1000 \ln \alpha = 24.844 \times 10^6 \, T^{-2} - 76.248 \times 10^3 \, T^{-1} + 52.612$$

(Majoube, 1971; Merlivat, 1969, in Gonfiantini, 1971), and for ice-vapour fractionation by (Merlivat and Nief, 1967):

$$1000 \ln \alpha^2 H = -94.5 + 16.289 \times 10^6 \, T^{-2}$$

tions such as:

$$R = R_0 f^{(\alpha-1)}$$

where R and R_0 are the isotope ratios at time t and zero respectively, f is the residual fraction of reactant at time t and α the isotope fractionation factor governing the distribution of isotopic species. Expressed in forms of δ-values

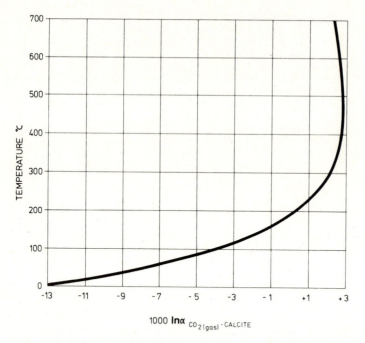

Fig. 2. The ^{13}C fractionation between carbon dioxide (gas) and calcite. The relationship shown corresponds to the following equation (Bottinga, 1968):

$$1000 \ln \alpha^{13}C_{CO_2\text{-calcite}} = -2.988 \times 10^6 \, T^{-2} + 7.6663 \times 10^3 \, T^{-1} - 2.4612 \, .$$

and ϵ one obtains:

$$\ln \frac{1+\delta}{1+\delta_0} = -\frac{\epsilon}{\epsilon+1} \ln f \, ,$$

and for values of δ, δ_0 and $\epsilon \ll 1$

$$\delta - \delta_0 \simeq -\epsilon \ln f$$

Calculated changes in isotopic compositions of a vapour reservoir during continuous condensation and the isotopic composition of the precipitation (rain and snow) are shown in Fig. 3. The initial vapour reservoir had a $\delta^{18}O \simeq -12\text{‰}$ which decreases rapidly as condensation proceeds. The importance of this process is further discussed in Chapter 1 by J.R. Gat. The Rayleigh equation describes systems in which the isotopic composition of the product is controlled by equilibrium or kinetic isotope separation factors but is removed instantaneously after its formation (e.g. rain from a cloud). In many natural systems, however, the product accumulates and iso-

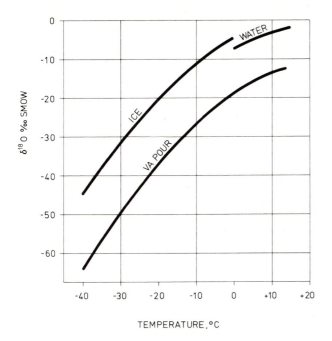

Fig. 3. Theoretical isotopic composition of vapour and condensed phases in a system which cools adiabatically and in which condensation started at 15°C from a vapour with $\delta^{18}O = -12‰$ (from Gonfiantini, 1971).

tope measurements are made on aliquots from the "product reservoir" as well as the "reactant reservoir". The isotopic difference between the two is not equal to the isotope separation factor ϵ_T which controls the instantaneous isotopic composition of the product at temperature T.

In such a closed system the following mass-balance equation is valid:

$$f\overline{\delta}_a + (1-f)\overline{\delta}_b = \delta_a^0$$

where $\overline{\delta}_a$ and $\overline{\delta}_b$ describe the isotopic compositions of the reactant and product reservoirs, δ_a^0 the initial isotopic composition of the reactant or the isotopic content of the total system, and f the residual fraction of the reactant. The isotopic difference between the two reservoirs is

$$\Delta = \overline{\delta}_a - \overline{\delta}_b$$

Assuming the homogeneity of the remaining fraction f of the reactant, the instantaneous isotopic composition of the product (δ_b) is given by:

$$\overline{\delta}_a + \epsilon \simeq \delta_b$$

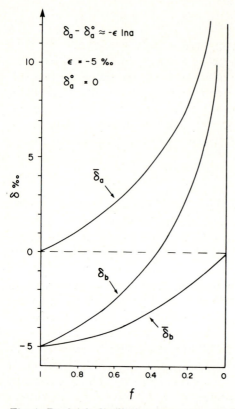

Fig. 4. Rayleigh distillation in an open and closed system. The open system describes the isotopic evolution of the instantaneously removed product (δ_b) and the continuously changing but well-mixed reservoir ($\overline{\delta}_a$). Under closed-system conditions the product accumulates and, if well mixed at all times, has an isotopic composition of $\overline{\delta}_b$. Its changing isotope composition is shown by the lower curve which intersects the composition δ_a^0 at $f = 0$.

and the isotopic composition of the changing reactant reservoir is at any instant described by:

$$\overline{\delta}_a \simeq \delta_a^0 - \epsilon \ln f$$

Combining these expressions one obtains for the product reservoir:

$$\overline{\delta}_b = \overline{\delta}_a - \Delta = \overline{\delta}_a + \epsilon \ln f/(1-f)$$

The different relationships are graphically shown in Fig. 4.

STANDARDS

The selection of standards for reporting stable isotope analyses is a very important problem in isotope geochemistry because their definition and availability will depend whether it is possible to compare the results of different laboratories. The International Atomic Energy Agency (IAEA) in Vienna, Austria, has therefore twice undertaken to coordinate the preparation, calibration and distribution of internationally acceptable standards. A summary of the latest effort has been presented by Gonfiantini (1978).

Oxygen

At present two internationally accepted ^{18}O standards exist: V-SMOW and PDB. The V-SMOW (Vienna Standard Mean Ocean Water) is a water standard the isotopic composition of which is close to the average ocean water, therefore, it is used in all hydrologic investigations. PDB is a carbonate standard derived from the rostrum of *Belemnitella americana* from the Pee Dee Formation of South Carolina (U.S.A.) and has its place primarily in paleoenvironmental studies. It is not used as reference for ^{18}O analyses of igneous and metamorphic rocks and is less and less used for sedimentary carbonates.

The V-SMOW standard has been artificially prepared and is distributed by the IAEA. Its ^{18}O content is practically identical to the SMOW standard previously defined by Craig (1961a) but which is not available. However, it had been defined with reference to a water standard of the National Bureau of Standards (Gaithersburg, Md, U.S.A.) which is called NBS-1 where:

$$(^{18}O/^{16}O)_{SMOW} = 1.008 \, (^{18}O/^{16}O)_{NBS-1}$$

Craig (1961a) evaluated the isotope ratio of SMOW and found:

$$(^{18}O/^{16}O)_{SMOW} = (1993.4 \pm 2.5) \times 10^{-6}$$

and Baertschi (1976) measured for V-SMOW:

$$(^{18}O/^{16}O)_{V\text{-}SMOW} = (2005.2 \pm 0.45) \times 10^{-6}$$

Besides V-SMOW and NBS-1 ($\delta^{18}O_{NBS-1/SMOW} = -7.94$; Craig, 1961a) there are a few other ^{18}O standards distributed by the IAEA. Of importance is specially a ^{18}O-poor sample which is called SLAP (Standard Light Arctic Precipitation). Its $\delta^{18}O$ value vs. V-SMOW has been determined to be:

$$\delta^{18}O_{SLAP/V\text{-}SMOW} = -55.5\text{‰}$$

The adoption of these two standards together with the intermediate NBS-1A ($\delta^{18}O_{\text{NBS-1A/SMOW}} = -24.33‰$; Craig, 1961a) permits the calibration of working standards over a wide range.

For technical reasons water samples have to be prepared by equilibrating CO_2 with the water. Therefore, if a water standard is to be used for other than water samples the CO_2–H_2O fractionation factor has to be known. Friedman and O'Neil (1977) summarized the experimental determinations of this critical value and suggest that for equilibration at 25°C the following value be adopted:

$$\alpha_{CO_2-H_2O} = R_{CO_2}/R_{H_2O} = 1.0412 \; (25°C)$$

The carbonate standard PDB has long been exhausted but was calibrated by Craig (1957) against the still available NSB-20, a limestone from Solenhofen in Southern Germany (available from National Bureau of Standards, Gaithersburg, Md., U.S.A.). He found:

$$\delta^{18}O_{\text{NBS-20/PDB}} = -4.14‰$$

These carbonates have to be prepared with 100% H_3PO_4 in order to obtain the CO_2-gas suitable for mass spectrometric analyses. Because out of the

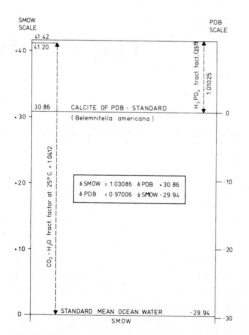

Fig. 5. Relationship between PDB and SMOW oxygen standards (modified from Friedman and O'Neil, 1977).

three oxygen atoms in the carbonates only two are extracted in the CO_2, this reaction has to be carried out at constant temperature (usually 25°C) in order to keep the isotope fractionation effects constant. The accepted value at 25°C is (Sharma and Clayton, 1965):

$$\alpha_{CO_2(H_3PO_4)\text{-calcite}} = 1.01025$$

The CO_2 equilibrated with SMOW and CO_2 extracted with H_3PO_4 from PDB have been measured against one another and a difference of 0.22‰ has been found between the two gases, the latter being enriched in ^{18}O.

The relationship between SMOW and PDB is shown in Fig. 5. From these a conversion formula from PDB to SMOW and vice versa can be determined where:

$$\delta^{18}O_{x/\text{V-SMOW}} = 1.03086\, \delta^{18}O_{x/\text{PDB}} + 30.86$$

Hydrogen

Only water standards (V-SMOW, NBS-1, NBS-1A, SLAP) are used for deuterium analyses. Friedman and O'Neil (1977) provide various inter-comparisons demonstrating the discrepancies that exist. However, the following definition is at present accepted (Craig, 1961a):

$$(^2H/^1H)_{\text{SMOW}} = 1.050(^2H/^1H)_{\text{NBS-1}}$$

Hageman et al. (1970) measured:

$$(^2H/^1H)_{\text{V-SMOW}} = (155.76 \pm 0.05) \times 10^{-6}$$

the δ^2H values for SLAP were defined following an inter-comparison organized by the IAEA (Gonfiantini, 1978) as:

$$\delta^2H_{\text{V-SMOW/SLAP}} = +428‰$$

As Friedman and O'Neil point out the analytical problems with deuterium analyses are such that relatively large analytical errors are possible, especially if the "parasitic" H_3^+ contribution in the mass spectrometer is incorrectly assessed. Therefore, many laboratories calibrate their mass spectrometers to internationally accepted standards and apply a blanket correction to their instrumental data. Especially data presented by Craig (1961a) have been used, accepting the following δ^2H values:

$$\delta^2H_{\text{NBS-1/SMOW}} = +428‰$$

$$\delta^2H_{\text{NBS-1A/SMOW}} = -183.3‰$$

Gonfiantini (1978) points out that "the adoption of V-SMOW as zero of the δ-scale and of prefixed values for SLAP corresponds in principle to a modification of the δ-scale which becomes:

$$\delta = [(R_{sample} - R_{V\text{-}SMOW})/R_{V\text{-}SMOW}]\delta_{SLAP}/[(R_{SLAP} - R_{V\text{-}SMOW})/R_{V\text{-}SMOW}]''$$

There is no reference standard for ^3H (tritium) measurements since the results are presented in terms of ^3H concentrations by direct measurement converting measured activities into absolute concentrations. The concentration unit is the tritium unit which corresponds to an abundance of 10^{-18} atoms of ^3H for one atom of hydrogen. One tritium unit (TU) is equivalent to 7.2 dpm per litre of water or 3.2 pCi/l. Working standards with known ^3H activities are available from the IAEA in Vienna.

Carbon

The universally accepted carbon standard is PDB. Since during the CO_2 extraction all carbon is transferred to the CO_2 few analytical problems do exist. The National Bureau of Standards distributes two reference standards, whose ^{13}C contents have been determined by Craig (1957); NBS-20 (Solenhofen limestone) and NBS-21 (spectrographic carbon):

$\delta^{13}C_{NBS\text{-}20/PDB} = -1.06$

$\delta^{13}C_{NBS\text{-}21/PDB} = -27.79$

There is some doubt about the latter value and it is possible that it is closer to −28.10 (Friedman and O'Neil, 1977).

In the future the IAEA will coordinate the preparation, calibration and distribution of new carbonate standards.

Differing from tritium measurements ^{14}C contents in natural materials are also referred to a standard which corresponds to wood grown during 1890 in an environment free of fossil carbon dioxide. Its ^{14}C content has been defined as representing the activity (A) of modern carbon and all measured samples are thus expressed in percent of modern carbon (pmc):

$$A = \frac{(^{14}C/^{12}C)_{sample}}{(^{14}C/^{12}C)_{standard}} \times 100 \text{ (pmc)}$$

For laboratory calibrations the National Bureau of Standards distributes (now in very limited quantities) a universally accepted oxalic acid standard (Olsson, 1970) whereby:

^{14}C activity modern carbon = 0.95 ^{14}C activity NBS oxalic acid in 1950

An activity of 100 pmc is close to the "steady-state" activity of tropospheric CO_2 and corresponds to 13.56 dpm/g of carbon. However, a difference exists between the carbon isotopic composition of atmospheric carbon dioxide and recent wood because of fractionation effects related to photosynthesis. The $\delta^{13}C$ of "modern" atmospheric CO_2 (free of combustion CO_2) is close to $-6.4‰$ vs. PDB (Craig and Keeling, 1963) and the corresponding "modern" wood has a $\delta^{13}C \simeq -25‰$. (The oxalic acid distributed by NBS has a $\delta^{13}C = -19.3‰$; Craig, 1961b.) If one accepts the assumption (Craig, 1954) that in any isotope exchange reaction the ^{14}C enrichment is approximately double the ^{13}C enrichment i.e.:

$\epsilon^{14}C \simeq 2\,\epsilon^{13}C\,‰$ or $0.2\,\epsilon^{13}C\,\%$

then there would be a difference of 3.7 pmc between "modern" atmospheric CO_2 and "modern" wood. This is schematically shown in Fig. 6 for atmospheric CO_2.

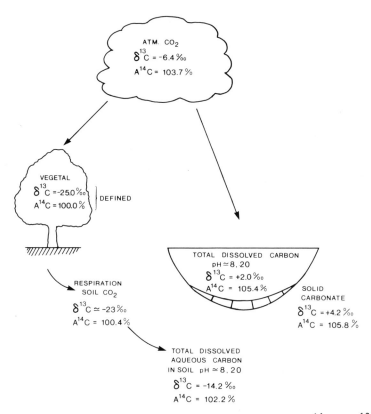

Fig. 6. Schematic representation of isotope effects for ^{14}C and ^{13}C in the atmospheric CO_2 system. In this case 1% variation in ^{14}C is equivalent to about 83 years.

Although small, such differences may be of importance in many natural systems, especially if one attempts to date very young materials. For example subrecent carbonates deposited in isotopic equilibrium with atmospheric CO_2 in Lake Abbè, Republic of Djibouti (Fontes and Pouchan, 1975), show ^{14}C contents of up to 102 pmc. Without the above correction such values would indicate post-nuclear deposits, which is impossible because they are found above present-day water levels and are at least a few hundred years old. Applying the correction such ages are found and paleoclimatic evidence can be derived from them.

Nitrogen

Atmospheric nitrogen is used as working and reference standard by most laboratories. The National Bureau of Standards distributed a nitrogen-gas sample (NBS-14) for calibration purposes, however, the supplies are nearly exhausted and two new reference standards have to be prepared. One will be an ammonium sulphate and the other should be a nitrate sample. The IAEA is at present preparing the calibration of the first and should begin shortly with its distribution.

Sulphur

The accepted reference standard is the troilite phase (FeS) from the Cañon Diablo meteorite. It has a ratio $^{32}S/^{34}S = 22.22$. If the isotope analyses are done with SO_2 then the ^{18}O contribution to mass 66 can bring about rather large correction factors. This is overcome by the SF_6 method (Puchelt and Kullerud, 1970) and new reference standards will have to be calibrated using this preparation technique. At present a NBS-120 is available but little used. However, the IAEA should shortly distribute a seawater sulphate ($\delta^{34}S \approx +20‰$) and coordinate the preparation and calibration of at least one sulphide standard.

Strontium

Strontium occurs with four stable isotopes ^{84}Sr, ^{86}Sr, ^{87}Sr and ^{88}Sr. The isotopes 84, 86, and 88 are not parts of any decay series. However, the most refined techniques of mass spectrometry do not yet permit an investigation of the slight (if any) isotope fractionations undergone by these species within the geochemical cycles of strontium. Thus, at present, the abundances of ^{84}Sr, ^{86}Sr and ^{88}Sr appear quite constant (see Table 1) and without geochemical interest.

The isotope ^{87}Sr, although naturally occurring, is also a daughter product of ^{87}Rb decay. Thus ^{87}Sr will exhibit variations due to: (a) amount of rubidium, and (b) residence time in the reservoir.

Because rubidium is not very frequent and because of its very long half-life (50×10^9 years), the production rate of ^{87}Sr in a reservoir is extremely low. However, the natural variations of this isotope are measurable and referred to ^{86}Sr concentrations as ^{87}Sr/^{86}Sr ratios. ^{86}Sr is chosen as a reference because it can be measured with a good accuracy and because its natural content is close to that of ^{87}Sr. Analyses are done by solid-source mass spectrometry. A reference standard for calibration is distributed by the National Bureau of Standards: NBS-987, a strontium carbonate with ^{87}Sr/^{86}Sr = 0.701.

Uranium

Three naturally occurring isotopes are known: ^{234}U, ^{235}U and ^{238}U. They are radioactive and their half-lives and average terrestrial abundances are listed in Table 1.

^{234}U is a daughter product in the decay series of ^{238}U and if a secular equilibrium is established between them the rate of the decay of the parent will equal the rate of decay of the daughter: their activity ratios will be one. In most known natural systems such is not the case because the geochemical behaviour of these two isotopes is slightly different. This uranium disequilibrium is thus a reflection of hydrogeologic and geochemical processes and time effects which affect the abundance of uranium and its isotopes in water and minerals. Their relationship is expressed in ^{234}U/^{238}U activity ratios and hence no standard is required. However, specific analytical techniques are necessary and are presented as an appendix to Chapter 7 in this volume.

REFERENCES

Baertschi, P., 1976. Absolute ^{18}O content of Standard Mean Ocean Water. *Earth Planet. Sci. Lett.*, 31: 341—344.

Bottinga, Y., 1968. Calculation of fractionation factors for carbon and oxygen exchange in the system calcite—carbon dioxide—water. *J Phys. Chem.*, 72: 800—808.

Bottinga, Y. and Craig, H., 1969. Oxygen isotope fractionation between CO_2 and water and the isotopic composition of the marine atmosphere. *Earth Planet. Sci. Lett.*, 5: 285—295.

Clayton, R.N. and Mayeda, T.K., 1963. The use of bromine pentafluoride in the extraction of oxygen from oxides and silicates for isotope analysis. *Geochim. Cosmochim. Acta*, 27: 43—52.

Craig, H., 1953. The geochemistry of the stable carbon isotopes. *Geochim. Cosmochim. Acta*, 3: 53—92.

Craig, H., 1954. Carbon-13 in plants and the relationship between carbon-13 and carbon-14 variations in nature. *J. Geol.*, 62: 115—149.

Craig, H., 1957. Isotopic standards for carbon and oxygen and correction factors for mass-spectrometric analysis of carbon dioxide. *Geochim. Cosmochim. Acta*, 12: 113—149.

Craig, H., 1961a. Standard for reporting concentrations of deuterium and oxygen-18 in natural water. *Science*, 133: 1833—1834.

Craig, H., 1961b. Mass-spectrometer analyses of radiocarbon standards. *Radiocarbon*, 3: 1—3.
Craig, H. and Keeling, C., 1963. The effects of atmospheric NO_2 on the measured isotopic composition of atmospheric CO_2. *Geochim. Cosmochim. Acta*, 27: 549—551.
Epstein, S. and Mayeda, T.K., 1953. Variations of the $^{18}O/^{16}O$ ratio in natural waters. *Geochim. Cosmochim. Acta*, 4: 213.
Epstein, S., Yapp, C.J. and Hall, J.H., 1976. The determination of the D/H ratio of non-exchangeable hydrogen in cellulose extracted from aquatic and land plants. *Earth Planet. Sci. Lett.*, 30: 241—251.
Fontes, J.Ch. and Pouchan, P., 1975. Les cheminées du lac Abbè (T.F.A.I.): stations hydroclimatiques de l'Holocène. *C.R. Acad. Sci. Paris, Sér. D*, 280: 383—386.
Friedman, I. and Hardcastle, K., 1970. A new technique for pumping hydrogen gas. *Geochim. Cosmochim. Acta*, 34: 125—126.
Friedman, I. and O'Neil, J.R., 1977. Compilation of stable isotope fractionation factors of geochemical interest. In: M. Fleischer (Editor), *Data of Geochemistry. U.S. Geol. Surv., Prof. Paper*, 440-KK: 1—12 (6th ed.).
Gonfiantini, R., 1971. Notes on isotope hydrology (manuscript).
Gonfinantini, R., 1978. Standards for stable isotope measurements in natural compounds. *Nature*, 271: 534—536.
Godfrey, J.D., 1962. The deuterium content of hydrous minerals from east central Nevada and Yosemite National Park. *Geochim. Cosmochim. Acta*, 26: 1215—1245.
Hageman, R., Nief, G. and Roth, E., 1970. Absolute isotopic scale for deuterium analysis of natural waters. Absolute D/H ratio for SMOW. *Tellus*, 22: 712—715.
Hoefs, J. and Schidlowski, M., 1967. Carbon isotope composition of carbonaceous matter from the Precambrian of the Witwatersrand system. *Science*, 155: 1096—1097.
Holt, B., 1975. Determination of stable sulphur isotope ratios in the environment. *Prog. Nucl. Energy, Ser. 9*, 12: 11—26.
Kreitler, C.W., 1975. Determining the source of nitrate in ground water by nitrogen isotope studies. *Bur. Econ. Geol., Univ. Texas, Rep.*, 83: 57 pp.
Longinelli, A. and Craig, H., 1967. Oxygen-18 variations in sulphate ions in seawaters and saline lakes. *Science*, 146: 56—59.
Longinelli, A. and Nuti, S., 1965. Oxygen isotopic composition of phosphate and carbonate from living and fossil marine organisms. In: E. Tongiorgi (Editor), *Stable Isotopes in Oceanographic Studies and Paleotemperatures, Spoleto 1965*. CNR, Rome, pp. 183—197.
McCrea, J.M., 1950. On the isotopic chemistry of carbonates and a paleotemperature scale. *J. Chem. Phys.*, 18: 849—857.
Majoube, M., 1971. Fractionnement en oxygène-18 et en deutérium entre l'eau et la vapeur. *J. Chim. Phys.*, 10: 1423—1436.
Merlivat, L. and Nief, G., 1967. Fractionnement isotopique lors des changements d'état solide-vapeur et liquide-vapeur de l'eau à des températures inférieures à 0°C. *Tellus*, 19: 122—127.
Ohmoto, H., 1972. Systematics of sulphur and carbon isotopes in hydrothermal ore deposits. *Econ. Geol.*, 67: 551—578.
Olsson, I.U., 1970. The use of oxalic acid as a standard. In: I.U. Olsson (Editor), *Radiocarbon Variations and Absolute Chronology, Proceedings of the 12th Nobel Symposium*. Wiley, New York, N.Y., p. 17.
O'Neil, J., 1968. Hydrogen and oxygen isotope fractionation between ice and water. *J. Chem. Phys.*, 72: 3683—3684.
Puchelt, H. and Kullerud, G., 1970. Sulphur isotope fractionation in the Pb-S system. *Earth Planet. Sci. Lett.*, 7: 301—306.

Rafter, T.A., 1957. Sulphur isotopic variations in waters, I. The preparation of sulphur dioxide for mass-spectrometer examination. *N.Z. J. Sci. Technol. Sect. B*, 38: 843—857.
Sakai, H. and Krouse, H.R., 1971. Elimination of memory effects in $^{18}O/^{16}O$ determinations in sulphates. *Earth Planet. Sci. Lett.*, 11: 369—373.
Schiegl, W.E. and Vogel, J., 1970. Deuterium content of organic matter. *Earth Planet. Sci. Lett.*, 15: 232—238.
Sharma, T. and Clayton, R.N., 1965. Measurement of $^{18}O/^{16}O$ ratios of total oxygen of carbonates. *Geochim. Cosmochim. Acta*, 29: 1347—1354.
Sofer, Z. and Gat, J., 1972. Activities and concentrations of ^{18}O in concentrated aqueous salt solutions: analytical and geophysical implications. *Earth Planet. Sci. Lett.*, 15: 232—238.
Taylor, H.P. and Epstein, S., 1962. Relation between $^{18}O/^{16}O$ ratios in coexisting minerals of igneous and metamorphic rocks, I. Principles and experimental results. *Geol. Assoc. Am. Bull.*, 73: 461—480.
Thode, H.G., Cragg, C.B., Hulston, J.R. and Rees, C.E., 1971. Sulphur isotope exchange between sulphur dioxide and hydrogen sulphide. *Geochim. Cosmochim. Acta*, 35: 35—45.
Tudge, A.P., 1960. A method of analysis of oxygen isotopes in orthophosphate and its use in the measurement of paleotemperatures. *Geochim. Cosmochim. Acta*, 18: 81—93.

Chapter 1

THE ISOTOPES OF HYDROGEN AND OXYGEN IN PRECIPITATION

JOEL R. GAT

INTRODUCTION

The oxygen isotopes ^{17}O and ^{18}O were discovered in natural oxygen in 1929 (Giauque and Johnston, 1929) and deuterium was identified as a component of natural hydrogen in 1932 (Urey et al., 1932). Although stable to radioactive decay, it soon transpired that there were variations in the natural abundance of the isotopic species of both oxygen and hydrogen among different chemicals or, for that matter, among water derived from various natural sources; primarily the result of fractionations caused by phase transitions, chemical or biological reactions and transport processes. The abundance variations due to the presently occurring radiogenic production of these isotopes is, by comparison, much smaller and quite negligible in the terrestrial water cycle. (This is not true, however, for materials of cosmic and lunar origin, where particularly the H/D ratios are much higher than on earth, due to solar wind influx.) These differences opened up the possibility of utilizing the abundance variations as tracers in the water cycle.

The stable isotope composition is usually measured with reference to a standard. In the case of isotopes of both oxygen and hydrogen in water, the standard commonly accepted is SMOW — the mean composition of ocean water * (Craig, 1961a). It is customary to express the stable isotope abundance as the relative difference in the ratio of heavy to the light (more abundant) isotope in the sample (R) with reference to that of SMOW. The greek δ, defined as $\delta = (R - R_{SMOW})/R_{SMOW}$ denotes the isotope abundance (data are usually reported in the form of δ‰ units $\equiv \delta \times 1000$). The value of R_{SMOW} is given as $R_{18_O} = (2005.2 \pm 0.45) \times 10^{-6}$ (Baertschi, 1976) and $R_D = (155.76 \pm 0.05) \times 10^{-6}$ (Hageman et al., 1970).

---

* SMOW has been defined rigorously in terms of an actual water standard, NBS-1 by Craig (1961a) as follows: $(R^{18O}_{SMOW}) = 1.008\, R^{18O}_{NBS-1}$; $R^{D}_{SMOW} = 1.050\, R^{D}_{NBS-1}$; where R is the atom ratio of the isotopic water species $H_2^{18}O/H_2^{16}O$ and HDO/H_2O respectively. Nowadays a barrel of water (Vienna-SMOW), whose composition matches that of the defined SMOW as closely as could be measured, is deposited at the IAEA at Vienna for reference.

Radioactive oxygen isotopes are too short-lived to be of use as geophysical markers. However, the radioisotope of mass 3 of hydrogen (tritium, with a half-life of 12.26 years) was detected in natural waters in 1950 (Falting and Harteck, 1950; Grosse et al., 1951). It is being produced through interaction of cosmic radiation with the atmosphere, as predicted by Libby (1946). Since 1952, the tritium from nuclear weapon tests has overshadowed the natural production by orders of magnitude. Both natural and bomb-produced tritium appears to a large extent in the form of tritiated water and has been very useful as a water tracer in the hydrologic cycle. The unit used to measure natural concentration levels is the tritium unit (TU); 1 TU corresponds to an atom ratio of tritium to hydrogen of 10^{-18}, being equivalent to a radioactivity of about 3.2×10^{-12} Ci/g of water.

Our picture of the global distribution of stable isotopes and of tritium in atmospheric waters and precipitation has been obtained through the IAEA-WMO network of monthly precipitation samples. This network had been initiated in 1961 and includes maritime (island and weather ships) stations, sites in coastal areas and some inland continental stations; 144 sampling sites distributed throughout the globe. The coverage of the IAEA network has been supplemented by a number of local surveys. Previous to this, tritium was monitored in precipitation from a limited number of sites: Chicago (Kaufman and Libby, 1954; Von Buttlar and Libby, 1955 and Begemann and Libby, 1957) and Ottawa (Brown and Grummit, 1956) have the longest record of observation.

Data are regularly published by the International Atomic Energy Agency (IAEA, 1969, 1970, 1971, 1973, 1975).

TRITIUM IN ATMOSPHERIC WATERS

Relatively few reliable measurements on natural tritium levels in the atmosphere were made before the atmospheric test of a thermonuclear device in 1952 swamped the atmosphere with artificial tritium. These early measurements sufficed, however, to establish natural tritium levels in precipitation, ranging for different locations from about 4 to 25 TU and enabled a crude estimate of the natural mean global decay rate of 0.5 ± 0.3 tritons/cm^2, which is of the same order of magnitude as the estimated production in the atmosphere by cosmic rays (Nir et al., 1966). Details of the distribution in time and space are sketchy and our picture of natural tritium geophysics relies to a large extent on additional information from the post-1952 era, based on the assumption that the atmospheric fate of cosmic-ray- and bomb-produced tritium was similar (with the exception of the interhemispheric distribution).

The pattern of tritium levels in precipitation at Ottawa, Canada (the station with the longest continuous record), is essentially repeated at other

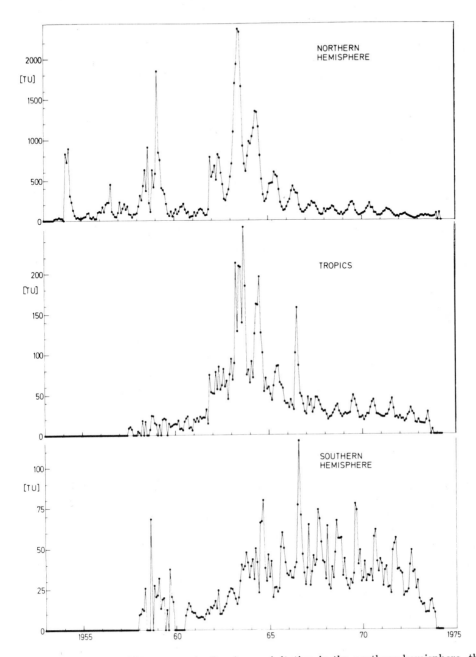

Fig. 1-1. Average tritium concentration in precipitation in the northern hemisphere, the tropical stations and the southern hemisphere, based on records of IAEA network (from Groeneveld, 1977).

northern hemisphere stations, albeit with varying amplitudes and some significant phase shifts. Fig. 1-1 shows the average northern hemisphere values and the notable feature of this curve is a yearly cycle of maximum concentrations in spring and summer and a winter minimum, with typical concentration ratios of 2.5—6 between maximum and minimum values. The annual cycle is superimposed upon long-term changes which have ranged over three orders of magnitude since 1952.

There is a marked latitude dependence: concentrations are highest north of the 30th parallel, with values lower by a factor of 5 or so at low-latitude and tropical stations (Athavale et al., 1967). In the southern hemisphere the yearly cycle is displaced with the season by half a year, and the mean tritium levels in atmospheric waters are lower than at comparable north-latitude stations (Taylor, 1966). This is a reflection of the predominant northern location of weapon testing sites and the slow inter-hemispheric transport of tracers. It is only in recent years (the seventies), following a long period of limited nuclear test activity and declining atmospheric tritium levels, that northern and southern hemisphere tritium concentrations are becoming comparable.

Tritium geophysics, like that of other atmospheric components of stratospheric origin such as ozone or "fallout" isotopes, is dominated by the timing, location and intensity of exchange of tropospheric and stratospheric air masses, as well as, of course, the tritium concentration in the stratosphere at the time that such an exchange takes place. Exchange occurs predominantly during late winter and in spring (the so-called spring leak of the tropopause) in the region of baroclinic zones and tropopause discontinuities of the mid-latitudes (Newell, 1963).

The changing tritium inventory of the stratosphere of recent years reflects the massive injections by weapon tests in 1954, 1955, 1958 and again during 1961—1962, mostly in the northern hemisphere. At any time, the inventory decreases by 5.5% per year through radioactive decay and some of the tritium leaks into the troposphere from where it is lost into the ocean or groundwaters, both of which can be considered a sink for the stratospheric tritium. Estimates of the residence time of tritiated water vapour in the lower stratosphere vary between one and 10 years (Martell, 1970; Lal and Peters, 1967). Inter-hemispheric mixing in the stratosphere seems to occur on a similar time scale.

The residence time of water in the lower troposphere, on the other hand, is of the order of 5—20 days. This is a short period relative to large-scale north-to-south mixing in the troposphere, but within the time scale of horizontal atmospheric motions (Eriksson, 1967). As a result tritium is deposited onto the face of the earth within the latitude band of its penetration from the stratosphere or more precisely of its distribution on top of the so-called moist layer, which extends to 500 mbar approximately. This explains the latitudinal distribution of tritium levels in the precipitation (Eriksson, 1966). 1966).

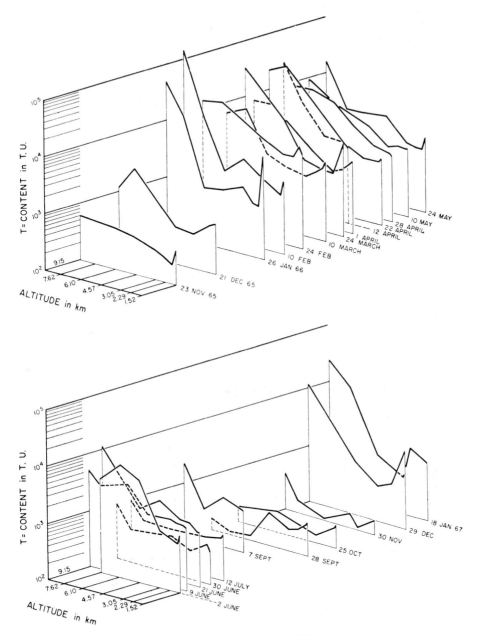

Fig. 1-2. Tropospheric tritium profiles (from Ehhalt, 1971).

Ehhalt (1971, 1974) reported the most comprehensive measurements of tritium levels in the troposphere over both land and sea, to which we owe the confirmation of the transport models outlined above. From his measured profiles (Fig. 1-2) the stratospheric source of tritium is abundantly clear: very high concentrations in the upper troposphere appear toward the end of December (up to 10^5 TU) and indicate the timing of the stratospheric injection; these upper tropospheric levels decline considerably by mid-June. At that time there seems to be almost no gradient of the TU values throughout the troposphere from 2 km upwards to the tropopause: the HTO mixing ratio (HTO molecules/kg of air) then decreases with altitude in parallel with that of the water vapour. Above the 2-km level the tritium content is surprisingly uniform over both land and sea, along each latitude band. Over the continents during summer, there is an increase in tritium amounts in the lowest 2 km, apparently due to re-evaporation of part of the winter and spring precipitation. In contrast, tritium levels are low over the sea throughout the year, a result of the uptake of tritium by molecular exchange into the oceans. (Moisture evaporated from the ocean is very low in tritium content due to the long residence time of water in the ocean.) The delay in the appearance of the annual tritium peak in precipitation relative to the time of its injection (June vs. the late winter months) is attributed by Ehhalt to this re-evaporation of moisture from the continents, which provides an additional source of tritium to the atmosphere during summer.

The atmospheric moisture system has been treated as a box model by Bolin (1958) and Begemann (1961) (Fig. 1-3), and graphically illustrated by Eriksson (1967). In these models tritium is added from aloft (F_T); loss occurs through rainout and exchange with the tritium deficient surface waters over the ocean while over land the loss is to the groundwater systems or through runoff. The soil, which returns most of the precipitated tritium through evapotranspiration acts as a buffer which maintains high continental tritium levels.

In discussing the model we have to distinguish whether the tritium concentration is expressed in terms of "HTO mixing ratios" or in relative terms as the tritium content of moisture, expressed in tritium units. We note as a peculiarity of the latter way of expressing data that rainout does not change the tritium content of the residual moisture of the atmosphere, except for a minor depletion by a few percent due to the isotope effect in the liquid-vapour equilibrium *. Rainout, however, decreases the precipitable mass of tritium in the atmosphere. The buildup of the tritium content inland (in tritium units) is inversely proportional to the moisture content in the atmosphere for any given flux F_T, and therefore, paradoxically, rainout increases

---

* Bolin (1958) has estimated a value of $f = 0.75$ for the tritium fractionation factor between the atmosphere-soil transition. Bigeleisen (1962) has shown on theoretical grounds that $\ln \alpha_{T-H}/\ln \alpha_{D-H} \sim 1.3\text{—}1.5$ for the type of transitions considered. Taking then the deuterium fractionation in the atmospheric water cycle as a yardstick, where $\Delta \delta D$ between vapour and water is typically $\Delta = 70\text{‰}$ one then obtains a more realistic tritium isotope effect of 10%, so that $f \simeq 0.9$.

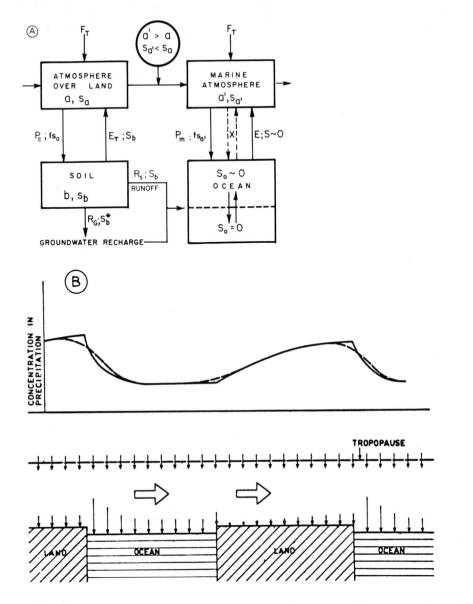

Fig. 1-3. A. The tritium budget of the atmosphere: box model representation (adapted from Bolin, 1958, and Begemann, 1960). a, a' and b = the water content in continental and marine atmosphere and soil column respectively. s_a, $s_{a'}$, s_b, s_0 = tritium content of these reservoirs and of the oceanic surface layers, respectively (in TU); s_b^*, the tritium content of recharging waters, may differ from s_b because of holdup of water in the soil column. F_T = tritium influx from aloft (in moles/s). P_c, P_m, E, E_T, R_s, R_G and X = the precipitation amounts (continental and marine), evaporation and evapotranspiration fluxes, surface and groundwater runoff and the molecular exchange flux between ocean and atmosphere, respectively.
B. Schematic picture of deposition of tritium and concentration of tritium in precipitation in a westerly zonal air current (from Eriksson, 1967). The length of the arrows is proportional to the rate of vertical transports. The dashed line indicates the smoothing in the concentrations which will take place due to longitudinal eddy mixing.

the inland tritium-unit gradient, except where this water loss is compensated for by evaporation.

From the model we can see that tritium levels in marine air are a measure of air-sea exchange processes, in particular the precipitation-evaporation balance. As first proposed by Suess (unpublished lecture) and in greater detail by Craig and Gordon (1965), the relative tritium loss to the ocean through precipitation (F_p) and molecular exchange (F_x) is given by:

$$\frac{F_x}{F_p} \simeq \frac{E}{P} \cdot \frac{h}{1-h}$$

(the formula gives a good approximation as long as ocean tritium levels are negligibly small relative to atmospheric ones); h is the relative humidity in the marine atmosphere and $h = 0.75$ is a reasonable value for marine air. We then find $F_x/F_p \simeq 3$ (E/P being close to 1 on a global average), indicating molecular exchange to be the dominant process of tritium loss to the ocean.

Tritium buildup over the continent comes about as a result of the cutting off of the supply of low activity oceanic vapour, while influx from aloft continues. The inland gradient could be expected to be highest during late winter, spring and summer when F_T is at its peak value. This effect is, however, somewhat balanced by the lower content of vapour in the winter atmosphere.

As stated, re-evaporation of moisture from the continent during summer acts to extend the spring maximum of tritium concentrations into the summer, but does not affect the inland buildup of tritium content (expressed in tritium units) except when there is a holdup (delay) of water, so that the re-evaporated moisture has a noticeably different tritium age than the atmospheric moisture. Ages of a few years are quite typical for soil moisture or for waters in sizeable inland lakes. During periods of rising tritium levels, such as the decade of 1952—1963, the continental water reservoirs were relatively low in tritium content and their evaporation reduced the continental gradient. During recent years with declining atmospheric tritium levels, the continental reservoirs may retain the memory of high tritium levels of the past and reverse the normal tritium flux, contributing tritium to the atmosphere. In this case the inland tritium gradient is increased.

Buildup of tritium concentrations over continents is quite gradual; for example, tritium-unit levels double over Central Europe over a distance of 1000 km (Eriksson, 1966). The interaction over the ocean, on the other hand, becomes effective over very short distances, especially when the continental air is very unsaturated relative to ocean surface waters. An extreme case is found on the leeside of Europe in the Mediterranean Sea (Fig. 1-4), where the intense sea-air interaction is mirrored in the extreme drop in tritium activity, to 10% of the continental value. The inland gradient on the

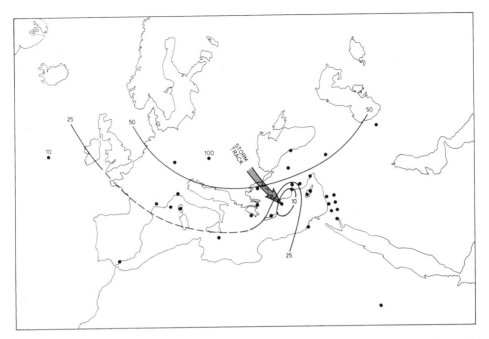

Fig. 1-4. Tritium in precipitation in the Mediterranean Sea area (March 1964; from Gat and Carmi, 1970). Dots indicate data points, all measurements are normalized relative to precipitation in Vienna (100) as a model of continental station.

Eastern Mediterranean shore also is more abrupt than usual, the tritium-unit doubling length being about 100 km (Carmi and Gat, 1973) due to the limited extent and intra-continental position of the Mediterranean Sea.

Precipitation is the main mechanism for removing tritium from the atmosphere over the continent. Moreover it is the vehicle for the downward transport of tritium within the troposphere. As a result of the rapid exchange of isotopes between the rain droplets and ambient vapour (Bolin, 1958; Booker, 1965), the falling rain drops contribute tritium to the lower troposphere during the period where strong vertical tritium gradients exist. Indeed, Ehhalt (1971) has noticed an additional source of tritium in the lower atmosphere, at a height just below the freezing level. Only the frozen phases, i.e. hail and snow, are not subject to this exchange and can carry the high tritium levels all the way to the ground.

STABLE ISOTOPE DISTRIBUTION IN ATMOSPHERIC WATERS: DATA

The stable isotope composition of precipitation samples (monthly averages, as obtained through the precipitation collection network described

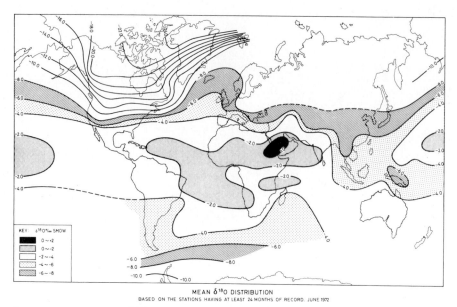

Fig. 1-5. Mean annual $\delta^{18}O$ values in precipitation: summary of IAEA precipitation network data (from Yurtsever, 1975).

above) at any one station generally exhibits seasonal variations; as a rule these are correlated with the temperature changes so that the precipitation during winter months is depleted in the heavy isotope species relative to the summer rains. The annually averaged worldwide distribution for each of these stations, as illustrated in Fig. 1-5, shows that atmospheric waters are depleted relative to the ocean and other surface water bodies, with marine rain closest to the ocean water composition. The degree of depletion has been related by Dansgaard (1964) phenomenologically to geographic parameters, such as latitude, altitude, distance from the coast and the amount of precipitation (the so-called altitude effect, amount effect, etc.). The latitude effect over the North American continent is roughly at 0.5‰ $\delta^{18}O$/degree latitude (Yurtsever, 1975). The altitude effect depends on local climate and topography, but gradients of 0.15—0.5‰ $\delta^{18}O$/100 m (1.2—4‰ δD/100 m) are typical. The isotope composition of mean worldwide precipitation was estimated by Craig and Gordon (1965) to be $\delta^{18}O = -4‰$, $\delta D = -22‰$.

A multiple linear regression analysis between the mean isotopic composition of precipitation at network stations and latitude, altitude, precipitation amounts and temperature respectively (Yurtsever, 1975), shows that the isotope variations can be described primarily by the temperature variations (the partial coefficient of correlation being 0.815) and that the other parameters do not significantly improve this correlation. Using 363 monthly values of precipitation from stations Thule, Groenedel, Nord and Vienna, Yurtsever (1975) found the following linear relationship between monthly $\delta^{18}O$ values

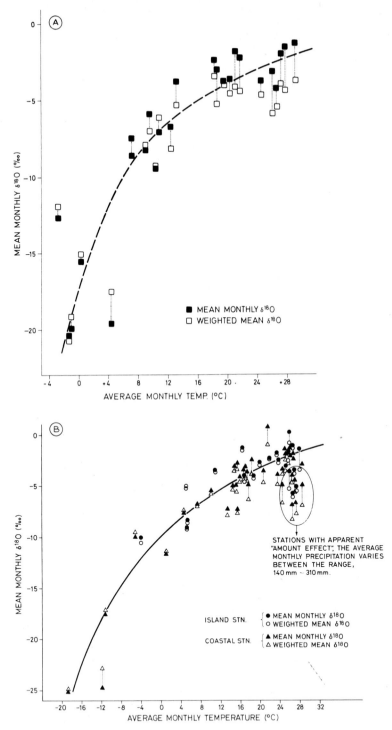

Fig. 1-6. $\delta^{18}O$-temperature correlation for average annual data from continental (A) and marine and coastal stations (B) (from Yurtsever, 1975).

and the mean monthly surface temperature:

$$\delta^{18}O = (0.521 \pm 0.014)\, T(°C) - (14.96 \pm 0.21)$$

Dansgaard (1964) gave the relationship of $\delta^{18}O = 0.695\, T(°C) - 13.6$ for Northern Atlantic coastal stations. On a worldwide basis the relationship is far from linear, however, but is represented by a set of curves for different geographic areas, which all approach values close to $\delta = 0$ at the higher-temperature range (Fig. 1-6).

The change of the ^{18}O concentration in precipitation is correlated with a commensurate change in the concentration of deuterium, as was first noted by the Chicago group (Friedman, 1953). A very good linear relationship exists between the (amount weighted) mean δD and $\delta^{18}O$ values of all precipitation samples from the IAEA network:

$$\delta D = (8.17 \pm 0.08)\, \delta^{18}O + (10.56 \pm 0.64)\, ‰$$

with a correlation coefficient of $r = 0.997$ and a standard error of estimate of δD of $\pm 3.3‰$. This line closely resembles the "meteoric water line" (MWL), defined by Craig (1961b) as the locus of the isotope composition of worldwide freshwater sources:

$$\delta D = 8\, \delta^{18}O + 10‰$$

The δD vs. $\delta^{18}O$ relationship for precipitation in any given region, however, often differs from the global equation. Eight continental stations in North America, for example, show the relationship of $\delta D = 7.95\, \delta^{18}O + 6.03‰$ which has a similar slope but a lower intercept value than the meteoric water line. On the other hand, data from stations in regions such as Australia, the Middle East and Central Africa fall above the meteoric water line with a definite deuterium excess.

In order to relate the composition of any water sample to the MWL, Dansgaard (1964) has defined the "d-parameter" (deuterium excess parameter) as $d = \delta D - 8\delta^{18}O$. The value of d of any sample can be interpreted as the intercept with the δD axis (for $\delta^{18}O = 0$) of the line with slope $\Delta\delta D/\Delta\delta^{18}O = 8$ which passes through that point; such a line presumably would be the locus of all precipitation samples which are derived from that particular air mass by rainout; as will be discussed, the "d-parameter" according to this view relates to the vapour-forming process whereas the position of any point on the slope = 8 line is determined by the rainout process. Lately some reservations have, however, been raised concerning this approach.

In some of these areas, the linear correlation between $\delta^{18}O$ and δD values of precipitation samples (to the extent that it exists at all) is not that of a slope = 8 line. This is particularly the case in arid regions where $\Delta\delta D/\Delta\delta^{18}O <$

8 and typically has values of 5—6. This also applies to fifteen tropical island stations (Yurtsever, 1975) where the best fit line was found to be $\delta D = 6.17 \delta^{18}O + 3.97‰$, with a slope significantly lower than that of the MWL.

Much less is known concerning the isotope composition of atmospheric moisture, mainly because sampling of atmospheric moisture has to be done without fractionation of the isotopes; this requires special techniques (Horibe, reported by Craig and Gordon, 1965; Roether and Junghans, 1966) which cannot be activated easily on a routine basis.

In continental areas the surface air seems close to isotopic equilibrium with local precipitation (Craig and Horibe, 1967), except in the vicinity of a lake (Fontes and Gonfiantini, 1967). In coastal areas the situation is more variable, due to the interplay of continental and oceanic air masses, the latter being closer to equilibrium with the ocean water than with the local rain. However, even marine air is not at isotopic equilibrium with surface waters; a survey of water vapour over major oceans has shown it to be depleted in ^{18}O relative to equilibrium vapour (Craig and Gordon, 1965), the deviation from equilibrium increasing with the saturation deficit in the oceanic air. Over wide regions of the Pacific Ocean, for example, one finds values of $\delta^{18}O$ between —10.5 and —14‰.

Upper air data reported by Taylor (1972) and Ehhalt (1974) show extreme depletion of the heavy isotopes in the upper troposphere, with values as low as $\delta D = -500‰$ at the 500-mbar level. This supposedly is the result of depletion in large cloud systems. Rain, however, never carries such an isotopic composition to the ground, as it forgets it through exchange with the ascending air masses in the cloud and beneath it. Only in hailstones can one find some record of the upper air composition, as outlined below.

MODELS OF THE ISOTOPE FRACTIONATION DURING EVAPORATION AND CONDENSATION OF WATER IN THE ATMOSPHERE

The first attempts at interpreting the isotope composition of atmospheric waters viewed the system as a global distillation column fed by evaporation from the ocean, with moisture condensing as a result of the cooling of the air masses at higher latitudes and altitudes. Provided that equilibrium exists between the condensed phases and the vapour at all times, then the degree of depletion of the heavy isotopic water species with respect to both hydrogen and oxygen is correlated in the residual waters. The actually observed slope of 8 of the MWL was close enough to the enrichment expected in such a column on the basis of the vapour pressure differences of the isotopic molecules at ambient temperatures to suggest that precipitation forms under phase equilibrium conditions. On the other hand, the MWL does not pass through the datum point of surface ocean water, as it should do in a proper distillation column which is fed by the ocean; evidently evaporation from the ocean does not produce vapour which is in isotopic equilibrium with it.

The simplest model to describe changes in the isotope composition during a phase transition is the classical Rayleigh model, which requires the material formed to be removed in instantaneous isotopic equilibrium with the phase from which it is derived. The equations which describe the changing isotopic composition of the vapour from which water condenses is then given by:

$$\frac{dR_V}{dN} = \frac{R_V}{N}(\alpha_e - 1); \quad \frac{d\ln R_V}{d\ln N} = (\alpha_e - 1) \tag{1}$$

where:

$$\alpha_e \equiv \frac{dN_i/dN}{N_i/N} = \frac{R_C}{R_V}$$

the unit separation factor at equilibrium, is a function of temperature. N and N_i are the number of moles of the abundant and heavy isotopic species, respectively; $R = N_i/N$; subscript V stands for the vapour phase, C for the liquid condensed phase and 0 for initial condition; $f = N/N_0$.

The best values for α_e are those given by Majoube (1971); for ^{18}O:

$$\ln \alpha = \frac{1.137}{T^2} \times 10^3 - \frac{0.4156}{T} - 2.0667 \times 10^{-3};$$

and for D:

$$\ln \alpha = \frac{24.844}{T^2} \times 10^3 - \frac{76.248}{T} + 52.612 \times 10^{-3}.$$

Equation (1) in integrated form and δ nomenclature gives:

$$\delta_c = \delta_0 + (\bar{\alpha}_e - 1) \ln f \tag{1a}$$

Dansgaard (1964) used such a model to describe the process of rainout. Since the process is one of cooling, the value of parameter α changes and integration must be carried out numerically, with α varying in the fashion prescribed by the cooling rate of the air mass. Dansgaard could fit such calculations to the observed temperature effect of $d\delta^{18}O/dT = 0.7‰/°C$ for the case where an air mass with a dew point of 20°C cools adiabatically.

S. Epstein (unpublished lecture notes) and Craig and Gordon (1965) elaborated on the simple Rayleigh model by taking into account the liquid water content of a cloud L, which is in isotopic equilibrium with the vapour. In this, the two-phase Rayleigh model, the evolution of the isotopic composition depends then on both L and the degree of rainout measured by f, as

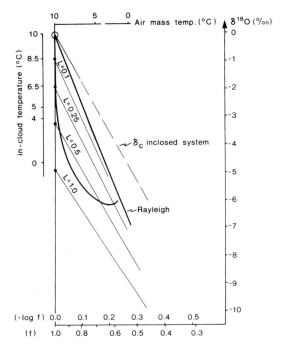

Fig. 1-7. Change in ^{18}O isotopic composition of precipitation as a result of rainout as predicted by various models, for an air mass with dew point of $10°C$ (adapted from Gat and Dansgaard, 1972). f = remaining fraction of atmospheric moisture, except for dashed line where f is remaining vapour phase. Dashed line indicates the composition of liquid water in a closed system (with no rainout). Solid line gives the δ-values in precipitation for the classical Rayleigh process according to the equation: $\delta_c = \delta_0 + (\alpha - 1) \ln f$. The lines marked L refer to precipitation in the two-phase model according to equation (2a); L is liquid water content of the air mass. Thick curved line represents the hypothetical evolution of precipitation on passage over a mountain chain with buildup and then dissipation of clouds. Temperature scale on left refers to the in-cloud temperature which causes the L-values as marked.

shown in equation (2):

$$d(NR) = R_c \cdot dN \tag{2}$$

$$\frac{\partial \ln R}{\partial \ln f} = \frac{(\alpha_e - 1)}{1 + \alpha_e L} \; ; \quad \frac{\partial \ln R}{\partial L} = \frac{(\alpha_e - 1)}{(1 + \alpha_e L)(1 + L)} \tag{2a}$$

The changes of δ with rainout predicted from these models are shown in Fig. 1-7.

Unlike the in-cloud processes, the evaporation into humid (yet unsaturated) air differs from a Rayleigh process in two respects (Craig et al., 1963): the separation factor is larger than the one dictated by the difference in vapour pressure of the isotopic water molecules and, secondly, a limiting enrichment is reached in the residual evaporated water body as it dries up. The first of these effects reflects the favoured transport of the lighter isotopic species in the diffusion-controlled stage of the evaporation process; the second effect is the result of the backflow of ambient moisture to the surface which finally balances the outflowing isotope flux.

The evaporation process has been modelled by Craig and Gordon (1965) as a series of steps which transport water from a virtually saturated sublayer near the liquid surface, through a region in which molecular diffusion plays a role, into the fully turbulent atmosphere of humidity h (normalized with respect to saturated vapour at the temperature of the surface) in which no further isotope fractionation occurs. A mathematical treatment for the case of an isolated water body without inflow leads to the equation:

$$\frac{d \ln R}{d \ln N} = \frac{h(R_s - R_a)/R_s - (1 - 1/\alpha_e) - \Delta\epsilon}{1 - h + \Delta\epsilon} = \frac{h(R_s - R_a)/R_s - \epsilon}{1 - h + \Delta\epsilon} \quad (3)$$

$$1 - \frac{1}{\alpha_e} + \Delta\epsilon \equiv \epsilon$$

where subscripts s and a stand for the surface waters and atmospheric moisture respectively. According to this equation R_s approaches a limiting steady-state (ss) value when $R_{ss} = R_a (1 - \epsilon/h)$ or, in δ nomenclature, when $\delta_{ss} \simeq \delta_a + \epsilon/h$. This is achieved only in the final stages of the dessication of the water body.

A more rigorous treatment takes into account possible differences between surface water and the bulk properties, due to incomplete mixing. For practical purposes this effect is negligibly small. Note furthermore that in the literature which is concerned exclusively with surface waters, the fractionation factor is defined as $\alpha = R_V/R_C$; this is the reciprocal value of the α_e defined in our treatment of the in-cloud processes. For the sake of consistency of presentation in this review, $\alpha_e = R_C/R_V$ will be used throughout.

A water body open to inflow (I) may also reach a steady-state isotope composition when at hydrologic steady state. This will be achieved once the isotope composition of the evaporation flux (δ_E) equals that of the (net) inflow: $\delta_E = \delta_I$. This steady-state concentration is given by the expression (Gat and Levy, 1978):

$$\delta^*_{ss} = \left(\frac{\epsilon}{h} + \delta_a + \frac{I}{E} \cdot \frac{(1-h)}{h} \delta_I\right) \bigg/ \left(1 + \frac{1-h}{h} \cdot \frac{I}{E}\right) \quad (4)$$

which for the case of a terminal lake, where $I = E$, simplifies to:

$$(\delta_s - \delta_I) = \epsilon + h(\delta_a - \delta_I) \quad (4a)$$

Here, once more, as is the case in the isolated dessicating water body, the isotope enrichment (relative to input) depends on atmospheric humidity and on the isotope fractionation factor.

In these equations $\Delta\epsilon$ — the kinetic excess separation factor — is the difference between the effective and thermodynamic separation factor; its magnitude depends on the detailed evaporation mechanism, but to the best of present knowledge it can be approximated by the expression of $15 \cdot (1-h)$‰ for ^{18}O and $13 \cdot (1-h)$‰ for deuterium (Merlivat, 1978). The numerical value of $\Delta\epsilon$ for the two isotopic species for any given humidity is rather similar. As far as ^{18}O is concerned this kinetic effect nearly doubles the enrichment compared to the equivalent one of the equilibrium process when $h \sim 0.5$. The value of $(1 - 1/\alpha_e)$ for deuterium, however, is larger by almost one order of magnitude than either the kinetic effect or the equivalent ^{18}O enrichment factor and the effect of $\Delta\epsilon$ is thus not pronounced as far as the deuterium enrichment is concerned. As a result of this the value of $\Delta\delta D/\Delta\delta^{18}O$ during evaporation is less than in the equilibrium column, so that the slope of the evaporation lines, i.e. the locus of the isotope compositions of residues of evaporating surface waters, has values of between 3 and 6. The detailed dependence of the slope of the evaporation line on the ambient parameters and on the separation factor is given by Gat (1971).

Enrichment of isotopes also occurs in raindrops falling through unsaturated air beneath the cloud base. This enrichment also usually proceeds along "evaporation lines" of slopes less than 8, and is most noted in rains of the arid zone or in the case of "slight" rain amounts. The process was studied in the laboratory by Stewart (1975) who found that the enrichment in drops which evaporate into a dry atmosphere fitted a Rayleigh-type equation with an exponent of $\alpha_e(D/D_i)^{0.58}$ *. This is close to the one predicted by theory for the gas-kinetically controlled diffusion process, taking the peculiar aerodynamic structure around the drop (so-called ventilation factor) into account.

IN-STORM VARIATION OF ISOTOPIC COMPOSITION, CLOUD MODELS AND HAILSTONE STUDIES

There are surprisingly few studies on stable isotope variations in discrete meteorological events except for the early measurements by Dansgaard (1953) and Epstein (1956). Bleeker et al. (1966) observed showers associated with different synoptic situations and found characteristic patterns

---

* D, D_i = the value of the gas kinetic diffusion coefficient through air of H_2O and of the isotopic water molecules, respectively. There is some disagreement between different authors as to their value, but measurements of Merlivat (1978) are the most recent ones.

for the passages of a cold front, warm fronts, etc. The first shower of many storms was found to be enriched in the heavy isotopic species compared to later rains at that site; this effect is attributed to the partial evaporation of raindrops in the yet undersaturated surface air. In a classical study, Ehhalt and Ostlund (1970) collected both rain and vapour during a flythrough of Hurricane Faith and determined the degree of disequilibrium dϵ between them (for the case of deuterium). dϵ_D was found to be usually negative, with a minimum value of $-20\%_o$. Very close to the eye of the hurricane dϵ, however, is positive. The authors interpreted the data by assuming that rain condensing at high levels (above the sampling points) has a low deuterium content and will therefore exhibit a δD value which is lower than the one corresponding to isotopic equilibrium with the local environment, more so in the case of larger raindrops for which the isotopic adjustment time (i.e. approach to equilibrium by means of molecular exchange) is of the order of a few minutes. Conversely, in very strong updraft regions in which most of the liquid water is carried in the form of large drops, we expect positive dϵ values. It is thus concluded from the observations that the inner edge of the rainband cell (the eye of the hurricane) is a region of updrafts with a zone of heavy rains from high altitudes around it. The vertical gradient of the isotope content and isotopic exchange between droplets and vapour are the key elements in the interpretation of the data.

The occurrence of fast isotopic exchange between droplets, water vapour and falling rain (Friedman et al., 1962) has been the basis for a number of attempts to model clouds as multi-stage vertical distillation columns. Kirschenbaum (1951) apparently was the first to take this approach. More recently Tzur (1971) modelled a cloud as a system of air and droplets updrafting counter-currentwise to the falling larger drops. His prediction is for isotopic equilibrium between rain and vapour to be established at some height above the cloud base; the height depends on the exchange kinetics and in particular on the drop size, as the rate of approach to equilibrium with the ambient air by molecular exchange depends critically on it. For example for raindrops falling at terminal free fall velocities a relaxation distance of about 5 m applies to drops of radius of 0.1 mm and 1 mile for 1-mm drops (Bolin, 1958).

Ehhalt (1967 and unpublished report, 1972) has presented a most elaborate model of this kind in order to derive the height distribution of the isotopic composition in the cloud to be used for the study of hailstone growth. According to Ehhalt the vertical local in-cloud gradient of isotopic composition is described rather well by a Rayleigh law with the fractionation factor chosen appropriate for the presence of solid phases. At cloud top levels the depletion in the heavy isotopes in atmospheric water becomes extreme. However, as far as the isotope composition of the rain under the cloud itself is concerned, all these models lead to the conclusion that rain is close-to-equilibrium with the moisture in surface air.

Only snow or hail keep a record of the upper air isotopic composition and, therefore, have been favourite objects for research as probes of a cloud's internal structure (Facy et al., 1963; Bailey et al., 1969; Macklin et al., 1970; Jouzel et al., 1975). The approach in all these cases was to assume a vertical isotope profile as predicted by a Rayleigh-type law for an adiabatically cooling updraft column. The measured stable isotope content in different parts of the hailstone is then matched to this profile, to determine the height of formation. The amount of material accreted at each level is related to the length of sojourn of the hailstone at that height or to the local velocity. For example, in the hailstone described in Fig. 1-8 one can identify two ascents of the hailstone through the cloud.

Such a model predicts a rather constant tritium content in the updraft core of the cloud (Ehhalt, 1967). This feature was used by Jouzel et al. (1975) as a test of the growth of each hailstone within the updraft zone. Variations in the tritium content within some hailstones were interpreted as due to the entrainment of ambient air into the ascending boundary layer of the cloud; in the latter case one can no longer assume a Rayleigh profile and the stable isotope data cannot then be assigned to an uniquely determined

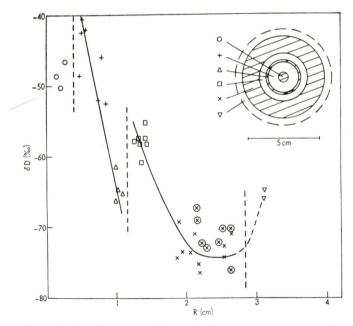

Fig. 1-8. Distribution of δD in a hailstone, as a function of its radius (from Macklin et al., 1970). The equivalent spherical model of the stone is depicted in the upper right-hand corner and the symbols used for samples from the various layers are as indicated. The opaque layers are hatched. The encircled crosses are the δD values for samples comprised in part of the clear ice layer lying within the thick outer opaque layer.

height. It must be noted that a vertical "Rayleigh distribution" of the stable isotopes in an updraft zone is an assumption which has not been put to the test of factual confirmation. More measurements of the vapour composition in and around cloud systems would be desirable.

These studies have illustrated the absence of a simple or unique correlation between the tritium concentration and the stable isotope content in atmospheric waters. In convective storms (Groeneveld, 1977) as in hail these parameters are not correlated at all since tritium levels are rather constant and stable isotope contents vary widely. The vertical distribution of tritium and heavy isotopes in the free atmosphere is, however, inversely related and this may express itself in an apparent correlation between them in some instances, especially where different precipitation samples represent the composition at different air strata. Such a situation may arise when comparing rain and snow samples. Tritium levels within individual rain showers in Israel were found to vary by a factor of two or so over periods of hours (Gat et al., 1962). Such variations seem to be more typical of coastal regions, where marine and continental air masses interplay.

STABLE ISOTOPE DISTRIBUTION IN ATMOSPHERIC WATERS: THE GLOBAL MODEL

The classical hypothesis for explaining the observed stable isotope distribution was that atmospheric vapour is formed through non-equilibrium fractionation by evaporation from the oceans, thereby generating the deuterium excess of $d = 10‰$ of the meteoric waters. Rainout was viewed as a Rayleigh fractionation process which operates on this primary vapour, with condensation occurring under conditions of equilibrium between the phases (and thus invariant with respect to the value of the "d-parameter"). The latitude, altitude and amount effects are then viewed simply as a measure of the degree of rainout.

Craig and Gordon (1965) showed that the marine atmosphere is neither simply constituted of the evaporation flux from the ocean nor is it in local equilibrium with the surface ocean waters. As shown schematically in Fig. 1-9 the evaporative flux (which from material-balance considerations is equal to the mean worldwide precipitation) mixes in the marine atmosphere with depleted residual vapour originating from the marine cloud layers. According to this model the isotope composition of the marine vapour, the precipitation condensed from it and the mean worldwide precipitation, all fall on the meteoric water line.

Local deviations from the mean vapour composition over parts of the ocean result from particular local air-sea interaction patterns and the air mixing regime. Probably there is a vertical gradient of the isotope composition as well, so that rain associated with stagnant low-altitude air, marine

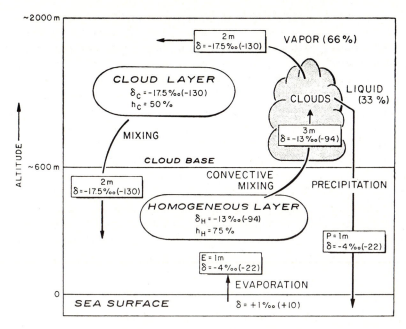

Fig. 1-9. Stable isotope composition of the marine atmosphere (from Craig and Gordon, 1965). δ-values are given for ^{18}O and, in parentheses, for deuterium.

drizzle and orographic precipitation is as a rule enriched in the heavy isotopes relative to precipitation from deep convective systems. The value of the "d-parameter" over some marine areas deviates considerably from the worldwide average value of $d = 10‰$. Extreme deviations are found in the Eastern Mediterranean Sea area, where the deuterium excess parameter in precipitation shows values of up to $d = 37‰$, presumably the result of evaporation occurring into very dry (continental) air (Gat and Carmi, 1970). Other local oceanic vapour sources, such as the Persian Gulf, the Indian Ocean or the Black Sea, also generate precipitation whose isotopic composition deviates from the worldwide meteoric water line. The latter dominates conditions over large continents and according to this view the continental precipitation originates from vapour which represents an average of the major mid-latitude oceanic source regions for atmospheric moisture.

The re-evaporation of water from over continents or lakes can also affect the composition of atmospheric moisture, as shown at Lake Geneva by Fontes and Gonfiantini (1970). Evapotranspiration, however, returns the groundwaters to the atmosphere essentially unchanged with respect to their isotopic composition (Zimmerman et al., 1967) as does a closed drainage system in which all the inflowing waters are finally evaporated. This effect occurs notwithstanding the enrichment of isotopes in the plant leaves

(Gonfiantini et al., 1965), since there is a constant throughflow of the water through the plants.

The recycling of water into the atmosphere, the most conspicuous example of which occurs in the Amazon basin (Salati et al., 1979) compensates for the rainout to some degree and restores the atmospheric water content, thereby reducing the inland isotopic gradient of depletion of the heavy isotopes. Re-evaporation, however, does not change the value of the "d-parameter" except by evaporation from a partially drained surface water body. In that case the evaporation flux increases the d-value of the ambient moisture just as it does over the ocean; this being the mirror-image effect to the enrichment of the residual surface waters along an evaporation line.

The view that the worldwide meteoric water line represents precipitation formed by a continuous process of cooling from an air mass seems a simplistic one. Yet, the composition of the most depleted precipitation samples in Antarctica, with a measured isotope content of $\delta^{18}O = -50‰$ (Epstein et al., 1965) is correctly predicted by the simple Rayleigh law (equation 1a): $\delta_c = \delta_0 + (\alpha_e - 1) \ln f$, when f corresponds to the ratio of the moisture content of air cooled to about $-50°C$ (which is the mean temperature of the area according to Loewe, 1967) with air saturated at $20°C$. In spite of this apparent agreement there are a number of disturbing features:

(1) The meteorological pattern of air movement is not consistent with a gradual poleward drift of low-latitude surface air masses, as implied by the simple Rayleigh model.

(2) The simultaneous integration of Rayleigh equations for both ^{18}O and deuterium under conditions of cooling, i.e. changing α_e values, results in a curved line on the $\lambda^{18}O$ vs. λD plot. * This curve is straightened out to some extent on translation of λD to δD values (Taylor, 1972) but not sufficiently to account for the really remarkable straight line dependence of δD on $\delta^{18}O$ in the meteoric water line.

(3) The latitudinal "distillation column" is certainly not a closed one since additional evaporation from over the middle- and high-latitude ocean occurs. This could be expected to reduce the latitude gradient of stable isotope content in proportion to the amount of secondary vapour added to the system.

The excellent correlation found between $\delta^{18}O$ and the ground temperature also requires comment, as one would have rather expected a correlation with the temperature of condensation in the cloud. This feature is now understood to be the result of the isotope equilibrium which is established between the vapour and raindrops; precipitation is thus removed at close to isotopic equilibrium with the surface air at a rate determined roughly by the

---

* $\lambda \equiv \ln R = \ln(1 + \delta)$; integration of formulas such as equations (1), (2) or (3) results in an isotopic composition expressed in terms of the λ values; the approximate relation $d\lambda \simeq d\delta$ is valid only when δ values are not too far removed from the zero of the scale.

average cooling rate of the whole air mass, as required by a Rayleigh model. Such a model applies, however, only along a storm path during a continental traverse. In this case the slope described on the $\delta^{18}O$ vs. δD plot would have a value of between 7 and 9, depending on the temperature range and the initial isotope composition.

In contrast with the phenomena here described which apply over a limited geographical extent, H. Craig (unpublished communication) proposed the worldwide meteoric water line to be a mixing line, possibly between marine moisture on the one hand and drier air from the top of large cloud systems, on the other hand. Mixing lines can be shown to be perfectly straight lines on the δD vs. $\delta^{18}O$ diagrams. According to this view then, the basic global fractionation step occurs vertically rather than horizontally across latitudes, with the clouds functioning as multi-stage fractionation columns. The worldwide isotope pattern is dictated by the mixing pattern of the upper tropospheric air with the oceanic vapour, predominantly in areas of widespread air mass subsidence. The slope of 8 of the meteoric water line, according to this view, does not represent a continuous fractionation process, but rather the relation between the mean isotope composition of the upper air and surface air masses. Locally one should be able to identify families of meteoric water lines, each of which extends over a limited range of isotope compositions. Each of those is related to a limited system of precipitating air masses mostly along a latitude band. The worldwide mixing and circulation patterns of tropospheric air and the buildup and dissipation of cloud fields thus assumes the central role in determining the worldwide distribution pattern of stable isotopes in the atmosphere. Details of this model have, however, not been worked out and we cannot identify the air masses involved.

In this global model one still considers the altitude or amount effect to be a manifestation of the moisture loss from an air mass by cooling; the latitude effect has a more complicated geophysical structure.

There is, however, another effect which modifies the simple picture of the distribution of the stable isotopes in atmospheric waters. This is the enrichment of the heavy isotopic species as a result of evaporation from falling raindrops beneath the cloud base. This effect, which was first described by Ehhalt et al. (1963) appears as the dominant reason for the clustering of data in the arid regions along evaporation lines of slope less than 8. The same effect has been recognized by Dansgaard (1964) in his analysis of worldwide precipitation samples as a shift between the weighted and unweighted mean composition of the precipitation. The "evaporated" rain samples are characterized by a smaller value of the "d-parameter" than the in-cloud moisture and as a result the meteoric water line loses its diagnostic value for characterizing the vapour source. One would expect, however, that snow or ice are not prone to the processes of exchange and enrichment and represent the in-cloud composition with greater fidelity than the rain samples collected by the precipitation network. As an example, snow in Jerusalem was reported by Gat and Dansgaard (1972) with a deuterium excess of 37‰, which is higher than the "d-parameter" of any of the rain samples in that area at that time. This arouses the suspicion that data of rain samples are all shifted to some extent relative to the composition at the cloud base, where rain is assumed to be in isotopic equilibrium with the atmospheric moisture. Taylor (1972) found indeed a difference of almost 10‰ between the "d-parameter" of high-altitude vapour over Europe and that of the precipitation in that region.

The secondary enrichment of the raindrops as a result of partial evaporation during their fall to the ground is responsible for some of the geographic effects on the isotope composition of rain; it contributes to the altitude effect, especially in valleys or in the mountain shadow, where cloud base levels are high so that there is a lengthy path for the falling raindrops through unsaturated air. Appropriately, this feature was named a pseudo-altitude effect (Moser and Stichler, 1971). The amount effect is also related to this phenomenon: the initial portions of a shower are enriched in the heavy isotopic species as a result of the evaporation (Dansgaard, 1964); as the shower continues, humidity beneath the cloud base increases and the continuing molecular exchange brings its composition closer to equilibrium with the incloud isotope composition. Because of this effect the duration of a shower may appear as an additional parameter on which the isotopic composition of rain depends.

From the foregoing discussion it is evident that one still lacks a satisfactory global model of the isotope distribution in atmospheric waters and more information, especially more atmospheric moisture data, are required. On the other hand, the extensive precipitation surveys have established in reasonable detail the meteoric isotope input into the local hydrological systems, as a basis for ground- and surface water studies.

REFERENCES

Athavale, R.N., Lal, D. and Rama, S., 1967. The measurement of tritium activity in natural waters, II. Characteristics of global fallout of ^{3}H and ^{90}Sr. *Proc. Indian. Acad. Sci.*, 65: 73—103.
Baertschi, P., 1976. Absolute ^{18}O content of Standard Mean Ocean Water. *Earth Planet. Sci. Lett.*, 31: 341—344.
Bailey, I.H., Hulston, J.R., Macklin, W.C. and Stewart, J.R., 1969. On the isotopic composition of hailstones. *J. Atmos. Sci.*, 26: 689—694.
Begemann, F., 1960. Natural tritium. In: *Nuclear Geology, Varenna 1960*. Laboratorio di Geologia Nucleare, Pisa, pp. 109—128.
Begemann, F. and Libby, W.F., 1957. Continental water balance, groundwater inventory and storage times, surface ocean mixing rates and world wide water circulation pattern from cosmic ray and bomb tritium. *Geochim. Cosmochim. Acta*, 12: 277.
Bigeleisen, J., 1962. Correlation of tritium and deuterium isotope effects. In: *Tritium in the Physical and Biological Sciences, 1*. IAEA, Vienna, pp. 161—168.

Bleeker, W., Dansgaard, W. and Lablans, W.N., 1966. Some remarks on simultaneous measurements of particulate contaminants including radioactivity and isotopic composition of precipitation. *Tellus*, 18: 773—785.

Bolin, B., 1958. On the use of tritium as a tracer for water in nature. *Proc. 2nd U.N. Conf. Peaceful Uses of Atomic Energy, Geneva*, 18: 336—344.

Booker, D.V., 1965. Exchange between water droplets and tritiated water vapour. *Q. J. R. Meteorol. Soc.*, 91: 73—79.

Brown, R.M., 1970. Distribution of hydrogen isotopes in Canadian waters. In: *Isotope Hydrology*. IAEA, Vienna, pp. 3—21.

Brown, R.M. and Grummitt, W.E., 1956. The determination of tritium in natural waters. *Can. J. Chem.*, 34: 220—226.

Carmi, I. and Gat, J.R., 1973. Tritium in precipitation and fresh water sources in Israel. *Isr. J. Earth Sci.*, 22: 71—92.

Craig, H., 1961a. Standard for reporting concentrations of deuterium and oxygen-18 in natural waters. *Science*, 133: 1833—1834.

Craig, H., 1961b. Isotopic variations in meteoric waters. *Science*, 133: 1702—1703.

Craig, H. and Gordon, L., 1965. Deuterium and oxygen-18 variation in the ocean and the marine atmosphere. In: E. Tongiorgi (Editor), *Stable Isotopes in Oceanographic Studies and Paleotemperatures, Spoleto 1965*. pp. 9—130.

Craig, H. and Horibe, Y., 1967. Isotopic characteristics of marine and continental water vapour. *Trans. Am. Geophys. Union.*, 48: 135—136.

Craig, H., Gordon, L. and Horibe, Y., 1963. Isotopic exchange effects in the evaporation of water: low-temperature experimental results. *J. Geophys. Res.*, 68: 5079—5087.

Dansgaard, W., 1953. The abundance of ^{18}O in atmospheric water and water vapour. *Tellus*, 5: 461—469.

Dansgaard, W., 1964. Stable isotopes in precipitation. *Tellus*, 16: 436—468.

Ehhalt, D., 1967. Deuterium and tritium content of hailstones: additional information on their growth. *Trans. Am. Geophys. Union*.

Ehhalt, D., 1974. Vertical profiles of HTO, HDO and H_2O in the troposphere. *NCAR Tech. Note*, NCAR-TN/STR-100: 131 pp.

Ehhalt, D.H., 1971. Vertical profiles and transport of HTO in the troposphere. *J. Geophys. Res.*, 76: 75—84.

Ehhalt, D. and Östlund, G., 1970. Deuterium in Hurricane Faith 1966. *J. Geophys. Res.*, 75: 2323—2327.

Ehhalt, D., Knott, K., Nagel, J.F. and Vogel, J.C., 1963. Deuterium and oxygen-18 in rain water. *J. Geophys. Res.*, 68: 3775—3780.

Epstein, S., 1956. Variations of the $^{18}O/^{16}O$ ratios of fresh water and ice. *Natl. Acad. Sci. Nucl. Sci. Ser., Rep.*, 19: 20—28.

Epstein, S., Sharp, R. and Gow, A.J., 1965. Six year record of oxygen and hydrogen isotope variations in South Pole firn. *J. Geophys. Res.*, 70: 1809—1814.

Eriksson, E., 1966. Major pulses of tritium in the atmosphere. *Tellus*, 17: 118—130.

Eriksson, E., 1967. Isotopes in hydrometeorology. In: *Isotopes in Hydrology*. IAEA, Vienna, pp. 21—33.

Facy, L., Merlivat, L., Nief, G. and Roth, E., 1963. The study of the formation of a hailstone by means of isotopic analysis. *J. Geophys. Res.*, 68: 3841—3848.

Falting, V. and Harteck, P.Z., 1950. Der Tritium Gehalt der Atmosphäre. *Z. Naturforsch.*, 5a: 438—439.

Fontes, J. and Gonfiantini, R., 1967. Composition isotopique et origine de la vapeur d'eau atmospherique dans la région du Lac Léman. *Earth Planet. Sci. Lett.*, 7: 325—329.

Friedman, I., 1953. Deuterium content of natural water and other substances. *Geochim. Cosmochim. Acta*, 4: 89—103.

Friedman, I., Machta, L. and Soller, R., 1962. Water vapour exchange between a water droplet and its environment. *J. Geophys. Res.*, 67: 2761—2766.

Gat, J.R., 1971. Comments on the stable isotope method in regional groundwater investigation. *Water Resour. Res.*, 7: 980—993.

Gat, J.R. and Carmi, I., 1970. Evolution of the isotopic composition of atmospheric waters in the Mediterranean Sea area. *J. Geophys. Res.*, 75: 3039—3048.

Gat, J.R. and Dansgaard, W., 1972. Stable isotope survey of the fresh water occurrences in Israel and the Jordan Rift Valley, *J. Hydrol.*, 16: 177—211.

Gat, J.R. and Levy, Y., 1978. Isotope hydrology of inland sabkhas in the Bardowil area. Sinai. *Limnol. Oceanogr.*, 123: 841—850.

Gat, J.R., Karfunkel, V. and Nir, A., 1962. Tritium content of rainwater from the Eastern Mediterranean area. In: *Tritium in the Physical and Biological Sciences, 1*. IAEA, Vienna, pp. 41—54.

Giauque, W.F. and Johnston, H.L., 1929. An isotope of oxygen, mass 18. *J. Am. Chem. Soc.*, 51: 1436—1441.

Gonfiantini, R., Gratsiu, S. and Tongiorgi, E., 1965. Oxygen isotopic composition of water in leaves. In: *Isotopes and Radiation in Soil-Plant Nutrition Studies*. IAEA, Vienna, p. 405.

Groeneveld, D.J., 1977. *Tritium Analysis of Environmental Water*. Ph. D. Thesis, University of Groningen, Groningen, 131 pp.

Grosse, A., Johnston, W.M., Wolfgang, R.L. and Libby, W.F., 1951. Tritium in nature. *Science*, 113: 1—2.

Hageman, R., Nief, G. and Roth, E., 1970. Absolute isotopic scale for deuterium analysis of natural water: absolute D/H ratio for SMOW. *Tellus*, 22: 712—715.

International Atomic Energy Agency (IAEA), 1969, 1970, 1971, 1973, 1975. *Environmental Isotope Data, No. 1, 2, 3, 4 and 5, World Survey of Isotope Precipitation From 1953 Onwards*. IAEA, Tech. Rep. Ser., Nos. 96, 117, 129.

Jouzel, J., Merlivat, L. and Roth, E., 1975. Isotopic study of hail. *J. Geophys. Res.*, 80: 5015—5030.

Kaufman, S. and Libby, W.F., 1954. The natural distribution of tritium. *Phys. Rev.*, 93: 1337—1344.

Kirschenbaum, I., 1951. Physical properties and analysis of heavy water. *Natl. Nucl. Energy Ser.*, Div. III, 4A.

Lal, D. and Peters, B., 1967. Cosmic ray produced radioactivity on the earth. In: S. Flugge (Editor), *Encyclopedia of Physics, 46*. Part 2, pp. 551—612.

Libby, W.F., 1946. Atmospheric helium-3 and radiocarbon from cosmic radiation. *Phys. Rev.*, 69: 671—672.

Loewe, F., 1967. Antarctic meteorology. In: R.W. Fairbridge (Editor), *Encyclopedia of Atmospheric Science*. Reinhold, New York, N.Y., pp. 19—26.

Macklin, W.C., Merlivat, L. and Stevenson, C.M., 1970. The analysis of a hailstone. *Q. J. R. Meteorol. Soc.*, 96: 472—486.

Majoube, M., 1971. Fractionnement en oxygène 18 et en deuterium entre l'eau et sa vapeur. *J. Chim. Phys.*, 10: 1423—1436.

Martell, E.A., 1970. Transport patterns and residence times for atmospheric trace constituents vs. altitude. *Adv. Chem.*, 93: 138—157.

Merlivat, L., 1978. Molecular diffusivities of water $H_2^{16}O$, $HD^{16}O$ and $H_2^{18}O$ in gases. *J. Chim. Phys.*, 69: 2864—2871.

Merlivat, L. and Coantic, M., 1975. Study of mass transfer at the air-water interface by an isotopic method. *J. Geophys. Res.*, 80: 3455—3464.

Moser, H. and Stichler, W., 1971. Die Verwendung des Deuterium und Sauerstoff-18 Gehalts bei hydrologischen Untersuchungen. *Geol. Bavarica*, 64: 7—35.

Nir, A., Kruger, S.J., Lingenfelder, R.E. and Flamm, E.J., 1966. Natural tritium. *Rev. Geophys.*, 4: 441—456.

Newell, R.S., 1963. Transfer through the tropopause and within the stratosphere. *Q. J. R. Meteorol. Soc.*, 89: 167—204.
Roether, W. and Junghans, H.G., 1966. Apparatur zur kontinuierlichen Gewinnung von Luftwasserdampfproben. *Jahresber. Phys. Inst., Univ. Heidelberg*, II.
Salati, E., Dal Ollio, A., Matsui, E. and Gat, J.R., 1979. Recycling of water in the Amazon basin: an isotopic study. *Water Resour. Res.* (in press).
Stewart, M., 1975. Stable isotope fractionation due to evaporation and isotopic exchange of falling water drops: application to atmospheric processes and evaporation of lakes. *J. Geophys. Res.*, 80: 1138—1146.
Taylor, C.B., 1966. Tritium in southern hemisphere precipitation, 1953—1964. *Tellus*, 18: 105—131.
Taylor, C.B., 1972. The vertical variations of the isotopic concentrations of tropospheric water vapour over continental Europe and their relationship to tropospheric structure. *N.Z. Dep. Sci. Ind. Res., Inst. Nucl. Sci., Rep.*, INS-R-107: 45 pp.
Tzur, Y., 1971. *Isotopic Composition of Water Vapour*. Ph.D. Thesis, Feinberg Graduate School at the Weizmann Institute, Rehovot.
Urey, H., Brickwedde, I.G. and Murphy, G.M., 1932. A hydrogen isotope of mass 2 and its concentration. *Phys. Rev.*, 39: 1—15.
Von Buttlar, H. and Libby, W.F., 1955. Natural distribution of cosmic-ray-produced tritium, II. *J. Inorg. Nucl. Chem.*, 1: 75—91.
Yurtsever, Y., 1975. Worldwide survey of stable isotopes in precipitation. *Rep. Sect. Isotope Hydrol., IAEA, November 1975*, 40 pp.
Zimmerman, U., Ehhalt, D. and Münnich, K.O., 1967. Soil water movement and evapotranspiration: change in the isotopic composition of the water. In: *Isotopes in Hydrology*. IAEA, Vienna, pp. 567—585.

Chapter 2

CARBON-14 IN HYDROGEOLOGICAL STUDIES

W.G. MOOK

INTRODUCTION

The radioactive isotope of carbon, ^{14}C, is continuously being produced in the earth's atmosphere. Through a series of exchange, assimilation and other processes of carbon fixation, all living organic matter contains ^{14}C, as well as the carbonate shells of marine and terrestrial molluscs. In the course of time ^{14}C decays at a certain rate according to the law of radioactive decay. This allows carbonaceous compounds to be dated by measuring the remaining ^{14}C.

The radiocarbon dating technique has become a standard tool for the archaeologist, the Quaternary geologist and the paleobotanist (Mook, 1980). Tens of thousands of age determinations have routinely been produced by tens of ^{14}C laboratories all over the world. The hydrogeological application, the age determination of groundwater, is, although technically a straightforward procedure, more complicated from a geochemical point of view. It will be the main aim of this chapter to state and discuss these problems and to approximate the meaning of a ^{14}C age of groundwater as closely as possible.

THE ABUNDANCE OF ^{14}C

The origin of ^{14}C

The primary source of ^{14}C is the transitional region between the stratosphere and the troposphere. ^{14}C is formed through a nuclear reaction between the secondary cosmic ray neutrons and nitrogen nuclei (Libby, 1965):

$^{14}N + n \rightarrow {}^{14}C + p$

The ^{14}C atoms oxidize to form $^{14}CO_2$ molecules. These become mixed with the inactive atmospheric carbon dioxide and subsequently enter the bio- and hydrosphere.

The natural abundance of ^{14}C

Although the ^{14}C production is not uniformly distributed over all latitudes — the highest production is at mid-latitudes — the ^{14}C content of the troposphere over one hemisphere is very much uniform. This has been shown by ^{14}C measurements on tree rings (Lerman et al., 1970). It also appeared that the tropospheric ^{14}C content over the southern hemisphere is about 4—5‰ lower than over the northern. This is probably caused by a larger uptake of CO_2 by the vast ocean surface of the southern hemisphere, while the mixing rate between both hemispheric tropospheres is limited.

From the atmospheric production of ^{14}C to the final radioactive decay mainly in marine and continental sediments, there is a continuous stream of ^{14}C (cf. Fig. 2-1). This stationary state maintains a more or less constant ^{14}C concentration in natural carbonaceous compounds as tropospheric carbon dioxide, oceanic bicarbonate, living terrestrial plants and animals. This does not necessarily mean that all compounds have equal ^{14}C concentrations. Differences up to 5% between the compounds mentioned do occur, because of different exchange rates between the carbon reservoirs and isotope fractionation effects.

Also within a reservoir small variations are observed. Through ^{14}C measurements on dendrochronologically dated wood several small- and large-frequency effects were established. This leads to calibrated ^{14}C time scales presently used by archeologists. For hydrogeological applications, however, this phenomenon of secular variations is of minor importance, considering dominating hydrochemical uncertainties.

^{14}C concentrations are generally given as specific activities, i.e. the radio-

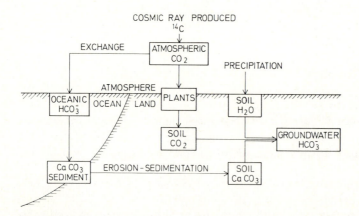

Fig. 2-1. A part of the natural carbon cycle, as far as relevant for the origin of ^{14}C in groundwater.

activity in ^{14}C disintegrations per minute per gram of carbon. The natural specific activity of atmospheric CO_2, oceanic bicarbonate and living plants and animals is about: $A_0 \simeq 14$ dpm/g C. According to an international convention specific activities are compared with a standard activity: A_{ox} = 0.95 times the specific activity of NBS oxalic acid (= 0.95 × 13.56 dpm/g C in the year A.D. 1950). The measured activity of a sample A is then given as a percentage of this standard activity:

$$a^{14} = A/A_{ox} \times 10^2 \text{ (pmc)} \tag{1}$$

(pmc = percentage of modern carbon). Deviations of a measured natural activity A_{0m} or a_{0m}^{14} from the standard activity are presented as delta values:

$$\delta^{14} = (A_{0m} - A_{ox})/A_{ox} \times 10^3 = (a_{0m}^{14} - 100) \times 10 \; (‰) \tag{2}$$

As we mentioned before, during the transition of carbon from one reservoir to another isotope fractionation occurs. The magnitude of this effect in a specific sample, which might be of any age, is still represented by the stable carbon isotopic ratio $^{13}C/^{12}C$. By measuring the δ^{13} value we can correct the original ^{14}C content of the material for fractionation effects occurred during the carbon fixation. Theory predicts (Bigeleisen, 1952; Craig, 1954) that the relative influence of a fractionation, whether equilibrium or kinetic, is twice as large for ^{14}C as it is for ^{13}C:

$$(A_c/A_m) - 1 = 2(R_c^{13}/R_m^{13} - 1)$$

where the subscripts c and m refer to the corrected and measured values respectively; R^{13} is the isotopic ratio $^{13}C/^{12}$ of the sample. Dividing A and R by the respective standard values: A_{ox} and R_{PDB}^{13} gives:

$$a_c^{14}/a_m^{14} = 1 + 2(\delta_c^{13} - \delta_m^{13})/(1 + \delta_m^{13} \times 10^{-3}) \tag{3}$$

By convention the ^{14}C activities are corrected to a δ_c^{13} value of $-25‰$ vs. PDB.

A definition, similar to equation (2), for relative deviations from the standard activity is in use for ^{13}C corrected activities:

$$\Delta^{14} = (A_{0c} - A_{ox})/A_{ox} \times 10^3 = (a_{0c}^{14} - 100) \times 10 \; (‰) \tag{4}$$

where A_0 and a_0^{14} apply to directly measured natural activities as well as to measured (tree ring) activities A_m and a_m^{14} which are corrected for the known material age T according to:

$$a_0^{14} = a_m^{14} \, e^{\lambda T} \tag{5}$$

From the equations (2), (3) and (4) we derive:

$$\Delta^{14} = \delta^{14} - 2(\delta^{13} + 25)(1 + \delta^{14} \times 10^{-3})/(1 + \delta^{13} \times 10^{-3}) \quad (6)$$

Generally the last factor and in many cases also the second last factor may be taken equal to unity:

$$\Delta^{14} \simeq \delta^{14} - 2(\delta^{13} + 25)$$

The influence of man on the ^{14}C abundance

During the last century man has influenced the natural ^{14}C content of the atmosphere in two ways. First the combustion of fossil fuel, containing no ^{14}C, has reduced the tropospheric ^{14}C level probably by about 10%. This is simply a linear dilution effect:

$$a^{14} = (1 - x)a_0^{14}$$

where a_0^{14} refers to the natural abundance and x is the fraction of fossil-fuel-derived CO_2. This "Suess effect" is completely equivalent to the changes in ^{13}C content of atmospheric CO_2 as described by Keeling (1961).

Since 1963 this effect, however, is surpassed by the injection of ^{14}C into the atmosphere by the nuclear explosion tests. The course of the tropospheric ^{14}C content is given in Fig. 2-3 for the northern and the southern hemisphere. Temporarily the ^{14}C content has been a factor of 2 above normal. The exchange with the ocean causes a gradual decrease towards natural conditions. In hydrogeology this nuclear bomb effect is very useful in dating recent groundwaters.

THE ^{14}C AGE DETERMINATION

The radioactive decay of ^{14}C

Radiocarbon decays according to:

$$^{14}C \rightarrow {}^{14}N + \beta^-$$

A ^{14}C activity is measured by detecting the β^- particles. Because the β^- particle energy (E_{max} = 156 keV) and the natural activities (<16 dpm/g C) are low, the detection efficiency has to be high. For this reason ^{14}C laboratories use proportional gas counters with internal sample or liquid scintillation counters.

The disintegration obeys the radioactive decay law:

$$A = A_0\, e^{-\lambda t}$$

where A and A_0 are the measured and original (natural) specific radioactivity of the material, t is time and λ is the decay constant. This is related to the half-life by $\lambda = \ln 2/t_{1/2}$. The half-life used for archaeological and geological dating is the "Libby" half-life of 5568 years. In hydrology we prefer to use the more recently established value of 5730 (±40) years.

The age determination

Provided A_0 is a known value, the age of carbonaceous material can be determined from the revised relation:

$$T = -(t_{1/2}/\ln 2) \ln a^{14}/a_0^{14} \tag{7}$$

where the coefficient is 8270 years. In routine ^{14}C dating the assumption of A_0 (and a_0^{14}) being known works well. Moreover, the resulting ^{14}C time scale can easily be checked by analysing dendrochronologically dated tree rings. This calibration has been obtained for the last 8000 years.

In hydrology the situation is much more complicated, because the origin of the carbon in water is less certain than it is for plant or animal carbon. Moreover, there is no way of checking the assumption with certainty. We can only try to obtain indications about the origin of the carbon from the stable carbon isotopic composition and from hydrochemical data.

^{14}C DATING IN GROUNDWATER

Throughout this chapter we use the following notation:

a, b, c	molar concentrations of dissolved CO_2, HCO_3^- and CO_3^{2-}, respectively
Σ	molar concentrations of dissolved total inorganic carbon ($= a + b + c$)
g	gaseous CO_2
$a_{g0}^{14}, \delta_{g0}^{13}$	^{14}C activity (in pmc) and ^{13}C content (in ‰) of soil CO_2
$a_{a0}^{14}, \delta_{a0}^{13}$	^{14}C activity (in pmc) and ^{13}C content (in ‰) of dissolved CO_2
$a_{l0}^{14}, \delta_{l0}^{13}$	^{14}C activity (in pmc) and ^{13}C content (in ‰) of marine $CaCO_3$
$a_{l}^{14}, \delta_{l}^{13}$	^{14}C activity (in pmc) and ^{13}C content (in ‰) of actual soil $CaCO_3$
$a_{b0}^{14}, \delta_{b0}^{13}$	^{14}C activity (in pmc) and ^{13}C content (in ‰) of dissolved HCO_3^- prior to exchange
$a_{bu}^{14}, \delta_{bu}^{13}$	^{14}C activity (in pmc) and ^{13}C content (in ‰) of dissolved HCO_3^- after (partial) exchange with "g0"
$a_{be}^{14}, \delta_{be}^{13}$	^{14}C activity (in pmc) and ^{13}C content (in ‰) of dissolved HCO_3^- after complete exchange
$a_{bs}^{14}, \delta_{bs}^{13}$	^{14}C activity (in pmc) and ^{13}C content (in ‰) of actual dissolved HCO_3^-
$a_{\Sigma}^{14}, \delta_{\Sigma}^{13}$	^{14}C activity (in pmc) and ^{13}C content (in ‰) of total dissolved carbon of recent recharge
$a_m^{14}, \delta_{\Sigma}^{13}$	^{14}C activity (in pmc) and ^{13}C content (in ‰) of total dissolved carbon in the actual sample

$\epsilon_g, \epsilon_a, \epsilon_c$ ^{13}C fractionation (in ‰) of gaseous CO_2, dissolved CO_2 and dissolved CO_3^{2-} with respect to dissolved HCO_3^-

ϵ_{ag} ^{13}C fractionation (in ‰) of dissolved with respect to gaseous CO_2

ϵ_s ^{13}C fractionation (in ‰) of solid calcite with respect to dissolved HCO_3^-

ϵ_{sa} ^{13}C fractionation (in ‰) of solid calcite with respect to dissolved CO_2

The origin of ^{14}C in groundwater

Several possible sources for the dissolved inorganic carbon content of groundwater have been indicated. We devote one section to each process. The first processes occur *in the upper soil zones*.

(1) CO_2 produced by the decomposition of *the organic material* in the soil (humus) or by root respiration in the unsaturated zone can dissolve *soil carbonates* (Münnich, 1957):

$$CaCO_3 + CO_2 + H_2O \rightarrow Ca^{2+} + 2\,HCO_3^- \qquad (8)$$

Starting from a ^{14}C content in the organics of 100 pmc (natural recent ^{14}C level) and in the limestone of 0 pmc (old marine carbonates), the dissolved bicarbonate is expected to have an original ^{14}C content a_{b0}^{14} (to be defined in section "The dissolution-exchange model") of 50 pmc. To simplify our discussion about the ^{13}C content in this section, we will mention the expected δ_{b0}^{13} for two alternative δ^{13} values of the soil CO_2: -25‰ for temperate climatic vegetations, and -15‰ for semi-arid regions (see section "The isotope dilution correction"). The δ^{13} value of soil limestone is assumed to be $+1$‰. The δ_{b0}^{13} value matching these conditions will be about -12 or -7‰ respectively. During the last years many examples have been given of a^{14} values of groundwater samples exceding 100 pmc. This is caused by the fact that the recent vegetation contains nuclear bomb ^{14}C. A nice example of the correlation between high tritium and (with some retardation) high ^{14}C values in groundwater has been given by Hufen et al. (1974).

(2) The above dissolution process might also be caused by *atmospheric CO_2* dissolved in the rain water, instead of the organic soil CO_2. The concentration of atmospheric CO_2 (0.03%) is in most cases negligibly small as compared to that of the CO_2 in the soil air (in the order of several percent).

The relatively small direct contribution of atmospheric CO_2 to the carbonate dissolution is confirmed by the slow response of the groundwater ^{14}C content to the sudden rise of the atmospheric ^{14}C level around 1963 (Münnich et al., 1967).

However, in areas of little or no vegetation, where the groundwater is consequently very soft, the contribution of the atmospheric CO_2 must be considered seriously. Because this has a δ^{13} value of about -7 to -9‰, the bicarbonate will then show relatively large δ^{13} values around -3.5‰.

(3) Especially in regions free of carbonates (Pearson and Friedman, 1970)

the soil CO_2 also contributes to *rock erosion* (Vogel and Ehhalt, 1963):

$$CaAl_2(SiO_4)_2 + 2\ CO_2 + H_2O \rightarrow Ca^{2+} + Al_2O_3 + 2\ SiO_2 + 2\ HCO_3^- \qquad (9)$$

The respective a_{bo}^{14} and δ_{bo}^{13} values in this case are 100 pmc and −25 and −15‰ respectively. The silicate weathering is considered to be a very slow process. Therefore, the actual contribution of this source of CO_2 to the groundwater carbon inventory usually is negligibly small.

(4) Part of the H^+ ions in the acid soil water can originate from *humic acids* (Vogel and Ehhalt, 1963):

$$2\ CaCO_3 + 2\ H(Hum) \rightarrow Ca(Hum)_2 + Ca^{2+} + 2\ HCO_3^- \qquad (10)$$

showing a_{bo}^{14} and δ_{bo}^{13} values of 0 pmc and +1‰ respectively. According to Vogel and Ehhalt the humic acid content of soil water is generally very small. However, we have serious evidence (Vogel and Ehhalt, 1963; Mook, 1970) that in areas with peat bogs or peat layers a considerable contribution of CO_2 is due to this process.

The next processes refer to subsequent alterations of the ^{14}C and ^{13}C content of the groundwater due to *underground generation of CO_2* in the aquifer.

Concerning this, a point of general interest has been made by Oeschger (see Winograd and Farlekas, 1974, p. 91). The fact is that a $^{14}C/H_2O$ ratio is not affected by the injection of "dead" carbon into the aquifer. Therefore, in some cases the consideration of these ratios would lead to better conclusions than the use of the values $^{14}C/C$.

(5) During a later stage of the groundwater movement in an aquifer carbon can be added through *oxidation of organic material* (peat) (Brinkman et al., 1959; Pearson and Hanshaw, 1970):

$$6n\ O_2 + (C_6H_{10}O_5)_n + n\ H_2O \rightarrow 6n\ H^+ + 6n\ HCO_3^-$$

followed by:

$$m\ H^+ + m\ CaCO_3 \rightarrow m\ Ca^{2+} + m\ HCO_3^- \qquad (11)$$

a_{bo}^{14} depends on the age of the organic compound. Moreover, during this addition of bicarbonate to the water, the water itself might already be of a considerable age. If $m = 6n$, δ_{bo}^{13} will generally be about −12‰.

(6) In the absence of free oxygen *sulphate reduction* might occur in the aquifers (Matthes et al., 1969; Pearson and Hanshaw, 1970):

$$SO_4^{2-} + CH_4 \rightarrow HS^- + H_2O + HCO_3^- \qquad (12)$$

where a_{b0}^{14} completely depends on the age of the methane. In this case δ_{b0}^{13} can have strongly deviating values (down to very negative), because already CH_4 generated by bacterial decomposition of organic matter has a low ^{13}C content (down to $-70‰$).

(7) If the Ca^{2+} concentration in the aquifer decreased due to *calcium-sodium exchange*, additional carbonate can be dissolved, thus maintaining saturation with calcite (Pearson and Swarzenki, 1974).

(8) Another source of CO_2 might be *volcanic activity*. This magmatic CO_2 is generally derived from limestone and consequently contains no ^{14}C. The original ^{14}C activity of the water may therefore be affected seriously by the additional solution of limestone as well as by isotopic exchange. There is no way by which the effect on the δ^{13} value can be predicted.

(9) Another process might affect the ^{14}C (and ^{13}C) content of groundwater. Through some kind of *exchange with the soil carbonate* or limestone in the unsaturated zone or in the aquifer the ^{14}C content of the dissolved bicarbonate would decrease. This problem has been discussed by several authors.

We should clearly distinguish between two different processes which can essentially occur without affecting the average dissolved carbon content of the water:

(a) *Isotopic exchange*, involving a solid state diffusion of CO_3^{2-} ions through the calcite crystals. This process has been observed and calculated (Urey et al., 1951) to be extremely slow in massive carbonates. In soil carbonates, however, a much larger surface area is exposed to exchange. The fact that a thin surface layer of the carbonate grains will soon have reached a state of isotopic equilibrium with the dissolved carbon (Münnich et al., 1967) does not mean that no further exchange takes place (Münnich and Roether, 1963). The radioactive decay of ^{14}C inside the carbonate grains causes a continuing diffusion of ^{14}C to further inside (Münnich, 1968). This does not apply to the stable ^{13}C. Results from artificial spike experiments by the Heidelberg laboratory (Münnich et al., 1967; Thilo and Münnich, 1970) have shown that the influence of the exchange on the ^{14}C age of groundwater is limited, depending on the grain size, on the carbonate content of the aquifer relative to the dissolved carbonate concentration of the water and presumably on the pH of the water. In high-carbonate aquifers especially at temperatures above normal, high ^{14}C ages are considered to be questionable. Data have been reported, however, which contradict this prediction (Pearson et al., 1972). These authors observed no differences in δ^{13} as well as a^{14} between a series of cold (15°C) and hot (60°C) springs in one system.

(b) *Dissolution-precipitation* of carbonates in the soil (Ingerson and Pearson, 1964; Geyh, 1972; Salomons and Mook, 1976) seems to be a frequently occurring process in regions subject to long periods of dryness and high surface temperatures. In regions of temperate climate this process can be induced by the seasonal temperature changes (Wendt et al., 1967). The reprecipitation is mostly confined to a small soil zone with relatively low carbonate content. By a rough comparison of the masses of solid calcium carbonate involved and the dissolved bicarbonate seeping down annually we feel that generally this process will be of minor importance. For the fact that part of the soil carbonate dissolved by the infiltrating water might contain some ^{14}C and that the δ^{13} value is somewhat below the normal marine value is taken care of in our equations of section "The dissolution-exchange model".

The above conclusions do not generally apply to regions with hydrothermal activity; at high temperatures (above 100°C) the isotopic exchange might be a process of considerable importance (Craig, 1966).

Considering the number of possible carbon sources, the situation seems little encouraging. However, the chemical composition of groundwaters indicate that equation (8) in most cases is by far the most dominating process, although the possible disturbing factors should be borne in mind. Therefore, before a ^{14}C analysis of groundwater is performed, it is essential that a chemical analysis of the sample is available.

The CO_2—$CaCO_3$ concept

The method of dating groundwater by means of ^{14}C was first proposed by Münnich (1957). A schematic diagram of the origin and the pathway of ^{14}C is presented by the small-scale carbon cycle of Fig. 2-1. In order to clarify the origin of ^{14}C in a groundwater sample more specifically, we have to rewrite equation (8) in a more complete form:

$$(a + 0.5b)\, CO_2 + 0.5b\, CaCO_3 + H_2O \rightarrow 0.5b\, Ca^{2+} + b\, HCO_3^- + a\, CO_2 \qquad (13)$$

In the soil $CaCO_3$ is dissolved by an equal molar amount of biogenic CO_2, whereas an additional amount of CO_2 stabilizes the solution chemically. For the ^{14}C analysis of groundwater all CO_2 is extracted from the solution after acidification.

As was mentioned in the previous section, process 1, the dissolved bicarbonate is, according to equation (13), expected to have a ^{14}C content of 50 pmc while an additional amount of CO_2 shifts the ^{14}C content to a somewhat higher value. The early experience was that in many cases the recent groundwater values are up to 85 pmc (Brinkman et al., 1959). In several publications (Vogel, 1967, 1970; Evin and Vuilaume, 1970) no statement is used about the original ^{14}C content. In some cases (Salati et al., 1974; Gonfiantini et al., 1974a) definite conclusions are drawn because of the fact that the ^{14}C content of the groundwater sample is very low (<5 pmc) or very high (>100 pmc). Definite conclusions, however, about flow directions and velocities of the groundwater can be deduced from age differences within the aquifer. The ages are calculated after defining the upper, supposedly recent sample as being of zero age:

$$T = -8270 \ln a^{14}/a^{14}_{recent} \tag{14}$$

This certainly is the most careful way of handling ^{14}C data on groundwater. As a matter of fact there are many hydrogeologists who feel that ^{14}C will never provide us with more accurate groundwater ages.

On the other hand, cases exist where knowing the absolute age even of a single groundwater sample is very important. It frequently occurs, for instance, that the isotope hydrologist is asked whether in specific cases drinking water supplies might have a chance of carrying recent pollution. In the next section we will therefore try to approximate the answer as closely as possible.

In our discussion of the CO_2—$CaCO_3$ concept we have to distinguish between the closed-system and the open-system model (Garrels and Christ, 1965; Deines et al., 1974). In the first the carbonate is dissolved by infiltrating water which has initially been in contact with a CO_2 reservoir (the soil CO_2). In the open system the carbonate is dissolved by water in continuing contact with the CO_2 reservoir at a fixed partial CO_2 pressure. In nature the carbonate dissolution will often take place under mixed conditions.

The closed dissolution system

In a closed system with respect to CO_2 the water reaches equilibrium with the gaseous CO_2 in the absence of solid calcite. The calcite dissolution subsequently occurs in the absence of gaseous CO_2.

The chemical dilution correction

Attempts have been made to correct the measured ^{14}C content or, which comes to the same, to calculate the original ^{14}C content of the water from other data. Assuming equation (13) be the only process taking place during groundwater recharge, the ^{14}C content of the total dissolved inorganic carbon (Σ) will be (Münnich, 1957):

$$(a + b)a^{14}_{\Sigma} = 0.5ba^{14}_1 + (0.5b + a)a^{14}_0 \tag{15}$$

Assuming the ^{14}C activity of the soil carbonates (a^{14}_1) to be equal to zero — an assumption which might not be valid (process 9, previous section, and section "The dissolution-exchange model") — the ratio between the original and the natural ^{14}C content is:

$$a^{14}_{\Sigma}/a^{14}_0 = (a + 0.5b)/(a + b) = (\Sigma - 0.5b)/\Sigma \tag{16}$$

where Σ is the total inorganic carbon content of the sample.

The resulting age is (equation 7):

$$T = -8270 \ln a^{14}/a_\Sigma^{14} = -8270[\ln a^{14}/a_0^{14} - \ln(a + 0.5b)/(a + b)] \qquad (17)$$

The factor a_Σ^{14}/a_0^{14} (mentioned as F, Q or q in literature) is often referred to as the *dilution factor* — incorrectly, because with high dilution (of natural ^{14}C active carbon by inactive carbon from limestone) the factor is low. It has been introduced by Ingerson and Pearson (1964) and Geyh and Wendt (1965), and later discussed and used by Tamers (1967a, b, 1975; Tamers and Scharpenseel, 1970). The use of this factor implies the measurement of the dissolved CO_2 and bicarbonate concentrations in the sample. In practice, b can simply be determined by acid titration in the field or in the laboratory (the CO_3^{2-} concentration is negligibly small in most cases); $a + b$ results from an alkali titration; as an alternative a can be calculated from a pH measurement (preferentially in the field), using known values of the first apparent dissociation constant of carbonic acid.

Although the literature (publications by M.A. Tamers; Pearson and Hanshaw, 1970, p. 277) reports some cases, where the chemical dilution correction seems to present reasonable results from a hydrogeologic point of view, some fundamental objections can be raised. The addition of carbon from other sources to the groundwater mass is being ignored (Winograd and Farlekas, 1974), as well as later changes in pH in the aquifer.

Occasionally, especially in German literature (cf. Willkomm and Erlenkeuser, 1972), the fact that part of the dissolved carbon in groundwater originates from carbonates with low or zero ^{14}C content is referred to as the *hard water effect*. By this is meant the age difference between conventional dating (a_0^{14} = 100 pmc) and the corrected ages (50 pmc $< a_\Sigma^{14} <$ 100 pmc). Assuming a_Σ^{14} to be 85 pmc, the hard water effect is 1350 years. Theoretically, however, the effect is not necessarily larger, the harder the water is.

The isotopic dilution correction

An alternative procedure to correct ^{14}C ages from groundwater has been introduced by Ingerson and Pearson in 1964 (Pearson, 1965). The idea is that the chemical origin of the total dissolved inorganic carbon determines the stable carbon isotopic composition as it does the ^{14}C content. From equation (13) we read:

$$(a + b)\delta_\Sigma^{13} = 0.5b\delta_1^{13} + (0.5b + a)\delta_0^{13} \qquad (18)$$

where δ_1^{13}, δ_0^{13} and δ_Σ^{13} refer to the limestone, the biogenic CO_2 and the total carbon content respectively. From equation (18) we derive:

$$(\delta_\Sigma^{13} - \delta_1^{13})/(\delta_0^{13} - \delta_1^{13}) = (a + 0.5b)/(a + b) \qquad (19)$$

Comparison between equations (16) and (19) shows that the "dilution factor" can as well be written in terms of δ^{13} values:

$$a_\Sigma^{14}/a_0^{14} = (\delta_\Sigma^{13} - \delta_1^{13})/(\delta_0^{13} - \delta_1^{13}) \qquad (20)$$

The choice of the proper value of the ^{13}C content of soil carbonate (δ_1^{13}) is not very critical. Marine limestone values range from 0 to +2‰. However, due to dissolution-reprecipitation processes (process 9a, p. 56) negative δ_1^{13} (to −2‰) values have been observed and used (Smith et al., 1976).

Larger natural variations are observed in δ_0^{13} depending on the type of vegetation in the recharge area. According to the photosynthetic process of CO_2 assimilation, three groups of terrestrial plant material contributing to the soil CO_2 generation are distinguished: the Calvin group, showing δ_0^{13} of −27 ± 5‰, the Hatch Slack group having δ_0^{13} values of −13 ± 4‰ and the Crassulacean Acid Metabolism type showing a large spread around −17‰ (Lerman, 1972a; Troughton, 1972); wood has a δ_0^{13} of −25 ± 5‰.

In moderate climates with mixed vegetations the plant decay in the soil is expected to generate CO_2 having an average δ^{13} of −25‰. Direct measurements have shown δ_0^{13} values of about −25‰ (Galimov, 1966). In semi-arid regions, however, where plants often obey the HS cycle, much higher δ_0^{13} (and consequently, δ_b^{13}) values are observed (Rightmire, 1967: arid West Texas, −17‰; Lerman, 1972b: −14 ± 4‰; Wallick, 1976: arid Tucson Basin, −12‰; Mazor et al., 1974; Gonfiantini et al., 1974b).

The conclusion is that in regions of temperate climate we can safely assume a δ_0^{13} value around −25‰. In semi-arid regions great caution is necessary in applying the isotopic dilution correction for obtaining true groundwater ages. In this respect, the chemical dilution correction will cause less uncertainties.

In cases where the actual δ^{13} values are reasonably well known, the closed-system concept should provide equal dilution factors, whether they are chemically or isotopically derived (Bedinger et al., 1974).

The open dissolution system

In an open system with respect to CO_2 calcite dissolution occurs, while the solution remains in contact with the gaseous CO_2.

Isotopic exchange in the unsaturated zone

The closed dissolution system presumably applies to (semi-arid) regions with little generation of soil CO_2 and where recharge occurs during short heavy rainfall, washing the biogenic CO_2 out of the soil zone. Recently, Atkinson (1977) has pointed out that even under these circumstances the system might act as an open system. The fact is that CO_2 might also be generated in the zone of percolation by decay of organic matter washed down by the percolating water. Contrary to the common soil air CO_2, this CO_2 is referred to as ground air carbon dioxide. In most cases, at least in temperate climates, the inorganic carbon formed according to equation (8) can easily exchange isotopes with the gaseous CO_2 in the unsaturated zone. The possible occurrence of this process (Münnich, 1957) which was discussed by Vogel and Ehhalt (1963) changes the δ^{13} value of the dissolved bicarbonate, without affecting the chemical composition of the solution. Therefore, we have to be cautious with simply applying the isotopic dilution correction. Extremely high δ^{13} values in groundwater bicarbonate (up to +9‰) have been imputed to exchange between bicarbonate and CO_2 produced by the process of equation (10) (Vogel, 1967; Mook, 1970).

The isotopic exchange is assumed to explain the fact that a value of 85 ± 5 pmc for the ^{14}C content of recent groundwater frequently occurs (Vogel, 1970). This conclusion was drawn from a histogram of many ^{14}C data. However, recent waters are found having lower ^{14}C content. The model (see for instance Pearson and Hanshaw, 1970, p. 280; Mook, 1976, p. 222), considering the dissolution and exchange processes as determining the chemical and isotopic composition of the groundwater, can be treated rigorously. We used a somewhat different starting point as was applied by Wendt et al. (1967) who considered the gaseous CO_2 in the unsaturated zone in an overall mass balance.

The dissolution-exchange model

The qualitative basis for this dissolution-exchange model was given by Vogel and Ehhalt (1963). For a detailed quantitive treatment we refer to Mook (1972, 1976). This approach has principally been the comparison of the starting and final conditions in a mass balance, while the δ^{13} value accounts for the isotopic exchange in the open or partly open system. Deines et al. (1974), on the other hand, describe the continuous evolution of the groundwater in both an open and a closed system. The conclusions are essentially the same. However, incorrect values for the isotope fractionations were applied.

Suffice it to state the conditions and give a brief outline and the resulting equations. The dissolution-exchange model only applies to cases where: (1) equation (13) is the only source of dissolved carbon; (2) no exchange between the solution and the solid soil carbonate occurs; (3) the pH is solely determined by the dissolution process; and (4) the water sample does not consist of a mixture of waters of different age. A schematic diagram of the ^{14}C and ^{13}C concentrations according to the dissolution-exchange model is presented in Fig. 2-2. The essential steps in the formation of the dissolved carbon are:

Step 1. The soil carbonate reacts with CO_2 dissolved in the infiltrating rain water (equation 8). Depending on the actual conditions the a^{14} and δ^{13}

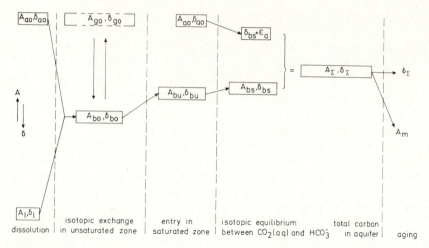

Fig. 2-2. Schematic representation of the dissolution-exchange model for determining the initial ^{14}C activity of groundwater and of the use of symbols. This model applies to a groundwater system which is (partly) open to exchange with soil CO_2 during the formation, but does not account for any additional dissolution or precipitation of calcite or other mineral in the aquifer. The vertical scale is merely qualitative.

values of the gaseous $CO_{2(g0)}$ or of the CO_2 dissolved in isotopic equilibrium ($CO_{2(a0)}$) should be taken ($\epsilon_{ag} \simeq -1‰$; equation (25)). The resulting equations for ^{14}C and ^{13}C are equivalent to those obtained earlier (equation 15):

$$a_{b0}^{14} = 0.5(a_{a0}^{14} + a_1^{14})$$

and:

$$\delta_{b0}^{13} = 0.5(\delta_{a0}^{13} + \delta_1^{13}) \tag{21}$$

Concerning the actual a^{14} and δ^{13} values of the soil carbonate, these might be altered from "marine values" to mixed values because of additional formation of secondary carbonate in the soil (Geyh, 1970; Hendy, 1971; Salomons and Mook, 1976; Plummer, 1977). If the actual soil carbonate consists of a mixture of old marine limestone and recently formed calcium carbonate (Münnich, 1957), an equation can be derived (Mook, 1976) to relate the values of a_{10}^{14} and δ_{10}^{14} with a_1^{14} and δ_1^{13}:

$$a_1^{14} - a_{10}^{14} = [a_{a0}^{14}(1 + 2\epsilon_{sa}/10^3) - a_{10}^{14}] \frac{\delta_1^{13} - \delta_{10}^{13}}{\delta_{a0}^{13} + \epsilon_{sa}(1 + \delta_{a0}^{13}/10^3) - \delta_{10}^{13}} \tag{22}$$

or if $a_{10}^{14} = 0$ pmc, approximately:

$$a_1^{14} \simeq a_{g0}^{14} \cdot \frac{\delta_1^{13} - \delta_{10}^{13}}{\delta_{g0}^{13} + \epsilon_{sa} - \delta_{10}^{13}}$$

where $\epsilon_{sa} \simeq 10‰$, slightly depending on temperature (cf. equation 25). If the secondary carbonate, however, is of considerable age, the calculated value of a_Σ^{14} will become too high and consequently also the resulting groundwater age.

Step 2. The bicarbonate exchanges isotopically with the gaseous CO_2 in the unsaturated zone, causing a parallel shift of ^{14}C and ^{13}C concentrations towards an isotopic equilibrium situation (equal to calcite saturation in an open system; Deines et al., 1974):

$$\frac{a_{bu}^{14} - a_{b0}^{14}}{a_{be}^{14} - a_{b0}^{14}} = \frac{\delta_{bu}^{13} - \delta_{b0}^{13}}{\delta_{be}^{13} - \delta_{b0}^{13}} \tag{23}$$

where the complete exchange is represented by:

$$a_{be}^{14} = a_{g0}^{14}(1 - 2\epsilon_g/10^3)$$

and:

$$\delta_{be}^{13} = \delta_{g0}^{13} - \epsilon_g(1 + \delta_{g0}^{13}/10^3) \tag{24}$$

The equilibrium fractionations are given by:

$$\epsilon_g = -9.483 \times 10^3/T + 23.89‰$$

$$\epsilon_a = -9.866 \times 10^3/T + 24.12‰$$

$$\epsilon_{ag} = -0.373 \times 10^3/T + 0.19‰ \tag{25}$$

(Vogel et al., 1970; Mook et al., 1974). T refers to the absolute temperature ($= t + 273.15°K$). From these relations and from an extensive evaluation of Rubinson and Clayton (1969) and Emrich et al. (1970) we further deduce:

$$\epsilon_s = -4.232 \times 10^3/T + 15.10‰ \text{ (calcite)}$$

Step 3. An additional amount of CO_2 (a_0) is considered to stabilize the bicarbonate solution chemically as dissolved CO_2, while furthermore the total carbon content is extracted from the sample for the isotope analyses:

$$ba_{bu}^{14} + a_0 a_{a0}^{14} = \Sigma a_\Sigma^{14}$$

and:

$$b\delta^{13}_{bu} + a_0\delta^{13}_{a0} = \Sigma\delta^{13}_\Sigma \tag{26}$$

The resulting initial ^{14}C activity of the groundwater then is:

$$a^{14}_\Sigma = \left[(a + 0.5b)a^{14}_{a0} + 0.5ba^{14}_1 + \{a^{14}_{g0}(1 - 2\epsilon_g/10^3) - 0.5(a^{14}_{a0} + a^{14}_1)\}\right.$$

$$\left.\times \frac{\Sigma\delta^{13}_\Sigma - a\delta^{13}_{a0} + 0.5b(\delta^{13}_{a0} + \delta^{13}_1)}{\delta^{13}_{g0} - \epsilon_g(1 + \delta^{13}_{g0}/10^3) - 0.5(\delta^{13}_{a0} + \delta^{13}_1)}\right]/\Sigma \tag{27}$$

The first two terms refer to the closed-system dissolution process, the third represents the isotopic exchange with the soil CO_2. As we mentioned before, a^{14}_{g0} and δ^{13}_{g0} may be taken instead of a^{14}_{a0} and δ^{13}_{a0} (δ^{13}_{a0} is about 1‰ more negative than δ^{13}_{g0}). It should be noted however, that *from a theoretical point of view* the a^{14}_{g0} value depends on the δ^{13}_{g0} value chosen, since $a^{14}_{g0} = 100$ pmc only if δ^{13}_{g0} is -25‰ (by definition). For $\delta^{13}_{g0} = -24$‰ the a^{14}_{g0} has to be raised to 100.2 pmc, at $\delta^{13}_{g0} = -12$‰, a^{14}_{g0} automatically is 102.6 pmc. In practical applications this is only a minor adjustment. Moreover, in nature larger secular fluctuations in a^{14}_{g0} have been detected (section "The influence of man on the ^{14}C abundance"). The values of a/Σ and b/Σ can be calculated using the measured pH and known values for the apparent dissociation constants of carbonic acid (Harned and Davis, 1943; Harned and Scholes, 1941; Jacobsen and Langmuir, 1974; Mook and Koene, 1975). The chemical composition of the groundwater samples can be revealed in a more rigorous manner, determining the Ca^{2+}, Mg^{2+}, SO_4^{2-}, Na^+, K^+, Cl^-, NO_3^- and other ionic concentrations. Inserting the proper equilibrium constants, computer programs calculate the carbonic compounds (Truesdell and Jones, 1974; Plummer et al., 1976; Reardon and Fritz, 1978; Fontes and Garnier, 1977). Instead of a carbon content of inorganic origin Σ of $0.5b$, we have in case of dissolution of gypsum as well as leaching of chloride and nitrate:

$$\Sigma = [Ca^{2+}] + [Mg^{2+}] - [SO_4^{2-}] + 0.5([Na^+] + [Cl^-] + [K^+] - [NO_3^-])$$

Neglecting the $[H^+]$ and $[OH^-]$ concentrations the inorganic carbon concentration amounts to:

$$\Sigma \simeq 0.5CA = 0.5(b + 2c),$$

which in most practical cases is about equal to $0.5\,b$.

Equation (27) is exact; a practical approximation is given by:

$$a_\Sigma = \left[(a + 0.5b)a_{g0}^{14} + 0.5ba_1^{14} + 0.5(a_{g0}^{14} - a_1^{14})\right.$$

$$\left. \times \frac{\Sigma\delta_\Sigma^{13} - (a + 0.5b)\delta_{g0}^{13} - 0.5b\delta_1^{13}}{0.5(\delta_{g0}^{13} - \delta_1^{13}) - \epsilon_g}\right]\bigg/\Sigma \qquad (28)$$

The most striking point about this relation is, that its validity does not depend on the actual sequence of the steps 1, 2 and 3 (Mook, 1972). Equation (27) is also valid where: (1) the dissolution-exchange mechanism consists of any number of alternating dissolution and (partial) exchange steps; consequently, it also applies to a continuous process, i.e. simultaneous dissolution and exchange; (2) the water consists of a mixture of any number of unequal portions of different chemical and isotopic composition, provided they all consist of *recent* recharge.

The main advantage of equation (27) is that ^{14}C data on groundwater can be compared with an exact model. It has been shown (Mook, 1976) that in several cases the model gives more reasonable adjustments than the simpler correction procedures or assumptions. Frequently, limits can be assigned to the true ^{14}C ages.

Isotopic changes in the groundwater body

Once the chemical and isotopic composition of the groundwater has been composed in the unsaturated zone, further changes in the flow direction below the phreatic surface might occur. In order to describe these changes we will use the model of Wigley et al. (1977). Although this description is meant to include the primary generation of the dissolved carbon as well as the evolution of CO_2 in the later stage, we will restrict our applications. The latter process will be dealt with in section "Chemical exchange with atmospheric CO_2". The reason for excluding the original dissolution process is, that it does not account for incomplete isotopic exchange between dissolved bicarbonate and gaseous carbon dioxide in the unsaturated zone.

The model of Wigley and coworkers is based on the presence of additional carbon sources and sinks in the aquifer. Assuming incremental changes of the carbon content of the water, the related changes in the isotopic composition are described by differential equations, based on a mass balance for ^{12}C as well as ^{13}C:

$$d\Sigma = \sum_{i=1}^{n} dF_i - \sum_{i=1}^{m} dQ_i$$

and:

$$d(R\Sigma) = \sum_{i=1}^{n} R_i \, dF_i - \sum_{i=1}^{n} \alpha_i R \, dQ_i \tag{29}$$

where Σ is the total carbon content of the water, dF_i and dQ_i respectively are the carbon input (source) and output (sink) during step i respectively; the second equation contains the isotopic compositions of these quantities. The use of α_i requires the assumption of a state of isotopic equilibrium between the solution and the output carbon. This fractionation factor α_i is not simply related to the fractionations presented by equation (25), but depends on the chemical composition (pH) of the sample and has to be computed for each step.

Depending on the actual circumstances and processes occurring in the aquifer, equations (29) can be solved to deliver equations similar to the result of a Rayleigh distillation.

The advantage of this model is that it allows in principle to correct δ^{13} and a^{14} values measured in an actual series of samples to values originated by the unsaturated zone processes (δ_Σ^{13} in equation 27), provided the processes responsible for the changes are known. These should be deduced from a chemical analysis of the sample results.

Wigley and coworkers present some examples for qualitatively realistic system conditions. The equations consist of reaction steps, so that principally the over-all result of the calculations depends on the reaction path chosen. Nevertheless, in several cases the result turns out not to be very sensitive for the choice made. The analysis of a groundwater system is so complicated that the use of extensive computer programs is required (Plummer et al., 1976).

Relative ^{14}C ages of groundwater

In the foregoing sections we have primarily dealt with attempts to obtain *absolute* groundwater ages. Although these are important to know in several cases, the isotope hydrologist is frequently confronted with the determination of a groundwater flow direction and velocity. For solving this problem it is generally sufficient to know the *relative* ages of a series of water samples. Assuming that the water mass has a recharge area with a homogeneous vegetation and that this plant cover did not change over the period of time covered by the ^{14}C dates, a_Σ^{14} may be taken equal for all samples. Then the relative age of sample $k + 1$ relative to sample k is simply:

$$\Delta T = -8270 \ln a_{k+1}^{14}/a_k^{14} \tag{30}$$

As we mentioned before (section "The CO_2-$CaCO_3$ concept"), this has been applied to several cases.

A simple refinement in cases where the carbon content has increased during the course of the water in the aquifer is presented by Pearson and Hanshaw (1970). In our notation the (age difference Δt corrected) initial ^{14}C activity at $k + 1$ has become:

$$a^{14}_{k+1} e^{\lambda \Delta t} = \frac{(\Sigma_{k+1} - \Sigma_k)a^{14}_1 + \Sigma_k a^{14}_k}{\Sigma_{k+1}}$$

If the additional carbon is assumed to be of zero a^{14}_1 (limestone), a^{14}_{k+1} has to be corrected by a factor Σ_{k+1}/Σ_k, so that the relative age becomes:

$$\Delta T = -8270 \ln \frac{a^{14}_{k+1}}{a^{14}_k} \cdot \frac{\Sigma_{k+1}}{\Sigma_k} \tag{31}$$

If no correction is applied for additional carbonate dissolution, too large age differences are found. The flow velocities therefore will be overestimated. The flow direction of the water, however, can be established beyond doubt.

Several different types of aquifers have been discussed with respect to the ^{14}C age distribution of the groundwater (Vogel, 1970; Bredenkamp and Vogel, 1970).

Exchange with the atmosphere

If water is in open contact with the atmosphere, the dissolved carbon is subjected to two processes: (1) a chemical adjustment to the atmospheric CO_2 pressure; and (2) an isotopic adjustment to the isotopic composition of atmospheric CO_2. These two processes can occur, under special circumstances, in the soil during dry periods (semi-arid regions). In some cases they are to be hold responsible for the formation of calcrete (Salomons et al., 1977).

More frequently the hydrogeologist is confronted with these processes during sampling in the field. The chemical and isotopic changes in the water in open contact with the atmosphere are fairly rapid. Consequently, water samples collected should be sealed as soon as possible. On the other hand, one should be cautious in collecting water from open wells. In this case it is necessary that the well contains fresh groundwater or that the sample is taken from several meters below the water table. We will discuss the processes briefly in the next sections.

Chemical exchange with atmospheric CO_2

If a water sample having a specific dissolved carbon content is exposed to the atmosphere, the partial CO_2 pressure (p_{CO_2}) rapidly approaches an equilibrium. Because the p_{CO_2} of groundwater (10^{-3} to 10^{-4} atmosphere) generally is larger than the atmospheric pressure ($\sim 3 \times 10^{-4}$ atmosphere), the excess of CO_2 will escape from the solution. This will give rise to a metastable state with a high pH in which the solution is oversaturated for calcite. In due course $CaCO_3$ is likely to precipitate. A calculation shows that fresh water with pH = 7.4, carbonate alkalinity ($CA = b + 2c$, section "The dissolution-exchange model") = 3.00 meq/l and total carbon content Σ = 3.32 mmole/l loses CO_2 until pH = 8.8 and Σ = 2.94 mmole/l while CA is unaffected before precipitation occurs. In a water mass of 1 m deep this situation is reached in a period of the order of days. For accurate pH measurements on groundwater it is therefore essential that the measurement is carried out immediately or that the sample is sealed properly.

The change in the isotopic composition of a solution in which $CaCO_3$ precipitates because of the evasion of CO_2, can be calculated quantitatively from a mass balance (section "Isotopic changes in the groundwater body"; Wigley et al., 1977). The similarity of this process to a Rayleigh distillation with two carbon sinks (evolution of CO_2 and precipitation of calcite) is apparent from the equation:

$$1 + \delta^{13} = (1 + \delta_0^{13}) \left(\frac{N^{12}}{N_0^{12}}\right) \left(\frac{\Delta g \cdot \epsilon'_g + \Delta s \cdot \epsilon'_s}{\Delta g + \Delta s}\right) / 1000$$

where δ^{13} and δ_0^{13} refer to the isotopic composition of the solution N^{12} and N_0^{12} are the (molal) quantities of ^{12}C after and before the reaction step respectively ($N^{12}/N_0^{12} \simeq \Sigma/\Sigma_0$); Δg and Δs are the (molal) quantities of CO_2 loss and calcite precipitation during the step, ϵ'_g and ϵ'_s the fractionation (in ‰) of the CO_2 and calcite with respect to the total solution. These numbers depend on the carbon chemical composition of the sample and have to be calculated from pH and known ϵ_g and ϵ_s numbers (equation 25).

According to Wigley et al. (1977) this process might alter δ^{13} of the solution appreciably, without much effect on a^{14}.

Isotopic exchange with atmospheric CO_2

The process of isotopic exchange with gaseous CO_2 has been dealt with in section "Isotopic exchange in the unsaturated zone". However, in this case the carbon dioxide has a δ^{13} value of —7 to —9‰, depending on the "purity" of the atmosphere. A state of isotopic equilibrium with CO_2 having a δ^{13} of —8‰ requires a δ^{13} value for the dissolved bicarbonate (δ_b^{13}) of $-8 - \epsilon_g \simeq +1$‰, depending on the temperature. In nature (Broecker and Walton, 1959) as well as in laboratory experiments (Mook, 1970) the exchange rate has been shown to be in the order of 10 mole/m² year without turbulence.

Groundwater dating using ^{14}C and tritium

It is essential that, whenever the ^{14}C content is analysed in a groundwater sample, also a tritium analysis is performed. If the sample is shown to have

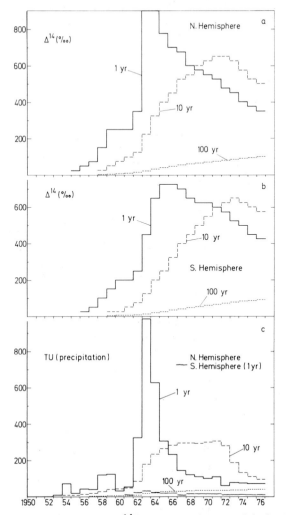

Fig. 2-3. A, B. The ^{14}C content of terrestrial plant material as deduced from measured atmospheric ^{14}C concentrations. The values are estimated averages from several stations on the northern and southern hemisphere. Considering the variations in the periods of plant growth and the differences between stations, uncertainties of ±5‰ should be assigned to the high $\Delta^{14}C$ values. The figures indicate single-year values and averages over the decade and century ending with the year indicated (data from Trondheim, Groningen, Los Angeles, Argentina, Pretoria, New Zealand and other stations, mostly as published in *Radiocarbon*. C. The tritium content of rainfall over the northern and southern hemispheric continents. The values are calculated from the IAEA computer tape on precipitation data. Considering the differences between the continents (northern hemisphere: Europe and U.S.A.; southern hemisphere: South America and Australia) and the fact that the data are non-weighted averages, uncertainties of up to ±25 TU should be assigned to the high tritium values. For the northern hemisphere averages have been given over the decade and century ending with the year indicated. All tritium data are corrected for radioactive decay until 1-1-1977, using $T_{1/2}$ = 12.25 years.

a significant tritium content (above 1 TU), the water is of subrecent age (younger than 50 years) or, more likely, consists of a mixture of older and recent water. The ^{14}C and tritium data in combination with hydrogeological evidence might enable us to decide about the possible water sources as well as about the significance of the ^{14}C ages as such and the validity of certain correction procedures (Mazor et al., 1974).

For dating (sub)recent groundwaters (last 50 years) a phenomenological model has been introduced by Eriksson (1962), later discussed and applied by Geyh (1972), using a combination of ^{14}C and tritium data. The original ^{14}C input ($a_\Sigma^{14} = a_0^{14} \times$ dilution factor) is calculated by comparing the ^{14}C and tritium output at the discharge with the known ^{14}C and tritium input functions since the nuclear test period. For a discussion of this *exponential model* we refer to Geyh et al. (1971).

Without detracting from the importance of tritium data, we do not feel that it is realistic to seek a generally applicable procedure by which absolute groundwater ages can be obtained from a combination of ^{14}C and tritium data. Too many variables are unknown, cf. the mixing ratio of old and young water, the tritium content of the precipitation to which the sample refers, the ^{14}C content of the humus layer effective in producing the soil CO_2.

As stated before the combined ^{14}C and tritium data can provide age limits. For instance, an a_m^{14} value of 130 pmc (Δ^{14} = +300‰) together with a tritium content of 100 TU certainly point to recent water. Values below 50 pmc and no tritium (<0.1 TU) refer to waters almost certainly hundreds if not thousands of years old. The principle of Fig. 2-3 might be useful in assigning age limits to (sub)recent waters. The atmospheric ^{14}C and tritium contents have been averaged over different periods to account for the age span covered by the vegetation and by the infiltration, as reflected by the water sample.

SUMMARY

Through a continuous production in the atmosphere all living plants contain ^{14}C, within a certain range in equal concentrations.

Biogenic processes in the top soil generate CO_2 which, incorporated in infiltrating rainwater, dissolves soil carbonates.

Depending on the actual chemical processes, groundwater contains a certain amount of ^{14}C during recharge. This presents a possibility of dating groundwater.

Under certain conditions and within certain limits absolute groundwater ages may be obtained applying correction procedures using the chemical and stable carbon isotopic composition of the sample.

The effect of carbon sources and sinks in the aquifer can be calculated by the use of Rayleigh distillation equations.

Relative ages can be determined with more certainty. These provide groundwater flow direction and velocities.

For being able to appreciate ^{14}C data on groundwater samples, it is essential that some chemical analyses are available, as well as the δ^{13} value and the tritium content. The latter is especially important when dealing with young groundwaters.

REFERENCES

Atkinson, T.C., 1977. Carbon dioxide in the atmosphere of the unsaturated zone: an important control of groundwater hardness in limestone. *J. Hydrol.*, 35: 111—123.

Bedinger, M.S., Pearson, F.J.Jr., Reed, J.E., Sniegocki, R.T. and Stone, C.G., 1974. The waters of hot springs national park, Arkansas — their origin, nature and management. *U.S. Geol. Surv., Rep.*, 102 pp.

Bigeleisen, J., 1952. The effects of isotopic substitutions on the rates of chemical reactions. *J. Phys. Chem.*, 56: 823—828.

Bredenkamp, D.B. and Vogel, J.C., 1970. Study of a dolomite aquifer with carbon-14 and tritium. In: *Isotope Hydrology*. IAEA, Vienna, pp. 349—371.

Brinkman, R., Münnich, K.O. and Vogel, J.C., 1959. ^{14}C-Altersbestimmung von Grundwasser. *Naturwissenschaften*, 46: 10—12.

Broecker, W.S. and Walton, A., 1959. The geochemistry of ^{14}C in fresh water systems. *Geochim. Cosmochim. Acta*, 16: 15.

Craig, H., 1954. Carbon-13 in plants and the relationships between carbon-13 and carbon-14 variations in nature. *J. Geol.*, 62: 115—149.

Craig, H., 1966. Isotopic composition and origin of the Red Sea and Salton Sea geothermal brines. *Science*, 154: 1544—1548.

Deines, P., Langmuir, D. and Harmon, R.S., 1974. Stable carbon isotope ratios and the existence of a gas phase in the evolution of carbonate groundwaters. *Geochim. Cosmochim. Acta*, 38: 1147—1164.

Emrich, K., Ehhalt, D.H. and Vogel, J.C., 1970. Carbon isotope fractionation during the precipitation of calcium carbonate. *Earth Planet. Sci. Lett.*, 8: 363—371.

Eriksson, E., 1962. Radioactivity in hydrology. In: H. Israel and A. Krebs (Editors), *Nuclear Radiation in Geophysics*. Springer-Verlag, New York, N.Y., pp. 47—60.

Evin, J. and Vuilaume, Y., 1970. Etude par le radiocarbone de la nappe captive de l'Albien du Bassin de Paris. In: *Isotope Hydrology*. IAEA, Vienna, pp. 315—332.

Fontes, J.C. and Garnier, J.M., 1977. Determination of the initial ^{14}C activity of the total dissolved carbon: age estimation of waters in confined aquifers. *Proc. Water Rocks Interactions*, Strasbourg, August 17—25, pp. I.363—I.376.

Galimov, E.M., 1966. Carbon isotopes in soil CO_2. *Geochem. Int.*, 3: 889—898.

Garrels, R.M. and Christ, C.L., 1965. *Solutions, Minerals and Equilibria*. Harper and Row, New York, N.Y.

Geyh, M.A., 1970. Carbon-14 concentration of lime in soils and aspects of the carbon-14 dating of groundwater. In: *Isotope Hydrology*. IAEA, Vienna, pp. 215—222.

Geyh, M.A., 1972. On the determination of the dilution factor of groundwater. *Proc. 8th Int. Conf. Radiocarbon Dating, New Zealand, 1972*, pp. 369—380.

Geyh, M.A. and Wendt, I., 1965. Results of water sample dating by means of the model of Münnich and Vogel. *Proc. 6th Int. Conf. Radiocarbon and Tritium Dating, Pullman, U.S.A., 1965*, pp. 597—603.

Geyh, M.A., Merkt, J. and Müller, H., 1971. Sediment-, Pollen-, und Isotopenanalysen an jahreszeitlich geschichteten Ablagerungen im zentralen Teil des Schleinsees. *Arch. Hydrobiol.*, 69: 366—399.

Gonfiantini, R., Conrad, G., Fontes, J.Ch., Sauzay, G. and Payne, B.R., 1974a. Etude isotopique de la nappe du continental intercalaire et de ses relations avec les autres nappes du Sahara septentrional. In: *Isotope Techniques in Groundwater Hydrology, 1.* IAEA, Vienna, pp. 227—240.

Gonfiantini, R., Dinçer, T. and Derekoy, A.M., 1974b. Environmental isotope hydrology in the Hodna region, Algeria. In: *Isotope Techniques in Groundwater Hydrology, 1.* IAEA, Vienna, pp. 293—314.

Harned, H.S. and Scholes, S.R., 1941. The ionization constant of HCO_3^- from 0 to 50°C. *J. Am. Chem. Soc.*, 63: 1706—1709.

Harned, H.S. and Davis, R.Jr., 1943. The ionization constant of carbonic acid in water and the solubility of carbon dioxide in water and aqueous salt solutions from 0 to 50°C. *J. Am. Chem. Soc.*, 65: 2030—2037.

Hendy, C.H., 1971. The isotopic geochemistry of speleothems, I. The calculation of the effects of different modes of formation on the isotopic composition of speleothems and their applicability as palaeoclimatic indicators. *Geochim. Cosmochim. Acta*, 35: 801—824.

Hufen, T.H., Lan, L.S. and Buddemeier, R.W., 1974. Radiocarbon, ^{13}C and tritium in water samples from basaltic aquifers and carbonate aquifers on the island of Oahu, Hawaii. In: *Isotope Techniques in Groundwater Hydrology, 2.* IEAE, Vienna, pp. 111—126.

Ingerson, E. and Pearson, F.J., 1964. Estimation of age and rate of motion of groundwater by the ^{14}C method. In: *Recent Researches in the fields of Hydrosphere, Atmosphere and Nuclear Geochemistry*. Maruzen, Tokyo, p. 263.

Jacobson, R.L. and Langmuir, D., 1974. Dissociation constants of calcite and $CaHCO_3^+$ from 0—50°C. *Geochim. Cosmochim. Acta*, 38: 301—318.

Keeling, C.D., 1961. The concentration and isotopic abundances of atmospheric carbon dioxide in rural and marine air. *Geochim. Cosmochim. Acta*, 24: 277—298.

Lerman, J.C., 1972a. Carbon-14 dating: origin and correction of isotope fractionation errors in terrestrial living matter. *Proc. 8th Int. Conf. Radiocarbon Dating, New Zealand, 1972*, pp. 612—624.

Lerman, J.C., 1972b. Soil-CO_2 and groundwater: carbon isotope compositions. *Proc. 8th Int. Conf. Radiocarbon Dating, New Zealand, 1972*, pp. 612—624.

Lerman, J.C., Mook, W.G. and Vogel, J.C., 1970. ^{14}C- in tree rings from different localities. In: I.U. Olsson (Editor), *Radiocarbon Variations and Absolute Chronology*. Almquist and Wiksell, Stockholm, pp. 275—299.

Libby, W.F., 1965. *Radiocarbon Dating*. University of Chicago Press, Chicago, Ill., 175 pp.

Matthes, G., Fauth, H., Geyh, M.A. and Wendt, I., 1969. Anäerobe bakterielle Kohlenwasserstoffoxydation als mögliche Fehlerquelle bei ^{14}C-Altersbestimmungen von Grund-wässern. *Geol. Jahrb.*, 87: 45—50.

Mazor, E., Verhagen, B.T., Sellschop, J.P.F., Robins, N.S. and Hutton, L.G., 1974. Kalahari groundwaters: their hydrogen, carbon and oxygen isotopes. In: *Isotope Techniques in Groundwater Hydrology, 1*. IAEA, Vienna, pp. 203—223.

Mook, W.G., 1970. Stable carbon and oxygen isotopes of natural waters in the Netherlands. In: *Isotope Hydrology*. IAEA, Vienna, pp. 163—189.

Mook, W.G., 1972. On the reconstruction of the initial ^{14}C content of groundwater from the chemical and isotopic composition. *Proc. 8th Int. Conf. Radiocarbon Dating, New Zealand, 1972*, pp. 342—352.

Mook, W.G., 1976. The dissolution-exchange model for dating groundwater with ^{14}C. In: *Interpretation of Environmental Isotope and Hydro-chemical Data in Groundwater Hydrology*. IAEA, Vienna, pp. 213—225.

Mook, W.G. (Editor), 1980. *The Principles and Applications of Radiocarbon Dating*. Elsevier, Amsterdam (in press).

Mook, W.G. and Koene, B.K.S., 1975. Chemistry of dissolved inorganic carbon in estuarine and coastal brackish waters. *Estuarine Coastal Mar. Sci.*, 3: 325—336.

Mook, W.G., Bommerson, J.C. and Staverman, W.H., 1974. Carbon isotope fractionation between dissolved bicarbonate and gaseous carbon dioxide. *Earth Planet. Sci. Lett.*, 22: 169—176.

Münnich, K.O., 1957. Messung des ^{14}C-Gehaltes von hartem Grundwasser. *Naturwissenschaften*, 44: 32—34.

Münnich, K.O., 1968. Isotopen-Datierung von Grundwasser. *Naturwissenschaften*, 55: 158—163.

Münnich, K.O. and Roether, W., 1963. A comparison of carbon-14 and tritium ages of groundwater. In: *Radioisotopes in Hydrology*. IAEA, Vienna, pp. 397—404.

Münnich, K.O., Roether, W. and Thilo, L., 1967. Dating of groundwater with tritium and ^{14}C. In: *Isotopes in Hydrology*. IAEA, Vienna, pp. 305—319.

Pearson, F.J., Jr., 1965. Use of $^{13}C/^{12}C$ ratios to correct radiocarbon ages of materials initially diluted by limestone. *Proc. 6th Int. Conf. Radiocarbon Dating*, Pullman, U.S.A., pp. 357—366.

Pearson, F.J. and Friedman, I., 1970. Sources of dissolved carbonate in an aquifer free of carbonate minerals. *Water Resour. Res.*, 6: 1775—1781.

Pearson, F.J., Jr. and Hanshaw, B.B., 1970. Sources of dissolved carbonate species in groundwater and their effects on carbon-14 dating. In: *Isotope Hydrology*. IAEA, Vienna, pp. 271—286.

Pearson, F.J., Jr. and Swarzenki, W.V., 1974. ^{14}C evidence for the origin of arid region groundwater, Northeastern Province, Kenya. In: *Isotope Techniques in Groundwater Hydrology*, 2. IAEA, Vienna, pp. 95—108.

Pearson, F.J., Bedinger, M.S. and Jones, B.F., 1972. Carbon-14 ages of water from the Arkansas hot springs. *Proc. 8th Int. Conf. Radiocarbon Dating, New Zealand, 1972*, pp. 330—341.

Plummer, L.N., 1977. Defining reactions and mineral transfer along an apparent flow path in the Floridan aquifer. *Water Resour. Res.* (submitted).

Plummer, L.N., Jones, B.F. and Truesdell, A.H., 1976. WATEQF a Fortran IV version of WATEQ, a computer program for calculating chemical equilibrium of natural waters. *U.S. Geol. Surv., Water Resour. Invest.*, 76-13.

Reardon, E.J. and Fritz, P., 1978. Computer modelling of groundwater ^{13}C and ^{14}C isotope compositions. *J. Hydrol.*, 36: 201—224.

Rightmire, C.T., 1967. *A Radiocarbon Study of the Age and Origin of Galiche Deposits*. M.A. Thesis, University of Texas.

Rubinson, M. and Clayton, R.N., 1969. Carbon-13 fractionations between aragonite and calcite. *Geochim. Cosmochim. Acta*, 33: 997—1002.

Salati, E., Menezes Leal, J. and Mendes Campos, M., 1974. Environmental isotopes used in a hydrogeological study of northeastern Brazil. In: *Isotope Techniques in Groundwater Hydrology*, 1. IAEA, Vienna, pp. 259—282.

Salomons, W. and Mook, W.G., 1976. Isotope geochemistry of carbonate dissolution and reprecipitation in soils. *Soil Sci.*, 122: 15—24.

Salomons, W., Goudie, A. and Mook, W.G., 1977. The isotopic composition of calcrete deposits from Europe, Africa and India. *Earth Survey Processes* (submitted).

Smith, D.B., Downing, R.A., Monkhouse, R.A., Otlet, R.L. and Pearson, F.J., Jr., 1976. The age of groundwater in the chalk of the London Basin. *Water Resour. Res.*, 12: 392—404.

Tamers, M.A., 1967a. Radiocarbon ages of groundwater in an arid zone unconfined aquifer. In: *Isotope Techniques in the Hydrologic Cycle*. Am. Geophys. Union, Geophys. Monogr. Ser., 11: 143—152.

Tamers, M.A., 1967b. Surface-water infiltration and groundwater movement in arid zones of Venezuela. In: *Isotopes in Hydrology*. IAEA, Vienna, pp. 339—351.

Tamers, M.A., 1975. Validity of radiocarbon dates on groundwater. *Geophys. Surv.*, 2: 217—239.

Tamers, M.A. and Scharpenseel, H.W., 1970. Sequential sampling of radiocarbon in groundwater. In: *Isotope Hydrology*. IAEA, Vienna, pp. 241—256.

Thilo, L. and Münnich, K.O., 1970. Reliability of carbon-14 dating of groundwater: effect of carbonate exchange. In: *Isotope Hydrology*. IAEA, Vienna, pp. 259—269.

Troughton, J.H., 1972. Carbon isotope fractionation by plants. *Proc. 8th Int. Conf. Radiocarbon Dating, New Zealand, 1972*, pp. 420—438.

Truesdell, A.H. and Jones, B.F., 1974. WATEQ — a computer program for calculating chemical equilibria of natural waters. *J. Res., U.S. Geol. Surv.*, 2: 233—248.

Urey, H.C., Lowenstam, H.A., Epstein, S. and McKinney, C.R., 1951. Measurements of paleotemperatures of the Upper Cretaceous of England, Denmark and the southeastern United States. *Geol. Soc. Am. Bull.*, 62: 399—416.

Vogel, J.C., 1967. Investigation of groundwater flow with radiocarbon. In: *Isotopes in Hydrology*. IAEA, Vienna, pp. 355—369.

Vogel, J.C., 1970. Carbon-14 dating of groundwater. In: *Isotope Hydrology*. IAEA, Vienna, pp. 225—239.

Vogel, J.C. and Ehhalt, D., 1963. The use of carbon isotopes in groundwater studies. In: *Radioisotopes in Hydrology*. IAEA, Vienna, pp. 383—396.

Vogel, J.C., Grootes, P.M. and Mook, W.G., 1970. Isotopic fractionation between gaseous and dissolved carbon dioxide. *Z. Phys.*, 230: 225—238.

Wallick, E.I., 1976. Isotopic and chemical considerations in radiocarbon dating of groundwater within the semi-arid Tucson Basin, Arizona. In: *Interpretation of Environmental Isotope and Hydrochemical Data in Groundwater Hydrology*. IAEA, Vienna, pp. 195—212.

Wendt, I., Stahl, W., Geyh, M.A. and Fauth, F., 1967. Model experiments for ^{14}C water-age determinations. In: *Isotopes in Hydrology*. IAEA, Vienna, pp. 321—336.

Wigley, T.M.L., Plummer, L.N. and Pearson, F.J., Jr., 1977. Carbon isotope evolution of groundwater. *Geochim. Cosmochim. Acta* (submitted).

Willkomm, H. and Erlenkeuser, H., 1972. ^{14}C measurements on water, plants and sediments of lakes. *Proc. 8th Int. Conf. Radiocarbon Dating, New Zealand, 1972*, pp. 312—323.

Winograd, I.J. and Farlekas, G.M., 1974. Problems in ^{14}C dating of water from aquifers of deltaic origin. In: *Isotope Techniques in Groundwater Hydrology, 2*. IAEA, Vienna, pp. 69—91.

Chapter 3

ENVIRONMENTAL ISOTOPES IN GROUNDWATER HYDROLOGY

J.Ch. FONTES

INTRODUCTION

Classically, groundwater flow patterns are deduced from indirect investigations. For instance, flow directions are deduced from water potentials, transmissivities are calculated from pumping test data. In all hydrogeological studies the basic assumption is that water continuity is respected within the aquifer but no direct identifications can be obtained on the water itself. Isotope hydrology will partially fill the gap by providing information on type, origin and age of the water. For that purpose, constitutive isotopes of the water molecules (^{18}O, ^2H, ^3H) are well suited since they represent, in the present state of our knowledge, the best conceivable tracers of the water molecules.

Some general rules can be recognized for the distribution of these isotopes in groundwaters: if the isotope content does not change within the aquifer, it will reflect the origin of the water. If the isotope content changes along groundwater paths, it will reflect the history of the water. Origin deals with location, period and processes of the recharge. History deals with mixing, salinization and discharge processes.

The application of artificial isotope tracing can theoretically yield similar information and, as shown below, in some case studies even more than environmental isotopes (e.g. recharge studies in the unsaturated zone). But generally, the use of artificial labelling of water molecules is limited. The cost of artificial tracers, and its detection limits which have to be compatible with permissible concentrations at the input introduce constraints for a system in which the tracer will be thoroughly diluted. Furthermore, and above all, it is difficult to reach a steady-state concentration at the outlet and transitory phenomena are difficult to study.

In contrast, environmental techniques will allow to tackle any hydrological problem with practically no limit to the spatial and temporal scales.

The most commonly used in isotope hydrology are those which are a constituent of the water molecules (^{18}O, ^2H, ^3H) although some other environmental isotopes (^{14}C, ^{13}C, ^{34}S, ^{15}N), which occur in dissolved compounds, may be extremely valuable for studying groundwater cycles (see Chapters 2, 6 and 10, this volume).

As will be seen from the methodology, isotopic analyses can be undertaken as an independent approach to solve hydrogeological problems but studies including combined hydrogeological, hydrochemical and isotopic data will yield more detailed and, in some cases, safer conclusions.

BASIC PRINCIPLES

Causes and empiric laws of distribution of ^{18}O, ^{2}H and ^{3}H in natural water cycles will be found in Chapter 1. Basic principles used in groundwater studies are summarized below.

Stable isotopes

Stable isotope concentrations in waters are basically controlled by the number of condensation stages resulting in precipitations and by ambiental conditions of any subsequent evaporation.

The condensation process, which is isotope fractionating, depends on the temperature and to a lesser degree on pressure changes. For a given atmo-

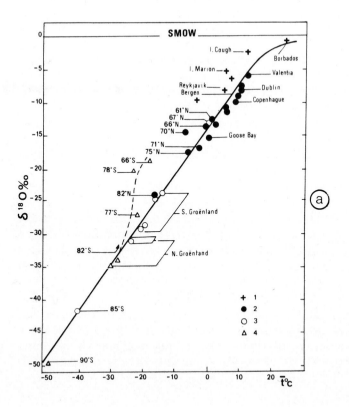

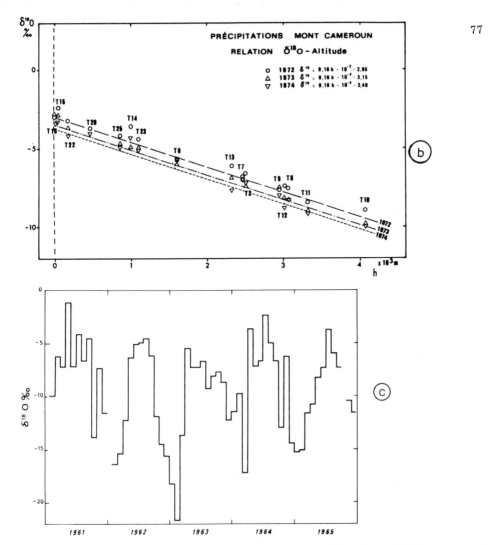

Fig. 3-1. Temperature dependence of the isotopic composition of precipitations: natural labelling of groundwater recharge in stable isotopes. (a) Worldwide relationship between $\delta^{18}O$ of rain and ground temperature at the collection site (annual average values) (after Dansgaard, 1964). The mean temperature gradient (approximately 0.7 $\delta^{18}O/°C$) integrates various types of climate (symbols 1—4) and is therefore not suitable for local studies. Symbols: 1 = island stations, $-4 < t < 25°C$; 2 = continental stations, $-17 < t < 11°C$; 3 = high-latitude stations, $-41 < t < -14°C$; 4 = Greenland and Antarctica stations, $-50 < t < -19°C$. (b) Isotopic gradient of heavy isotope content of precipitations in altitude. Profile on Mt. Cameroon (annual weighted mean values). The gradient of -0.16 $\delta^{18}O/100$ m is constant over 3 years of observation (from Fontes and Olivry, 1976). Typical values for the isotopic gradients in temperate zones are close to -0.3 $\delta^{18}O/100$ m. The altitude effect is the most suited labelling for groundwaters studies. (c) Seasonal variations in stable isotope content of precipitations. The amplitude between summer peak and winter valley (weighted monthly values) reaches 20‰ in the case of continental climate of Central Europe (here Vienna, Austria) but a difference of several permil is still noticeable even under the most temperate climates. Results from IAEA-WMO network (IAEA, 1969, 1970). This labelling can be useful for groundwater studies dealing with fast or poorly mixed systems.

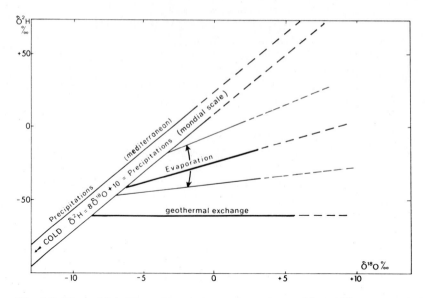

Fig. 3-2. Natural labelling of water in stable isotopes: δ^2H–$\delta^{18}O$ relationship. Precipitations lie on parallel straight lines with slope close to 8 which accounts for the unicity of the condensation process which always occurs in thermodynamic conditions (i.e. equilibrium between vapour and liquid phase) (see Chapter 1, this volume). The value of the intercept depends on the origin of the condensing vapour. A typical value for oceanic precipitations is +10 and the equation $\delta^2H = 8\, \delta^{18}O + 10$ (Craig, 1961) is generally used as a reference for precipitations when no local data are available. In regions where the vapour comes from closed seas or inland seas this intercept generally increases (see Chapter 1). In the eastern Mediterranean the value of the intercept is +22 (Gat and Carmi, 1970). The intercept is referred to as deuterium excess, $d = \delta^2H - 8\, \delta^{18}O$ (Dansgaard, 1964). Evaporating bodies of water lie also on straight lines. These lines have variable slopes (depending on local kinetic conditions of evaporation, i.e., on local values of evaporation rate) and variable intercept with the meteoric water line (depending on the initial stable isotope contents of the water before evaporation started). Generally the slope of an evaporation line is within the range 2—5.

Waters heated in rocks generally rich in ^{18}O show deviations in ^{18}O whereas the ^2H content remains unchanged. This property is used for geothermal studies (see Chapter 5, this volume).

spheric vapour, the more pronounced the cooling process the more depleted in ^2H and ^{18}O becomes the vapour phase and thus the subsequent liquid (or solid) phase. A multistage cooling gives a condensed phase (liquid or solid) progressively depleted in heavy isotopes.

For practical purposes the temperature effect is translated in term of (Fig. 3-1):

— Altitude — existence of a negative linear correlation of heavy isotope content of rains with altitude.

— Amount — negative tendency between amount of rainfall and isotope content.

— Distance to the source of vapour — continental precipitations are depleted in ^{18}O and ^2H as compared to marine and coastal rains.
— Paleoclimate — precipitation fallen under cooler climatic conditions are depleted in heavy isotopes as compared to those from warmer periods.
— Seasonal and short-term variations — winter precipitations are depleted in ^{18}O and ^2H with respect to summer rains.

The evaporation process tends to increase the isotope content of the remaining liquid phase. This enrichment is inversely correlated with the relative humidity, i.e. with the density of water vapour molecules depleted in heavy isotopes, whose condensation at the liquid surface counteracts the enrichment due to evaporation. However, on a $\delta^2H-\delta^{18}O$ diagram stable isotope contents of evaporated waters fall below the local meteoric water line giving thus a sensitive fingerprint of evaporation (Fig. 3-2).

Tritium (Fig. 3-3)

This radioactive isotope of hydrogen ($T_{1/2}$ = 12.35 years) behaves similarly to deuterium in its fractionation patterns. Because of the relative mass difference between deuterium and common hydrogen, this fractionation is large (about 15% at room temperature). However, variations due to isotopic fractionation are still of minor importance if compared to the fluctuations due to meteorological factors. Furthermore, tritium contents are determined as concentrations (tritium units, TU *) and not as permil or percent variation from a standard.

Natural tritium is produced by the impact of cosmic neutrons on nitrogen nuclei in the upper atmosphere resulting in steady-state concentrations ≤20 TU in precipitations. Since 1952, this natural background has been swamped by enormous amounts of man-made tritium which were injected into the stratosphere during open air thermonuclear tests. At its maximum level of 1963, the contribution of artificial tritium to precipitation reached 2 to 3 orders of magnitude above that of natural tritium.

Because of the features of the tropospheric circulations and the relationships between the stratospheric reservoir and the troposphere (see Chapter 1) the tritium rain-out is monitored in latitude and season. Tritium activity in rains increases towards mid and high latitudes with respect to low latitudes. At a given location the maximum activity of rain is observed in spring (spring peak) and corresponds to about three times the annual weighted mean. Then the winter valley gives a minimum value at about one half of this average.

Furthermore, the length of the transit of wet air masses over continents or

---

* One tritium unit (1 TU) corresponds to a concentration of one atom of ^3H in 10^{18} atoms of ^1H.

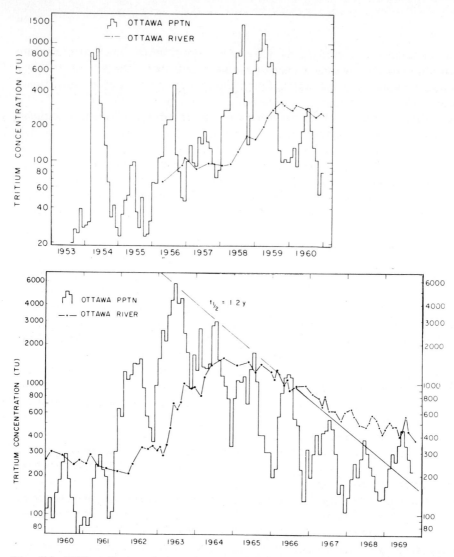

Fig. 3-3. Tritium-time variations in precipitations in the northern hemisphere (from Brown, 1970). The natural level (several TU) increased markedly at the end of 1963 after the beginning of the thermonuclear tests which injected enormous amounts of ^3H in the stratosphere. Then, in concordance with these tests the major peaks occur in 1958 and 1963 and are separated by a deep valley in 1960. From the moratorium of 1963 at the end of aerial tests the tritium level in precipitations decreased until 1968 with a time constant of 1.2 years which reflected the rate of tritium transit from the stratospheric reservoir.

For each single year one observes a strong seasonal variation with a maximum rain-out at the end of the spring (spring peak) and a minimum at the beginning of winter. This pattern which was studied in detail in Ottawa, is of large general value since tritium-time variations in precipitations are roughly homothetic within the northern hemisphere. It is referred to as the ^3H input function for hydrogeological studies. From the results given by the IAEA-WMO network (IAEA 1969—1978) and from the long record of Ottawa it is practically possible to estimate the ^3H input function at any location in the northern hemisphere. In the southern hemisphere, the patterns are more complicated because of the low ^3H level due to the lack of significant supply of ^3H in the austral stratosphere.

oceans leads to variations in the tritium content due to mixing with tritiated vapour and molecular exchange with free water surface, respectively. Schematically continental rains are enriched in tritium with respect to marine ones.

The results of these combined effects on tritium rain-out is rather complicated (see Chapter 1, this volume) but for practical purposes one would retain the following principles of labelling:

— Prenuclear levels (i.e. prior to 1952) corrected for decay are everywhere below 5 TU as an average value.

— Thermonuclear tritium in precipitations is presently above 5 TU on annual average basis at any location of the northern hemisphere (IAEA, 1978).

— Thermonuclear tritium rain-out when weighted with annual rainfall gives roughly proportional activities from location to location within the northern hemisphere (this proportionality accounts for the cumulative effects of tritium variations factors: latitude, distance to the open sea). This property is useful for the approximate evaluation of the tritium content in precipitation on a given region for which no continuous records are available. Extrapolation can be done on the basis of records from IAEA-WMO stations (IAEA, 1969, 1970, 1971, 1973, 1975, 1978).

In groundwater studies one would adopt as a guidance: (1) no tritium, i.e. background of gas counting technique after electrolytic enrichment, indicates waters older than 20—50 years; and (2) detectable tritium means mixing with recent (past 1952) waters (if one disregards the very special case of detectable tritium of prethermonuclear age).

The possible contribution of past-1952 waters in tritium-bearing systems can then be evaluated case by case on the basis of the expected local tritium activity of the annual recharge after 1952.

Care must be taken that the large decrease in ^{3}H activity of precipitations due to the destorage of stratospheric reservoirs after the end of aerial thermonuclear tests may lead to some ambiguity in interpreting low tritium content, say 5—30 TU, in groundwaters. Depending on the areas of study this could mean either waters of the year or mixing including tritium free waters. More comments on this subject are presented below.

Methodology of environmental isotopic studies of groundwaters

The interpretation of the isotope contents in groundwaters requires the investigation of several effects which may lead to discrepancies between the tracer input at the surface or subsurface and the aquifer where the environmental tracer is investigated:

— Evaporation and molecular exchange on stable isotope and tritium content of water which infiltrates; this effect can lead to an enrichment in

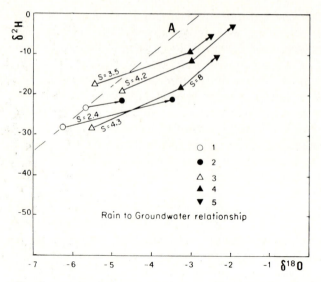

Fig. 3-4. Modification of the stable isotope content of rainwater due to evaporation before or during infiltration (studies in Israel, from Gat and Tzur, 1967). Symbols: A = local meteoric waters line; 1 = precipitations (weighted means); 2 = groundwaters supplied by 1; 3 = sprinkler water supply of irrigation plots; 4 = water collected in raingauges or irrigated plots; 5 = drainage of lysimeters in irrigated plots. Waters which undergo free air evaporation move on evaporation lines (slopes lower than 8). Drainage waters are also enriched in heavy isotopes as compared to supply waters but move on slope-8 lines suggesting that evaporation occurs at equilibrium.

stable isotopes of the recharge as compared to rains in dry regions (e.g. Israel, Gat and Tzur, 1967; Fig. 3-4).

— Effect of preferential seasonal seepage through the unsaturated zone; this effect will give more relative weight to surface waters of cold and or humid periods in the recharge.

Furthermore, it is now noteworthy that water displacement in soils and in aquifers is generally dispersive. Isotopic signals contained in the input tend to be smoothed out. Distinctions between systems thus imply that the respective isotope contents are significantly different. This is a common analytical problem. Moreover, isotope contents must be representative of the various components of the system. This is a conceptual problem for the sampling which, in any case, should be solved in close collaboration between hydrogeologist and isotope geochemist.

GROUNDWATER RECHARGE

Isotope techniques may contribute to solve the following problems dealing with groundwater recharge:

(1) Qualitatively, does the recharge occur? This problem exists in areas in which the occurrence of recent recharge is questionable, especially arid zones. Furthermore, for preservation of groundwater resources and evaluation of pollution risks, it is important to determine the areas of rapid seepage towards the aquifers.

(2) Quantitatively, what is the annual recharge rate through the unsaturated zone? This question is of high importance for the evaluation of the renewal of groundwater resources especially if one keeps in mind that evapotranspiration from the soil zone is the most difficult term of the water budget to estimate.

Qualitative evidence of recent recharge

To prove the occurrence of recent recharge is essential for the evaluation of groundwater resources in regions with heavy pluviometric deficit. Recharge occurs when pluviosity exceeds evapotranspiration plus run-off during a given period. Such excess, which extends over several months per year in temperate regions (e.g. November to March in the Paris Basin), is still observed during one or two months per year in dry tropical climates of monsoon type. For instance, in the Sahelian regions north of Lake Chad, recharge must take place in August although annual evapotranspiration exceeds precipitation by one order of magnitude (250 mm for rainfall and 2.5 m for evapotranspiration). Thus the question of occurrence of recharge only arises in regions where rainfall is not only low but where no clear seasonal excess of precipitation over evapotranspiration can be observed, i.e., areas in which average rainfall does not exceed 50 mm randomly distributed.

In arid areas such as Sahara or Kalahari deserts, recent recharge of some localized shallow aquifers has been demonstrated by their tritium content which is similar to that of recent precipitation. Obviously, because of global distribution patterns, the tritium activity is much higher in recent Sahara groundwaters (more than 20 TU, Conrad et al., 1975) than in those from Kalahari (more than 2 TU, Verhagen et al., 1974). (See Chapter 1 for discussion on ^{3}H distribution between the two hemispheres.)

From other studies performed in arid zones: Sinai (Gat and Issar, 1974) Djibuti, Oman (J.Ch. Fontes, unpublished data), Saudi Arabia, Qatar (IAEA, unpublished data) it appears that bomb tritium is also frequent in unconfined aquifers especially in underflows of intermittent wadis. This supply of recent water is generally attributed to exceptional pluvial events. However, in the management of groundwater resources of arid zones care must be taken of the fact that the occurrence of tritium does not prove that net recharge is taking place. Small amounts of recent water can reach the water table of old reservoirs whose water storage is globally decreased by evapotranspiration. For instance, the unconfined aquifer of the Grand Erg in the northwest Sahara discharges tritium-free waters at Béni-Abbès whereas tritium-rich waters are found in the shallower parts of the recharge area (Gonfiantini et al., 1974a; Conrad et al., 1975). Furthermore, it is also possible that shallow groundwater becomes tritiated by exchange with air moisture in the unsaturated zone.

A special case of arid zone is that of permanent frozen soils (permafrost) which are often assumed to be impervious since always below $0°C$ at atmospheric pressure. Theoretically, under these conditions the recharge would occur only if a liquid phase can exist, i.e. if a long improvement of the climate take place. However, preliminary results obtained in Canada suggest that the seepage still occurs since tritium is found at a depth of some meters below ground surface (Michel and Fritz, 1978).

In pollution studies high environmental tritium levels can indicate the areas where vertical permeability and thus pollution risks are high (Matthess et al., 1976). However, this attractive technique will become less easy to use in the future because of the decrease in ^3H activity in rains.

The problem of recent infiltrations into confined aquifers where the piezometric head is supposed to eliminate the possibility of a mixing of shallow and deep waters can also be of importance for pollution studies. The occurrence of some minor amounts of tritium diluted in very old waters indicates a shallow contribution. In fact, numerous isotopic studies of confined aquifers have shown that waters often contain small amounts of tritium (Bortolami et al., 1973; Smith et al., 1976). However, this does not necessarily imply a leakage from above in the system but could be due to a mixing with shallow water during the discharge, in the upper part of the borehole. Despite the difficulties of interpretation, studies dealing with vulnerability of confined aquifers to surface pollution must include tritium measurements.

Quantitative estimates of recharge

The amount of water which reaches the water table can be estimated if one knows: (1) the age of the water at a given depth in the unsaturated zone or below the water table; (2) the mechanism of dispersion of water within the porous medium; and (3) the total and the effective porosity.

The age of the waters, i.e. the time elapsed after precipitation can be evaluated using any kind of time-dependent variable. For instance, the time of the beginning of the use of fertilizers in a given region is generally well known. In that case, the depth of the front of infiltration of fertilizers provides an estimate of the apparent vertical permeability assuming that the behaviour of ions in aqueous solutions is representative of that of the waters. Such studies performed in the chalk of the London Basin (Young et al., 1976; Cole and Wilkinson, 1976) show that the nitrate front moves downward in the unsaturated zone at a rate of approximately 1 m/yr. The same kind of study has been undertaken using environmental and artificial isotopes, primarily tritium. The average stable isotope content of the fraction of annual rainfall effective in recharge is very similar from year to year and the pioneer studies by Brinkman et al. (1963) and Eichler (1965) demonstrate that the stable isotope content of soil water is also very constant in space and time because of a smoothing out of seasonal and interannual (if any) variations in the unsaturated zone. Environmental ^{18}O and ^{2}H are thus poorly suitable for studies on infiltration which requires an annual time scale. In contrast, the distribution of tritium in precipitations has strongly varied in time from year to year especially within the time interval 1960—1970 which includes the 1963 "peak". Furthermore, artificial labelling of waters with tritium is cheaper and easier to measure than artificial labelling in ^{18}O and ^{2}H. For these reasons, tritium has been more extensively used than the other isotopic tracers of water molecules to draw time dependencies in the vertical movement of groundwaters.

Breakthrough curves of artificially labelled waters in experimental set ups or field studies provide direct evidence for the type of water movement that might occur in the unsaturated zone.

Water movement in the unsaturated zone was first investigated, using isotope techniques, in Germany (Zimmermann et al., 1965, 1966, 1967). In these studies, both environmental and artificial hydrogen isotopes were used. Conclusions regarding soils, vegetal cover and climate from Central Europe (Rhine Valley) can be summed up as follows:

(1) Most summer rains do not contribute to groundwater recharge.

(2) Dispersion, mainly due to molecular diffusion, is very active during downward movement, and homogenizes the isotope content after some centimetres, or some tens of centimetres, of vertical movement (Fig. 3-5).

(3) Evaporation tends to increase the heavy isotope content of the upper layers of bare soils (or of the vegetal cover) down to some centimetres (Fig. 3-6), depending on the ratio between the diffusion coefficient and the effective upward velocity of water pumped out by evaporation (effective velocity = evaporation flux/porosity).

(4) Infiltration rate is within the order of 1 m/yr.

(5) Recharge, expressed as the fraction of precipitation which percolated, is the range of 25—75%, depending on season, amount of rain, and soil characteristics (Fig. 3-5).

Further studies on the use of artificial or environmental tritium in the estimation of groundwater recharge were then made on different soils under different climates.

In the Geneva Lake Basin, it was found (Blavoux and Siwertz, 1971) that at the basis of 1 m of reworked glacial material in a lysimeter, a single tritium pulse applied at the surface gave a bimodal distribution curve suggesting two types of seepage. One is rapid and attributed to movements on cracks, the other one is delayed and supposed to be due to the microporosity. This latter pulse was pushed down by a weekly application of tritium-free water with a constant stable isotope content. A significant increase in heavy

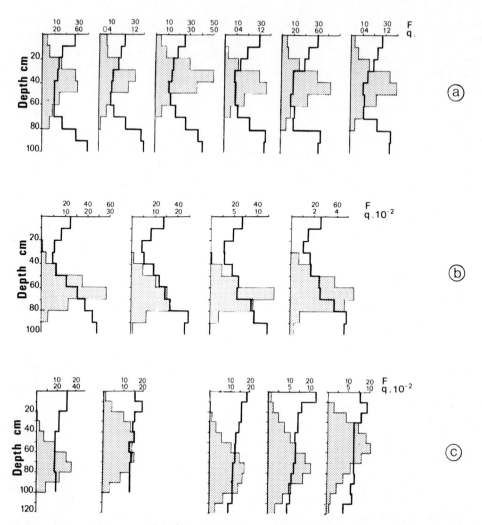

Fig. 3-5. Tritium and water profiles of different types of soils after tracing experiments in Western Germany (after Zimmermann et al., 1967). Symbols: F = water content in percent per volume; q (shaded) = tritium content in cpm/g of soil water. (a) Sand covered with grass. Tracing 19 April 1966, sampling of the 6 profiles 28 July 1966. Although the distance between each profile can reach 2 m, the displacement of the peak was very similar (width of the peak and depth of peak maximum) indicating homogeneous dispersive conditions. The total rainfall after the tracer input was 300 mm. (b) Loamy sand covered with grass. Tracing 21 December 1965, sampling of the 4 profiles 19 March 1966 at horizontal distance of 2 m. The total rainfall after the tracer input was 180 mm. (c) Loamy sand covered with grass. Tracing 21 December 1965. Profiles are cored at horizontal distances of 2 m. On left, 2 profiles taken on 9 March 1966, total rainfall after tracer input 130 mm. On right, 3 profiles taken on 26 April 1966, total rainfall after tracer input 210 mm. In both cases we observed significant dispersion and large variations in the depth reached by the peak of tracer.

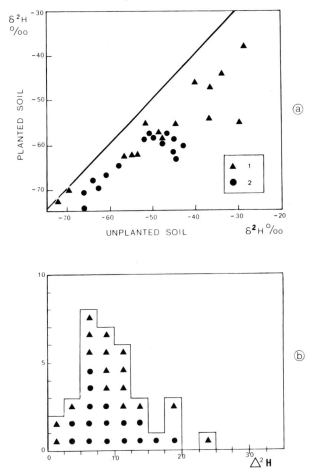

Fig. 3-6. Isotopic behaviour during evaporation from soils: Giessen, West Germany (after Zimmermann et al., 1967). (a) Deuterium content of samples taken at the same depth in bare soil (abscissa) and in planted soils (ordinate). All the samples lie below the line of slope 1 indicating a systematic enrichment in heavy isotopes for bare soil waters. This enrichment appears larger in the shallow layers as shown by the symbols: 1 = samples taken at depths 0—10 cm, 2 = samples taken at depths 10—20 cm. (b) Frequency distribution of the observed enrichment in heavy isotopes between evaporation from bare soil and evapotranspiration from planted soils. Same symbols as above. The most frequent and the maximum enrichment observed would correspond to about 2—5‰ in ^{18}O respectively assuming a slope of 4 for the δ^2H—$\delta^{18}O$ correlation during the evaporation process from bare soil.

isotope content was also observed during the summer months indicating an active enrichment by evaporation.

In a six-year record of natural tritium profile obtained from soil cores in Denmark, Andersen and Sevel (1974) showed that a glaciofluvial outwash

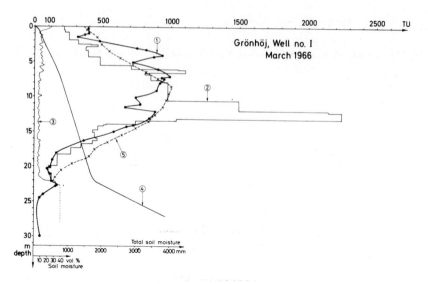

Fig. 3-7. Natural tritium profile on a soil made of a glaciofluvial outwash in Denmark (from Andersen and Sevel, 1974). *1* = measured values; *2* = displacement flow model (net recharge, i.e., rainfall minus evapotranspiration of a given month, displaces earlier stored soil moisture under a pure piston flow mechanism); *5* = displacement flow with dispersion (same recharge as above and assuming a dispersion described by $C(x, t) = C_0 \exp[-(x-x_0)^2/4Dt]$, C = concentration at time t and depth x, D = dispersion coefficient taken as 10^{-7} m^2/s); *3* = soil moisture content; *4* = cumulated soil moisture content.

of sand and gravels allowed a downward velocity of about 4.5 m/yr (Fig. 3-7). Calculations assumed that the fraction of precipitation which is not removed by evapotranspiration is well mixed with the existing soil water, within the first metre of soil. Dispersion is taken into account assuming homogeneous medium and isotropic conditions. Corrected for the effective porosity, this downward velocity gave an annual recharge rate of 358 mm for a mean annual precipitation of 700 mm and without any run-off. It is remarkable that in this case the downward velocity of the 1963 tritium peak was approximately 10 times slower than the seepage velocity of seasonal rainfall estimated from neutron gauge measurements. This discrepancy suggests that soil moisture profiles indicate "pressure waves" displacement rather than actual rainwater infiltration.

Under "Mediterranean-type" climate of the Gambier Plain in Australia, Allison and Hughes (1974) measured environmental tritium profiles in different types of soils (sandy to loamy and clay-rich). They calculated recharge rates from 40 to 140 mm/yr according to the type of soils. Calculations were made taking into account the diffusion of water in soil materials and using different models of "piston flow" and partial mixing during downward movement.

Under semi-arid conditions in the Transvaal, Bredenkamp et al. (1974) investigated the recharge rate of sandy and loamy soils cored for water extraction and tritium analyses. The assumption was made that downward movement occurred by "piston flow". They obtained values between 16 and 53 mm/yr for the recharge with an average annual rainfall of 560 mm and an effective porosity of 2.5%.

In arid soils from Washington State (U.S.A.), Isaacson et al. (1974) drilled wells in silty sands down to a depth of 90 m in the unsaturated zone. No thermonuclear tritium was detectable below a depth of 5 m in the soil profile. These observations were completed with a study of a giant lysimeter (diameter 3 m, depth 20 m). It appears from this study that although infiltration is rather fast and easy for heavy rains, the water content decreases rapidly downward to a value of approximately 6% per volume which remains constant below 5 m. The conclusion was that water moves rapidly downward in winter, but is taken up by evaporation and evapotranspiration during summer since "archaic" (i.e. prethermonuclear test) tritium occurred below 5 m. No direct vertical recharge seems to take place by rain infiltration under these conditions of aridity (160 mm of mean annual rainfall). However, one must note that in this area lateral recharge also occurs since significant amounts of tritium are found in the saturated zone (water table at 94 m depth).

A special case of unsaturated zone is provided by the Upper Cretaceous chalk of the London and Paris Basins. It is noteworthy that the microporosity is high (about 40%) in this sediment mainly composed of tests of micro-organisms like coccoliths and foraminifera. Smith et al. (1970) have extracted water from a chalk profile in the London Basin, down to a depth of 27 m into the unsaturated zone. The water content was very uniform, 0.210 ± 0.003 g per gram of wet chalk corresponding to 38% per volume. The distribution of the tritium content with depth showed a very sharp peak (600 TU) at about 4 m and a secondary peak between 7 and 9 m (Fig. 3-8). Interpreted as the peaks observed in precipitations in 1963 and 1958 respectively, this figure suggested: (1) essentially piston flow displacement, and (2) downward velocity of 0.88 m/yr. Using a dispersive model based on molecular diffusion (diffusion coefficient of water in porous media = 10^{-9} to 10^{-10} m^2/s), assuming that 85% of the total recharge goes through the interstitial porosity (and 15% through cracks) and that effective evapotranspiration is equal to 0.75 of Penman's figure, then the major peak of the profile can be accounted for. The calculated annual displacement of 0.88 m/yr, corrected for a volumetric water content of 38% signifies an annual mean recharge of 334 mm/yr. This value is in good agreement with the data of the water budget in the area which gives an annual infiltration rate of 280 mm/yr on the whole catchment area.

However, the authors point out that the calculation of tritium seepage based on monthly water balance (precipitation minus evapotranspiration)

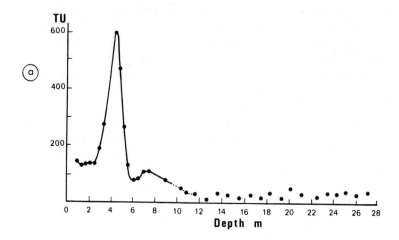

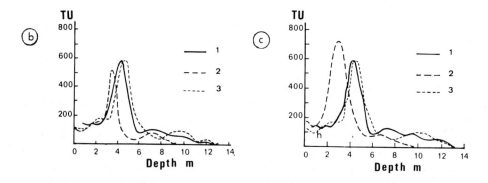

Fig. 3-8. Natural tritium profile in water extracted from the unsaturated zone of the chalk in the London Basin (after Smith et al., 1970). (a) Tritium variations with depth; the remarkably sharp peak at about 4 m is attributed to the 1963 ^{3}H maximum in precipitations (see Fig. 3-3). The lower diagrams represent modelizations of the observed values. (b) Model without molecular diffusion. 1 = observed profile, 2 = model assuming that the input is equal to 0.85 (rainfall − Penman's evapotranspiration), 0.15 of the input is lost downward by direct seepage through cracks; 3 = model assuming input equal to 0.85 (rainfall − 0.75 Penman's evapotranspiration), Penman's figure for evapotranspiration is reduced by 25% in order to match the total tritium content of the profile. (c) Model with molecular diffusion assuming that $C = C_0' \exp[-x^2/4Dt]$ (C = concentration at time t and depth x, from the plane, C_0 = concentration at the plane after time t calculated by normalizing the curve so that $\int_{-\infty}^{+\infty} C dx$ = the amount of activity per unit area of the plane source, D = diffusion coefficient of water in soil taken as 10^{-10} m^2/s); 1 = observed values; 2 = model assuming input = 0.85 (rainfall − 0.7 Penman's evapotranspiration); 3 = model assuming 30% of direct input without any loss by evapotranspiration. Although these treatments are not completely satisfactory (see the respective positions of the primary and the secondary peak at about 7 m depth), they provide a very instructive approach of how environmental isotopes can contribute to infiltration studies.

cannot take into account the influence of heavy rains which can infiltrate rapidly even during periods with pluviometric deficit. Because of the possibility of such fast seepage the soil water balance should be calculated on each single rainfall. If a significant (30%) contribution of summer rain with a relatively high tritium content can directly infiltrate below the level at which it can be taken up and removed by evapotranspiration, the tritium balance in the profile becomes more difficult to match with the observed concentrations in the input. It appears that more information is needed on the dispersion within the porous chalk. Particularly it will be important to investigate intergranular seepage in chalk which may be more dispersive than can be calculated on the basis of molecular diffusion only.

In conclusion of this review of isotopic studies in the unsaturated zone, it appears that more efforts are needed to investigate the following critical points.

— What is the influence of the first layers of the soil on the dispersion?

— To what extent is the layer-by-layer downward seepage representative of water movement?

— Which fraction of the rain can follow the cracks network and which possibilities of mixing can exist between circulation in cracks and through the pores?

Attempts to estimate the recharge have also been made by direct sampling in the saturated zone. In an unconfined aquifer of the Rhine Valley, Atakan et al. (1974) find a recharge rate of 164 mm/yr ± 25% for a sand with 30% of porosity. The average annual rainfall in this area is 686 mm. The observed stratification of successive recharges is corrected for the effect of dispersion in the porous medium, and an accurate evaluation of the various sources of errors is presented. The recharge rate calculated on the basis of environmental tritium stratification is in agreement with the estimates obtained from hydrological studies (150—225 mm/yr) but disagrees with the figure obtained from lysimeter studies (392 mm/yr).

Identification of recharge areas

The determination of areas where an aquifer is recharged is important for estimates of groundwater resources and the definition of the limits of the protection zone. Special attention must thereby be paid to complex aquifers whose sedimentological texture (e.g. multilayered) and structural frame work (e.g. leaking faults) might complicate the interpretation of hydrological data. Furthermore, the use of the isotope approach is particularly pertinent for areas with little hydrologic background data.

Many recharge studies based on environmental isotope take advantage of the fact that the isotope content varies as a function of altitude. The isotopic contents of groundwater corresponds to a distribution of the product of the isotope concentration of the recharge at a given altitude by the amount of rain which infiltrates there.

Methodology. The first study of this nature was attempted by Fontes et al. (1967) on the aquifer of Evian (Haute-Savoie, France). The isotopic gradient in precipitations was determined from the relationship between temperature and ^{18}O content on a monthly basis at the station of Thonon-les-Bains. This relationship was converted into an ^{18}O/altitude gradient through temperature/altitude gradient. The authors proposed an "average altitude of recharge" of 820 m which was compatible with local hydrogeological settings. No corrections were made for the altitude distribution of the catchment area, the variation of rainfall with altitude and the various amounts of precipitations which are removed by evapotranspiration according to the season.

In Switzerland (Siegenthaler et al., 1970) sampled precipitations at two altitudes on several sides of Alpes and Jura and obtained oxygen isotope gradients of —0.4 and —0.2‰ per 100 m respectively. These values were used to calculate the altitude of recharge of springs. Evaporation was invoked when the measured isotope content was higher than that expected on the basis of the isotope gradient, but the same limitations encountered in the previous example arose for the interpretation of the data.

Somewhat more complicated was the interpretation of data obtained in

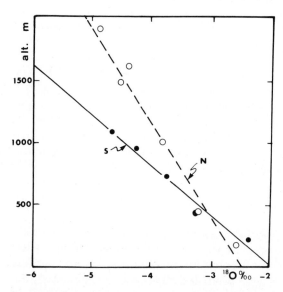

Fig. 3-9. Determination of recharge areas: isotopic gradient of groundwaters in altitude in Gran Canaria Island (after Gonfiantini et al., 1976). Samples of which the tritium content was greater than 5 TU have been selected to establish these correlations assuming that they are recent and thus representative of local infiltration. The difference between the correlation for the north of the island $h(m) = -777 \, \delta^{18}O - 1947$ and for the south of the island $h(m) = -405 \, \delta^{18}O - 810$ is due to the fact that on this latter area rains have undergone evaporation during rainfall. The corresponding gradients in altitude are —0.13‰ and 0.24‰ per 100 m on the northern and southern region respectively.

a study of Central Italy (Zuppi et al., 1974). The authors pointed out that the definition of the "isotopic altitudes of recharge" has to take into account seasonal variations of the isotopic gradient with altitude. On the Tyrrhenian side of the Appenines, the measured $\delta^{18}O$ values were -0.17 and -0.54% per 100 m in summer and winter precipitation respectively. The average annual gradient was -0.34% per 100 m. In this region of karstified and fractured limestones and dolostones, the best estimate for the altitude of recharge were obtained with the latter value of the gradient.

In a study performed in Canary Islands Gonfiantini et al. (1976) used groundwaters of local origin to establish the isotopic gradient in altitude. Recent, and thus local, groundwater samples were selected on the basis of their tritium content ($\geqslant 5$ TU for this area of pure marine precipitations very depleted in ^3H). Two different gradients were obtained for ^{18}O variations in altitude in Gran Canaria Island (-0.13% per 100 m on the northern side and -0.24 for the southern side). A comparison of ^2H with ^{18}O data demonstrated this is due to partial evaporation of rains during their fall on the southern side (Fig. 3-9).

Similar investigations were conducted in the Sperkhios Valley (Greece). Stahl et al. (1974) measured a gradient of -0.16% per 100 m on groundwaters whose local origin was determined on the basis of a geological survey. They calculated altitudes of recharge consistent with hydrogeological and topographical data for several springs.

Correction for the topographic effect. Payne and Yurtsever (1974) have reported on a study in Nicaragua where they estimated the location of recharge of deep groundwater in the Chinandega Plain. The plain covers an area of about 1100 km² between the Pacific Ocean and the drainage divide of the Cordillera Marrabios. Inland from the coast the topography rises gradually to an altitude of about 200 m at a distance of 20 km, after which the gradient becomes more steep with the maximum elevations at the crest of the Cordillera being 1745 m. Samples of precipitation and groundwater were collected at different elevations in a transverse strip extending inland from the coast (Fig. 3-10).

The mean $\delta^{18}O$ values for each collection site, weighted for the amount of precipitation, are plotted against the respective elevations and fall on line *A* in Fig. 3-10. The sampling period extended over almost two rainy seasons and the altitude effect (isotopic gradient) was determined individually for each rainy season. It appeared that these isotopic gradients in altitude differed by about 50% and there was a marked difference in the $\delta^{18}O$ values at a given elevation. Thus the best estimate of $\delta^{18}O$ at a given elevation was obtained from groundwater samples for which reasonable estimates of their origin of recharge could be made.

Bellavista spring has a $\delta^{18}O$ value defining the isotopic composition of recharge by precipitation falling above the elevation (800 m) of the spring.

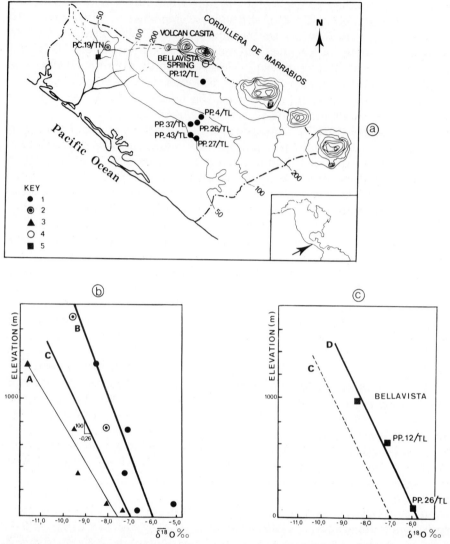

Fig. 3-10. Determination of recharge areas in Nicaragua (after Payne and Yurtsever, 1974). (a) Location map of sampling points: *1* = drilled well, *2* = dug well, *3* = precipitation station, *4* = spring, *5* = river sampling points, altitude contours are in metres. (b) Heavy isotope gradient of precipitations weighted for amount against altitude of collection. Line *A*: weighted mean values for 1969; line *B*: weighted mean values for 1970; line *C*: weighted mean values for 1969-70. The average gradient is —0.26‰ per 100 m. Circles correspond to annual averaged samples. (c) Heavy isotope content of groundwater plotted against the weighted altitude of the area of catchment above the discharge point for Bellavista and drilled well PP-12/TL. The sampling point PP-26/T6 was taken as representative of the recharge in the plain. Points describe a parallel *D* to the average gradient *C*.

Similarly, the $\delta^{18}O$ of the drilled well PP-12/TL is characteristic of recharge above 280 m. However, in both cases the mean altitude of recharge will be determined by any differences in the amount of precipitation and by any differences in land surface area at different elevations. Data available at the time indicated no major differences in precipitation, so only a weighting factor for different surface areas at different elevations estimated by planimetry was used to obtain mean altitudes of 1000 m for Bellavista springs and 625 m for PP-12/TL.

The drilled well PP-26/TL was quite distinct from other drilled wells in being the shallowest and having a tritium concentration comparable to that of current precipitation and furthermore the most enriched stable isotopic composition. This was strongly suggestive that water from this well originated as local recharge on the plain. Added confidence in this assumption was provided by sampling shallow groundwater ($\delta^{18}O = -5.8‰$) and the base flow of a stream ($\delta^{18}O = -5.9‰$) in the northwest of the area where there is no influence of recharge from the Cordillera. From Fig. 3-10 it will be seen that these points describe a line which has similar slope as that defined by the samples of precipitation covering the whole period of sampling.

The mean $\delta^{18}O$ values of the drilled wells, which individually did not vary significantly with time, was $-6.86‰$ which indicated that recharge was primarily from elevations above 280 m.

Modification of the isotopic composition of the recharge within the aquifer. The basic assumption in these studies is that the isotopic composition of the discharging groundwaters had not been modified during the subsurface passage. This is not always the case as illustrated in an example presented by Fontes and Zuppi (1976).

In Central Italy many areas with recent volcanism do exist and at one site, at Lavinio on the Tyrrhenian shore, a spring discharges which has a constant deuterium content throughout the year ($\delta^2H = -48.0‰$) whereas the $\delta^{18}O$ values vary between -3.6 and $-7.6‰$.

During summer the recharge rate is low. Groundwaters circulations reach a depth which allows the isotopic exchange with high-temperature material to take place. During winter the aquifer is recharged and the spring discharges waters which circulate at the upper and cooler part of the aquifer. An alternative could be that the circulation occurs at the same level but with variable water flow and thus variable amount of exchange.

Whatever the flow pattern it appears that the area of recharge cannot be determined only from ^{18}O contents, since a comparison of regional ^{18}O altitude gradients (Zuppi et al., 1974) with the average ^{18}O content ($-5.60‰$) of springwater would indicate an altitude of recharge of about 120 m (Fig. 3-11). The actual value of $-7.80‰$ is indicated by the intercept of the local meteoric water line with the exchange line (Fig. 3-11). This value gives an

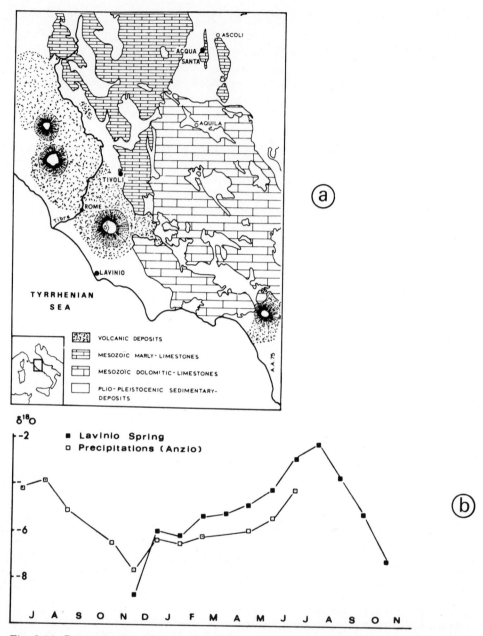

Fig. 3-11. Determination of recharge areas in a perivolcanic system in Italy (from Zuppi et al., 1974; Fontes and Zuppi, 1974). (a) Location map. (b) ^{18}O content at Lavinio spring compared to weighted mean values for precipitations at the nearby coastal station of Anzio. From these results one could conceive that (1) groundwater flow preserves seasonal variations in rains, i.e. is not dispersive, (2) the mean altitude of recharge is close to 120 m assuming that infiltration occurs during the entire year (see Fig. 11d). These two conclusions would be in contradiction with the high dispersivity of volcano-detrital

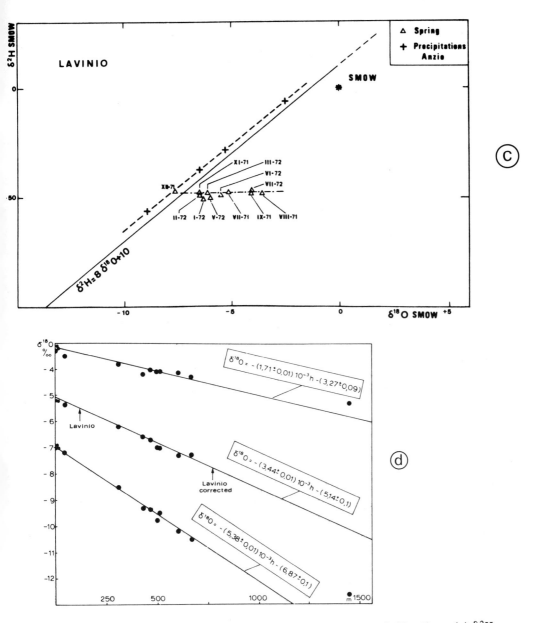

deposits which do not allow piston flow movement and summer infiltrations. (c) δ^2H–δ^{18}O relationship in Lavinio spring. The oxygen shift suggests a seasonal exchange with heated rocks. The intercept with the local meteoric water line established at Anzio gives the actual average ^{18}O content of the recharge. This value allows determination of the corrected altitude of recharge, i.e., 750 m if using the average gradient (see Fig. 11d) which appears suitable because this altitude corresponds to fractured lavas. (d) ^{18}O gradient in monthly weighted precipitations on the Tyrrhenian Basin. Upper curve: summer gradient, lower curve: winter gradient, middle curve: average gradient.

altitude of 650—700 m for the isotopic altitude of recharge assuming infiltration is taking place during the entire year (Fig. 3-11). This altitude corresponds to a volcanic cone of recent fractured lavas which may represent the zone of catchment of this geothermal system.

This example illustrates the fact that fictitious results can be obtained if ^{18}O is used as a conservative tracer in areas or in systems in which modifications of the initial composition of the recharge occur by geothermal exchange or evaporation.

RELATIONS BETWEEN SURFACE- AND GROUNDWATERS

The potential interconnection between surface and groundwaters can be deduced from the piezometric contours. But because of strong variations in horizontal permeabilities in the alluvial deposits which generally form the river bank it is sometimes difficult to ascertain to what extent pressure variations indicated by piezometric maps mean mass transfer of water. The identification of the zones of farthest penetration of surface water into aquifers is important for the localization of future areas of withdrawal and, of course, pollution risks.

Evaporation from lakes and permanent stagnant surface waters often results in an enrichment in heavy isotopes as compared to normal groundwaters. Enrichments are also possible in rivers and in any kind of surface waters even with short exposure time to the free atmosphere.

In areas with strong topographic and thus hydraulic gradient, rivers usually originate in high altitudes and therefore have low isotope contents. For sedimentological and hydraulic reasons the aquifers of any importance are located in the lowest parts of the basins. Thus a difference in heavy isotope content can be expected between the local infiltration (low-altitude precipitation) and the supply from the river.

Surface waters represent a complex distribution of groundwater, soil water and precipitation in the upper parts of the watershed which are drained by the river according to the respective local permeabilities. Therefore the use of the tritium contents as an index of contribution of surface water to groundwaters will require long-term series of measurements and sophisticated mathematical models.

Leakage from rivers

Numerous isotopic investigations conducted in orographic regions with various climates, hydraulic regimes and geological settings confirm the occurrence of leakage from rivers and provide an estimate of its spatial extension, provided well distributed sampling points are available. In most cases the river recharges its own alluvial deposits. This has been shown for Alpine torrents, for the Dranse, a tributary of Geneva Lake (Blanc and Dray, 1967) and for the low part of the Rhône Valley in the vicinity of the nuclear power plant of Pierrelatte, where studies included long-range observations of piezometry, stable isotopes and tritium (Bosch et al., 1974). In the plain of Venice recharge by rivers was recognizable by a difference of about —2‰ in ^{18}O with local water for the rivers Brenta and Piave with mid-altitude catchment areas and of about —4‰ for Adige which is coming from the central zone of the Alps (Bortolami et al., 1973). In South America Vogel et al. (1975) recognize the fingerprints of Andine precipitations, which are highly depleted in heavy isotopes ($\delta^{18}O$ down to —19‰), within the aquifers of the Pampa del Rosario, indicating that they are supplied by the torrents coming from the Cordillera.

The possible contributions of major floods of surface network to aquifers in regions of flat topography have been less investigated. In the centre of the Chad Basin along the Chari River groundwaters show isotopic content higher than local groundwater recharged by precipitations (IAEA, unpublished data, A. Chouret et al., unpublished data), indicating that infiltration occurs. Major floods have also been thought to be responsible for the replenishment of groundwaters underlying the extremely arid Pampa del Tamarugal in northern Chile. However, stable isotope data clearly show that this is not the case and that recharge must occur by subsurface flows in alluvial fills in essentially dry Andean river beds (Fritz et al., 1978b).

The two arms of the Nile River in the region of Khartoum (Sudan) are isotopically different. The White Nile reflects equatorial rains of Central Africa ($\delta^{18}O$ = +1.19‰; δ^2H = +16.7‰ whereas the Blue Nile reflects rains from high altitude of the Ethiopian plateau ($\delta^{18}O$ = —1.96‰; δ^2H = —1.56‰) both slightly increased by evaporation. The local aquifer is mainly containing waters ($\delta^{18}O$ = —9.8‰; δ^2H = —72‰) reflecting a more humid and cooler period of recharge (paleoclimatic effect). The different proportions of mixing of this old reserve of groundwaters with each type of recent Nile waters can be evidenced by their fit on mixing lines in a δ^2H—$\delta^{18}O$ diagram. Updip the confluence groundwaters are lying on the mixing line given by the two types of surface waters (IAEA, unpublished data).

In a recent study in the area of Bern (Switzerland), Siegenthaler and Schotterer (1977) estimate the contribution of surface to groundwaters on the basis of stable isotope contents.

Leakage from lakes

The pioneer work of Gonfiantini et al. (1962) shows that the evaporative enrichment in ^{18}O of the volcanic lake of Bracciano (Central Italy) permits the recognition of the contribution of lake waters to the shallow aquifer on the southeastern shore of the lake. The same type of labelling was used by Payne (1970) to demonstrate that Chala Lake (between Tanzania and

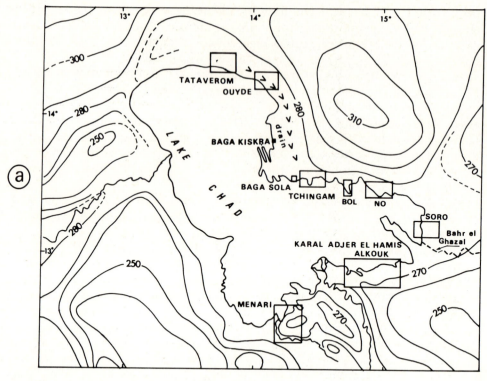

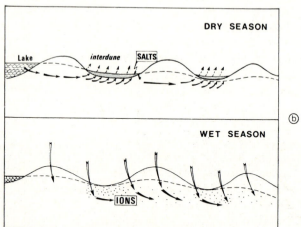

Fig. 3-12. Relationship between lakes and groundwaters. (a) Location map of Lake Chad (382 m) and piezometric contours. The lake is mainly supplied by rivers coming from the south. It has no surface outlet and represents a piezometric high for groundwaters. (b) Mechanism of salt regulation of Lake Chad. The flood of the surface network reaches the lake during dry season. The coastal aquifer is recharged by lake water which evaporates in topographic lows (interdunes). During the wet season salts are dissolved by rain waters and brought to the water table where they are removed by the landward hydraulic gradient.

Kenya) was not contributing by more than 6% to the discharge of springs located downdip the lake.

The case of Lake Chad illustrates the complexity of the relationship between surface and groundwaters as evidenced by a combined piezometric, geochemical and isotopic approach. Lake Chad is the terminal lake of the Chari river system (Fig. 3-12). The climate is highly evaporitic with about 2.30 m/yr of evaporation in free water bodies. Despite these conditions of internal drainage under arid climatic conditions, the waters in the lake keep a low salinity (about 350 ppm total dissolved solids). Thus salinity has remained low and constant for at least one century. Even during and after the drastic drought of 1972-73 no significant increase in salt content has been observed when the average lake area (20,000 km^2) was reduced to about one half. The salinity is thus kept at a low level by an active mechanism of regulation. Previous piezometric studies have shown that on its northeastern shore, the lake is connected to a phreatic aquifer whose level is generally lower than the average lake level (Schneider, 1965). It was thus thought that the saline regulation of Lake Chad was simply due to a leakage of saline waters towards the aquifer. Evaporation represents about 90—95% of the total input by the Chad River. One has to assume a steady state in which the underground leakage was about 1/10 to 1/20 of the annual input with 5 to 10 times its weighted mean salinity since roughly one half of the ionic content of the input remains entrapped in the clay minerals of the bottom sediments. Such a seepage would easily occur through the sandy shore (Quaternary erg) of the eastern part of the lake (Kanem). But this simple concept based only upon piezometry is not in agreement with all the observed data: (1) the salinity of groundwaters in the vicinity of the lake is very variable from 700 to 10,000 ppm but higher than on the shore (Roche, 1974); (2) the chemical facies change from calcium bicarbonate type to sodium bicarbonate type; and (3) the stable isotope content of the groundwater is generally much lower than in the lake (typical values are $\delta^{18}O =$ -2 to $-5‰$), whereas the lake has an average ^{18}O content of $+5‰$ with values as high as $+10$ to $+12‰$ on the presumed infiltration front (Fontes et al., 1970).

As no physico-chemical process exists which could decrease the heavy isotope content of a given water body (except exchange with CO_2 or carbonate precipitation for ^{18}O), one must admit that, at least, a strong dilution of lake water with water depleted in heavy isotopes is occurring within the aquifer.

Detailed studies including horizontal and vertical cross sections of the aquifer have led to the following interpretation of the salt regulation mechanism (Fontes, 1976):

— The lake is enriched in salt and heavy isotopes by evaporation, but waters are not well mixed, the maximum enrichment in ^{18}O and salt is reached on the northeastern shore.

— The flood of the lake shows a lag of about 5 to 6 months after the monsoon period which supplies the surface network of the Chari River several hundreds of kilometers southward; thus the highest level of the lake is reached during dry season (Fig. 3-12).

— At maximum lake level, waters infiltrate through the sandy shore; this infiltration, which is noticeable by the similarity of isotope contents, extends over a narrow fringe of land (some metres to some tenths of metres).

— The water containing dissolved salts moves away from the lake according to the hydraulic gradient and evaporates strongly when the water table becomes close to the topographic surface, i.e., in the interdunes.

— The salt content of the solution increases in this depression and sometimes saline crusts are formed.

— When the rainy summer monsoon occurs, rains dilute the saline solutions and dissolve the saline crust and salts are thus washed down to the water table.

Finally, it appears that the saline regulation of Lake Chad is the product of cascade processes of concentration and dilution with meteoric waters. After some stages, lake water is completely eliminated from the system and salts are only transported by meteoric waters infiltrated during the rainy season. The same mechanism can occur in the coastal zone itself where the salt deposits left behind by the lake flood are dissolved and transported to the aquifer by the precipitations of the rainy season which corresponds to the low level of the lake.

MECHANISM AND COMPONENTS OF THE RUN-OFF

Determination of the respective contributions of rainwater, groundwater and presumably soil water to a flood discharge in surface systems represents one of the most important problems of hydrology. The knowledge of the components of the flood can be used for applications in the field of the chemistry of surface water (see, e.g., Chapter 11, this volume) in the determination of the residence time of groundwaters in the aquifer, in the evaluation of snow melt to the surface run-off. These aspects are closely related to the general problems of surface water management.

Graphical analyses of flood hydrograph are generally used to evaluate the groundwater component to the flood. Physico-chemical methods including measurements of conductivity and temperature during the flood can also provide a positive input (Pinder and Jones, 1969; Andrieux, 1976), however, the shape of the flood hydrograph may change for each single episode and the difference in physico-chemical parameters of the initial components may not be strong enough to evaluate the mixing. These methods thus do not provide reliable estimates of rain and groundwater contributions. The same criticism is also valid for environmental isotope measurements.

Again, the hydrograph separation utilizes isotopic differences between different sources. But taking into account the interest of the question it would thus be highly recommendable that any investigation of flood episodes include temperature recording and a dense time series of sampling for hydrochemistry, stable isotope and tritium including base flow, groundwater and rainwater as reference.

First attempts in using the environmental isotopes in flood studies were done by Hubert et al. (1969) who measured the tritium content and major ions of the exceptional flood of the Dranse River in September 1968. At peak discharge the flow of this tributary of Geneva Lake was about 400 m^3/s and the tritium content reached about 265 TU whereas the base flow and precipitation responsible for the flood were measured within the range 200—250 TU and 100—150 TU respectively. Furthermore, it was possible to follow the dilution of Dranse water into Geneva Lake several months after the flood (Meybeck et al., 1970). Run-off coefficients of some percent were determined on the basis of ^{3}H analyses in several basins (5—90 km^2) in France (Crouzet et al., 1970).

The potential use of isotope techniques to the study of the run-off in snow melt processes was also illustrated by Dinçer et al. (1970) in the Modry Dul catchment in northern Czechoslovakia. The stream at the outlet of the basin was sampled daily during the snow melt period, precipitation was sampled close to the stream sampling station at 1030 m and also at 1410 m, and snow pack samples were taken at different depths at three points in the basin. Measurements of the tritium and stable isotopic composition were made although the major contribution to the study was provided by tritium.

The tritium content of the snow pack samples varied between 200 and 300 TU with a mean value of 250 TU. The tritium concentration of the base flow of the creek in winter had a mean value of 730 TU. After the onset of snow melt the values of the run-off decreased. This suggested that the run-off was a mixture of two types of water, one having the tritium concentration of winter base flow and the other one having the composition of meltwater. Indeed, this assumption was supported on two occasions after the onset of the snow melt period when the temperature fell below 0°C and melting stopped and the tritium concentration increased close to that of the winter base flow. The variations in tritium concentration of the run-off during the snow melt period were used to estimate the relative proportions of meltwater and baseflow in the total run-off from the basin. In this study, the direct contribution of precipitation was neglected in view of its minor importance during the snow melt period as compared to the water equivalent of the snow pack in the basin.

A joint study to the previous one and including ^{18}O and ^{3}H measurements has been performed in the Dischma Valley in the Swiss Alps (Martinec et al., 1974). The basin has an area of about 43 km^2 and extends over 1668—3145 m in altitude. Long-range run-off records were available which showed

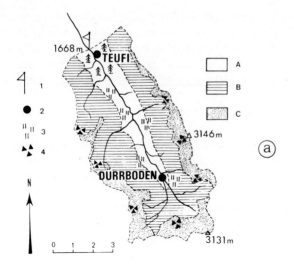
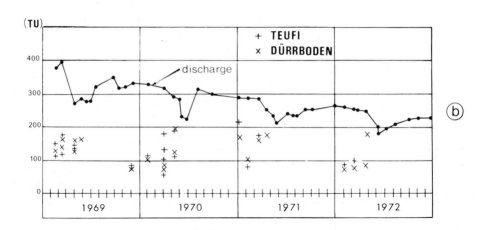
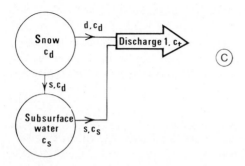

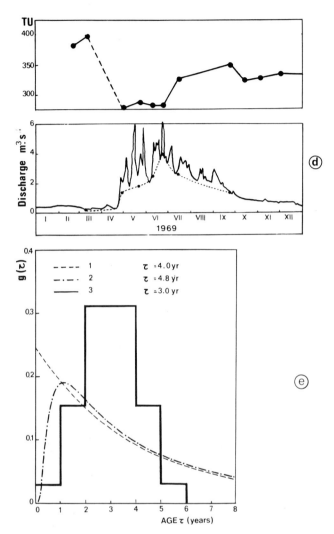

Fig. 3-13. Mechanisms and components of the run-off: Dischma Basin in Switzerland (after Martinec et al., 1974). (a) Physiographical scheme of the Dischma Basin: *1* = stream gauge station and isotope sampling, *2* = snow core sampling, *3* = weadows, *4* = outcropping rocks; *A, B* and *C* refer to the weighting areas of the isotopic composition of precipitations. (b) Tritium activity in the discharge (black dots) and in the snow cover (crosses). (c) Principle of the tritium mass balance in the discharge: d and s = relative proportions of direct and subsurface flow ($d + s = 1$); C = concentrations. (d) Determination of the respective contributions of surface and subsurface flows on the basis of ^3H data. Above: ^3H in the discharge, below: subsurface flow contribution to the hydrogram (black dots). (e) Residence time of water delivered by the subsurface according to: *1* = exponential (i.e., well-mixed model), *2* = dispersive, and *3* = binomial models. The function $g(\tau)$, where τ is the mean age or mean residence time, represents the fraction of outflow of the year $t - \tau$ appearing in the outflow at time t.

the seasonal influence of snow melt, which gives a discharge peak in June whose maximum 15.55 m³/s was approximately 10 times higher than average interannual discharge (1.67 m³/s). Fig. 3-13 shows the tritium concentration of the creek and the separation of the hydrograph from tracer mass balance computations. The tritium balance based upon measurements in flow, snow cover and precipitations suggested that the percentage of the direct meltwater to the run-off was about 50% (41—63%) at or just before peak discharge (Fig. 3-13). Different models of groundwater recharge and storage were proposed to calculate the residence time of groundwater discharged during winter baseflow. The best fit was obtained for an exponential distribution of the residence times of each annual contribution assuming that 2/3 of winter and 1/3 of summer water contribute to the base flow discharge. The mean residence time is about 4 years.

The run-off of the small (650 ha) basin of the Hupsele Beek in Eastern Netherlands was studied by Mook et al. (1974) using ^{18}O as natural tracer for groundwater and rainwater during the flood caused by a long period of moderate autumn rains in November 1972. Run-off and rains were automatically sampled at time intervals of 8 h. The run-off response appears much faster than the time interval of sampling, but it was still possible to propose that 87% of the precipitations were recharging the aquifer and 13% were rapidly drained.

The run-off process was extensively studied in a number of watersheds in Canada (Sklash et al., 1976; Fritz et al., 1976). In central and eastern Canada run-off arises from summer storms whose stable isotopic composition is normally markedly different from the annual weighted mean composition of precipitation. Consequently, the isotopic composition of these storms is quite different from that of groundwater and thus provides a characteristic label for these two components in a study of the run-off process. If the total rainfall and run-off are known, then the contribution of groundwater to the run-off may be estimated from the isotope mass balance.

In the Wilson Creek watershed (22 km²) in Manitoba (Fig. 3-14), a violent summer storm (40 mm in eight hours having a δ^{18}O of —19‰), fell in August 1973 giving rise to a flood having a minimum δ^{18}O value of —16‰. Prior to the storm the δ^{18}O content of the base flow was —14.5‰, which was not very different from groundwater (δ^{18}O = —15‰) in the basin. The isotope mass balance indicated that 90% of the run-off was pre-storm water and even at maximum discharge from the creek this component only dropped to 60%. Similar conclusions could be deduced from the chemical composition of the run-off and groundwater, but were much less quantitative.

Complementary studies were conducted in Kenora Big Creek and Big Otter Creek watersheds located on Quaternary sandy deposits in western and southern Ontario. In these basins, 25—50% of the total flood could be attributed to water of pre-storm origin.

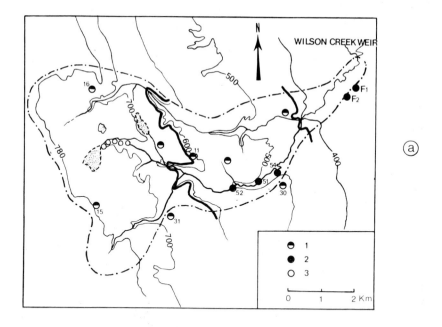

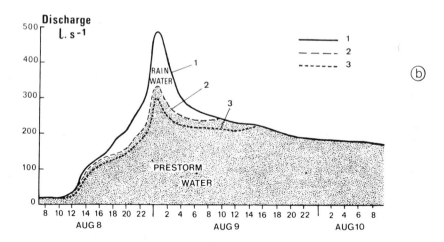

Fig. 3-14. Mechanism and components of the run-off: the Wilson Creek, Canada, experimental watershed (after Fritz et al., 1976). (a) Map of the Wilson Creek. 1 = recording rain gauge, 2 = piezometric nest, 3 = upstream sampling stations. Topographic contours are in metres above sea level. (b) Components of the run-off calculated from isotope mass balance. 1 = total stream discharge at Weir; 2 = separation using average base flow at Weir $\delta^{18}O = -14.5‰$; 3 = separation using a maximum expectable value for the base flow $\delta^{18}O = -14‰$. About 90% of the flow given by a storm of 48 mm in 8 hours are waters stored before the storm.

LEAKAGE BETWEEN AQUIFERS

The interconnection between two different aquifers separated by an impermeable layer can be shown on the basis of the respective piezometry, hydrochemistry, water balance and structural analysis of the area. In some cases, however, e.g. in carbonate aquifers, the interconnection may be difficult to prove because the piezometry cannot always be established. The knowledge of possible connections between aquifers will help in groundwater resources evaluations as well as in the forecasting of decrease of water quality by leakage from above or below. This kind of study can be done using specific chemico-physical parameters of the two water masses. But, because ^{18}O and ^{2}H are conservative when they have reached the water table under a reasonable thickness of sediments the simultaneous use of ^{2}H or ^{18}O often represents the most suitable tool to study the leakage, i.e., the mixing between two reservoirs of waters, provided the two reservoirs are isotopically different. When used simultaneously ^{2}H and ^{18}O contents define straight lines of mixing between the representative points of the two end members.

The Sahara provides some clear examples of mixing between aquifers. The aquifer contained in the continental sandstones of lower Cretaceous ("Continental intercalaire") extends from the border of Morocco to the Nile Valley where it is called aquifer of the Nubian sandstones (Fig. 3-15). In its eastern part (east of the Mzab structural rise) there is no outcrop of "Continental intercalaire". The aquifer is confined with a very stable ^{18}O content near −8‰ (Gonfiantini et al., 1974a). In its western part (west of the Mzab structural rise) the aquifer is unconfined and is recharged through outcrops of Gourara, Touat and Tiddikelt around the upper Cretaceous Tademaït Plateau. Other outcrops are located in the south Saharian Atlas. North of Gourara the "Continental intercalaire" is covered by the complex of the Grand Erg Occidental. Detailed isotope surveys show that the isotope contents of waters of the Grand Erg have typical value of $\delta^{18}O = -5‰$ observed in several locations (Conrad et al., 1966; Conrad and Fontes, 1970). The piezometric surface of the aquifer of the Grand Erg Occidental is several metres above that of the "Continental intercalaire" and leakage should occur since no impervious layer lies between the two aquifers (the base of the Grand Erg is made of probably fractured limestones). In the Gourara the "Continental intercalaire" is recharged by the southern edge of the Grand Erg (Fig. 3-15). Going southward the "Continental intercalaire" is buried under the Tademaït Plateau whose basement includes impervious levels of clays. There the aquifer of the "Continental intercalaire" shows a decrease in ^{18}O content towards values typical for the general circulation in the confined part of the aquifer (i.e. −8‰). This illustrates the decreasing contribution of the Grand Erg to the "Continental intercalaire" below the Tademaït Plateau. This is in agreement with the general picture of flow within the "Continental intercalaire" as indicated by the piezometry (circulation towards southwest in the western part of the Sahara). In the region of the Gabès Gulf in southern Tunisia, the "Continental intercalaire" is affected by a dense network of vertical faults. The piezometry indicates that this region is a discharge area. The "Continental intercalaire" is covered by impervious levels of marls and then by another aquifer itself confined in Cretaceous limestones, sandstones and dolostones ("Complexe terminal"). Differences in piezometric head suggest that the "Continental intercalaire" may leak into the "Complexe terminal" through the faults network. Isotope analyses confirm this hypothesis. In a $\delta^2H–\delta^{18}O$ diagram the points corresponding to the faulted region fall on a straight line between the typical values for "Continental intercalaire" and for the "Complexe terminal" (Fig. 3-15).

Gonfiantini et al. (1974b) investigated groundwater systems of the Hodna Plain, an almost closed depression in the North Africa plateau at its boundary with the Saharian region (200—300 mm of annual precipitations). The plain collects the intermittent floods which dry up and give rise to a salty chott in its central part (Fig. 3-16). Two aquifers are contained in the recent sediments and geological layers of Tertiary age. Detailed hydrogeological studies prove that connections are possible with leakage from below and from above because the deeper aquifer is semi-confined by discontinuous levels of Quaternary clay deposits of continental origin. This study was a demonstration of the usefulness of a combined approach which includes hydrochemistry. North and east of Chott-el-Hodna, the shallow aquifer is enriched in heavy isotopes with respect to the deeper one although it is assumed that the mechanism of recharge is the same for both deep and shallow aquifer (precipitations in the Hodna Mountains). Evaporation is not responsible for this difference since representative points are lying along the meteoric water line $\delta^2H = 8\ \delta^{18}O + 10$ (Fig. 3-16; see Chapter 1). Outside the area of artesian flow the deep aquifer shows a decrease in ^{18}O with increasing depth. The evolution of the stable isotope contents is thus interpreted in terms of mixing of two different bodies of water. In one area it was suspected on the basis of salt content and piezometry that leakage from below occurred. This was not confirmed by stable isotope and tritium contents which suggested local fast infiltration of recent water resulting in low salinity and a piezometric high.

South of the Chott, the deuterium control proves the evaporation from the shallow aquifer and back extrapolation of the evaporation line to the meteoric water line suggests that the water of the shallow aquifer could, in most parts, come from a leakage from below (Fig. 3-16).

Similar indications for deep (artesian) water leakage into shallow aquifer were noted in Hungary on the basis of deuterium and tritium content (Deak, 1974). No significant leakage could be found between two deep artesian aquifers in the Kaikoura Plain, New Zealand (Brown and Taylor, 1974). In Saudi Arabia it was possible to estimate the regions where shallow

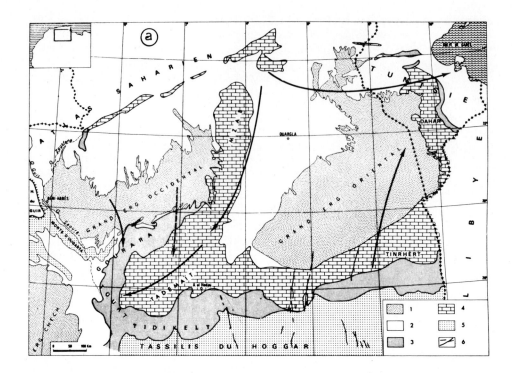

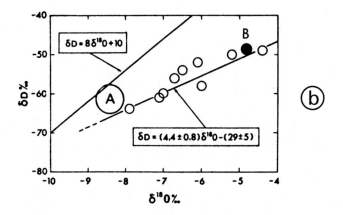

Fig. 3-15. Leakage between aquifers. (a) Aquifer of the "Continental intercalaire" in the northern Sahara (from Gonfiantini et al., 1974a). *1* = Ergs; *2* = Tertiary and Quaternary ("Complexe terminal"); *3* = "Continental intercalaire"; *4* = Upper Cretaceous and Eocene; *5* = Paleozoic and old crystalline rocks; *6* = Faults, solid arrows flow lines in the aquifers of the "Continental intercalaire", dotted arrow flow line in the aquifer of the "Grand Erg Occidental". The aquifer of "Continental intercalaire" is unconfined west of 3rd meridian and confined in eastern Sahara. (b) Northwestern Sahara: heavy isotope

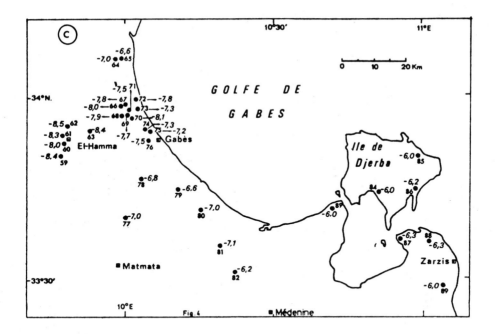

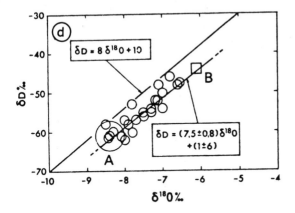

content of waters from the aquifer of the "Continental intercalaire" (here unconfined). A = average value for groundwaters of "Continental intercalaire" in its confined part. This trend can be interpretated as a mixing with waters of the Grand Erg aquifer (average value B). (c) Detailed study of the area of discharge of the aquifer of "Continental intercalaire" and "Complexe terminal" (Gulf of Gabes). (d) Mixing between waters of "Continental intercalaire" (symbol A) and "Complexe terminal" (symbol B) in the area of the Gulf of Gabes (south Tunisia).

112

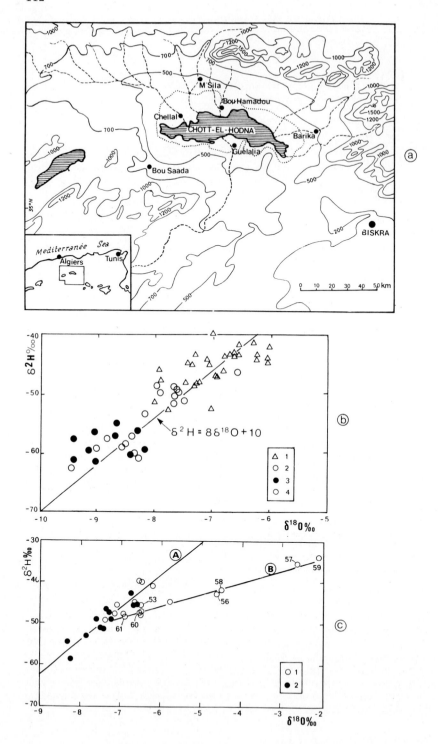

aquifers are recharged by leakage from the Umm-Er-Radhuma deep artesian aquifer (Dinçer et al., 1974).

ISOTOPE HYDROLOGY OF FRACTURED ROCKS

It is generally admitted (Suszczynski, 1972) that hydrogeology of fractured rocks refers to groundwater movement in rocks whose porosity is only due to cracks, fissures and fractures in any kind of solid (impermeable) rock. These rocks generally cannot provide aquifers in the proper sense of the word, i.e., a porous medium continuously saturated with water. For that reason, the basic principles of hydrogeologic studies cannot apply to water in fractured rocks. Until the recent past, these rocks were even considered as impervious at least as far as crystalline rocks were concerned (the role of karstic reservoirs for groundwater storage was obviously known). Thus, although fractured rocks cover approximately one third of the continents, the knowledge on their groundwater hydrology still remains empiric. Generally, the exploitation of groundwater resources is made in the weathered and hence porous parts of fractured rocks, where classical methods (drawing of piezometric maps and evaluation of hydraulic gradients, conductivity pumping tests and evaluation of transmissivities) can be applied.

In the fractured rocks themselves, the main questions that isotope hydrology may contribute in solving are:

— Does groundwater of a given catchment area mix well or poorly within the fractured system?

— Is there any significant flow and renewal of groundwater resources?

A negative answer to the first question by measuring large differences in stable isotope contents within a groundwater system leads to the most simple interpretation. In such a case, we are dealing with several independent groundwater systems, each of which has its own hydraulic characteristics. If stable isotopes suggest a good homogeneity of a groundwater body, it would not necessarily mean that the system can be considered as a unique aquifer. The homogeneous stable isotope content of groundwaters in frac-

---

Fig. 3-16. Leakage between aquifers (after Gonfiantini et al., 1974b). (a) Location map of the region of study in Algeria. (b) δ^2H—δ^{18}O relationship in samples from north and east of Chott-el-Hodna, showing the mixing between two water bodies. 1 = shallow groundwaters; 2, 3 and 4 = deep groundwaters (2: ^{14}C > 2%; 3: ^{14}C < 2%; 4: ^{14}C not measured). The global scale value (δ^2H = 8 δ^{18}O + 10) has been adopted for the meteoric water line. (c) δ^2H—δ^{18}O relationship in samples from south of the Chott. 1 = shallow groundwater; 2 = deep groundwater. Line A = meteoric water line (assumed) δ^2H = 8 δ^{18}O + 10; line B = evaporation line calculated from salt-bearing samples. Back extrapolation of line B to line A give an intercept which suggests that the shallow aquifer may be recharged by leakage from below.

tured rocks may be due to a supply from a unique episode or from a reservoir with already well-mixed water (flood from a river, leakage from a porous and well-mixed aquifer).

A positive answer to the second question can be easily provided by the occurrence of tritium which indicates that recent water is present. Differences in tritium content from location to location may sometimes help to prove the heterogeneity of systems in which stable isotope contents are constant (e.g. several distinct episodes of recharge in the same area). But, variable tritium content could also mean that, although homogeneous, the system has flow movements slow enough for age effect to be detected in tritium decay. The occurrence of homogeneous tritium contents (together with homogeneous stable isotope content) can be attributed to good mixing but also to a given episode of recharge (e.g. flood) in isolated channels. The lack of tritium will be an indication of a poor recharge and possibly of no recharge at all. If correlated with a low ^{14}C content of dissolved inorganic carbon, the lack of tritium will indicate that the system is discharging old waters inherited from an ancient humid period or from an old exceptional flood event, or from another aquifer. Variations in stable isotope content together with constant tritium content would be expected in fast systems in which evaporation or geothermal exchange would occur.

Within fractured rocks karstified limestones or dolostones represent a special case of limited spatial but of high economic importance. In these systems, large reservoirs can be found by carbonate dissolution which always takes place along pre-existing fractures. The same question about mixing or not arises for karstic as for any other kind of fractured rocks. In karstic systems, flow is generally proved by major springs and hence recharge must occur in the outcrop areas of karstified limestones or dolostones which are porous enough to generally eliminate surface runoff during rainfall. The problems which are most specific for karstic systems deal with (1) the importance of rainfall distribution upon the mechanism of recharge and discharge of the system; (2) the importance of storage; and (3) the possible occurrence in the system of inactive (i.e. "by passed") reservoir which are discharged (and recharged) during exceptional rainy episodes.

Environmental isotope analyses permit discussion on the origin and the age distribution of groundwaters and therefore are a well-suited tool for the study of karstic systems, especially if one considers that other hydrological tools can only be employed with difficulty. For these reasons, it is to be expected that isotope studies in karst hydrology will expand markedly in the near future.

Non-carbonate fractured rocks

Aquifers in fractured, non-carbonate rocks are generally assumed to be of limited economical importance. Therefore only few detailed hydrogeological investigations have been undertaken and only in exceptional cases have been combined with environmental isotope studies. An example is IAEA sponsored studies in the Canary Islands, where tritium data show that recharge of volcanic rocks occurs mainly in the central and highest part of Gran Canaria (Gonfiantini et al., 1976). In Tenerife, the general lack of tritium was interpreted as the result of a possible stratification of successive recharge episodes. In the groundwaters of the islands, the large variations which occur in the stable isotope content of precipitations in altitude were preserved in groundwaters, and correlated to their salt content (Fig. 3-17). The independence and the non-mixing of groundwater circulations are thus demonstrated.

Groundwater systems on Iceland belong to some extent to the group of fractured rock systems since the waters are contained in cracks and cavities of lava. The decrease in deuterium content with increasing altitude is indicative of local recharge without significant mixing (Arnason and Sigurgeisson, 1967). The occurrence of small amounts of tritium even in deep geothermal water shows that recent supply reaches rapidly great depths of more than

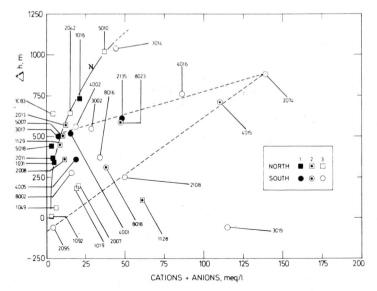

Fig. 3-17. Isotope hydrology of fractured rocks. Non-carbonate rocks: volcanic rocks of Gran Canaria, Canary Islands (from Gonfiantini et al., 1976). The difference in elevation between the sampling point and the altitude of recharge calculated on the basis of the ^{18}O gradient in altitude (see p. 93 and Fig. 3-9) is plotted against the total dissolved salt. Squares: north of the Island; circles: south of the island. Symbols *1, 2, 3* refer to the type and the number of analyses available. The length of groundwater circulations is roughly correlated with their total dissolved solids. This suggests that the mineralisation is obtained in independent groundwater paths.

2000 m in some cases (assuming that this shallow supply is not due to a simple mixing near the ground surface).

In the Central Austrian Alps, deuterium and tritium analyses of groundwaters collected in a 7-km long tunnel (Fig. 3-18) revealed slow and vertical seepage since no ^3H was found in the central part and since ^2H variations respected the altitude effect (Rauert and Stichler, 1974).

Recent studies of groundwater in fractured rocks have been performed under the Mont Blanc (Western Alps; Bortolami et al., 1977; Fontes et al.,

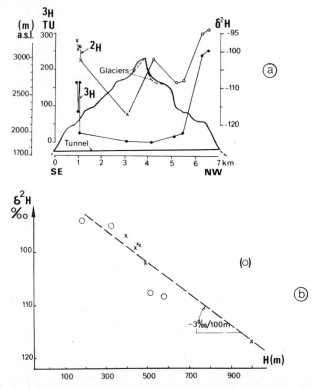

Fig. 3-18. Isotope hydrology of fractured rocks. Non-carbonate rocks: crystalline and metamorphic rocks from Austrian Central Alps (after Rauert and Stichler, 1974). (a) Cross section of the massif: location of sampling sites along the 7-km-long tunnel of Tuxer Hauptkamm; deuterium and tritium content of the seepage water. On both sides ^3H contents decrease to approximately background level after 1 km on the southeastern side and 2 km on the northwestern side. Thus in the center of the tunnel the seepage water was older than 18 years in 1972 (date of sampling). (b) Relationship between the stable isotope content of the seepage water and the height of rocks above the sampling point. The rather good linear correlation with a slope of —3.0‰ per 100 m describes a reasonable altitude effect suggesting that the infiltration is vertical. A point located beneath the glacier is anomalously enriched in ^{18}O. Symbols: circles, northwestern side; crosses, southeastern side of the massif.

1978). About 80 springs (about 0.5 m³/s of cumulated flow) were sampled in the tunnel which crosses the French-Italian border beneath the metamorphic and crystalline massif. Stable isotope contents indicate that water movement occurs along fractures where little or no mixing occurs. Regional values for stable isotope gradient of precipitations in altitude suggest that the recharge area extends over the entire topographic profile up to about 4000 m. Tritium concentrations are in agreement with these conclusions and indicate furthermore that groundwaters are recent (^{3}H concentrations between 50 and 250 TU). Thus recharge occurs rapidly through 200 m of low-temperatures ice ($-10°$C) and through an average of 2000 m of granite.

In the centre of the tunnel, waters are evaporated as indicated by their location below the meteoric water line in the δ^2H—δ^{18}O diagram. This evaporation may occur at the surface of the snow and ice cover by sublimation and even within the unsaturated cracks due to the difference in temperature between the inner part of the rock warmed by the geothermal gradient and the iced infiltration zone.

Karsts

In southern Europe (Mediterranean regions), the Middle East and in northern Africa, carbonate rocks are often the only potential source of good-quality water. But because of large variations in the discharge (frequently intermittent) or because of complicated tectonic features, it is difficult to establish whether the surface area of catchment is really representative of the drainage area. Thus, the first uses of isotopes were done in connection with dye studies to investigate the groundwater paths. For this purpose, injections of artificial tritium were made in grecian karsts (Burdon et al., 1963; Leontiadis and Dimitroulas, 1971). Connections between sink holes and springs were demonstrated and values were proposed for mean velocities and volumes stored in the systems. However, these experiments were not repeated because of the large amount of tritium needed (about 10^3 Ci).

Environmental isotope studies were first performed in southern Turkey (Dinçer and Payne, 1965) where the main water resources come from karstic reservoirs with huge springs of up to 50 m³/s of discharge. Although preliminary, this study was of high importance because it took place in 1963-64, when the tritium rain-out was at its maximum and the identification of recent and old waters was unambiguous. The springs contain generally large amounts of tritium which are sometimes close to the computed activity of the recharge for the previous years, which indicates short transit times and likely small storage. Seasonally the same spring (Homa EIE kampi) exhibits a decrease in ^{3}H content which indicates that older water could participate in the discharge.

This preliminary study was complemented by further measurements which demonstrated the role of recent rainfall (Fig. 3-19; Bakalowicz and

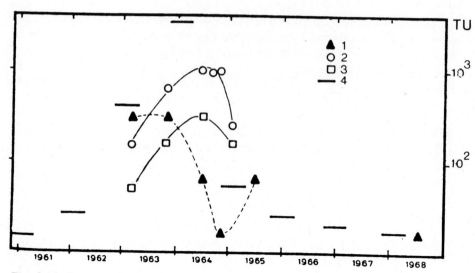

Fig. 3-19. Isotope hydrology of karstic systems in Turkey (after Bakalowicz and Olive, 1970). Tritium content of: (1) karstic spring of Homa; (2) Lake of Beysehir, (3) Lake of Egridir. The monthly average tritium activity of the recharge (4) has been calculated by comparison with the values measured in Europe. The peak in the spring of Homa is attributed to 1963 peak in precipitations showing the large participation of recent rainfalls in the discharge. The low tritium content of the spring at the end of 1964 is attributed to a base flow discharge of old waters stored in the deepest parts of the karst.

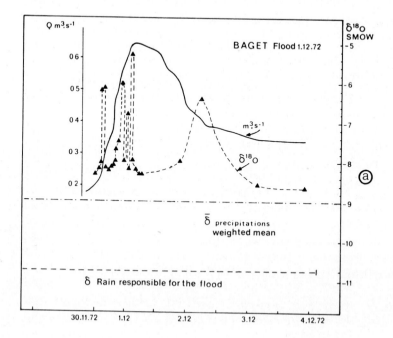

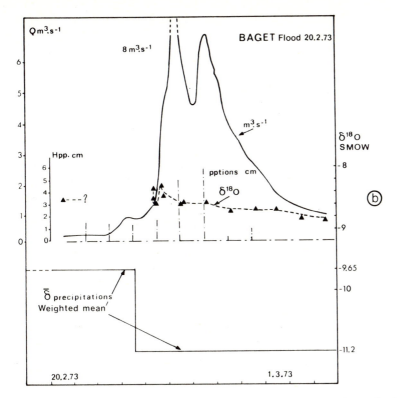

Fig. 3-20. Isotope hydrology of karstic systems. Flood studies at Le Baget, southwestern France (after Fontes, 1976). (a) Flood discharge after a dry period. The flood water is quite different either from the rain which generated the flood or from the monthly weighted mean value of precipitations before the flood. The karstic system is acting according to a piston flow mechanism. The discharge during flood rise corresponds to separated reservoirs with isotope content higher than that of base flow. This ^{18}O-rich water could correspond to summer rain waters stored in the lowest parts of the system. (b) Flood discharge following a rainy period. This major flood was sampled after and during several rainy episodes. Each rainfall (dotted lines) is referred to the scale Hpp in cm. The isotopic discharge is different from the isotopic content of the rains showing that the system still acts according to a piston flow mechanism. But the discharge is homogeneous suggesting that the different reservoirs of the system were filled by the same rainy episode. The slight fluctuations at the beginning of the flood reflect the contribution of waters collected at low altitude. The continuous decrease in ^{18}O during the flood is attributed to the increasing proportions of waters coming from the highest parts of the basin.

Olive, 1970), and established that groundwater recharge was partially due to leakage from lakes (Dinçer and Payne, 1971). Within underground reservoirs, it appeared from comparisons between theoretical models and output functions that the best fit was obtained by the so-called "well-mixed" model in which the contribution of each yearly rainfall is supposed to mix rapidly within the reservoir.

Other attempts to use environmental ^{3}H and ^{18}O in the investigation of a karstic system were carried out in the Swiss Jura (Burger et al., 1971). Because the springs exhibited a general decrease in ^{3}H content which parallels that in precipitations, it was assumed that the output was mainly water of the past year. This was in agreement with dye experiments which gave apparent velocities of some metres per hour to some hundreds of metres per hour over distances of 2—16 km. Because the stable isotope content was smoothed out at the outlet as compared to fluctuations in the precipitation, it was also proposed that the storage implied good mixing in a large reservoir. It was not discussed to which degree the smoothing out of seasonal variations was due to mixing in the unsaturated zone.

In the central part of the French Pyrenees, Bakalowicz et al. (1974) noted, on the basis of ^{18}O variations, transit times of about two months for the seasonal contributions to appear at the outlet of two small watersheds (12 and 83 km^2 respectively). In the same area the base flow of another watershed of 13 km^3, the Baget, had a rather constant ^{18}O content which was found to be very close to the annual recharge by precipitation. The authors argued that this would not necessarily imply that the saturated zone of the karst was homogenized, but could also mean that the successive discharges, randomly distributed, were representative of the mean input. Heavy showers occurring in summer produced slight but noticeable increases in the ^{18}O content at the outlet. These variations were interpreted either as a direct contribution of low-altitude rains to the discharge or as the discharge of the unsaturated zone of the karst which would represent a reservoir of waters enriched in heavy isotopes. The fact that even in winter the flood begins with an increase of the ^{18}O content led the authors to accept the second interpretation.

Flood episodes have been studied in detail in the same system (Fontes, 1976). On the basis of stable isotopes and tritium, the following interpretations are proposed:

— During the floods, the isotope content of the discharge is independent of that of the rainfall which caused it.

— During the flood rise, the system is thus discharging according to a piston flow mechanism.

— Summer rains accumulate in the lowest parts of the system which are the most karstified and which provide the greatest potential of storage.

— Heavy rains in autumn removed the summer waters accumulated in the reservoirs at low altitude. The discharge is mainly due to these waters enriched in heavy isotopes which reach the outlet as separated pulses (Fig. 3-20) corresponding to the discharge of isolated reservoirs.

— When heavy rains occur in winter, the reservoirs of the lowest part of the karst are filled with water of the previous rains.

— At the beginning of each flood, a small enrichment in heavy isotopes reflects first the discharge of waters stored at low altitude; then the isotope content decreases while waters coming from high altitudes reach the outlet.

Fontaine de Vaucluse, in southern France, is a major karstic spring (about 30 m³/s where the type of vauclusian (i.e. karstic) discharge was defined. It was found (Margrita et al., 1970) using ^{3}H measurements at the spring and in the rainfall upon the area of catchment that during spring, which is a rainy season, the residence time is short and flow occurs through by-pass systems of cracks. While summer proceeds, waters remaining from the previous period are mixed in decreasing proportions with (old?) waters stored in the system. Autumn rains give rise to a mixing of waters different from the rains themselves, according to a "piston flow" displacement.

MECHANISM OF SALINIZATION

The origin of the total dissolved solids (TDS) can generally be deduced from chemical studies. For instance, the variations of the characteristic ratios Cl^-/SO_4^{2-}, Mg^{2+}/Ca^{2+}, $Cl^-/Na^+ + K^+$, HCO_3^-/SO_4^{2-} and the evolution of saturation indices of the main solid phase components, can provide valuable information on the various mechanisms of salt concentration. However, in some cases, e.g. when the solubility product of the major dissolved salts is reached, it becomes difficult or impossible to follow the chemical evolution of a solution and thus to draw information on its origin. In the case of dilute solutions, it is often not possible to determine from the chemical composition whether the salt content is due to evaporation or to leaching of solid salts, e.g. in irrigated zones where evaporation is high. However, ^{18}O and ^{2}H contents will allow the distinction of leaching, without evaporation and thus without isotopic enrichment, from an increase in TDS due to evaporation. In ideal cases, it would be even possible to evaluate the respective contribution of leaching and evaporation (Fontes and Gonfiantini, 1967).

The Sebkha el Melah (northwestern Sahara), terminal lake of the intermittent Wadi Saoura, provides such an example. The Sebkha is usually dry and the bottom is covered with a salty crust mainly formed of halite. When a major flood occurs, it reaches the Sebkha and increases its TDS by evaporation and partial dissolution of the salt crust. After some months of evaporation and infiltration, the Sebkha dries again and a new salt crust is formed (Conrad, 1969). Below this crust, a layer of clays rich in organic matter acts as an impervious barrier for a confined aquifer saturated with respect to sodium chloride. The problem which cannot be solved by chemical analyses is: does saturation in the aquifer occur by leaching of salts in the sediments of the Sebkha bottom or is the solution already saturated by evaporation when it infiltrates and recharges the aquifer. In the latter case, high heavy isotope content is expected because of the large isotopic enrichment which accompanies evaporation and salt concentration under the arid climate of the Sahara (Fontes and Gonfiantini, 1967). In the former process of recharge of the aquifer, the isotopic enrichment must be variable but lower, depending on the residence time of the water in the Sebkha before infiltration.

Results of isotopic, chemical and piezometric measurements (Conrad et al., 1966; Conrad and Fontes, 1970) are:

Location	Salinity	Piezometric rise (cm)	$\delta^{18}O$ (‰ vs. SMOW)
Center of the Sebkha	saturated NaCl	105	−4.9
South of the Sebkha	saturated NaCl	20	+9.6
North of the Sebkha	saturated NaCl	0	+19.6

The interpretation is as follows:

— When the flood reaches the basin, a part of the water infiltrates immediately in the permeable margin of the basin and recharges the deepest part of the confined aquifer; the isotope content is the one of the flood water and halite saturation is then reached by leaching in the sediment; the piezometric head is maximum.

— South of the Sebkha, infiltration occurs during the enrichment process, a part of the salt content is due to leaching.

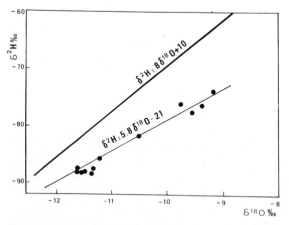

Fig. 3-21. Mechanism of salinization groundwater samples from the Juarez Valley in Mexico (after Payne, 1976). On the δ^2H–$\delta^{18}O$ diagram groundwater points are lying on an evaporation line. Samples located down dip from the city of Juarez are also the richest in salt and stable isotopes suggesting that evaporation by recycling of surface (irrigation) water is responsible for the salt content.

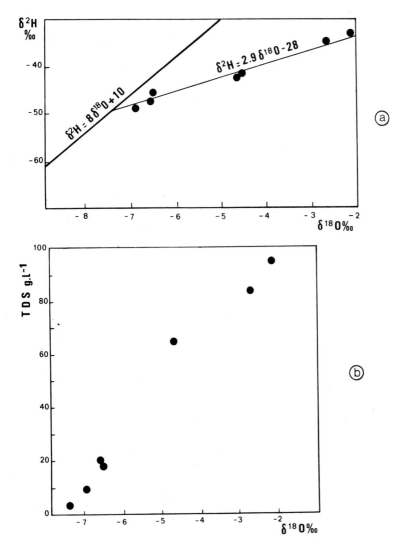

Fig. 3-22. Mechanism of salinization groundwaters from south of Chott-el-Hodna, Algeria (after Gonfiantini et al., 1974b). (a) Salt-rich waters are lying along an evaporation line (cf. Fig. 3-16c). (b) The enrichment in salt is well correlated to the increase in ^{18}O suggesting a mechanism of evaporation from the water table itself in this area. This effect would be isotope fractionating and thus would take place during rises of the water table.

— North of the Sebkha, infiltration occurs at the end of the concentration process, the heavy isotope content is high, the piezometric head is low.

The change in stable isotopic composition due to evaporation can also provide information on the mechanism causing salinity in irrigated areas. The data plotted in Fig. 3-21 represent the stable isotopic composition of water samples taken from irrigation wells in the Juarez Valley of the Rio Bravo in Mexico (Payne, 1976). The points fall on a line, the slope of which is characteristic of an evaporation process. The most depleted samples are for wells which were sampled close to the city of Juarez. The increase in isotopic enrichment of ^2H and ^{18}O corresponds both to the location of the sampled wells down the hydraulic gradient of the valley and the increase in salinity of the waters. Thus the most enriched and more saline waters are found farthest from the city of Juarez. The isotope data, therefore, indicate that the increase in salinity is due to evaporation. The gradual increase in isotopic enrichment moving down the valley suggests a re-cycling of the excess irrigation water with the resultant increase in salinity.

Stable isotope data also suggested that evaporation gave rise to increases in salinity of shallow groundwater south of the Chott-el-Hodna in Algeria (Gonfiantini et al., 1974b). The stable isotope content for shallow groundwater sampled from auger holes where the water table is less than 1 m from the surface is quite distinct from that of deeper groundwater in this area which plots on or close to the meteoric water line. The shallow groundwater samples are also characterized by high salinity of the order of tens of grams

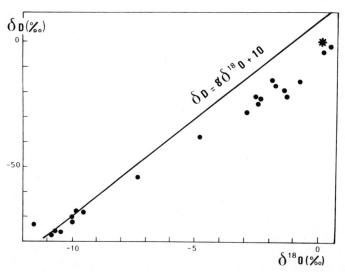

Fig. 3-23. Mechanism of salinization: coal mine waters from the Upper Silesian Basin (after Rozkowski and Przewlocki, 1974). Water seepage originated from direct infiltration of precipitation are lying on or close to the meteoric water line and they are slightly salty (TDS < 3 g/l). Seepage also results in actual brines (69—223 g/l) with heavy isotope content which can be as high as SMOW. All points lie on a line which suggests a mixing between local precipitations and synsedimentary (relictual) water of marine origin. Hydrogeological studies show that the heavy-isotope-rich waters are located in the deepest poorly drained parts of the system.

of total dissolved solids per litre. The linear relationship of salinity versus $\delta^{18}O$ shown in Fig. 3-22 suggests that the salinity is caused by evaporation of water from the shallow groundwater table.

The investigation of the origin of salts in water can be of practical significance: seawater intrusions are very critical for freshwater management in coastal areas. In such a case, environmental isotopes allow the calculation of the mixing ratio. The results can be compared to salt concentration measurements and give information on the diffusion of salts between the two water masses (Cotecchia et al., 1974).

The determination of the origin of mine waters and their dissolved salts is important for mining operations and assessments of the stability of clay horizons. In coal mines in Poland it has been shown on the basis of stable isotope content that salty water intrusions were due to a dilution of fossil brine entrapped in the sediments by circulating waters of meteoric origin (Fig. 3-23; Rozkowski and Przewlocki, 1974).

GROUNDWATER DATING

The radiometric age of a water is only the mathematical transcription of e.g. ^3H or ^{14}C activity in terms of time. These activities are the weighted averages of the respective contributions of numerous elementary flows each of it with its own ^3H or ^{14}C contents. The measured activities are activity distributions and radiometric ages are also ages distributions (logarithmic). To correlate radiometric ages of groundwaters with calendar ages one must deal either with pure "piston flow" systems or assume that the dispersion is restricted to recharge episodes corresponding to short periods of time as compared to the residence time in the system.

Unconfined aquifers.

The dispersion in unconfined aquifers is due to differences in elementary turbulent flows and also to continuous supply of water at any point of the water table. Thus the mixing occurs between supply from updip and vertical seepage.

For small aquifers of relatively high porosity one can assume that each annual recharge (R) is homogenized in the reservoir which discharges an aliquot equal to R in the same time. In that simple case the mean residence time within the reservoir is equal to the reciprocal of the annual recharge rate (Fig. 3-24). One can calculate that the concentration of a given annual contribution decreases exponentially with its age (Eriksson, 1962; Geyh and Mairhofer, 1970). This model is generally used with tritium data since the turn-over time of these small aquifers is generally short. Tritium activities at a discharge point are matched to tritium activities of the recharge (average

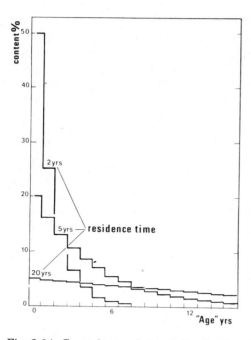

Fig. 3-24. Groundwater dating (after Geyh and Mairhofer, 1970). The exponential model (Ericksson, 1962; Geyh and Mairhofer, 1970) assumes that a fraction of the precipitations of the year enters into a good mixing within the reservoir which releases an equivalent amount of well-mixed water. Ordinate: fraction of the precipitations of a given year which participates to the storage; abscissa: corresponding years. In that model the reciprocal of the contribution of the last year is equal to the mean residence time, e.g. if at any time the reservoir contains 50% of water of the last year it means that the residence time is 2 years.

annual activity of the rainfall corrected for evapotranspiration) in order to find the best fit for R. Such an approach has been used by Hubert et al. (1970).

Another treatment has been proposed by Münnich and Roether (1963) for aquifers containing recent and possibly old (prethermonuclear) waters. They calculate the maximum contribution of the recharge of a given year which could account for the ^3H content observed at the outlet. The respective possible contributions increase with increasing ages and finally reach 100% for the contribution of a year which (corrected for decay) would have a ^3H activity equal or lower than the measured value. These values are plotted on an histogram and the age corresponding to the 50% value for the maximum contribution is the minimum age of the main part of the water.

But it is now established that groundwater bodies are generally not very well mixed. Thus these models are not easily suitable and can only give rough estimate for the mean age of the fraction of the aquifer in which the flow is rapid, i.e., the shallower part.

Radiocarbon measurements have been used in unconfined aquifer to investigate the possible occurrence of old waters. These old waters can exist in arid regions where recent recharge is small as compared to the discharge. Groundwater from the unconfined aquifer of Tertiary limestones and sandstones of the "Hamada du Guir" in northwestern Sahara was found to contain dissolved inorganic carbon (DIC) showing possible age effect (Conrad and Fontes, 1972). This age effect could account for the low ^{18}O content of the waters ($\delta^{18}O = -9.5‰$) attributed to precipitations from the last humid period of the Holocene.

In temperate regions the detection of old waters can prove the stratification within the aquifer. On the Island of Schiermonnikoog Vogel (1967) finds ^{14}C age stratification from recent on top to about 1000 years at 50—70 m depth. However, no significant variations were observed in ^{18}O content nor does the discussion include the possibility that chemical dilutions were important.

The chemical dilution consists in the mixing of soil-derived "active" carbon with "dead" carbon of the unsaturated zone and the aquifer. It is well known in areas where soils and aquifers contain significant amounts of carbonate minerals (Münnich and Vogel, 1959; Brinkman et al., 1959, 1960; Ingerson and Pearson, 1964; Wendt et al., 1967; Tamers, 1967; Münnich, 1968; Geyh, 1970, and others). The "chemical dilution factor" has been introduced by Ingerson and Pearson (1964). It was considered until now that no chemical dilution occurred in carbonate-free terrains. This conclusion was in agreement with low ^{13}C contents of total dissolved carbon generally found in these areas (Pearson and Friedman, 1970; Geyh, 1970; Hufen et al., 1974). But ^{14}C activities of shallow groundwaters on areas of the Canadian Shield virtually free of carbonates show dilutions which can approach 50 percent modern carbon (pmc) in waters which contains significant amounts of tritium (Fritz et al., 1978a).

Thus, in general, the interpretation of ^{14}C measurements in unconfined aquifers is more difficult than ^{3}H data because of the chemical and isotopic modelization which is involved in ^{14}C age calculation (see Chapter 2, this volume).

Confined aquifers

Because groundwater velocities are generally low in confined aquifers there is no other isotopic practical mean than ^{14}C to date the waters. Very interesting reconnaissance studies have been made using ^{39}Ar (Oeschger et al., 1974) or ^{32}Si (Lal et al., 1970) to date waters. But until now these techniques require samples of several cubic metres for one age determination and their perspectives of practical uses are necessarily limited to fundamental surveys.

It is generally admitted that groundwater circulations are laminar in confined aquifers. Thus, to some extent the problem of age determination is simpler than for unconfined aquifers.

Paleoclimatological and chemical limitations for age determinations. Until recently it was admitted that glacial periods in high latitudes and altitudes were correlated with low evaporation and/or high rainfalls. This popular concept is oversimplified and the following paleoclimatic picture is now proposed for both hemispheres for low and mid-latitudes (Rognon and Williams, 1977):

— From 40,000 to 20,000 years B.P. (before present): heavy rainfall and high lake levels.

— From 17,000 to 12,000 years B.P.: intertropical aridity, dune building and lake dessication.

— From 11,000 to 5000 years B.P.: high precipitations and very high lake level.

During the late Quaternary, the major glacial episode started at about 25,000 years B.P. and produced a major eustatic decrease with a minimum oceanic level of about —90 m to —120 m at 18,000 years B.P. (Mörner, 1971). Then the climatic set back produced the melting of ice caps and rise of sea level which was practically achieved at 7000 years B.P. For hydrological studies dealing with old groundwaters one must therefore consider that basins with external drainage were actively drained between 20,000 and 10,000 years B.P., and recharged between 10,000 and 5000 years B.P. For periods older than 20,000 years B.P., any paleohydrological reconstitution is very risky.

As discussed by W.G. Mook (Chapter 2) and Fontes and Garnier (1977, 1979), among others, the hydrochemistry of major ions and the stable isotope content of total dissolved inorganic carbon (DIC) and of solid carbonates from the aquifer must be known for ^{14}C interpretation. Because of the dilution of the active carbon from soil zones into the inactive carbon of the aquifer during carbon mineralization and because of pollution risks of the DIC which is not a closed reservoir, it appears that the ^{14}C method of age determination of DIC is limited to the past 20,000 to 25,000 years. As discussed before this time interval covers the last renewal of groundwaters in basins with external drainage.

For both these geochemical and paleohydrological reasons, apparent ages greater than, say, 25,000 years B.P. would not be considered as true ages. They can be the result of mixing between very old (^{14}C-free) waters with some contribute of ^{14}C-bearing waters especially in boreholes with large withdrawal. Furthermore one must keep in mind that even if a solution of carbon species may be in equilibrium with the solid phase for each chemical species and for stable isotope contents, it may not be in equilibrium for ^{14}C. Thus ^{14}C will tend to diffuse and to be lost in the solid phase. Presently no means exists which could allow investigation of this process on long time ranges. It is therefore possible that DIC with corrected ages of 30,000 or 35,000 years be actually much younger.

The stable isotope content of groundwaters is expected to reflect average local climatic conditions during the recharge. A significant climatic variation over a period of time which is long as compared to the turn-over of local aquifers will be marked on their stable isotope content. With respect to recent waters this paleoclimatic labelling may consist in differences in ^{18}O (or in ^2H) or in deuterium excess, or in both.

Stable isotopes and ^{14}C evidences for paleowaters in confined aquifers.
Because of the low flow velocities predicted from Darcy's law it is to be expected that very old waters be found after some kilometres or some tens of kilometres of confinement. Paleowater occurrence was thus first investigated in confined aquifers.

The pioneer work, of Münnich and Vogel (1962) and Degens (1962) dealt with the aquifer of the Nubian sandstones (lower Cretaceous) in western Egypt. They found stable isotope contents ($\delta^2H = -85‰$; $\delta^{18}O = -11‰$) much lower than expected for present-day precipitations in this area. Calculated ages, assuming an initial ^{14}C activity of 72.5% after the carbon mineralization fell in the range 18,000 to 40,000 years. The interpretation was that these artesian paleowaters were recharged in pluvial time after an episode of eustatic drainage. Because of the low dissolved salt contents (150—300 ppm) a local cold recharge through outcrops of Nubian sandstones was preferred rather than an hypothetical recharge in the Tibesti Mountains followed by a long underground transit, as also discussed.

The aquifer of Nubian sandstones was also investigated in Sinaï and Negev deserts (Issar et al., 1972; Gat and Issar, 1974). The dissolved inorganic carbon was dated from 13,200 to more than 31,000 years taking into account the ^{13}C content. The ^{18}O and deuterium content are substantially higher ($\delta^{18}O = -6.0$ to $7.5‰$, $\delta^2H = +30$ to $+65‰$) than in the Egyptian part of these sandstones (Knetsch et al., 1962).

But these waters were attributed to paleoclimatic recharge on the bases of ^{14}C ages and also because the deuterium excess ($d \simeq +10‰$) was lower than the present-day excess in the area ($+22‰$, Gat and Carmi, 1970).

In the region of Chott-el-Hodna at the border between the Atlas Plateau and the Sahara in Algeria, Gonfiantini et al. (1974b) investigated deep and shallow groundwaters around the Chott. On the northern and eastern sides of the Chott artesian groundwaters show a correlation between ^{14}C and stable isotopes contents (^{14}C = 10.9 to 0 pmc; $\delta^{18}O = -7.5$ to $-9.4‰$; $\delta^2H = -49$ to $-62‰$). In a $\delta^2H-\delta^{18}O$ diagram all these waters lie on a rough correlation line with a slope close to 8 indicating that no evaporation has occurred (see p. 109 and Fig. 3-16). This is interpreted as the result of a mixing between deep water depleted in heavy isotopes and ^{14}C with more recent waters with heavy isotopes content similar to that of local recharge. The authors explained the difference in stable isotopes between artesian deep water and local recharge by a paleoclimatic effect, i.e., a change in the isotopic composition of the recharge. However, in that case flow patterns are complicated because the aquifer is made of Tertiary and Quaternary continental deposits. These deposits are not continuous and the deep aquifer is therefore semi-confined. If mixing may occur between waters precipitated at different altitudes, it could account for the observed difference in stable isotopes.

The chalk of the London Basin (Smith et al., 1976) and the Lincolnshire limestone (Downing et al., 1977) in eastern England, exhibited the same tendency. Between the discharge areas and the outcrops, the ^{18}O content decreased by about 0.70‰. Ages were estimated using the mixing model presented by Ingerson and Pearson (1964). They ranged between recent and more than 25,000 years ago in both the aquifers. Since this range covers the last glacial epoch waters were expected to be more depleted in heavy isotopes. For instance, Dansgaard et al. (1969) claim that in Greenland the end of ice age (~10,000 years B.P.) is marked by an increase of about 13‰ in the ^{18}O content of precipitation. It was thus proposed that recharge took place during warm interstadials only.

A similar low ^{18}O variation between recent and old waters was observed by Fontes and Garnier (1977) in the aquifer of the "Calcaires carbonifères" in the north of France. The age of the older water collected in this system adjusted to about 15,000 years using an exchange-mixing model.

In an investigation of the calcareous aquifer to south Dobrogea in Roumania, Tenu et al. (1975) calculate ages from 1500 to 25,000 years using the correction proposed by Vogel (1970). Low ^{18}O (−11.0 to 12.8‰) and ^2H (−64.6 to −78.2‰) contents are attributed to paleoclimatic recharge. The deuterium excess d^2H is high due to the participation of vapour evaporated under conditions of low relative humidity. The observed range for d is +17.2 to +26.6‰. As this figure increases with the deficit in air moisture which is positively correlated to the temperature it is pointed out that d should be positively correlated with air temperature. Reporting apparent ages versus deuterium excess of groundwaters, the curve is close to that of Milankovitch on the estimation of air temperature. Evaluations of temperature differences between present and the time of recharge were proposed. This approach is undoubtedly promising but one must note that more efforts are needed to ascertain some critical points: (1) radiometric ages are generally undercorrected by Vogel's approach, (2) the calculation of the deuterium excess magnifies the uncertainty on ^{18}O measurement by a factor of 8, and (3) the Milankovitch curves still need experimental support.

In the Tulum Valley in central western Argentina Vogel et al. (1972) find variable ^{18}O content for confined groundwaters (−5.8 to −12.0‰). Apparent ages of DIC evaluated using Vogel's correction are ranging between 2000 and 13,300 years B.P. The ^{18}O content of shallow and recent (bomb-carbon-bearing) groundwater is very low due to their recharge by Andine rivers coming from high altitude. No explanation could be proposed in this preliminary work on the difference in stable isotopes contents between paleo- and recent waters. This is interesting since generally paleowaters are ^{18}O depleted with respect to recent ones.

Vogel and Van Urk (1975) point out that confined and ^{14}C age groundwaters from Kalahari do not show any significant ^{18}O decrease as compared to recent recharge.

In the plain of Venice Bortolami et al. (1973) do not observe a paleoclimatic effect on the stable isotope content of artesian waters. The dissolved inorganic carbon of these waters has a ^{14}C content of about 72 pmc in the areas of recharge and close to zero in the center of the plain. However, the average ^{18}O values of deep groundwaters are similar to the respective present-day surface waters which recharge the aquifer.

Radiometric flow rates in confined aquifers. Studies are not numerous which allow the estimation of radiometric flow rates, i.e., those velocities which are based upon radiometric time intervals. Basic requirements for this estimation are: the localization of recharge areas and piezometric contours and the certainty that the system was closed with respect to total dissolved carbon (i.e., no mixing of waters, no precipitation nor dissolution of carbonate, no supply of any other form of dissolved carbon).

In the aquifer of the Carrizo Sand (Texas) a comparison between ^{14}C flow rates and velocities calculated from hydrologic data showed a good agreement (Fig. 3-25, Pearson and White, 1967).

In the Venice plain it was assumed that the initial ^{14}C activity of total dissolved carbon remained the same during the infiltration in the recharge area (Bortolami et al., 1973). Flow velocities were thus deduced from differences in ^{14}C contents and ranged from 3 to about 1 m/yr (Fig. 3-26). A similar

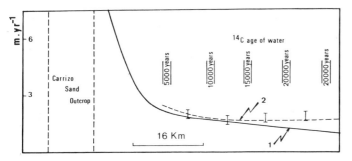

Fig. 3-25. Groundwater dating. Comparisons between rates of flow deduced from hydrologic data and isotopic data (after Pearson and White, 1967). Curve *1*: drawn from calculation of the actual Darcy's velocity at each site: $V = Ki/\rho$ (K = permeability, i = hydraulic gradient and ρ = effective porosity). Curve *2*: drawn from ^{14}C age calculation using Pearson's correction technique (assuming that the dilution of ^{14}C within the soil and the aquifer is reflected by changes in ^{13}C contents in terms of a two-component mixing: organogenic CO_2 and carbonate).

approach was used by Tenu et al. (1975) for a Roumanian aquifer where velocities of 6.9 to 2.9 m/yr were obtained.

In the London Chalk, calculated flow velocities fell close to 0.9 m/yr. But it was pointed out that the distribution of permeabilities was complicated which follows that the determination of actual velocities was difficult (Smith et al., 1976). The aquifer of the Lincolnshire limestones showed velocities of 0.5 to 1 m/yr (Downing et al., 1977).

In the "Calcaires carbonifères" of north France radiometric flow rates were in agreement with Darcy's velocities. But a large discrepancy appeared if the calculation took into account classical values of porosity for limestones:

Radiometric flow rates (m/yr)		Hydraulic velocities (m/yr)	
A_0 constant [1]	A_0 calculated [2]	filtration velocity [3]	actual velocity [4]
0.3—2.0	0.4—4.0	0.5—6.0	~100

[1] A_0 constant: same conditions of mineralization of carbon in the soils and in the aquifer extending on the whole circulation time.
[2] A_0 calculated from the model of Fontes and Garnier (1977, 1979).
[3] Velocity calculated according to Darcy's law from hydrogeologic parameters.
[4] Actual velocity calculated assuming a porosity of 2% and an average Darcy's velocity of 2 m/yr.

This discrepancy could be due (1) to a large underevaluation of porosity in these limestones, (2) to a diffusion process of ^{14}C from the aqueous carbon into the carbonate matrix which would not be at equilibrium, and (3) to a recent change in the hydraulic gradients due to the large withdrawal of groundwaters in this industrial area.

The methodological conclusions on this very important aspect of the use of environmental isotopes in the measurement of low flow rates are:

(1) More efforts are needed in selecting the aquifers in which hydrological data are available.

(2) A complete set of data including water chemistry, ^{13}C contents of the total dissolved carbon, of the solid carbonate of soil or of the aquifer, and possibly of the soil carbon dioxide, is required to determine the corrected ^{14}C age.

CONCLUSIONS

Since about 20 years, isotope hydrology has improved its methodology especially through meetings sponsored by the International Atomic Energy

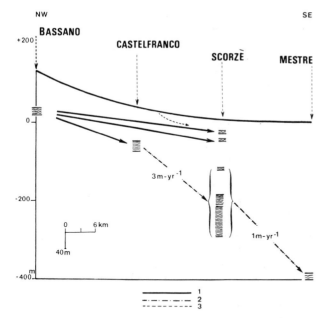

Fig. 3-26. Groundwater dating. Cross section of the flow patterns in the plain of Venice as deduced from radiometric data. The aquifer is unconfined in Bassano, becomes confined between Bassano and Castelfranco and multilayered downdip. Locations of screens (sampling points) are indicated for each borehole. Symbols: 1 = fast circulations (some 10^2 m/yr); 2 = low flow in the deep parts of the confined aquifer; 3 = recent (vertical or oblique) supplies indicated by tritium contents. The flow velocities were calculated assuming that the initial ^{14}C activity of total dissolved carbon was not altered by further isotopic exchange within the confined parts of the aquifer.

Agency in 1963, 1967, 1970, 1974 and 1978.

Basically the information which can be gained from an isotopic study is complementary to that which could be obtained from an hydrochemical study (identification of water bodies, length of groundwaters paths). Isotopic studies deal more particularly with:

(1) Origin of groundwater, if one uses the constitutive stable isotopes (^{18}O and ^2H) which are conservative within low-temperature aquifers; this leads to conclusions on quantitative study of recharge mechanisms, identification of recharge areas, relationships between surface waters and groundwater, identification of mixing and recharge-discharge mechanisms in fractured rocks.

(2) Transit or residence time, if one uses radioactive isotopes: tritium has the advantages of the constitutive isotopes (no other interactions with soils and rocks other than those undergone by the water itself), radiocarbon, to be interpreted, will require hydrochemical and ^{13}C data.

Generally stable constitutive isotopes can be used without any limitations of spatial scale (from the lysimeter to the largest aquifers); on the time scale, tritium will be suitable only for fast circulations whereas ^{14}C can be used either for unconfined or confined aquifers. However, there is still a gap in the age determination of groundwaters of some centuries because of the uncertainty on ^{14}C correction. This time range which is of primary importance for the study of the recharge of confined aquifers will be covered when ^{39}Ar and/or ^{32}Si methods will be suitable for practical uses.

ACKNOWLEDGEMENTS

The author is indebted to B.R. Payne head of the section of hydrology at the International Atomic Energy Agency who criticized helpfully the manuscript, made numerous suggestions and provided unpublished material from the Agency. Very special thanks are also due to P. Fritz who made a heavy editing work on the form and on the substance.

REFERENCES

Allison, G.B. and Hughes, M.W., 1974. Environmental tritium in the unsaturated zone: estimation of recharge to an unconfined aquifer. In: *Isotope Techniques in Groundwater Hydrology, 1.* IAEA, Vienna, pp. 57—72.
Andersen, L.J. and Sevel, T., 1974. Profiles in the unsaturated and saturated zones, Grønhøj, Denmark. In: *Isotope Techniques in Groundwater Hydrology, 1.* IAEA, Vienna, pp. 3—20.
Andrieux, C., 1976. Le système karstique du Baget, 2. Géothermie des eaux à l'exutoire principal selon les cycles hydrologiques 1974 et 1975. Actes deuxième colloque d'hydrologie en pays calcaire. *Ann. Sci. Univ. Besançon, Géol., Sér. 3,* 25: 1—26.
Arnason, B. and Sigurgeirsson, Th., 1967. Hydrogen isotopes in hydrological studies in Iceland. In: *Isotopes in Hydrology.* IAEA, Vienna, pp. 35—47.
Atakan, Y., Roether, W., Münnich, K.O. and Matthes, G., 1974. The Sandhausen shallow-groundwater tritium experiment. In: *Isotope Techniques in Groundwater Hydrology, 1.* IAEA, Vienna, pp. 21—43.
Bakalowicz, M. and Olive, Ph., 1970. Teneurs en tritium des eaux du karst du Taurus occidental et de Pisidie (Turquie). *Schweiz. Z. Hydrol.,* 32: 475—480.
Bakalowicz, M., Blavoux, B. and Mangin, A., 1974. Apports du traçage isotopique naturel à la connaissance du fonctionnement d'un système karstique — teneurs en oxygène 18 de trois systèmes des Pyrénées (France). *J. Hydrol.,* 23; 141—158.
Blanc, P. and Dray, M., 1967. La nappe du delta de la Dranse. Qualité chimique des eaux en rapport avec son alimentation. *Rev. Inst. Pasteur, Lyon,* 1: 277—295.
Blavoux, B. and Siwertz, E., 1971. Traçage isotopique de l'infiltration sur lysimètre. *C.R. Acad. Sci., Paris,* 273: 2056—2059.
Bortolami, G.C., Fontes, J.Ch. and Panichi, C., 1973. Isotopes du milieu et circulations dans les aquifères du sous-sol vénitien. *Earth Planet. Sci. Lett.,* 19: 154—167.
Bortolami, G.C., Fontes, J.Ch. and Zuppi, G.M., 1977. Hydrogeology and hydrochemistry of the Mont Blanc Massif. *Proc. ECOG V, Pisa, September 5—10, 1977* (abstract).
Bosch, B., Guégan, A., Marcé, A. and Siméon, C., 1974. Tritium et bilan hydrogéologique

de la nappe alluviale du Rhône entre Donzère et Mondragon (Drôme-Vaucluse). *Bull. B.R.G.M., Sér. 2, Sect. III*, 3: 245—260.
Bredenkamp, D.B., Schutte, J.M. and Du Toit, G.J., 1974. Recharge of a dolomitic aquifer as determined from tritium profiles. In: *Isotope Techniques in Groundwater Hydrology, 1.* IAEA, Vienna, pp. 73—95.
Brinkmann, R., Münnich, K.O. and Vogel, J.C., 1959. ^{14}C Altersbestimmung von Grundwasser. *Naturwissenschaften*, 46: 10—12.
Brinkmann, R., Münnich, K.O. and Vogel, J.C., 1960. Anwendung der ^{14}C-Methode auf Bodenbildung und Grundwasserkreislauf. *Geol. Rundsch.*, 49: 244—253.
Brinkmann, R., Eïchler, R., Ehhalt, D. and Münnich, K.O., 1963. Über den Deuterium-Gehalt von Niederschlags und Grundwasser. *Naturwissenschaften*, 50: 611—612.
Brown, L.J. and Taylor, C.B., 1974. Geohydrology of the Kaikoura plain, Marlborough, New Zealand. In: *Isotope Techniques in Groundwater Hydrology, 1.* IAEA, Vienna, pp. 169—189.
Burdon, D.J., Eriksson, E., Papadimitropoulos, T., Papakis, N. and Payne, B.R., 1963. The use of tritium in tracing karst groundwater in Greece. In: *Radioisotopes in Hydrology.* IAEA, Vienna, pp. 309—320.
Burger, A., Mathey, B., Marcé, A. and Olive, Ph., 1971. Tritium et oxygène-18 dans les bassins de l'Areuse et de la Serrière. *Ann. Sci. Univ. Besançon, Géol., Sér. 3*, 15: 79—87.
Chouret, A., Fontes, J.Ch. and Mathieu, Ph., 1977. La nappe phréatique à la périphérie du lac Tchad (République du Tchad). Etude complémentaire. *Rap. ORSTOM, Centre de N'djaména*, 67 pp.
Cole, J.A. and Wilkinson, W.B., 1976. Un aperçu des recherches récentes sur la pollution des eaux souterraines au Royaume-Uni. *C.R. XIV, Journées Hydraul., Rapp. 8, Q. IV. Soc. Hydrotech. Fr.*, pp. 1—8.
Conrad, G., 1969. *L'Evolution Post-Hercynienne du Sahara Algérien (Saoura, Erg-chech-Tanezrouft, Ahnet-Mouydir). CNRS, Publ. Centre Rech. Zones Arides, Sér. Géol.,* 10: 530 pp.
Conrad, G. and Fontes, J.Ch., 1970. Hydrologie isotopique du Sahara nord-occidental. In: *Isotope Hydrology.* IAEA, Vienna, pp. 405—419.
Conrad, G. and Fontes, J.Ch., 1972. Circulations, aires et périodes de recharge dans les nappes aquifères du Nord-Ouest saharien: données isotopiques (^{18}O, ^{13}C, ^{14}C). *C.R. Acad. Sci., Paris, Sér. D*, 275: 165—168.
Conrad, G., Fontes, J.Ch., Létolle, R. and Roche, M.A., 1966. Etude isotopique de l'oxygène dans les eaux de la Haute-Saoura (Sahara nord-occidental). *C.R. Acad. Sci., Paris,* 262: 1058—1061.
Conrad, G., Marcé, A. and Olive, Ph., 1975. Mise en évidence, par le tritium, de la recharge actuelle des nappes libres de la zone aride saharienne (Algérie). *J. Hydrol.*, 27: 207—224.
Cotecchia, V., Tazioli, G.S. and Magri, G., 1974. Isotopic measurements in research on seawater ingression in the carbonate aquifer of the Salentine Peninsula, Southern Italy. In: *Isotope Techniques in Groundwater Hydrology, 1.* IAEA, Vienna, pp. 441—463.
Craig, H., 1961. Standard for reporting concentrations of deuterium and oxygen-18 in natural water. *Science*, 133: 1833—1834.
Crouzet, E., Hubert, P., Olive, Ph., Siwertz, E. and Marcé, A., 1970. Le tritium dans les mesures d'hydrologie de surface. Détermination expérimentale du coefficient de ruisellement. *J. Hydrol.*, 11: 217—229.
Dansgaard, W., 1964. Stable isotopes in precipitation. *Tellus*, 16: 436—468.
Dansgaard, W., Johnsen, S.J., Moeller, J. and Langway, C.C., Jr., 1969. One thousand centuries of climatic record from Camp Century on the Greenland ice sheet. *Science*, 166: 377—381.
Deak, J., 1974. Use of environmental isotopes to investigate the connection between sur-

face and subsurface waters in the Nagykunsàg region, Hungary. In: *Isotope Techniques in Groundwater Hydrology, 1*. IAEA, Vienna, pp. 157—167.

Degens, E.T., 1962. Geochemische Untersuchungen von Wasser aus der ägyptischen Sahara. *Geol. Rundsch.*, 52: 625—639.

Dinçer, T. and Payne, T., 1965. Isotope survey of the karst region of Southern Turkey. *Proc. 6th Conf. Radiocarbon and Tritium Dating, Washington State Univ.*, pp. 671—686.

Dinçer, T. and Payne, B.R., 1971. An environmental isotope study of the southwestern karst region of Turkey. *J. Hydrol.*, 14: 307—321.

Dinçer, T., Payne, B.R., Florkowski, T., Martinec, D. and Tongiorgi, E., 1970. Snowmelt runoff from measurements of tritium and oxygen-18. *Water Resour. Res.*, 5: 438.

Dinçer, T., Noory, M., Javed, A.R.K., Nuti, S. and Tongiorgi, E., 1974. Study of groundwater recharge and movement in shallow and deep aquifers in Saudi Arabia with stable isotopes and salinity data. In: *Isotope Techniques in Groundwater Hydrology, 2.* IAEA, Vienna, pp. 363—378.

Downing, R.A., Smith, D.B., Pearson, F.J., Monkhouse, R.A., Otlet, R.L., 1977. The age of groundwater in the Lincolnshire limestone, England and its relevance to the flow mechanism. *J. Hydrol.*, 33: 201—216.

Eichler, R., 1965. Deuterium-Isotopengeochemie des Grund- und Oberflächenwassers. *Geol. Rundsch.*, 55: 144—155.

Eriksson, E., 1962. Radioactivity in hydrology. In: H.I. Israël and A. Krebs (Editors), *Nuclear Radiation in Geophysics, 42*. Springer-Verlag, Berlin-Heidelberg, pp. 47—60.

Fontes, J.Ch., 1976. *Isotopes du Milieu et Cycles des Eaux Naturelles: Quelques Aspects.* Thèse de doctorat ès Sciences. Université Pierre et Marie Curie, Paris, 218 pp.

Fontes, J.Ch. and Garnier, J.M., 1977. Determination of the initial ^{14}C activity of the total dissolved carbon, age estimation of waters in confined aquifers. *Proc. 2nd Int. Symp. Water-Rock Interactions, Strasbourg, August 17—25, 1977*, pp. I.363—I.376.

Fontes, J.Ch. and Garnier, J.M., 1979. Determination of the initial activity of the total dissolved carbon. A review of the existing models and a new approach. *Water Resour. Res.*, 12: 399—413.

Fontes, J.Ch. and Gonfiantini, R., 1967. Comportement isotopique au cours de l'évaporation de deux bassins sahariens. *Earth Planet. Sci. Lett.*, 7: 325—329.

Fontes, J.Ch. and Olivry, J.C., 1976. Premiers résultats sur la composition isotopique des précipitations de la région du Mont Cameroun. *Cah. ORSTOM, Sér. Hydrol.*, 13: 179—194.

Fontes, J.Ch. and Zuppi, G.M., 1976. Isotopes and water chemistry in sulphide bearing springs of Central Italy. In: *Interpretation of Environmental Isotope and Hydrochemical Data in Groundwater Hydrology*. IAEA, Vienna, pp. 143—158.

Fontes, J.Ch., Létolle, R., Olive, Ph. and Blavoux, B., 1967. Oxygène-18 et tritium dans le bassin d'Evian. In: *Isotopes in Hydrology*. IAEA, Vienna, pp. 401—415.

Fontes, J.Ch., Gonfiantini, R. and Roche, M.A., 1970. Deuterium et oxygen-18 dans les eaux du lac Tchad. In: *Isotope Hydrology*. IAEA, Vienna, pp. 387—404.

Fontes, J.Ch., Bortolami, G.C. and Zuppi, G.M., 1978. Hydrologie isotopique du massif du Mont-Blanc. In: *Isotope Hydrology 1978, 1*. IAEA, Vienna, pp. 411—440.

Fritz, P., Cherry, J.A., Weyer, K.U. and Sklash, M., 1976. Storm runoff analyses using environmental isotopes and major ions. In: *Interpretation of Environmental Isotope and Hydrochemical Data in Groundwater Hydrology*. IAEA, Vienna, pp. 111—130.

Fritz, P., Reardon, E.J., Barker, J., Brown, R.M., Cherry, J.A., Killey, R.W.D. and McNaughton, D., 1978a. The carbon-isotope geochemistry of a small groundwater system in northeastern Ontario. *Water Resour. Res.*, 14: 1059—1067.

Fritz, P., Silva, C., Suzuki, O. and Salati, E., 1978b. Isotope hydrology in northern Chile. In: *Isotope Hydrology 1978, 2*. IAEA Vienna, pp. 525—543.

Gat, J.R. and Carmi, I., 1970. Evolution of the isotopic composition of atmospheric waters in the Mediterranean Sea area. *J. Geophys. Res.*, 75: 3039—3048.

Gat, J.R. and Issar, A., 1974. Desert isotope hydrology; water sources of the Sinai Desert. *Geochim. Cosmochim. Acta*, 38: 1117—1131.

Gat, J.R. and Tzur, Y., 1967. Modification of the isotopic composition of rainwater by processes which occur before groundwater recharge. In: *Isotopes in Hydrology*. IAEA, Vienna, pp. 49—60.

Geyh, M.A., 1970. Carbon-14 concentration of lime in soils and aspects of the carbon-14 dating of groundwater. In: *Isotope Hydrology*. IAEA, Vienna, pp. 215—223.

Geyh, M.A., 1972. On the determination of the dilution factor of groundwater. *Proc. 8th Int. Conf. Radiocarbon Dating, Lower Hutt, N.Z.*, pp. 369—380.

Geyh, M.A. and Mairhofer, J., 1970. Der natürliche ^{14}C- und ^{3}H-Gehalt der Wässer. *Steir. Beitr. Hydrogeol.*, 22: 63—81.

Gonfiantini, R., Togliatti, V. and Tongiorgi, E., 1962. Il rapporto $^{18}O/^{16}O$ nell'acqua del lago di Bracciano e nelle falde a sud-est del lago. *Not. CNEN (Italy)*, 8(6): 39—45.

Gonfiantini, R., Conrad, G., Fontes, J.Ch., Sauzay, G. and Payne, B.R., 1974a. Etude isotopique de la nappe du Continental intercalaire et des relations avec les autres nappes du Sahara septentrional. In: *Isotope Techniques in Groundwater Hydrology, 1*. IAEA, Vienna, pp. 227—241.

Gonfiantini, R., Dinçer, T. and Derekoy, A.M., 1974b. Environmental isotope hydrology in the Hodna region — Algeria. In: *Isotope Techniques in Groundwater Hydrology, 1*. IAEA, Vienna, pp. 293—316.

Gonfiantini, R., Gallo, G., Payne, B.R. and Taylor, C.B., 1976. Environmental isotopes and hydrochemistry in groundwater of Gran Canaria. In: *Interpretation of Environmental Isotope and Hydrochemical Data in Groundwater Hydrology*. IAEA, Vienna, pp. 159—170.

Hubert, P., Marin, E., Meybeck, M., Olive, Ph. and Siwertz, E., 1969. Aspects hydrologique, géochimique et sédimentologique de la crue exceptionnelle de la Dranse du Chablais du 22 septembre 1968. *Arch. Sci. Genève*, 22(3): 581—604.

Hubert, P., Marcé, A., Olive, Ph. and Siwertz, E., 1970. Etude par le tritium de la dynamique des eaux souterraines. *C.R. Acad. Sci., Paris, Sér. D.*, 270: 908—911.

Hufen, T.S., Lau, L.S. and Buddemeir, R.W., 1974. Radiocarbon, ^{13}C and tritium in water samples from basaltic aquifers and carbonate aquifers on the island of Oahu, Hawaii. In: *Isotope Techniques in Groundwater Hydrology, 2*. IAEA, Vienna, pp. 111—127.

IAEA, 1969. Environmental Isotope Data, 1. World Survey of Isotope Concentration in Precipitation (1953—1963). *IAEA Tech. Rep. Ser.*, 96: 421 pp.

IAEA, 1970. Environmental Isotope Data, 2. World Survey of Isotope Concentration in Precipitation (1964—1965). *IAEA Tech. Rep. Ser.*, 117: 402 pp.

IAEA, 1971. Environmental Isotope Data, 3. World Survey of Isotope Concentration in Precipitation (1966—1967). *IAEA Tech. Rep. Ser.*, 129: 402 pp.

IAEA, 1973. Environmental Isotope Data, 4. World Survey of Isotope Concentration in Precipitation (1968—1969). *IAEA Tech. Rep. Ser.*, 147: 334 pp.

IAEA, 1975. Environmental Isotope Data, 5. World Survey of Isotope Concentration in Precipitation (1970—1971). *IAEA Tech. Rep. Ser.*, 165: 309 pp.

IAEA, 1978. Environmental Isotope Data, 6. World Survey of Isotope Concentration in Precipitation (1972—1975). *IAEA Tech. Rep. Ser.* 192: 187 pp.

Ingerson, E. and Pearson, F.J., Jr., 1964. Estimation of age and rate of motion of groundwater by the ^{14}C method. In: *Recent Researches in the Fields of Hydrosphere, Atmosphere and Nuclear Geochemistry*. Maruzen, Tokyo, pp. 263—283.

Isaacson, R.E., Brownell, L.E., Nelson, R.W. and Roetman, E.L., 1974. Soil-moisture transport in arid site vadose zones. In: *Isotope Techniques in Groundwater Hydrology, 1*. IAEA, Vienna, pp. 97—114.

Issar, A., Bein, A. and Michaeli, A., 1972. On the ancient waters of the Upper Nubian sandstone aquifer in central Sinai and southern Israël. *J. Hydrol.*, 17: 353—374.
Knetsch, G., Shata, A., Degens, E., Münnich, K.O., Vogel, J.C. and Shazly, M.M., 1962. Untersuchungen an Grundwässern der Ostsahara. *Geol. Rundsch.*, 52: 587—610.
Lal, D., Nijampurkar, V.N. and Rama, S., 1970. Silicon-32 hydrology. In: *Isotope Hydrology*. IAEA, Vienna, pp. 847—868.
Leontiadis, J. and Dimitroulas, Ch., 1971. The use of radioisotopes in tracing karst groundwater, 1. *Nucl. Res. Center, "Demokritos"*, DEMO 71/10E.
Margrita, R., Evin, J., Flandrin, J. and Paloc, H., 1970. Contribution des mesures isotopiques à l'étude de la Fontaine de Vaucluse. In: *Isotope Hydrology*. IAEA, Vienna, pp. 333—348.
Martinec, J., Siegenthaler, U., Oeschger, H. and Tongiorgi, E., 1974. New insights into the run-off mechanism by environmental isotopes. In: *Isotope Techniques in Groundwater Hydrology, 1*. IAEA, Vienna, pp. 129—143.
Matthess, G., Münnich, K.O. and Sonntag, C., 1976. Practical problems of groundwater model ages for groundwater protection studies. In: *Interpretation of Environmental Isotope and Hydrochemical Data in Groundwater Hydrology*. IAEA, Vienna, pp. 185—194.
Meybeck, M., Hubert, P., Martin, J.M. and Olive, Ph., 1970. Etude par le tritium du mélange des eaux en milieu lacustre et estuarien. Application au lac de Genève et à la Gironde. In: *Isotope in Hydrology*. IAEA, Vienna, pp. 523—541.
Michel, F.A. and Fritz, P., 1978. Environmental isotope in permafrost related waters along the Mackenzie Valley corridor. *Proc. 3rd Int. Conf. Permafrost, Edmonton, 1978*, 1: 207—211.
Mook, W.G., Groeneveld, D.J., Brown, A.E. and Van Ganswijk, A.J., 1974. Analysis of a run-off hydrograph by means of natural ^{18}O. In: *Isotope Techniques in Groundwater Hydrology, 1*. IAEA, Vienna, pp. 145—155.
Mörner, N.A., 1971. The position of the ocean level during the interstadial at about 30,000 B.P. A discussion from a climatic glaciologic point of view. *Can. J. Earth Sci.*, 8: 132—143.
Münnich, K.O., 1957. Messung des ^{14}C-Gehaltes von hartem Grundwasser. *Naturwissenschaften*, 44: 32—33.
Münnich, K.O., 1968. Isotopen-Datierung von Grundwasser. *Naturwissenschaften*, 55: 158—163.
Münnich, K.O. and Roether, W., 1963. A comparison of carbon-14 and tritium ages of groundwater. In: *Radioisotopes in Hydrology*. IAEA, Vienna, pp. 397—406.
Münnich, K.O. and Vogel, J.C., 1959. C-14 Altersbestimmung von Süsswasser-Kalkablagerungen. *Naturwissenschaften*, 46: 168—169.
Münnich, K.O. and Vogel, J.C., 1962. Untersuchungen an pluvialen Wässer der Ost-Sahara. *Geol. Rundsch.*, 52: 611—624.
Oeschger, H., Gugelmann, A., Loosli, H., Schotterer, U., Siegenthaler, U. and Wiest, W., 1974. ^{39}Ar dating of groundwater. In: *Isotope Techniques in Groundwater Hydrology, 2*. IAEA, Vienna, pp. 179—190.
Payne, B.R., 1970. Water balance of Lake Chala and its relation to groundwater from tritium and stable isotope data. *J. Hydrol.*, 11: 47—58.
Payne, B.R., 1976. Environmental isotopes as a hydrogeological tool. *Arbeitstag. "Isotope in der Natur" Gera, Zfl-Mitt.*, 5: 177—199.
Payne, B.R. and Yurtsever, Y., 1974. Environmental isotopes as a hydrogeological tool in Nicaragua. In: *Isotope Techniques in Groundwater Hydrology, 1*. IAEA, Vienna, pp. 193—202.
Pearson, F.J., Jr. and Friedman, I., 1970. Sources of dissolved carbonate in an aquifer free of carbonate minerals. *Water Resour. Res.*, 6: 1775—1781.
Pearson, F.J., Jr. and White D.E., 1967. Carbon-14 ages and flow rates of water in Carrizo Sand, Atascosa County, Texas. *Water Resour. Res.*, 3: 251—261.

Pinder, G.F. and Jones, D.F., 1969. Determination of the groundwater component of peak discharge from the chemistry of total runoff. *Water Resour. Res.*, 5: 438.
Rauert, W. and Stichler, W., 1974. Groundwater investigations with environmental isotopes. In: *Isotopes Techniques in Groundwater Hydrology, 2.* IAEA, Vienna, pp. 431—443.
Rognon, P. and Williams, M.A.J., 1977. Late Quaternary climatic changes in Australia and North-Africa: a preliminary interpretation. *Palaeogeogr., Palaeoclimatol., Palaeoecol.*, 21: 285—327.
Roche, M.A., 1974. *Traçage Naturel Salin et Isotopique des Eaux du Système Hydrologiques du Lac Tchad.* Thèse de doctorat ès Sciences Université Pierre et Marie Curie, Paris, 398 pp.
Rozkowski, A. and Przewlocki, K., 1974. Application of stable environmental isotopes in mine hydrogeology taking polish coal basins as an example. In: *Isotope Techniques in Groundwater Hydrology, 1.* IAEA, Vienna, pp. 481—502.
Schneider, J.L., 1965. Relation entre le lac Tchad et la nappe phréatique. *AIHS Symp. de Garde*, pp. 122—131.
Siegenthaler, U. and Schotterer, U., 1977. Hydrologische Anwendungen von Isotopenmessungen in der Schweiz. *Gaz-Eaux-Eaux Usées*, 7: 501—506.
Siegenthaler, U., Oeschger, H. and Tongiorgi, E., 1970. Tritium and oxygen-18 in natural water samples from Switzerland. In: *Isotopes in Hydrology.* IAEA, Vienna, pp. 343—385.
Sklash, M.G., Farvolden, R.N. and Fritz, P., 1976. A conceptual model of watershed response to rainfall, developed through the use of oxygen-18 as a natural tracer. *Can. J. Earth Sci.*, 13: 271—283.
Smith, D.B., Wearn, P.L., Richards, H.J. and Rowe, P.C., 1970. Water movement in the unsaturated zone of high and low permeability strata by measuring natural tritium. In: *Isotope Hydrology.* IAEA, Vienna, pp. 73—87.
Smith, D.B., Downing, R.A., Monkhouse, R.A., Otlet, R.L. and Pearson, F.J., 1976. The age of groundwater in the chalk of the London Basin. *Water Resour. Res.*, 12: 392—404.
Stahl, W., Aust, H. and Dounas, A., 1974. Original of artesian and thermal waters determined by oxygen, hydrogen and carbon isotope analyses of water samples from the Sperkhios Valley, Greece. In: *Isotope Technique in Groundwater Hydrology, 1.* IAEA, Vienna, pp. 317—339.
Suszczynski, E.F., 1972. L'hydrogéologie des roches fracturés et fissurés. *I.H.D. Working Groups on Underground Waters.*
Tamers, M.A., 1967. Surface-water infiltration and groundwater movement in arid zones of Venezuela. In: *Isotopes in Hydrology.* IAEA, Vienna, pp. 339—351.
Tenu, A., Noto, P., Cortecci, G. and Nuti, S., 1975. Environmental isotopic study of the Barremian-Jurassic aquifer in south Dobrogea (Roumania). *J. Hydrol.*, 26: 185—198.
Verhagen, B.Th., Mazor, E. and Sellschop, J.P.F., 1974. Radiocarbon and tritium evidence for direct rain recharge to groundwaters in the northern Kalahari. *Nature*, 249: 643—644.
Vogel, J.C., 1967. Investigation of groundwater flow with radiocarbon. In: *Isotopes in Hydrology.* IAEA, Vienna, pp. 355—369.
Vogel, J.C., 1970. Carbon-14 dating of groundwater. In: *Isotope Hydrology.* IAEA, Vienna, pp. 225—239.
Vogel, J.C. and Van Urk, H., 1975. Isotopic composition of groundwater in semi-arid regions of southern Africa. *J. Hydrol.*, 25: 23—36.
Vogel, J.C., Lerman, J.C., Mook, W.G. and Roberts, F.B., 1972. Natural isotopes in the groundwater of the Tulum Valley, San Juan, Argentina. *Hydrol. Sci. Bull.*, 17: 85—96.
Vogel, J.C., Lerman, J.C. and Mook, W.G., 1975. Natural isotopes in surface and groundwater from Argentina. *Hydrol. Sci. Bull.*, 20: 203—221.

Wendt, I., Stahl, W., Geyh, M. and Fauth, F., 1967. Model experiments for ^{14}C water-age determination. In: *Isotopes in Hydrology*. IAEA, Vienna, pp. 321—337.

Young, C.P., Hall, E.S. and Oakes, D.B., 1976. Nitrate in groundwater studies on the chalk near Winchester, Hampshire. *Water Res. Center, Tech. Rep.*, T.R. 31: 67 pp.

Zimmermann, V., Münnich, K.D., Roether, W., Kreutz, W., Schaubach, K. and Siegel, D., 1965. Downward movement of soil moisture traced by means of hydrogen isotopes. *Proc. 6th Int. Conf. C-14 and T-Dating, Pullman, Wash., June 7—11. USAEA Conf.*, 650652: 577.

Zimmermann, U., Münnich, K.O., Roether, W., Kreutz, W., Schaubach, K. and Siegel, O., 1966. Tracers determine movement of soil moisture and evapotranspiration. *Science*, 152: 346—347.

Zimmermann, U., Münnich, K.O. and Roether, W., 1967. Downward movement of soil moisture traced by means of hydrogen isotopes. *Am. Geophys. Union, Geophys. Monogr.*, 11: 28—36.

Zuppi, G.M., Fontes, J.Ch. and Létolle, R., 1974. Isotopes du milieu et circulations d'eaux sulfurées dans le Latium. In: *Isotope Technique in Groundwater Hydrology, 1*. IAEA, Vienna, pp. 341—361.

Chapter 4

ENVIRONMENTAL ISOTOPES IN ICE AND SNOW

H. MOSER and W. STICHLER

INTRODUCTION

The physical principles underlying the distribution of environmental isotopes in snow and ice, as well as their temporal and spatial variations, are basically the same as in the hydrosphere. Thus the contents of the stable isotopes deuterium (^{2}H) and ^{18}O are governed by fractionation phenomena which in snow and ice may occur not only at the phase boundary liquid-vapour, but also at the phase boundaries solid-liquid and solid-vapour. The known isotopic equilibrium fractionation factors have been compiled in Table 4-1. They show for example, that in an ice-water system in isotopic equilibrium the δ^2H value of the water is about 20‰ lower than that of the ice.

Isotope fractionations during phase changes occur at phase boundaries. They can be measured if physical processes insure that the original surface effects become volume effects. In the liquid and vapour phase, such a transition goes on during a relatively short time by convection, dispersion, and isotope exchange. In the compact solid phase (e.g. in ice), however, differences of isotope contents will be balanced only by molecular diffusion (Kuhn and

TABLE 4-1

The values of the isotopic equilibrium fractionation factor of deuterium and ^{18}O from ice and water.

$\alpha(^2$H$)$		$\alpha(^{18}$O$)$ experimental	Reference
theoretical	experimental		
1.0192 ± 0.002			Weston (1955)
1.0186	1.0171 ± 0.005	1.00048	Kuhn and Thürkauf (1958)
	1.0235		Merlivat and Nief (1967)
	1.0195	1.0031	O'Neil (1968)
		1.00265	Craig and Hom (1968)
	1.0208 ± 0.0007		Arnason (1969b)

Thürkauf, 1958) which takes a longer time in comparison with the processes above mentioned in vapours and liquids (see section "Isotope diffusion in compact glacier ice"). Therefore isotope fractionation effects normally cannot be observed in most cases during melting or sublimation of compact ice, whereas they do exist if the solid phase is in the form of porous snow or firn. This is due to the great surface and the small layer thickness of the grains which allow isotope equilibrium by convection within the pore volume. This has been confirmed through the determination of isotope effects during snow evaporation, snow melting, and meltwater percolation through snow layers. In addition, the mass transport within a snow cover occurring during evaporation and condensation processes within the pores as described by models (for instance, De Quervain, 1973) and measured by isotope contents (Moser and Stichler, 1975) documents the difference in isotopic behaviour between snow and ice during melting or sublimation.

If, on the other hand, the condensation of vapour as the forcing of water into a solid phase is considered then isotope effects close to those demanded by equilibrium conditions will occur. Investigations based on environmental isotopes in snow and ice so far have been primarily focussed on deuterium and ^{18}O measurements. Adopting special methods as described in the following sections it is possible to follow the processes during accretion and reduction of temporary snow covers, in the course of firning of the snow, and in the transition to ice. Related glaciological and hydrological problems can be studied successfully. To differentiate the phenomena with respect to time, measurements of the tritium contents and of precipitation fall-out have been utilized to a certain extent. Long-term dating of information stored in the ice using radioisotopes has been carried out only in isolated cases due to great difficulties associated with methodology and measuring techniques. As a result, many conceivable applications in the field of climatology, geology, volcanology, meteorology, atmospheric chemistry, and cosmic physics have to date been only partly explored or not at all through environmental isotope studies in snow and ice.

ISOTOPE CONTENT OF A SNOW COVER IN ACCRETION

The isotope content of a snow cover results primarily from the isotope contents of those snowfalls, which have contributed to the snow cover accretion. In addition, the snow cover may contain hoarfrost from the water vapour of the air and snow drifted by wind from other locations. As a whole, the snow cover reflects therefore the complex picture of the foregoing meteorological situations in its accretion area.

Record of isotope contents of precipitations in the snow cover

The amounts of the stable isotopes deuterium (^2H) and ^{18}O in snowfalls depend as in rain, essentially on the origin of the water vapour and on condensation temperatures. The tritium (^3H) contents of precipitations in the form of snow, however, are governed mainly by the admixture of bomb-tritium-bearing water vapour from the troposphere, and is thus influenced by meteorological phenomena (see e.g. Ehhalt, 1971). Isotope exchange with atmospheric water vapour, for example, and evaporation during precipitation events seems to be of minor importance in snowfall, even for stable isotopes whose changes in ratio can be measured with a greater degree of accuracy than is possible for the absolute measurement of tritium (Friedman and Smith, 1970).

The isotope contents of successive snow layers can frequently give information on short- and long-term isotopic variations in precipitations. Picciotto et al. (1960) discovered, on snow samples from Antarctica, a relation between the δ-value and the temperature measured by radiosondes in the clouds during the precipitation event. However, further measurements and calculations (for instance, Dansgaard, 1964; Aldaz and Deutsch, 1967; Lorius et al., 1969; Dansgaard et al., 1973; Merlivat et al., 1973; Lorius and Merlivat, 1975; Lorius, 1976) revealed that it is only possible to establish a relation between the δ-value and temperature in small regions for which the meteorological conditions are rather uniform (Fig. 4-1). These differ-

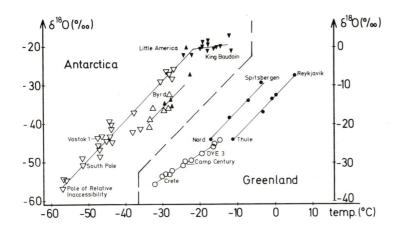

Fig. 4-1. Mean of δ-values of precipitation plotted against mean air temperature at ground level (from Dansgaard et al., 1973). Stations: circles = northern hemisphere; triangles = southern hemisphere; filled symbols indicate lower than 1000 m a.s.l., open symbols, higher than 1000 m a.s.l.

ences are thus basically due to the diverse thermodynamic histories of the condensing water vapours. For example, Aldaz and Deutsch (1967) took advantage of these findings to estimate under simple conditions the origin of snow in Antarctica. The isotopic altitude effect in the deuterium and ^{18}O contents known for rain and produced as a result of the temperature dependence of isotope fractionation during condensation has also been recorded for snow samples as documented by measurements on samples along an altitude profile (see also Chapter 1). Examples from various regions of the world (Fig. 4-2) show that the δ-values differ not only in their absolute values as a function of the climatic conditions prevailing at the respective

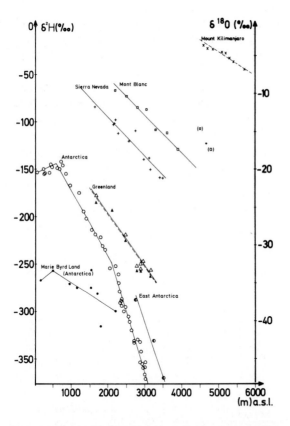

Fig. 4-2. Isotopic altitude effect in snow precipitation. Values from Mont Blanc measurements in brackets originate from the ridge and the peak. The deviation is probably caused by snow drift. The solid lines connect deuterium measurements, the dashed lines ^{18}O measurements. Data references: Mount Kilimanjaro, Gonfiantini (1970); Mont Blanc, Moser and Stichler (1970); Sierra Nevada, Friedman and Smith (1970); Greenland, Renaud (1969); Antarctica, East Antarctica, and Marie Byrd Land, Lorius et al. (1969).

geographic location of the sampling site, but also in the slopes of the consistently linear relation between δ-value and altitude differential. Investigations in the Alps disclosed variations in the altitude effect in fresh snow of 2 to 6 δ^2H per 100 m of difference in evaluation (Moser and Stichler, 1970). Furthermore, it should be noted that in some areas, other isotopic effects (the continental effect, for instance) overlay the altitude effect. The lack of altitude effect in the δ-values suggests that the snowfall derives from a high and horizontal cloud formation (Lorius and Merlivat, 1977; Ambach et al., 1968). An increase of the ^3H contents with altitude as measured by Merlivat et al. (1973, 1977) in Greenland and in Antarctica can be explained by an increased contact with upper atmosphere water vapour rich in tritium.

Similar to rain, snowfall exhibits seasonal variations of its ^2H, ^{18}O and ^3H contents. They remain detectable in the snow cover for a considerable time, especially in polar regions (see section "Isotope variations in the transition of snow to glacier ice"). The snow cover of a temperate glacier is used as an example in Fig. 4-3 (Moser and Stichler, 1975), and shows that the isotope content of the precipitation of the previous three months is preserved.

Changes in large-scale weather patterns likewise become visible in the isotope content of a snow cover. Thus, in Antarctica, the relatively high ^{18}O contents of the snow layers of the summer of 1957-58 were traceable to the

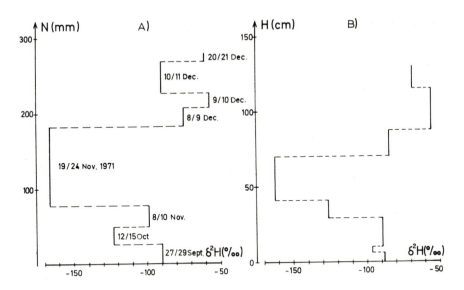

Fig. 4-3. A. δ^2H values of separate successive snowfalls on the Zugspitzplatt (Bavarian Alps) in relation to the water equivalents N (mm) of the single snow falls and respective precipitation dates. B. δ^2H values of a snow profile sampled on 2 January 1972; H (cm) = snow height (from Moser and Stichler, 1975).

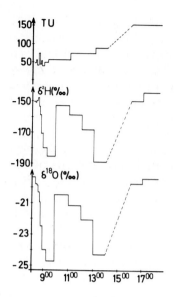

Fig. 4-4. Variations of the ^{2}H, ^{3}H and ^{18}O content in snow samples taken during a warm-front in Innsbruck (Austria), 16 January 1973 (after Ambach et al., 1975). The decrease of stable isotope content is due to continual condensation. The abrupt increase with following decrease in the stable isotope content can be explained by a second warmfront with following rain-out. The last increase of the isotope contents is caused by a new cold air stream with high ^{2}H, ^{18}O and ^{3}H contents.

above-average snow accumulation accompanying the warm weather (Picciotto et al., 1966; Epstein et al., 1963, 1965). In the Sierra Nevada (U.S.A.), Friedman and Smith (1972) found average δ^2H values for winter snow blankets that were about 20‰ lower in the winter of 1968-69 than in the following winter semesters of 1969-70 and 1970-71. They believed that dry cold air from the northeast has been overriding the moist and warm Pacific air. In the same observation area, δ^2H-measurements in the snow cover led to the local determination of a climatological divide (Friedman and Smith, 1970).

Epstein et al. (1963) and Ambach et al. (1975) observed a close correlation between variations of the δ^2H values during a particular snowfall and meteorological conditions of the precipitation event (cold front, warm front, occlusion) (Fig. 4-4).

Isotope content and wind drift

The time pattern of the isotope contents of precipitation in a snow cover may undergo changes through wind drift, which becomes particularly effective in the case of dry snow, in areas with little accumulation and exposed to wind (such as in Antarctica), but as well on steep slopes in connection with a widely varying isotope content. This disturbance of the original isotope distribution may lead to increased scattering in the δ-value vs. temperature relation and to a distortion of the isotopic altitude effect (Lorius et al., 1969). A drifting of snow from higher to lower altitudes results in a diminution of the δ-values. In contrast, a snow drift to higher levels, forming snow cornices, for example, will cause elevated δ-values on ridges and peaks (see Fig. 4-2, Mont Blanc, as an illustration). In Antarctica the pronounced influence of snow drifts complicates all the interpretations of the isotope contents from snow profiles (see, for instance, Arnason, 1980). Nevertheless Epstein et al. (1963) were able to prove on the basis of ^{18}O measurements on snow samples, that this snow had been wind-drifted for

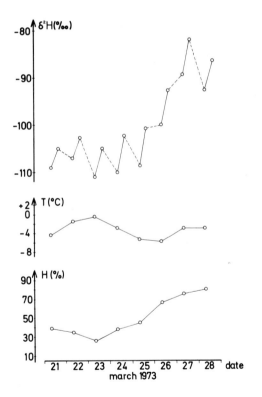

Fig. 4-5. δ^2H values of surface snow samples, air temperature (T) and air humidity (H) from a test series at Weissfluhjoch, Davos (Switzerland) (from Moser and Stichler, 1975). The daily fluctuations of the deuterium content (δ^2H) of surface snow samples are due to the antagonist effects of diurnal evaporation and nightly condensation of air moisture.

a distance of some 500 km, assuming that the wind drift had not caused a separation of the snowflakes according to isotope content.

Isotope content and formation of hoarfrost

Studies on the isotopic effects occurring when atmospheric moisture condenses on the ground or on the surface of snow and ice (formation of hoarfrost) have been scant so far, although valuable indications about the transport processes of the water vapour in the snow layer, and associated problems of snow metamorphosis, might be derived from this. Fig. 4-5 shows the results of examinations carried out on a snow surface (depth 1—2 cm) at the Weissfluhjoch, Davos (Switzerland) (Moser and Stichler, 1975). From the curve of the δ^2H values of the samples taken respectively in the morning and in the afternoon, it does appear that the increase in the δ^2H value caused by evaporation of the snow at the surface is offset by condensation of air moisture at the snowline. However, the isotope content of the snow surface underwent a significant change, when new air masses with higher air moisture and different δ^2H values were transported into the area. After a general increase of the δ^2H values the day-night alternation of evaporation and condensation then takes its course at higher isotope content level.

ISOTOPE DISTRIBUTION DURING THE REDUCTION OF A TEMPERATE SNOW COVER

The isotope distribution resulting from the accretion of a temperate snow cover remains qualitatively intact during the ablation period and therefore is particularly suited to follow the reduction of the snow blanket as it melts. Naturally, certain changes in isotope content occur: at the snow surface, isotope fractionations accompanying evaporation processes may increase the ^2H and ^{18}O contents, whereas in the interior of the snow cover the percolation of water tends, through mixing and fractionation effects, to produce a more or less pronounced homogenization of the isotope contents of the separate snow layers, connected with a general enrichment in ^2H and ^{18}O.

Isotope variations as a result of evaporation from the snow surface

Hydrological and radiation energy balances of snow covers commonly provide fairly low gross evaporation rates (De Quervain, 1951). This is mainly due to the fact that the evaporated mass is offset to a large degree by condensation of water vapour on the snow surface. Such a concept has also been confirmed qualitatively through measurements of the isotope content

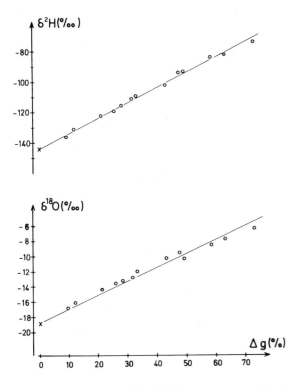

Fig. 4-6. Changes in δ^2H, $\delta^{18}O$ values and weight loss Δg of snow samples evaporating in a cold chamber of about $-10°C$ (from Moser and Stichler, 1975). Daily net evaporation rate about 0.01 g/cm². The isotope content was measured in bulk samples remaining after varying losses in weight.

in the snow surface (see Fig. 4-5). Thus one always observes an enrichment in ²H and ¹⁸O in the surface layer of a snow pack (Moser and Stichler, 1975; Martinec et al., 1977), but it is impossible to distinguish quantitatively between the evaporation from a liquid or the porous solid phase of the snow cover (see "Introduction"). On the other hand, the total enrichment in heavy isotopes in a snow blanket may contribute to an evaluation of the evaporation percentage in the hydrological balance (see section "Isotope mass balance of a snow cover"). Preliminary measurements with respect to the relationship between variations in isotope contents and evaporation rates of snow samples have been performed in a cold room ($-10°C$) (Moser and Stichler, 1975). Largely independent from the amount and the surface of the exposed snow sample a linear increase is observed between both δ^2H and $\delta^{18}O$ values and the relative weight loss (Fig. 4-6). Furthermore, in this evaporation process which dominates over condensation, the isotope

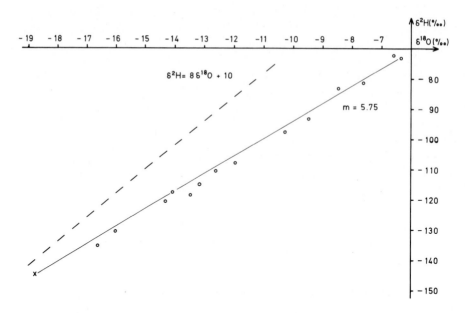

Fig. 4-7. δ^2H—$\delta^{18}O$ relationship of the samples of Fig. 4-6 (from Moser and Stichler, 1975). m = slope of the evaporation line. The dashed curve corresponds to the precipitation line. × = snow originally introduced.

content changes in the snow cover depend on the isotope content of the water vapour above the snow (W. Stichler, unpublished results). This indicates mass exchange phenomena between gaseous and solid H_2O phases. In the δ^2H—$\delta^{18}O$ relationship of the results represented in Fig. 4-7 the slope m = 5.75; this means that the snow evaporation in the cold room does not take place at a thermodynamical phase equilibrium.

Percolation of melt and rain water and their effects on the isotope profile of a temperate snow cover

The variation with time of the deuterium and tritium contents of snow profiles at the Weissfluhjoch, Davos (Switzerland) is shown in Fig. 4-8 (from Martinec et al., 1977). This and other isotope content profile measurements on temperate snow covers (Ambach et al., 1972; Moser et al., 1975; Moser and Stichler, 1975; Herrmann and Stichler, 1976) show that despite percolating rain and meltwater, the layering of the isotope content in a snow cover undergoes only a slight change during the ablation period in spring and summer. Theoretically, attainment of the isotope equilibrium leads to a fractionation in the δ^2H value of about 20‰ and in the $\delta^{18}O$ value of about 3‰ between solid and liquid phases (see "Introduction"). Therefore

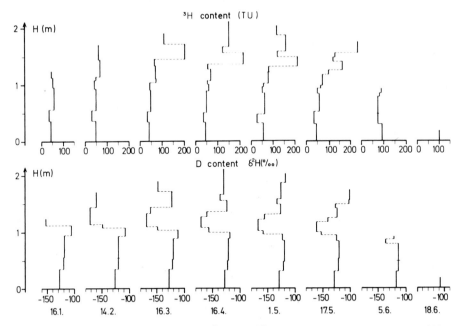

Fig. 4-8. Variations with time of the ^{3}H and δ^2H values of snow profiles (1973) on the Weissfluhjoch, Davos for the days indicated (after Martinec et al., 1977). H = snow height. The slight shift of the major peaks towards the base is due to mechanical compaction. The slight smoothening out of the peaks is due to melting, percolation and isotope exchange. With the onset of spring, now ablation occurs from the top of the profile.

a change of the isotope content in the snow cover occurs when a significant amount of water with a δ-value out of equilibrium percolates into the snow pack. Thus, for instance, an enrichment in heavy isotopes sets in whenever rain, with relative high δ-values, or meltwater from surface layers, which has been enriched in the isotope content by evaporation or other effects, infiltrates the snow cover. In partially melted layers such an enrichment may be observed as well (Judy et al., 1970). All in all, the isotope exchange between liquid and solid phases associated with the percolating water causes a certain homogenization of the isotope contents if the residence time is long enough, and, in general, a gross increase of the δ-value of the entire snow cover. These fractionation effects may be obscured to a large degree in regions with widely varying isotope contents of precipitation by modifications and mass exchange processes in the free water content of the snow cover. The formation of ice lenses could be cited as an example.

Studies of isotope fractionation by percolating meltwater have been carried out particularly in Iceland (Arnason, 1969a, 1980), which offers good conditions for such investigations, due to its small seasonal temperature vari-

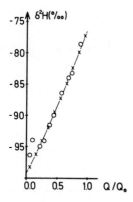

Fig. 4-9. Comparison of results from laboratory snow melt experiment (○) and calculated values (×) (from Buason, 1972). The deuterium concentrations in drain water samples are shown versus the fraction of melted snow. Q_0 is the amount of snow initially in the column. Q is the amount of snow that had melted at each measurement. Agreement is excellent except for the first three observations. This discrepancy could have been caused by a certain portion of the first melt water flowing down the inner surface of the glass tube instead of percolating down the snow column.

ations and associated limited changes of the δ-values of precipitation. Samples taken before the summer ablation period in pits and shallow cores on Icelandic glaciers showed in the firn layers of the preceding summer a rise in the deuterium content wherever a heavy rate of percolation could be assumed. To explain this phenomenon, a model experiment was performed on a homogeneous snow column which was melted from the top. The δ^2H value of the melt water running off was measured continuously and is shown in Fig. 4-9. The pattern of the δ^2H values in the meltwater was fairly accurately described theoretically assuming in the model a partial isotopic equilibrium (Buason, 1972). This theoretical model was tested against measurements obtained at the Weissfluhjoch, with snow lysimeter in 1970 and 1971 (Arnason et al., 1973). Based on the assumption of a constant content of free water of 10% and 20% in the snow cover, it was possible through a time parameter describing the dynamics of the isotope exchange in relation to the total melting time to arrive at a satisfactory adjustment of the theoretical values to the measurement results.

Considerable changes in the isotope content of a snow blanket were noted by Krouse and Smith (1973) after a heavy rain. The variation of the density profile during the precipitation period was pursued by profile recordings at close intervals (about twice daily), and the respective water contents of the separate layers, measured. In addition, the isotope content was used to distinguish between melt water and rain water, allowing a detailed observation of the water movement (obstruction of flow by ice lenses, for instance). A complete quantitative evaluation of isotope content measurements could not be made, however, since the δ-value of precipitation had not been registered. In a study by Herrmann and Stichler (1976) on a snow cover at the outskirts of the Upper Bavarian Alps an isotopic balance between precipitation and snow blanket could not satisfactorily explain the increase in isotopic content of the snow cover after a heavy contribution of meltwater from the upper layers following an onset of foehn, or rainfall. It was also noted that in this case isotope content changes took place the causes of which remain to be determined.

Mechanism of the reduction of a snow cover

The example given in Fig. 4-8 of the isotope content layering at the Weissfluhjoch clearly shows that the reduction of snow height occurs by two processes: One is due to the mechanical settling indicated by coloured threats identifying the single snow layers. The other reduction of the snow cover occurs by the melting of the uppermost layer in each case, whereas the layers below merely suffer a slight reduction in thickness. In contrast to this mechanism confirmed by almost all other isotope content profile recordings in the Alpine area (Fig. 4-10) reveals for the example of a snow profile recording in La Parva (Chile) (Moser et al., 1975) that there apparently the snow cover starts melting from the ground at the beginning of ablation. The upper-

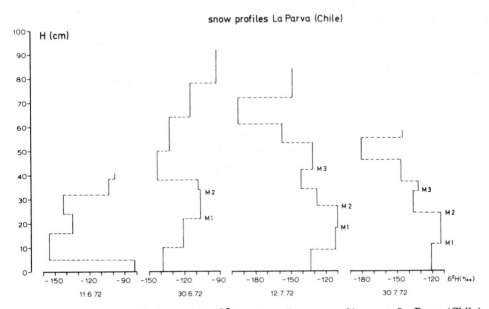

Fig. 4-10. Variation with time of the $\delta^2 H$ values of snow profiles near La Parva (Chile) (after Moser et al., 1975). H = snow height. The single snow layers have been identified by coloured threads ($M1, M2, M3$).

most layers, on the other hand, were reduced only toward the end, and then from the top downward. This may be related to the low rate of cooling of the ground during the short frost period in this region.

ISOTOPE VARIATIONS IN THE TRANSITION OF SNOW TO GLACIER ICE

Early studies by Epstein et al. (1963), Dansgaard (1964), Friedman et al. (1964) and others demonstrated that the information obtainable from the determination of isotope distribution in glaciers can contribute to the solution of a wide variety of glaciological problems. This has since been confirmed by numerous other studies.

The original, precipitation derived isotope distributions are subject to changes and two basic questions arise: what isotopic effects occur close to the surface as the snow cover ages, what modifications of the isotope content distribution are observed or can be expected in the porous firn, and, finally, in the compact ice. Note, that in this case a distinction must be made between polar (cold) glaciers, as well as polar ice sheets, and temperate glaciers. For the latter the occurrence during the ablation period of percolating melt water, with the isotopic effects discussed on pp. 150—153 has to be taken into account.

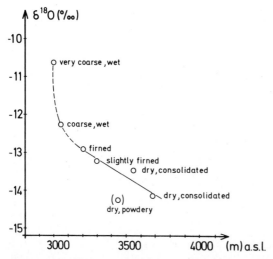

Fig. 4-11. $\delta^{18}O$ values of snow samples from the Vernagtferner region (Oetztal Alps, Austria) in relation to the altitude of the sampling sites (after Moser and Stichler, 1975). The snow type is indicated for each sampling site. The measuring points along the solid line exemplify the isotope altitude effect. Other effects (e.g. melting, evaporation) intensify the isotopic enrichment (dashed line).

Isotope variations in near-surface snow layers

The local isotope content distribution of the near-surface layers resulting from each preceding snowfall may, as a consequence of exposure and albedo of the glacier surface, undergo substantial changes through nonuniform enrichment of the remaining snow in ^{2}H and ^{18}O contents due to evaporation and melting processes. In this manner the isotopic altitude effect in the sur-

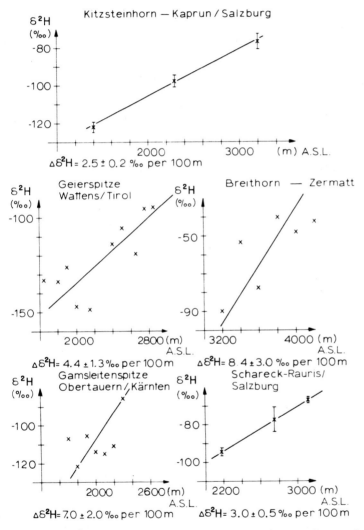

Fig. 4-12. δ^2H values of samples of pack snow from various regions of the Eastern and Western Alps in relation to the altitude of the sampling sites (from Moser and Stichler, 1970).

face layer of the snow cover, for example, may be intensified (see Fig. 4-11) or mitigated, even overcompensated (Fig. 4-12) (Moser and Stichler, 1970, 1975), depending on where a preferential energy absorption takes place on the glacier surface. These effects, which in general are to be expected only in temperate glaciers may partially or totally mask the seasonal pattern of the isotope contents of precipitation which should become visible in the vertical profile of the glacier firn. This may be a reason among others why it was not possible in some cases to detect seasonal variations of the δ-values in drill cores from temperate glaciers (Deutsch et al., 1966, for instance).

In polar glaciers, where generally no isotope content changes are produced as a result of melting and evaporation phenomena at the surface, considerable modifications of original isotope distributions by precipitation may, however, occur through wind drift as discussed previously.

Isotope profile in the porous firn zone of glaciers

If the isotope contents of precipitation conform to a time-dependent periodic function with the cycle of one year, and if these isotope contents impress a record in the snow cover of the glacier following the sequence of precipitation events, then an annual layering of the isotope contents in the vertical profile of the glacier will appear. This annual layering will be the more pronounced the greater the differences are in the isotope contents of precipitation in the course of a year, and the more undisturbed they are allowed to leave an imprint in the snow profile. In the case of a polar glacier, i.e. without percolating melt water, and under the supposition that glacier movements do not influence the annual layering, homogenization can only occur through a vertical isotope exchange. In the porous firn zone examined here, this could be realized in first line by a vertical transport of water vapour. A contributing factor in the near-surface layers may be variations in atmospheric pressure; in deeper firn layers down to the critical density of $\rho_c \approx 0.55$ g/cm^3, however, the main cause can only be diffusion in the water vapour phase (Dansgaard et al., 1973). Fig. 4-13 shows the diminution of the differences between summer and winter $\delta^{18}O$ values as a function of the residence time of the respective layer in the glacier (Dansgaard et al., 1973). Noteworthy in this case is the marked attenuation in annual layering of the isotope contents for the Antarctic stations. This is traceable to the already not very pronounced primary differences of the $\delta^{18}O$ values in the snow cover (caused by wind drift and a low rate of accumulation, especially in summer), but as well as by increased air motion within the firn due to the wide variations in atmospheric pressure produced by weather conditions. The Greenland stations represented in Fig. 4-13 differ principally in the accumulation rate. Small rates of accumulation lead to an increased effectiveness of diffusion during homogenization. In this sense, the greater compaction of the layer with growing depth results in a higher degree of homo-

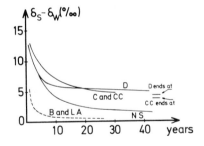

Fig. 4-13. Measured reduction of summer to winter difference of δ-values, $\delta_s - \delta_w$, in accumulated snow as a function of the time since deposition (from Dansgaard et al., 1973). Seasonal δ variations, and thus age and accumulation rates, can be determined as long as $\delta_s - \delta_w$ remains higher than 2‰. Stations: D = Dye III, C = Crête, CC = Camp Century, NS = North Site, B = Byrd, LA = Little America. (——— = Greenland, - - - - - - = Antarctica).

genization of isotope contents. On the basis of available results of isotope measurements on cores of deep drillings in Greenland, there resulted an empirical accumulation rate of 24 g/cm²/yr as a lower limit for the measurable survival of the annual layering of the δ-values in the firning process up to ice formation. Because of the unfavorable basic conditions in Antarctica, this accumulation rate ought to be higher there, attaining a value of about 34 g/cm²/yr (Dansgaard et al., 1973).

In temperate glaciers, homogenization of isotope content variations in the firn profile depend largely on the amount of percolating melt water (Arnason, 1974). Ice lenses formed in the firn by the refreezing of small amounts of melt water may obstruct waterflow and thus shield subjacent layers against the effects of percolation (Stichler and Herrmann, 1977). Furthermore, such ice lenses constitute vapour barriers for the vertical isotope exchange through the vapour phase and therefore possibly lessen the effects of diffusion (Johnsen et al., 1972).

Isotope diffusion in compact glacier ice

In the compact ice of a polar glacier, a homogenization of the variations of the isotope content introduced from the firn can take place only through the slow process of molecular diffusion in the solid state. Owing to the long periods to be examined it is no longer possible, however, to neglect ice movement, particularly in the vertical.

The diffusion flow model theory for diffusion in ice and glacier movement developed by Johnsen (1977) on the basis of earlier theoretical approaches, describes the change with time of the isotope content gradient in firn and ice. Fig. 4-14 shows, for the example of an ice core from Camp Century (Greenland) (see section "Dating firn and glacier ice"), how, accord-

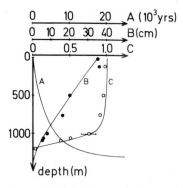

Fig. 4-14. Depth relationships of age (A), annual layer thickness (B), and relative amplitude of annual δ cycles (C) at the Camp Century deep ice core (from Johnsen, 1977). Open circles: field data; closed circles: results from diffusion flow model calculations.

ing to this theory, the age of the ice, the annual layer thickness and the relative amplitude of the variations of the $\delta^{18}O$ values in an annual layer become modified. The test points plotted confirm the very general theoretical approach in its application to these relatively simple conditions of ice dynamics at Camp Century. If one takes into account the diffusion in the porous firn zone it appears, according to this theory, that the diffusion length in the ice is about 7—8 cm. This diffusion length is largely independent of accumulation rate and temperature and hence supports the empirical statements made in the previous section on the critical rate of accumulation.

In temperate glaciers with their not very thick ice layers, diffusion processes of the type mentioned above play no significant part, since the residence time of the ice in the glacier is generally not more than a few thousand years. Nevertheless, in such a body of ice, homogenization is produced by an isotope exchange traceable to the melt water produced in zones of increased pressure (planes of sliding, for example) (Arnason, 1980).

Total beta activity in firn and ice profiles

Total beta activity of the fall-out of precipitation originates, as does its tritium content, from the nuclear weapon tests that have been carried out since 1953. Conformingly, the distribution in time and space of total beta activity of precipitations is very similar to that of tritium; and, for instance, the seasonal maximum also occurs in early summer (see Chapter 1). Somewhat different in comparison to tritium however is the deposition of total beta activity in the firn profile, in particular with regard to temperate glaciers. Whereas tritium as component of the water molecule labels the respective snow layers as "ideal" tracer because of its varying concentra-

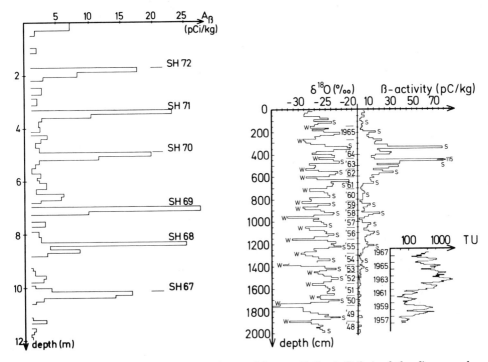

Fig. 4-15. Vertical profile of the specific total beta activity (pCi/kg) of the firn samples in the accumulation area of Kesselwandferner (Oetztal Alps, Austria) (after Ambach et al., 1976). Location 3.225 m a.s.l.; date, July 1973; activity measurements, October 1973 to January 1974. SH = summer horizons; values <1 pCi/kg have not been plotted.

Fig. 4-16. Profiles of total beta activity and $\delta^{18}O$ values along a 20-m firn core at Carrefour on the Greenland ice sheet (from Ambach and Dansgaard, 1970). For comparison, the graph at the lower right shows the tritium concentration in monthly samples of precipitation collected in Ottawa, Canada (Brown, 1970). The layers interpreted as summer and winter precipitation are indicated by S and W, respectively.

tion in the associated precipitation, the beta activity nuclides in rain water produce a vertical distribution profile in the firn influenced by adsorption at dust particles within summer ablation horizons (Ambach et al., 1968). In this way it becomes possible, also in the case of temperate glaciers, to register, in contrast to the profile of ^2H, ^{18}O and ^3H contents, a pronounced annual layering of total beta activity. Ambach et al. (1976) were able to demonstrate with the aid of measurements of the half-life of the beta-active residues of firn samples from a test core of the Kesselwandferner (Oetztal Alps, Austria) that the original deposition horizons of total beta activity are reactivated with the fall-out of later years through subsequently percolating melt water. Thus a marked annual layering remains conserved also in layers assigned to years of a relatively low deposited total beta activity (Fig. 4-15).

Similar results have been found by Jouzel et al. (1977) in the Mont Blanc area. This effect does not appear in polar glaciers; there, the profile of total beta activity reproduces horizons originating from precipitation with a high fall-out rate especially in the years 1963-64. The result is a different distribution with respect to maximum values and precipitation years on the northern and on the southern hemisphere (Arctic, Greenland, Antarctic), and depends on the point of the respective nuclear weapon test and on meteorological exchange phenomena (Ambach and Dansgaard, 1970; Clausen and Dansgaard, 1977; Reeh et al., 1977). To demonstrate this the distribution of total beta activity in the firn profile of a drill core from Carrefour (Greenland) is compared (Fig. 4-16) to that of the $\delta^{18}O$ values and that of the tritium content in precipitation, whose geographical distribution is detailed in Chapter 1.

Measuring accumulation rates

An uninterrupted sequence of annual layers each with characteristic isotope contents offer an excellent guide to determine the accumulation rate on glaciers. Similar to conventional stratigraphic determinations, the advantage of isotope methods over gauge measurements on glacier surfaces is that accumulation rates can be ascertained over longer periods, especially if age dating is carried out simultaneously. However, difficulties associated with stratigraphic methods in the identification of annual horizons apply in principle also to the isotope method, and have been discussed in the preceding sections. A favourable factor is that a homogenization in the stratigraphic profile does not necessarily lead to a disappearance of the isotope layering, and vice versa. A combination of both methods may thus furnish an optimum of information in order to calculate the rate of accumulation (Hattersley-Smith et al., 1975).

Especially in compact ice the variations of the δ-values offer the only possibility of detecting the accumulation rates of past periods. Johnsen (1977) was able to show on a $\delta^{18}O$ profile of the drilling at Camp Century (Greenland), that given a sufficient measuring density in the vertical profile, the Fourier analysis of a diffusion distorted isotope profile may still reveal annual layerings to a depth of 1160 m (Fig. 4-17). Reeh et al. (1978) transferred the annual layer thickness profiles along three 400-m ice cores from central Greenland (Dye III, Milcent, Crête) into accumulation rate records. The short-term (30—120 years) and intermediate-term (>120 years) deviations of the accumulation rate from the mean values measured during a time interval of 800 years and 1400 years, respectively, are shown in Fig. 4-18. To determine the accumulation rate during the past twenty years the characteristic horizons of fall-out deposition, which can be charted by measurements of tritium content and total beta activity, provide an additional guideline (Ambach et al., 1976; Merlivat et al., 1977). Age determinations based on

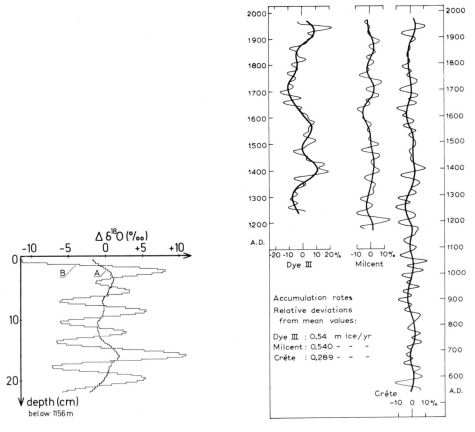

Fig. 4-17. Curve A: $\delta^{18}O$ profile from the Camp Century deep ice core smoothed by diffusion in firn and solid ice, measured on approximately 12,000-year-old ice. Curve B: The same profile deconvoluted by $L = 1.8$ m (diffusion length) and $\sigma = 0.1‰$ (standard deviation) (from Johnsen, 1977).

Fig. 4-18. Accumulation rate records from three Greenland ice-sheet stations, determined mainly from seasonal $\delta^{18}O$ cycles (from Reeh et al., 1978). The annual accumulation data series have been smoothed by filters with cut-off periods of 120 years (heavy curves) and 30 years (thin curves).

radioisotopes (e.g. ^{210}Pb) are sometimes suited to ascertain the accumulation rate of longer periods. Half-life and concentration of the nuclide examined (Picciotto et al., 1968, compare also section "Glacier changes and movements") are the limiting factors.

The experiences gained to date in the course of numerous studies on the accumulation rates in polar regions reveal that in particular in Antarctica a statement on accumulation can be made with certainty only in those cases where as many independent methods as possible are employed (Epstein et

TABLE 4-2

Comparison of hydrological balance and isotopic balance in snow lysimeter studies

Lysimeter area (km^2)	Observation period	Reference	Hydrological balance *				Isotopic balance *			
			ΔW (mm)	P (mm)	R (mm)	loss by $(E-C)$ (%)	isotope	I_{W+P} (TU or ‰ vs. SMOW)	I_R	loss (%)
Weissfluhjoch, Davos										
5.2	5/17 to 6/28, 1973	Martinec et al. (1977)	799.2	144.8	881.3	6.6	^3H	103.8	123.2	?
							^2H	−118.5	−103.8	9
							^{18}O	−16.27	−14.88	≈10.5
Lainbach Valley										
25	3/18 to 3/24, 1974	Herrmann and Stichler (1976)	175	3.5	176	1.4	^2H	−85.9	−83.7	2.0
							^{18}O	−12.20	−11.92	2.4

* ΔW = water equivalent of melted snow cover; P = precipitation; R = discharge; E = evaporated water equivalent within the observation period; C = condensed water equivalent within the observation period; I_{W+P} = weighed average value of isotope contents of snow cover and precipitation; I_R = weighed average value of isotope contents in discharge.
The losses resulting in the isotopic balance have been calculated from the difference between I_R and I_{W+P} by the Raleigh formula for isotope fractionation using the equilibrium factor (see text).

al., 1965; Picciotto et al., 1966, 1968; Johnsen et al., 1972; Arnason, 1980; Clausen and Dansgaard, 1977; Hammer et al., 1978).

SNOW AND ICE ISOTOPE HYDROLOGY

Measurements of isotope contents have also been used for the solution of hydrological problems in catchment areas with a temporary snow cover or with glaciers. The isotope labelling of the snow cover in its single layers, of the precipitations, and of the discharge concerning its origin gives possibilities for water balances and for a time differentiation between precipitation and discharge.

The following examples of applications describe experiments to estimate the net evaporation rate of a snow cover during the ablation period, the fractions of groundwater and surface water discharge from a snow covered catchment area, the fractions of snow meltwater and ice meltwater in a glacier discharge, and the residence time of water between precipitation and discharge in a glaciated area. One can foresee that isotope analyses will find in future an increasing application in snow and ice hydrology.

Isotope mass balance of a snow cover

An isotopic balance of snow cover and discharge can be established during the ablation period if different isotope contents are present in the snow cover, in precipitation and in the discharge. Studies in this field have been undertaken on snow lysimeters at the Weissfluhjoch by Arnason et al. (1973), Martinec (1974), Martinec et al. (1977), and in the Lainbach Valley near Benediktbeuren (Upper Bavaria; Herrmann and Stichler, 1976), the isotopic mass balance (equation 1) being compared to the hydrological balance (equation 2):

$$I_R R = I_W \Delta W + I_P P - I_{E-C}(E-C) \tag{1}$$

$$R = \Delta W + P - (E-C) \tag{2}$$

where R represents discharge; ΔW, the water equivalent of the melted snow cover; P, precipitation; E, the evaporated, and C, the condensed water equivalent within the observation period; the isotope contents I_R, I_W and I_p are the corresponding weighed average values during the ablation period reviewed. I_{E-C} is the isotope content of the combined water equivalent $(E-C)$.

In general, it is possible to measure the quantities I_R, I_W, I_p, R, W and P directly. Difficulties arise with respect to both the isotopic balance and the hydrological balance due to the substantial errors associated specially with

TABLE 4-3

Investigation areas to determine subsurface discharge (after Martinec, 1972)

Drainage basin	Area (km²)	Difference in elevation (m)	Forested area (%)	Average discharge (m³/s)
Modry Dul	2.65	1000—1554	30	0.1
Dischma	43.3	1668—3146	3	1.7

precipitation measurements of snow. A quantitative evaluation is given for the investigations represented in Table 4-2. The net evaporation rate for the isotopic balance was calculated using the values of the isotope fractionation during evaporation. In this case the Rayleigh formula with the equilibrium fractionation factor was adopted. Since in this procedure the surely existing kinetic isotope fractionation has been neglected, the loss rates indicated must be taken as maximum values.

Determination of discharge proportions from isotope content measurements

Discharges composed of waters with differing isotope contents can be analyzed quantitatively with respect to their separate shares by measuring the individual isotope contents for an isotopic balance. To establish the mixing equations:

$$I_t R_t = \sum_{i=1}^{n} I_i R_i$$

it is necessary to know I_t (isotope content in total discharge R_t) and all I_i (isotope contents in the partial discharges R_i). To determine the absolute values of the partial discharges, one of the quantities R_i or R_t must be known as well. Isotopic balances of this kind are the more accurate the more the separate isotope contents in the respective drainage area differ. Regions with temporary snow cover and glaciers offer favourable conditions in this respect, since wide variations of the isotope content in area and time occur there, in particular during melting processes.

Discharge proportions during the snow melt in drainage areas with temporary snow cover. To determine the direct runoff, that is, the share of total discharge flowing off superficially in the course of the ablation period of the snow cover, detailed and comparative isotope-hydrological and hydrological investigations have been conducted in the drainage areas of Modry Dul (northern Czechoslovakia) and Dischma near Davos (Switzerland) (Dinçer

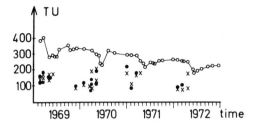 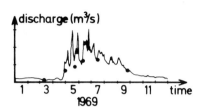

Fig. 4-19. Tritium content in the discharge (○) and in the snow cover of Dischma Valley near Davos (Switzerland) (from Martinec et al., 1974). The snow samples were collected from two locations: Teufi (●), near the discharge measuring station; Dürrboden (×), upper part of the valley. For further discussion see text.

Fig. 4-20. Separation of direct and subsurface flow by tritium data (from Martinec et al., 1974). Solid dots represent calculated subsurface flow from tritium data.

et al., 1970; Martinec et al., 1974; Martinec, 1975a). Some characteristic data for the two observation areas are listed in Table 4-3.

For the Dischma region Fig. 4-19 shows the variations of tritium concentrations in total discharge, and in the snow cover at two sampling points in the upper and the lower part of the observation area, over a period of four years. It can be noted that the tritium content in runoff is generally higher than in the snow samples. This suggests that it originates from precipitation of past years with high tritium content. This applies in particular to the winter months during which precipitation does not flow off directly.

In the course of the ablation period in early summer the tritium concentrations drop, but do not attain those of the snow samples. One can thus conclude that also during the period of snow melt runoff, water from past years accounts for a significant portion of discharge, and reversely, a substantial percentage of snow melt water seeps into the ground. For 1969, Fig. 4-20 shows the shares of subsurface flow in total discharge calculated from the tritium balance. Overall, computations proved that during ablation of the snow cover only 10—50% of the melt water runs off directly, i.e. on the surface. Between the years 1969 and 1972 the proportion of direct snow melt run off amounts on the average to 36%. Corresponding measurements for the Modry Dul region in 1966 gave 37%. Hydrological regression analyses agree well with these results (Martinec, 1975b).

In two drainage basins in the north of the Colorado Front Range (U.S.A.), Meiman et al. (1973) examined the $\delta^2 H$ values of discharges, of the snow cover and of some precipitation samples. From the pattern of the $\delta^2 H$ values in runoff during the snow melt it was possible to conclude for the one of the two areas that a heavy infiltration of melt water took place into the subsurface; furthermore, it was estimated that in the course of the entire year snow melt water accounts for about 60% of discharge.

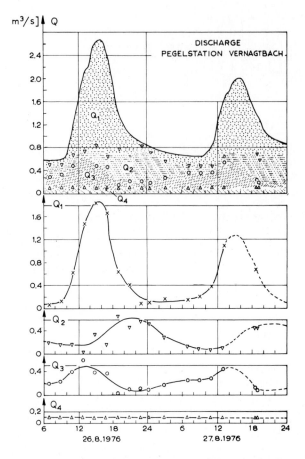

Fig. 4-21. Daily curve of total runoff Q as well as of discharge proportions of ice melt water (Q_1), and snow melt water running off directly (Q_2), longer retained melt water within the glacier (Q_3) and groundwater (Q_4), calculated from measurements of the tritium content, the deuterium content, electrolytic conductivity and total runoff for 26—27 August 1976 (Vernagtbach, Oetztal Alps, Austria) (from Behrens et al., 1979).

Discharge proportions in glaciered drainage areas during the ablation period. Glacier discharge components may differ significantly in their isotope contents: old glaciers ice melt water which is generally originated in precipitations fallen prior to the time of nuclear weapon tests is practically free of tritium, whereas snow melt water approximate the tritium content of precipitations (Ambach et al., 1973). The deuterium and ^{18}O contents allow a distinction between the melt water with its relatively high δ-values that flows off superficially during the ablation period and the intra- and subglacial glacier discharge with low δ-values. The latter two components can be distinguished by their mineralization measured by electrolytic conductivity.

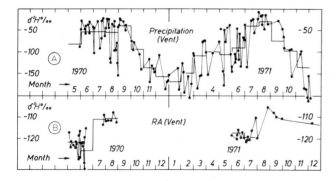
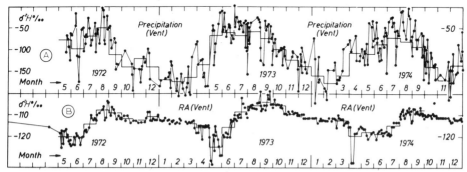

Fig. 4-22. Seasonal fluctuations of the deuterium concentration at Vent (Oetztal Alps, Austria) (from Ambach et al., 1976). A. Precipitation (Vent): dots correspond to individual precipitations, the step function giving the weighted monthly mean value. B. Water samples Rofenache Vent (RA): dots correspond to individual samples, the step function gives the unweighted monthly mean value.

Necessary conditions are homogenization and enrichment effects of the δ-values in the snow and the firn of temperate glaciers (Moser et al., 1972; Ambach et al., 1976).

As an example, Fig. 4-21 presents an analysis of the discharge components in the glacier creek of the Vernagtferner (Oetztal Alps, Austria) which in the course of the ablation period vary with the time of the day. The calculation was based on the deuterium, tritium and electrolytic conductivity mixing equations and the discharge hydrograph (Behrens et al., 1979; see also Behrens et al., 1971). In the annual progression of the δ^2H values in the discharge of the 98-km², 44% glaciered drainage basin of the Rofenache (Oetztal Alps, Austria) (Fig. 4-22) the beginning of the ablation period is marked by a substantial drop in δ^2H values caused by a melting of the winter snow cover. Subsequently the deuterium contents rise because of isotopically heavy melt water. The wide variations during the summer ablation period are the result of the variable contents of ice and snow melt water as well as of subglacial spring water. On the basis of the deuterium contents it thus becomes pos-

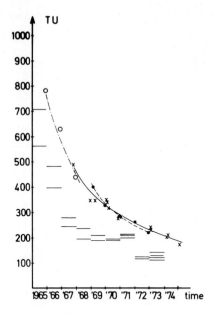

Fig. 4-23. Decrease in tritium contents in the winter base flow in the Vent Valley, Austria (×; Ambach et al., 1976), Dischma Valley, Switzerland (●; Martinec et al., 1974), Modry Dul, Czechoslovakia (○; Dinçer et al., 1970), and in mean annual precipitation of various European locations.

sible to indicate the sequence of the principal discharge components. The quantitative examination of the annual $\delta^2 H$ graph of the discharge of the individual glaciers of the region under investigation permitted the identification of the mean altitude of the separate drainage basins and average shares of the spring water and melt water during ablation (Ambach et al., 1976).

In the very much larger drainage basin of the North Saskatchewan River and the Mistay (Rocky Mountains, Canada), which cover an area of 1900 km² and are glaciered to 15%, tritium measurements were performed in creeks, rivers, firn and ice samples over a period of three years (Prantl and Loijens, 1977). The tritium values of the four components contributing to discharge in the area (summer precipitation, temporary snow cover and firn, ice and groundwater) were used in tritium balance equations and, adopting simplifying assumptions, proved possible to quantify the respective contribution to total discharge for a certain period. For example, for the middle of July, a discharge composition of 10% firn meltwater, 20% ice meltwater, and 70% storage water from the subsurface was determined. These values are in good agreement with data obtained from hydrological glacier discharge models.

As an additional example one might mention that for four drillings on two Icelandic glaciers an isotopic balance of the deuterium contents in the firn and in precipitation gave the proportion of the snow of the preceding winter that had melted during the ablation period (Arnason, 1980).

Residence time of water in the subsurface or in glaciers

Input-output comparisons of the variations of the tritium contents in a given hydrological system allow a calculation based on model assumptions of the average residence time of the water in the respective system (Siegenthaler, 1971). The tritium data from the discharge of the Modry Dul and Dischma drainage areas were introduced, with tritium measurements of precipitations, into a mathematical model used for the calculation of the groundwater storage. Residence times of about 2.5 years (Dinçer et al., 1970) and 4.5 years (Martinec et al., 1974) were thus determined for the respective basins. In both cases, dispersion models proved to be suitable mathematical propositions although different dispersion functions were used. From the average residence times thus obtained it was then possible to determine the underground storage volumes at 2.6×10^6 m^3 and 130×10^6 m^3, respectively.

The pattern of the tritium contents in the winter discharge from the glacier area of the Rofenache (Oetztal Alps, Austria) is represented in Fig. 4-23 and compared to the annual average tritium content in precipitation. The decrease in tritium content corresponds approximately to the values from the Dischma Valley (see above) indicating a residence time of the water of 4—5 years (Ambach et al., 1976).

HISTORICAL GLACIOLOGY

The information stored in the ice sheets of the polar regions comprise, in particular in Greenland and in Antarctica, periods of hundred thousands of years and thus allow a retrospect into the Pleistocene. Because of the correlation between δ-values and temperatures measurements of deuterium and ^{18}O content are suited, to study climatic variations in the past. From such information it might be possible to forecast climatic developments. The determinations of the concentration of radionuclides contribute to the development of the time scale.

Further possibilities so far not intensively exploited for isotopic investigations consist in recording long-term variations of the production rate of radioactive isotopes (for example, ^{14}C) and their causes, the intensity of cosmic radiation as well as the deposition rate of cosmic and terrestrial dust. Especially the latter subject should gain in significance in view of the increasing anthropogenic pollution of the atmosphere.

Finally, the isotope content distribution in glaciers allows certain conclusions to be drawn about their movement and mass balance.

Glacier changes and movements

The normal case of glacier movement is shown in longitudinal section in Fig. 4-24. It follows herefrom that ice surfaces the deeper in the ablation zone the higher it was formed in the accumulation zone. If such a flowage condition exists, then, conformingly, a reduction of the δ-values, caused by the isotopic altitude effect, must take place in the ablation zone from the firn line toward the glacier tongue. This was the conclusion of measurements performed in Canada on Saskatchewan Glacier and on Malaspina Glacier (Epstein and Sharp, 1959) and on Rusty Glacier (West and Krouse, 1970). Local deviations in these cases indicate a particular flowage behaviour of individual glacier arms.

To study the flowage of ice in the ice cap of Devon Island (75°N, 82°W), Bucher et al. (1976) undertook age determinations, based on the ^{14}C content, at different depths (down to 300 m) of four borings near the center of the ice cap. For this investigation they used the borehole probe mentioned in the next section. The age values increase with depth, reach about 6000 years and are in agreement with results calculated from flow models (Dansgaard and Johnsen, 1969). A ^{32}Si determination on a sample from a depth of 135 m also indicated an age (~700 years) that agrees well with measurements of vertical velocities within the period of the ice cap.

Finally, measurements of the δ-values in glaciers whose tongue protrudes into the ocean as an ice shelf provide information about what part of the shelf consists of glacier ice, of frozen seawater or of snow precipitation along the coast (Gow and Epstein, 1972; Morgan, 1972).

Dating firn and glacier ice

The long-term conservation of seasonal variations of isotope concentrations in the glacier ice (compare Figs. 4-14 and 4-17) permits a much better identification of annual ice layers than is possible by conventional stratigraphic methods. This applies in particular to the Greenland ice sheets. Johnsen et al. (1972) were able to show on the basis of ^{18}O measurements that on an undisturbed ice profile of Camp Century (Greenland) dating over

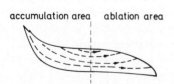

Fig. 4-24. Flow pattern of a normal glacier (from Reid, 1896).

TABLE 4-4

Radionuclides for ice dating

Nuclide and occurrence in ice	Natural concentration in recent ice	Necessary sample amount (kg)	Half-life (years)	Dating period (years)	Reference
^{3}H as ^{1}H^{3}HO in ice	ca. 40 dpm/kg of ice *	1	12.3	<100	Theodorsson (1977)
^{210}Pb as fall-out	some dpm/kg of ice	1	22	<100	Picciotto et al. (1968)
^{32}Si in solution or adsorbed on solid particles	0.4 dpm/10^3 kg of ice	10^3	295 ± 25	<1000	Clausen (1973)
^{39}Ar in air bubbles	0.1 dpm/10^3 kg of ice **	5×10^3	270	<1000	Oeschger et al. (1976)
^{14}C in CO_2 of air bubbles	0.2 dpm/10^3 kg of ice **	$1-5 \times 10^3$	5730	<25,000	Oeschger et al. (1976)
^{36}Cl in solution	10^{-2} dpm/10^3 kg of ice (estimated for regions with low accumulation rate)	10^4	3.1×10^5		Oeschger et al. (1976)

* Excluding "bomb tritium", i.e. only tritium produced by cosmic radiation.
** For an air content of the ice of about 100 dm^3 of air per 10^3 kg of ice (compare Raynaud and Delmas, 1977).

8300 years would be feasible. Thus in Central Greenland, given favourable conditions, an annual layering ought to be observable deep into the last ice age. The best dating by δ-values known at present was made on the drilling Dye III (Greenland) and fixes A.D. 1230 with an accuracy of ±5 years (Dansgaard et al., 1973) and on the Gisp ice core from Milcent (Greenland) giving the exact value of A.D. 1177 for the deposition of the bottom layer (Hammer et al., 1978). The criterion for a still measurable annual layering of δ-values in ice is that at a depth with a density of 0.55 g/cm^3 seasonal variations in the ^{18}O content are close to $\delta^{18}O \approx 2‰$, if eight samples per annual layer are evaluated.

Longer-term variations of the δ-values in ice can be traced for considerably greater periods than seasonal fluctuations. A corresponding analysis indicated that 10-year cycles, for example, should be observable in the ice cores of Camp Century or Dye III still after ca. 19,000 years; 100-year cycles still after some 40,000 years (Dansgaard et al., 1973). A combination of age determinations with flow models (see next section) should permit a considerable extension of the isotopic-stratigraphic dating even in regions where for glaciological or meteorological reasons the annual layering of δ-values is weak and/or not continuous (Hammer et al., 1978). Such a situation may be encountered in Antarctica.

Table 4-4 gives a survey of the radionuclides employed or considered for determining ice ages. Adoption of tritium generated by cosmic radiation has been limited to a significant degree by "bomb tritium" in the atmosphere and precipitations. The distribution of bomb tritium in time marks firn and ice horizons up to about 1953 (Fig. 4-16). Similar to the fall-out deposition (compare section "Total beta activity in firn and ice profiles"), characteristic horizons result which can be used to evaluate the accumulation rate and, consequently, also for "dating" purposes (Ambach et al., 1976). In the northern hemisphere, one of the most active horizons is the summer layer of 1963. Theodorsson (1977) determined the "prebomb" tritium in the ice of Iceland (Vatnajökoll Glacier) and of Greenland (Dye III). From the decrease in ^{210}Pb activity within a firn profile in the Antarctica, Picciotto et al. (1968) deduced an age of about 80 years. They calculated an accumulation rate, which agreed well with that determined on the basis of gauge and beta activity measurements.

Other nuclides listed in Table 4-4 require sample amounts that cannot be obtained conventionally from core drillings. In the last few years Oeschger et al. (1976) have developed a method by which the necessary sample amounts are melted in situ in an ice boring and the desired nuclides extracted from the water. Both dissolved ions and solid particles can be obtained in situ from the meltwater. However, the probing tool can be used only to a depth of about 400 m in dry boreholes where a stabilizing liquid is not required for the mechanical stability of the boring. First measurements on samples that were obtained with this probe in Antarctica and in Greenland proved the usefulness of the method and furnished ^{39}Ar and ^{32}Si ages that showed satisfactory agreement with the δ^{18}O stratigraphy (Oeschger et al., 1977).

^{18}O paleoclimatology on ice cores

Climatic changes cause only a shifting of the ^{18}O values in precipitations in those cases where the difference between condensation temperature in the precipitation cloud and the temperature at the beginning of the condensation process is influenced thereby. The records of climatic variations in ice are represented by the data of the 1387-m-deep core drilling at Camp Century in Greenland (Fig. 4-25, Dansgaard et al., 1969). The pattern of the δ^{18}O values in the vertical profile gives a detailed account of cold periods (low δ-values) and climatic optima (high δ-values). A direct conversion of the δ-values to average temperatures is not possible, since other effects produced in part by climatic conditions may also have influenced the isotope contents of precipitation at the time. Thus, for instance, an increase in ice thickness during cold periods leads to an additional reduction of δ-values

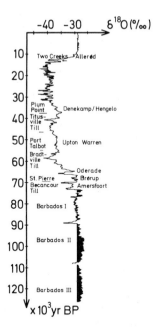

Fig. 4-25. The continuous δ^{18}O record for the Camp Century ice core. The time scale is calculated from a flow model (Dansgaard and Johnsen, 1969) for the Greenland ice and correlated with the major climatic events of the upper Pleistocene (from Dansgaard et al., 1974).

as a result of the isotopic altitude effect. On the basis of gas content measurements in ice Raynaud and Lorius (1977) concluded that more than half of the isotope variations shown in Fig. 4-25 between Holocene and Wisconsin had been caused by an elevation change in the ice surface but also in the ice formation site.

The time scale in Fig. 4-25 was calculated, in the absence of other dating possibilities, from a flow model for the Greenland ice (Dansgaard and Johnsen, 1969). This treatment allows a satisfactory correlation with known ^{14}C data, pollen dating and dated δ^{18}O values for deep-sea sediments. It was also shown that beyond all known climatic phenomena such as interstadials supplementary unknown informations are stored in the ice. Fourier analyses of successive δ-values indicated cycles of varying annual frequences which may be assigned in part to sun activity, and permit forecasts of future climatic developments: according to these data, a temperature decline in the next decade, and, around the middle of the 21st century, an optimum climate, may be expected, provided that anthropogenic influences will not disturb the natural climate pattern (Dansgaard et al., 1973).

Corresponding studies on the ice core of the 2164-m drilling at Byrd Station in Antarctica, gave a similar pattern of the δ^{18}O profile. However, it can only be related with reservations to the δ^{18}O curve for Camp Century, because of the considerably greater complexity introduced by ice movement which makes dating with simple flow models more difficult (Johnsen et al., 1972).

REFERENCES

Aldaz, L. and Deutsch, S., 1967. On a relationship between air temperature and oxygen isotope ratio of snow and firn in the South Pole region. *Earth Planet. Sci. Lett.*, 3: 267—274.

Ambach, W. and Dansgaard, W., 1970. Fallout and climate studies on firn cores from Carrefour, Greenland. *Earth Planet. Sci. Lett.*, 8: 311—316.

Ambach, W., Dansgaard, W., Eisner, H. and Möller, J., 1968. The altitude effect on the isotopic composition of precipitation and glacier ice in the Alps. *Tellus*, 20: 595—600.

Ambach, W., Eisner, H. and Pessl, K., 1972. Isotopic oxygen composition of firn, old snow and precipitation in alpine regions. *Z. Gletscherkd. Glazialgeol.*, 8: 125—135.

Ambach, W., Eisner, H. and Url, M., 1973. Seasonal variations in the tritium activity of run-off from an Alpine glacier, Kesselwandferner (Oetztal Alps). *IASH Publ.*, 95: 199—204.

Ambach, W., Elsässer, M., Moser, H., Rauert, W., Stichler, W. and Trimborn, P., 1975. Variationen des Gehalts an Deuterium, Sauerstoff-18 und Tritium während einzelner Niederschläge. *Wetter Leben*, 27: 186—192.

Ambach, W., Eisner, H., Elsässer, M., Löschhorn, U., Moser, H., Rauert, W. and Stichler, W., 1976. Deuterium, tritium and gross-beta-activity investigations on Alpine glaciers (Oetztal Alps). *J. Glaciol.*, 17: 383—400, see also in: *Isotopes and Impurities in Snow and Ice, Proc. Grenoble Symp., August-September 1975. IASH Publ.*, 118: 285—288, 1977.

Arnason, B., 1969a. The exchange of hydrogen isotopes between ice and water in temperate glaciers. *Earth Planet. Sci. Lett.*, 6: 423—430.
Arnason, B., 1969b. Equilibrium constant for the fractionation of deuterium between ice and water. *J. Phys. Chem.*, 73: 3491—3494.
Arnason, B., 1980. Stable isotopes in hydrology. In: J.R. Gat (Editor), *Ice and Snow Hydrology*. IAEA, Vienna (in press).
Arnason, B., Buason, Th., Martinec, J. and Theodorsson, P., 1973. Movement of water through snowpack traced by deuterium and tritium. In: *The Role of Snow and Ice in Hydrology*, Proc. Banff Symp., September 1972. UNESCO-WMO-IASH, pp. 299—312.
Behrens, H., Bergmann, H., Moser, H., Rauert, W., Stichler, W., Ambach, W., Eisner, H. and Pessl, K., 1971. Study of the discharge of Alpine glaciers by means of environmental isotopes and dye tracers. *Z. Gletscherkd. Glazialgeol.*, 7: 79—102.
Behrens, H., Moser, H., Rauert, W., Stichler, W. and Kirchlechner, P., 1979. Models for the run-off from a glaciated catchment area using measurements of environmental isotope contents. In: *Isotope Hydrology 1978*. IAEA, Vienna, pp. 829—846.
Brown, R.M., 1970. Distribution of hydrogen isotopes in Canadian waters. In: *Isotope Hydrology*. IAEA, Vienna, pp. 3—21.
Buason, Th., 1972. Equations of isotope fractionation between ice and water in a melting snow column with continuous rain and percolation. *J. Glaciol.*, 11: 387—405.
Bucher, P., Möll, M., Oeschger, H., Stauffer, B. and Paterson, W.S.B., 1976. Radiokohlenstoffdatierung polaren Eises in einem Bohrloch durch die Eiskappe des Devon Island (N.W.T., Kanada). *D. Ges. Polarforsch.*, 10.
Clausen, H.B., 1973. Dating of polar ice by ^{32}Si. *J. Glaciol.*, 12: 411—416.
Clausen, H.B. and Dansgaard, W., 1977. Less surface accumulation on the Ross Ice shelf than hitherto assumed. In: *Isotopes and Impurities in Snow and Ice*, Proc. Grenoble Symp., August-September 1975. IASH Publ., 118: 172—176.
Craig, H. and Hom, B., 1968. Relationship of deuterium, oxygen-18, and chlorinity in formation of sea ice. *Trans. Am. Geophys. Union*, 49: 216—217.
Dansgaard, W., 1964. Stable isotopes in precipitation. *Tellus*, 16: 436—468.
Dansgaard, W. and Johnsen, S.J., 1969. A flow model and a time scale for the ice core from Camp Century, Greenland. *J. Glaciol.*, 8: 215—223.
Dansgaard, W., Johnsen, S.J., Moeller, J. and Langway, C.C., Jr., 1969. One thousand centuries of climatic record from Camp Century on the Greenland ice sheet. *Science*, 166: 377—381.
Dansgaard, W., Johnsen, S.J., Clausen, H.B. and Gundestrup, N., 1973. Stable isotope glaciology. *Medd. Grønl.*, 197(2): 1—53.
Dansgaard, W., Johnsen, S.J., Reeh, N., Gundestrup, N., Clausen, H.B. and Hammer, G.U., 1974. Climatic changes, Norsemen and modern man. *Nature*, 255: 24—28.
De Quervain, M., 1951. Zur Verdunstung der Schneedecke. *Arch. Meteorol., Geophys. Bioklimatol., Ser. B*, 3: 47—64.
De Quervain, M., 1973. Snow structure, heat, and mass flux through snow. In: *The Role of Snow and Ice in Hydrology*, Proc. Banff Symp., September 1972. UNESCO-WMO-IASH, pp. 203—226.
Deutsch, S., Ambach, W. and Eisner, H., 1966. Oxygen isotope study of snow and firn on an Alpine glacier. *Earth Planet. Sci. Lett.*, 1: 197—201.
Dinçer, T., Payne, B.R., Florkowski, T., Martinec, J. and Tongiorgi, E., 1970. Snowmelt runoff from measurements of tritium and oxygen-18. *Water Resour. Res.*, 6: 110—124.
Ehhalt, D.H., 1971. Vertical profiles and transport of HTO in the troposphere. *J. Geophys. Res.*, 76: 7351—7367.
Epstein, S. and Sharp, R.P., 1959. Oxygen isotope variations in the Malaspina and Saskatchewan glaciers. *J. Geol.*, 67: 88—102.

Epstein, S., Sharp, R.P. and Goddard, I., 1963. Oxygen isotope ratios in Antarctic snow, firn and ice. *J. Geol.*, 71: 698—720.

Epstein, S., Sharp, R.P. and Gow, A.J., 1965. Six years record of oxygen and hydrogen isotope variations in South Pole firn. *J. Geophys. Res.*, 70: 1809—1814.

Friedman, I. and Smith, G., 1970. Deuterium content of snow cores from Sierra Nevada area. *Science*, 169: 467—470.

Friedman, I. and Smith, G.I., 1972. Deuterium content of snow as an index to winter climate in the Sierra Nevada Area. *Science*, 176: 790—793.

Friedman, I., Redfield, A.C., Schoen, B. and Harris, J., 1964. The variation of deuterium content of natural waters in the hydrologic cycle. *Rev. Geophys.*, 2: 177—244.

Gonfiantini, R., 1970. Discussion. In: *Isotope Hydrology*. IAEA, Vienna, p. 56.

Gow, A.J. and Epstein, S., 1972. On the use of stable isotopes to trace the origins of ice in a floating ice tongue. *J. Geophys. Res.*, 77: 6552—6557.

Hammer, C.V., Clausen, H.B., Dansgaard, W., Gundestrup, N., Johnson, S.J. and Reeh, N., 1978. Dating of Greenland ice cores by flow models, isotopes, volcanic debris, and continental dust. *J. Glaciol.*, 20: 3—26.

Hattersley-Smith, G., Krouse, H.R. and West, K.E., 1975. Oxygen isotope analysis in accumulation studies on an ice cap in Northern Ellesmere Island, N.W.T. In: *Snow and Ice, Proc. Moscow Symp. August 1971. IASH Publ.*, 104: 123—128.

Herrmann, A. and Stichler, W., 1976. Messungen des Gehalts an stabilen Isotopen in einer temperierten randalpinen Schneedecke. *Mitt. Geogr. Ges. München*, 61: 169—180.

Johnsen, S.J., 1977. Stable isotope homogenization of polar firn and ice. In: *Isotopes and Impurities in Snow and Ice, Proc. Grenoble Symp., August-September 1975. IASH Publ.*, 118: 210—219.

Johnsen, S.J., Dansgaard, W., Clausen, H.B. and Langway, C.C., Jr., 1972. Oxygen isotope profiles through the Antarctic and Greenland ice sheets. *Nature*, 235: 429—434.

Jouzel, J., Merlivat, L. and Pourchet, M., 1977. Deuterium, tritium and β-activity in a snow core taken on the summit of Mont Blanc (French Alps). Determination of accumulation rate. *J. Glaciol.*, 18: 465—470.

Judy, C., Meiman, J.R. and Friedman, I., 1970. Deuterium variations in an annual snowpack. *Water Resour. Res.*, 6: 125—129.

Krouse, H.R. and Smith, J.L., 1973. $^{18}O/^{16}O$ abundance variations in Sierra Nevada, seasonal snowpack and their use in hydrological research. In: *The Role of Snow and Ice in Hydrology, Proc. Banff Symp., September 1972*. UNESCO-WMO-IASH, pp. 24—38.

Kuhn, W. and Thürkauf, M., 1958. Isotopentrennung beim Gefrieren von Wasser und Diffusionskonstanten von D und ^{18}O im Eis. *Helv. Chim. Acta*, 41: 938—971.

Lorius, C., 1976. Antarctica: Survey of near surface mean isotope values. In: *Cambridge Workshop Monograph on Isotopic and Temperature Profiles in Ice Sheets*. Cambridge University Press, Cambridge.

Lorius, C. and Merlivat, L., 1975. Distribution of mean surface stable isotope values in East Antarctica; observed changes with depth in a coastal area. In: *Isotopes and Impurities in Snow and Ice, Proc. Grenoble Symp., August-September 1975. IASH Publ.*, 118: 127—137.

Lorius, C., Merlivat, L. and Hageman, R., 1969. Variation in the mean deuterium content of precipitation in Antarctica. *J. Geophys. Res.*, 74: 7027—7031.

Martinec, J., 1972. Tritium and Sauerstoff-18 bei Abflussuntersuchungen in repräsentativen Einzugsgebieten. *Gas Wasser, Abwasser*, 6: 2—8.

Martinec, J., 1974. Untersuchung der Schneeschmelze mit Umweltisotopen. *Österr. Wasserwirtsch.*, 26: 57—61.

Martinec, J., 1975a. Subsurface flow from snowmelt traced by tritium. *Water Resour. Res.*, 11: 496—498.

Martinec, J., 1975b. New methods in snowmelt-runoff studies in representative basins. In: *Hydrological Characteristics of River Basins and the Effect on these Characteristics of Better Water Management, Proc. Tokyo Symp., 1975. IASH Publ.*, 117: 99—107.

Martinec, J., Siegenthaler, U., Oeschger, H. and Tongiorgi, E., 1974. New insights into the run-off mechanism by environmental isotopes. In: *Isotope Techniques in Groundwater Hydrology, 1*. IAEA, Vienna, pp. 129—143.

Martinec, J., Moser, H., De Quervain, M.R., Rauert, W. and Stichler, W., 1977. Assessment of processes in the snowpack by parallel deuterium, tritium and oxygen-18 sampling. In: *Isotopes and Impurities in Snow and Ice, Proc. Grenoble Symp., August-September 1975. IASH Publ.*, 118: 220—231.

Meiman, J., Friedman, I. and Hardcastle, K., 1973. Deuterium as a tracer in snow hydrology. In: *The Role of Snow and Ice in Hydrology, Proc. Banff Symp., September 1972.* UNESCO-WHO-IASH, pp. 39—50.

Merlivat, L. and Nief, G., 1967. Fractionnement isotopique lors de changements d'état solide-vapeur et liquide-vapeur de l'eau à des températures inférieures à 0°C. *Tellus*, 19: 122—127.

Merlivat, L., Ravoire, J., Vergnaud, J.P. and Lorius, C., 1973. Tritium and deuterium content of the snow in Greenland. *Earth Planet. Sci. Lett.*, 19: 235—240.

Merlivat, L., Jouzel, J., Robert, J. and Lorius, C., 1977. Distribution of artificial tritium in firn samples from East Antarctica. In: *Isotopes and Impurities in Snow and Ice, Proc. Grenoble Symp., August-September 1975. IASH Publ.*, 118: 138—145.

Morgan, V.I., 1972. Oxygen isotope evidence for bottom freezing on the Amery Ice Shelf. *Nature*, 238: 393—394.

Moser, H. and Stichler, W., 1970. Deuterium measurements on snow samples from the Alps. In: *Isotope Hydrology*. IAEA, Vienna, pp. 43—57.

Moser, H. and Stichler, W., 1975. Deuterium and oxygen-18 contents as index of the properties of snow blankets. In: *Snow Mechanics, Proc. Grindelwald Symp., April 1974. IASH Publ.*, 114: 122—135.

Moser, H., Rauert, W., Stichler, W., Ambach, W. and Eisner, H., 1972. Messungen des Deuterium- und Tritiumgehaltes von Schnee-, Eis- und Schmelzwasser proben des Hintereisferners (Ötztaler Alpen). *Z. Gletscherkd. Glazialgeol.*, 8: 275—281.

Moser, H., Silva, C., Stichler, W. and Stowhas, L., 1975. Variation in the isotopic content of precipitation with altitude. *Final Rep., IAEA, Res. Contract*, No. 813/RB.

Oeschger, H., Stauffer, B., Bucher, P. and Moell, M., 1976. Instruments and methods. Extraction of trace components from large quantities of ice in bore holes. *J. Glaciol.*, 17: 117—127.

Oeschger, H., Stauffer, B., Bucher, P. and Loosli, H.H., 1977. Extraction of gases and dissolved and particulate matter from ice in deep boreholes. In: *Isotopes and Impurities in Snow and Ice, Proc. Grenoble Symp., August-September 1975. IASH Publ.*, 118: 307—311.

O'Neil, J.R., 1968. Hydrogen and oxygen isotope fractionation between ice and water. *J. Phys. Chem.*, 72: 3683—3684.

Picciotto, E., De Maere, X. and Friedman, I., 1960. Isotope composition and temperature of formation of Antarctic snows. *Nature*, 187: 857—859.

Picciotto, E., Deutsch, S. and Aldaz, L., 1966. The summer 1957—1958 at the South Pole: an example of an unusual meteorological event recorded by the oxygen isotope ratios in the firn. *Earth Planet. Sci. Lett.*, 1: 202—204.

Picciotto, E., Cameron, R., Crozaz, C., Deutsch, S. and Wilgain, S., 1968. Determination of the rate of snow accumulation at the Pole of Relative Inaccessibility, Eastern Antarctica: a comparison of glaciological and isotopic methods. *J. Glaciol.*, 7: 273—287.

Prantl, F.A. and Loijens, H.S., 1977. Nuclear techniques for glaciological studies in Canada. In: *Isotopes and Impurities in Snow and Ice, Proc. Grenoble Symp., August-September 1975. IASH Publ.*, 118: 237—241.

Raynaud, D. and Delmas, R., 1977. Composition des gaz contenus dans la glace polaire. In: *Isotopes and Impurities in Snow and Ice, Proc. Grenoble Symp., August-September 1975. IASH Publ.*, 118: 377—381.

Raynaud, D. and Lorius, R., 1977. Total gas content in polar ice; climatic and rheological implications. In: *Isotopes and Impurities in Snow and Ice, Proc. Grenoble Symp., August-September 1975. IASH Publ.*, 118: 326—335.

Reeh, N., Clausen, H.B., Gundestrup, N. and Johnsen, S.J., 1977. $\delta^{18}O$ and accumulation rate distribution in the DYE-3 area, South Greenland. In: *Isotopes and Impurities in Snow and Ice, Proc. Grenoble Symp., August-September 1975. IASH Publ.*, 118: 177—181.

Reeh, N., Clausen, H.B., Dansgaard, W., Gundestrup, N., Hammer, C.U. and Johnsen, S.J., 1978. Secular trends of accumulation rates at three Greenland stations. *J. Glaciol.*, 20: 27—30.

Reid, H.C., 1896. The mechanics of glaciers. *J. Geol.*, 4: 912—928.

Renaud, A., 1969. Etudes physiques et chimiques sur la glace de l'indlandsis du Groenland 1959. In: Section 3.1 by W. Dansgaard, L. Merlivat and E. Roth. *Medd. Grønl.*, 177: 62—76.

Robin, G. de Q., 1976. Reconciliation of temperature depth profiles in polar ice sheets with past surface temperatures deduced from oxygen-isotopes profiles. *J. Glaciol.*, 16: 9—22.

Siegenthaler, U., 1971. *Sauerstoff-18, Deuterium und Tritium im Wasserkreislauf, Beiträge zu Messtechnik, Modellrechnung und Anwendungen*. Dissertation, Universität Bern, Bern.

Stichler, W. and Herrmann, A., 1977. Variations of isotopes in snow covers as input of temperate glaciers. *Z. Gletscherkd. Glazialgeol.*, 13: 181—191.

Theodorsson, P., 1977. 40 years profiles in a polar and a temperate glacier. In: *Isotopes and Impurities in Snow and Ice, Proc. Grenoble Symp., August-September 1975. IASH Publ.*, 118: 393—398.

West, K.E. and Krouse, H.R., 1970. H_2O^{18}/H_2O^{16} variations in snow and ice of the St. Elias Mountain ranges (Alaska-Yukon-Border). In: *Recent developments in Mass Spectrometry, Proc. Int. Conf. Mass Spectroscopy, Kyoto (Tokyo)*, pp. 728—734.

Weston, R.E., Jr., 1955. Hydrogen isotope fractionation between ice and water. *Geochim. Cosmochim. Acta*, 8: 281—284.

Chapter 5

ISOTOPIC EVIDENCE ON ENVIRONMENTS OF GEOTHERMAL SYSTEMS

ALFRED H. TRUESDELL and JOHN R. HULSTON

INTRODUCTION

Isotope hydrology and the scientific study of geothermal systems have matured together. In early studies of hot-spring waters in Iceland, New Zealand, and Yellowstone and of fumarolic steam at Lassen, The Geysers, and Larderello, it was not possible to determine an origin for the water and steam. Some workers considered that both heat and water were derived from magmatic steam; others, noting seasonal changes in flow and temperature, suggested that at least part of the water was meteoric. This problem was largely resolved when Craig and his coworkers (Craig et al., 1956; Craig, 1963) showed that geothermal fluids are similar in deuterium content to local precipitation, which suggests that these fluids are dominated by meteoric water acquiring heat and solutes during deep circulation. The use by Craig (1953) of the carbon isotope fractionation between carbon dioxide and methane to indicate subsurface temperatures at Yellowstone, The Geysers, and Larderello preceded the application of chemical geothermometers to geothermal systems.

The distinctive magmatic origin of certain chemical components in geothermal fluids was disproved by Ellis and Mahon (1964), who showed that these components can be leached from volcanic rocks at reasonable geothermal temperatures and in amounts typical of geothermal fluids. Direct magmatic contributions to some systems have been inferred from the isotopic compositions of carbon and sulfur and from the presence of noble gases of mantle origin (Kononov and Polak, 1976), although this interpretation remains controversial.

This paper reviews the application of isotope chemistry to the study of geothermal systems, with the exception of tracer studies. We discuss at length the use of oxygen and hydrogen isotopes to indicate the sources of fluids and the nature of subsurface processes in geothermal systems, and the use of isotopic geothermometry to indicate subsurface temperatures. We review less thoroughly the application of natural radioactive isotopes in the dating of geothermal fluids, isotopic evidence for the origin of fluid constituents, and the use of isotopes in the petrologic study of geothermal systems. Specific chemical methods employed in the isotopic study of geothermal fluids are discussed in an appendix.

The application of isotope studies to geothermal systems has also been reviewed by Panichi and Gonfiantini (1978) and is the subject of a symposium volume edited by Gonfiantini (1977). The methods and background of isotope hydrology are well discussed by Bradley et al. (1972).

ISOTOPE HYDROLOGY OF GEOTHERMAL SYSTEMS

In early studies of geothermal systems (e.g., Allen and Day, 1935), it was assumed that the heat and much of the water entered as magmatic steam. This hypothesis could not be tested until the development of isotopic methods permitted comparison of the hydrogen- and oxygen-isotope composition of thermal waters with that of possible source waters.

Origin of thermal waters — meteoric and magmatic

Variations in the isotopic composition of surface waters and reasons for their occurrence are reasonably well known. Because ^{18}O and deuterium are fractionated at low temperatures into condensed phases, clouds originating from the evaporation of tropical seawater become progressively depleted in the heavier isotopes with precipitation during movement toward the poles, inland over landmasses, and to higher altitudes (see Chapter 1). By these processes the average precipitation at each point reaches an isotopic composition related to its latitude, altitude, and distance from the ocean. Fresh-water compositions were found by Craig (1961) to lie close to a "meteoric water line" described by δD (‰) * $= 8\, \delta^{18}O + \Delta$, where Δ is the "deuterium excess", which usually is equal to 10 but may be higher in certain coastal areas and lower in dry inland areas. (See Chapter 1 of this book for a discussion of variations in Δ.) The compositions of closed-basin lakes lie to the right of this line, along lines with $\delta D/\delta^{18}O$ slopes of about 5, reflecting non-equilibrium fractionation during evaporation. Thus deuterium isotope composition effectively "tags" rainwater, so that the recharge area of deeply circulating groundwater may be indicated by a comparison of its isotopic composition with that of shallow groundwater and directly collected precipitation from possible recharge areas.

---

* The permil (parts per thousand, ‰) notation is used throughout. Differences from isotopic standards are expressed as $\delta(‰) = [(R_x/R_s) - 1)] \times 10^3$ where $R_x = {}^{18}O/{}^{16}O$, D/H, etc., in the unknown and R_s = the same ratios in a standard. The standard for water is standard mean ocean water (SMOW); for carbon the Pee Dee belemnite (PDB); and for sulfur, the Canyon Diablo troilite (CDT).

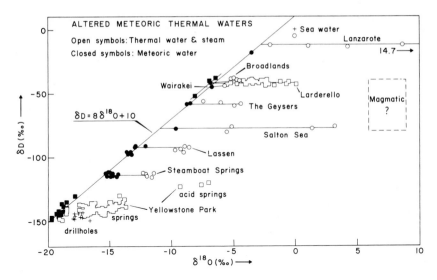

Fig. 5-1. ^{18}O and deuterium compositions of hot-spring, fumarole, and drill-hole fluids derived from meteoric water (open symbols) and of meteoric water local to each system (closed symbols). Data sources: Lanzarote — Arana and Panichi (1974); Broadlands — Giggenbach (1971); Wairakei — Stewart (1978); Larderello — Panichi et al. (1974); The Geysers — A.H. Truesdell (unpublished data, 1975); Salton Sea — Craig (1966); Lassen, Steamboat Springs — Craig (1963); Yellowstone — R.O. Rye and A.H. Truesdell (unpublished data, 1976).

The isotopic composition of magmatic water, considered in early studies to be a major component of thermal waters, is not well known and may even be variable. The ^{18}O content of this water must be higher than that of seawater as a result of high-temperature (minimum fractionation) exchange with silicates, which are relatively rich in ^{18}O ($\delta^{18}O > +5‰$). The deuterium content of magmatic water can be estimated from analysis of water in fluid inclusions, rapidly chilled submarine basaltic glass, volcanic emanations, and hydrous minerals of igneous rocks (reviewed by White, 1974, and Taylor, 1977). These studies suggest that the range in the deuterium content of magmatic water is about -40 to $-80‰$. The ranges of $\delta^{18}O = +6$ to $+9‰$ and of $\delta D = -40$ to $-80‰$ for magmatic water, shown in Fig. 5-1, are those proposed by White (1974).

Altered meteoric thermal waters

The isotopic composition of most deep geothermal waters (Figs. 5-1, 5-4, and 5-8; references in captions) is related to that of local meteoric water but does not in general lie on the meteoric water line. Hydrogen isotope compositions are similar in each individual case, but those for oxygen

isotopes are generally displaced towards higher ^{18}O content. If displacement from the meteoric water line resulted from a mixture of meteoric with magmatic waters, constant deuterium content would not be expected, and each mixing line should connect the composition of local meteoric water with that of magmatic water. This is possible in some areas (Larderello, The Geysers, Broadlands, Wairakei, El Tatio, and the Salton Sea), but clearly impossible in others (Lanzarote, Steamboat Springs, Yellowstone, and Lassen). As pointed out by Craig (1963), an observed constant deuterium content of meteoric and geothermal waters could be achieved by mixing with magmatic water only if the deuterium composition of magmatic water in each locality were equal to that of local meteoric water — an obvious absurdity. The correct explanation, first discussed by Craig et al. (1956), is that in geothermal systems oxygen isotopes are exchanged between hot rocks and deeply circulating meteoric water to produce an "oxygen isotope shift" to higher ^{18}O content in the water and a reverse shift to lower ^{18}O content in the rock. *
A comparable hydrogen isotope shift does not occur because rock minerals contain little hydrogen and because the water/rock ratios of geothermal systems are seldom so low that the hydrogen of rock minerals is a significant fraction of the total hydrogen. Combined oxygen- and hydrogen-isotope shifts have been described by Ohmoto and Rye (1974) and by Taylor (1977).

The magnitude of the oxygen isotope shift depends on the original δ^{18}O for both water and rock, the mineralogy and texture of the rock, the temperature, the water/rock ratio, and the time of contact. Waters from systems containing carbonate rocks (with original δ^{18}O values from +20 to +30‰), high temperatures, and low water/rock ratios (Lanzarote, the Salton Sea) exhibit large oxygen isotope shifts. Large oxygen isotope shifts from exchange with carbonate rocks can also occur at temperatures as low as 10°C in oilfield formation waters, where water/rock ratios are exceptionally low and contact times very long (Clayton et al., 1966). Systems with maximum temperatures below 150°C, moderate water/rock ratios and igneous rocks of an original δ^{18}O $\simeq$ +5‰ (Raft River, Idaho, and some Icelandic systems) may show little or no isotopic shift. Although the magnitude of oxygen isotope shift has been proposed as a qualitative geothermometer, it may prove misleading because surface isotopic composition can be affected by other processes (see below) and because systems with large water/rock ratios may show little apparent shift owing to extensive rock alteration. This is the case

---

* Some reverse shifts in water ^{18}O are possible. Interaction of high-^{18}O waters (seawater, connate water) with low-^{18}O rocks (igneous, not sedimentary) at moderate temperatures, at which ^{18}O fractionation exceeds the original isotopic difference between water and rock, will produce a reverse shift. Basalt-seawater interaction on spreading centers is a possible example.

at Wairakei, New Zealand, where the $\delta^{18}O$ value of reservoir silicates was +7.5 to +9.7‰ but has been altered to +2.6 to +5‰ (Clayton and Steiner, 1975) by near equilibrium with water of $\delta^{18}O$ = —5.7‰ (Stewart, 1978) at temperatures of 260—280°C. The resulting ^{18}O shift of Wairakei waters is small and was estimated to be zero by Craig (1963); restudy by Stewart (1978) showed it to be +1.1‰ (see Fig. 5-1). The apparent ^{18}O shift of Yellowstone thermal waters exceeds +5‰ (Fig. 5-1), but consideration of drill-hole water compositions indicates that the actual shift is less than +2‰ and that near-surface boiling has caused most of the apparent shift.

Extensive studies of ^{18}O and deuterium in minerals from igneous plutons by Taylor (1974, 1977) have shown that pervasive exchange with meteoric water occurs in many cases. Circulation of meteoric water by thermal convection through these plutons probably occurred during cooling after crystallization. These old igneous rocks exposed by erosion are undoubtedly fossil analogs of the recent intrusions that provide heat in modern high-temperature geothermal systems.

Altered oceanic thermal waters

A second large class of thermal waters consists of those originating from modern or ancient seawater that has been altered by reaction with rock minerals, by dilution with meteoric water, and in some cases by membrane filtration. The oceanic origin of these waters is indicated by their occurrence in marine sedimentary and volcanic rocks that do not contain evaporites, by their salinity near that of seawater, and in many cases by their coastal location. These saline waters have a high ^{18}O content from their seawater origin or from high-temperature reaction with high-^{18}O sedimentary rocks which has produced a positive oxygen isotope shift (Fig. 5-2; references in caption). Increasing degrees of oxygen isotope shift are exhibited by the thermal waters of Ibusuki and the green tuff formation in Japan, and of the Coast Range in California. The deuterium-chloride relations of these waters indicate only simple mixture of seawater with meteoric water (Fig. 5-3).

Although they issue from igneous rocks, the concentrated thermal brines of Arima, Japan (Figs. 5-2, 5-3), may be related to deep oilfield waters in which an extensive oxygen isotope shift, resulting from reaction with carbonate minerals, is accompanied by an increase in chloride and a decrease in deuterium that possibly are caused by membrane filtration, reaction with clay minerals, or exchange with meteoric waters (Clayton et al., 1966; Kharaka et al., 1973). The formation of these waters is poorly understood; they may be metamorphic waters, as defined by White (1957), with a hydrogen isotope shift due to an exceptionally low (<0.01?) water/rock ratio. Sakai and Matsubaya (1977) have shown that mixing of Arima brine with meteoric water occurs near the surface and can be traced by the tritium content.

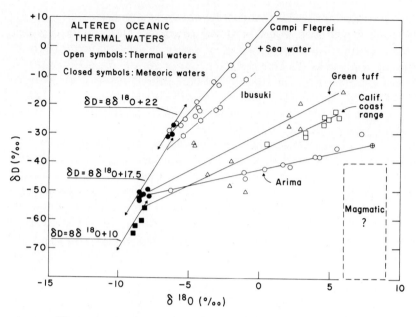

Fig. 5-2. ^{18}O and deuterium compositions of hot-spring waters derived from oceanic and connate water. Data sources: Campi Flegrei — Baldi et al. (1976); Ibusuki, green tuff, and Arima — Matsubaya et al. (1973) and Sakai and Matsubaya (1977); California Coast Range — White et al. (1973).

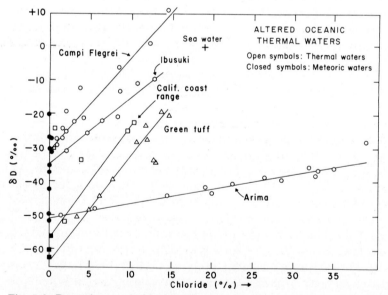

Fig. 5-3. Deuterium and chloride compositions of hot-spring waters derived from oceanic and connate water. Same data sources as in Fig. 2.

The coastal thermal waters of Campi Flegrei, Italy, are quite distinct isotopically (Figs. 5-2, 5-3) and appear to require a generative process that would enrich seawater in ^{18}O and deuterium while depleting it in chloride. The ^{18}O content of the most concentrated Campi Flegrei water may result from equilibrium with rocks at temperatures below 300°C, but the high deuterium content cannot be similarly explained, and it is possible that some of these waters results from non-equilibrium surface evaporation (see below) of seawater-rainwater mixtures.

Physical alteration of ascending thermal waters

Use of the isotopic composition of thermal waters to indicate their origin may be complicated by physical alteration occurring in the upflow zone of geothermal systems. In this zone, waters above surface boiling temperature must cool by conduction, by boiling and steam separation, or by mixing with cooler water. Only waters that cool by conduction retain their original isotopic composition. In drilled systems, waters may be collected before they boil and before they mix with near-surface cold water.

Analysis of steam and water from drill holes in hot-water systems (Broadlands, Wairakei, the Salton Sea, and Yellowstone in Fig. 5-1) makes possible a reconstruction of the isotopic composition of deep waters. In vapor-dominated systems, deep water is totally vaporized to steam, so that both fumaroles (Lanzarote, Lassen) and drill holes (Larderello, The Geysers) produce steam of the same isotopic composition as deep water.

The reconstruction of deep thermal water composition from that of surface thermal waters requires consideration of the effects of mixing with non-thermal waters, subsurface boiling and steam separation, and non-equilibrium surface evaporation.

Mixing with non-thermal water

Mixing with cooler water commonly occurs in the upper parts of geothermal systems. Cold water is denser than hot water and may be under sufficient pressure to invade shallow thermal aquifers. Mixture with cooler water may be demonstrated by larger variations in dissolved salt content than can be explained by steam loss, or by correlation between concentration changes in isotopes (^{18}O and deuterium) and in dissolved substances such as chloride that (in rocks without leachable salts) are not affected by near-surface rock reactions. Most thermal waters are higher in chloride than cold mixing waters (although exceptions occur), and the isotopic compositions of the two waters are likely to be quite different. Commonly the cold diluting water derives from precipitation falling close to the thermal area at a relatively low elevation and is isotopically distinct from the deep thermal water, which also is meteoric in origin but from higher elevations and greater distances from

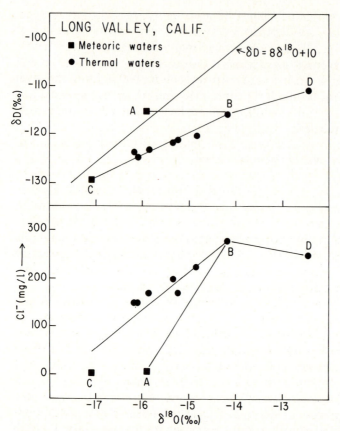

Fig. 5-4. ^{18}O, deuterium, and chloride compositions of hot-spring and shallow drill-hole waters from Long Valley, California; data from Mariner and Willey (1976). A and C are meteoric waters; B is possible unmixed thermal water; D is hot-spring water derived from B by non-equilibrium evaporation; other hot-spring waters (circles) appear to be mixtures of B and C. See text for further discussion.

the thermal area. In addition, the oxygen isotope shift of most thermal waters makes their ^{18}O content distinct from that of diluting waters. Recently studied geothermal systems with extensive mixing include: El Tatio, Chile (Cusicanqui et al., 1976; Giggenbach, 1978); Icelandic thermal areas (Arnason, 1977b); Long Valley, California (Mariner and Willey, 1976); Wieser and Bruneau-Grandview, Idaho (Rightmire et al., 1976); and Yellowstone Park, Wyoming (Truesdell et al., 1977a). Mixing relations are shown for Long Valley, Yellowstone, and El Tatio in Figs. 5-4, 5-6, and 5-8.

In the Long Valley, California, geothermal system, thermal waters are extensively mixed with cold meteoric water (Fig. 5-4, from Mariner and Willey, 1976). The thermal water probably originates as high-altitude precipitation on the Sierra Nevada with an isotopic composition similar to point A, then circulates as deep as 4 km, where it gains chloride and undergoes an oxygen isotope shift of about +1.5‰ to a composition near B (from a hot, shallow well). This water mixes with shallow, dilute, cold groundwater (similar to C) that originates as precipitation on mountains to the west in the rain shadow of the Sierras, and is relatively depleted in ^{18}O and deuterium, to produce the hot-spring waters. Processes affecting the isotopic and chemical composition of Long Valley waters must be more complicated than indicated in Fig. 5-4; although most hot springs issue at temperatures below boiling, some do boil, and so subsurface steam separation must have occurred. Only the composition of spring D shows clear evidence of steam heating and non-equilibrium evaporation; this may also have occurred in other springs. Finally, the correspondence in deuterium content between compositions A and B may be fortuitous, and it is probable that B is not the composition of end-member thermal water.

The absence of surface samples of undiluted thermal water is a general problem. If no unmixed thermal water reaches the surface, the subsurface temperature and the chemical and isotopic properties of the thermal water must be deduced from mixed samples. This can be done only (1) if a constituent such as tritium or ^{14}C can be assumed present in the cold component and absent in the hot component, (2) if this cold-water constituent is not removed after mixing by radioactive decay or by rock reactions, and (3) if only two homogeneous bodies of water are involved in the mixture. These assumptions were apparently successful at Hammat Gader, Israel (Mazor et al., 1973); at Manikaran, India (Gupta et al., 1976); and at Arima, Japan (Sakai and Matsubaya, 1977). At Yellowstone, although springs containing up to 800 ppm chloride occur, most mixed waters containing over 400 ppm chloride are free of ^{14}C and tritium, and so the end-member thermal water is not uniquely indicated (F.J. Pearson, unpublished data, 1977).

Effects of equilibrium liquid-vapor separation

High-temperature geothermal systems (such as at Yellowstone, The Geysers, Broadlands, El Tatio, and others) tend to be near boiling at all but the shallowest depths (White et al., 1971, 1975; Giggenbach, 1971, 1978). Upward movement of fluid before or during exploitation is accompanied by boiling and phase separation. Dissolved salts partition into the liquid phase and gases mainly into the vapor, according to their solubilities; isotopes also are strongly partitioned.

The light isotope of oxygen (^{16}O) is concentrated in the vapor at all temperatures below the critical point (374°C for pure water, higher if salts are present); the isotopic difference between vapor and liquid disappears at the critical point and increases with decreasing temperatures (Fig. 5-5). Below 221°C the light isotope of hydrogen (^{1}H) is increasingly concentrated in the vapor with decreasing temperature; between 221°C and the critical point it

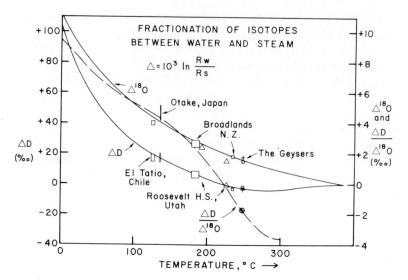

Fig. 5-5. Equilibrium fractionation of ^{18}O and deuterium between steam and water, and the ratios of deuterium and ^{18}O fractionations. Water-steam fractionations are given as values of $\Delta \equiv 1000 \ln[(1000 + \delta_w)/(1000 + \delta_s)]$, where δ_w and δ_s refer to $\delta^{18}O$ or δD of water and steam. Also shown are observed isotopic differences for steam and water from separators on geothermal wells and for steam samples collected at low and high flow rates from wells at The Geysers, California (see text).

is concentrated in the liquid, showing a maximum at 280°C (Bottinga, 1968a; Bottinga and Craig, 1968; other data reviewed in Truesdell et al., 1977a). Confirmation of experimental liquid-vapor fractionation factors and evidence for rapid isotopic equilibration between liquid and vapor is provided by analyses of water and steam from separators in geothermal fields (Fig. 5-5).

Because of these fractionations, water boiling during ascent from a hot aquifer will undergo changes in isotopic composition, the degree of change depending on the quantity of steam produced and the extent of fractionation between steam and water at the temperature of steam separation. Arnason (1977b) has calculated the effect on isotopes in liquid water of boiling when steam remains with the liquid and separates at a single temperature (single-stage separation). Single-stage steam separation results in changes in the isotopic composition of water that can be calculated from the equations:

$$\delta_{ws} - \delta_{wi} = (10^3 + \delta_{ws})(1 - \theta)$$

and:

$$\theta = (10^3 + \delta_{wi})/(10^3 + \delta_{ws}) = \eta + (1 - \eta)/\alpha$$

where δ_{ws} = $\delta^{18}O$ or δD of water at the separation temperature
δ_{wi} = $\delta^{18}O$ or δD of water at the initial temperature
η = liquid fraction = $(H_{vs} - H_{li})/(H_{vs} - H_{ls})$
H_{vs} = enthalpy of steam at the separation temperature
H_{ls} = enthalpy of water at the separation temperature
H_{li} = enthalpy of water at the initial temperature
α = R_l/R_v in equilibrium at the separation temperature
R = $^{18}O/^{16}O$ or D/H.

Truesdell et al. (1977a) have calculated the effects of continuous steam separation in which steam separates as soon as it is formed. Continuous steam separation results in changes in the isotopic composition of the waters expressed by the equations:

$$\delta_{ws} - \delta_{wi} = (10^3 + \delta_{ws})(1 - \theta)$$

and:

$$\theta = (10^3 + \delta_{wi})/(10^3 + \delta_{ws}) = \exp \int_{H_{ls}}^{H_{li}} (1 - 1/\alpha) \frac{dH_l}{H_l - H_v}$$

where symbols are as before. Most steam separation in geothermal systems is probably by multistage processes intermediate between the calculated models.

Yellowstone Park, Wyoming, provides extreme examples of boiling and mixing processes. Maximum subsurface water temperature is probably greater than 350°C, and, unless extensive mixture with cooler water occurs, more than 50% by weight of ascending deep thermal water must vaporize before reaching the surface. Fig. 5-6 shows surface and subsurface chloride and deuterium compositions of thermal waters from Norris Geyser basin, Yellowstone (Truesdell et al., 1977a).

At Norris, the chloride and deuterium compositions of most hot-spring waters can be explained by the subsurface mixture (Fig. 5-6, line C) of high-temperature thermal water (360°C, 310 ppm Cl⁻, and δD = −149‰) with local cold meteoric water, followed by boiling and multistage steam separation. Surface compositions of spring waters may therefore fall between those calculated for boiling of the mixed waters of line C to surface temperature with single-stage (line D) or continuous (line E) steam separation. Because little liquid-vapor deuterium fractionation occurs above 200°C, deuterium content is very sensitive to the mechanism of steam separation. Steam separating above 200°C has little effect on the deuterium composition of the remaining water (note on curve B the small difference between the deuterium content of 238°C drill-hole water and the calculated composition of 360°C

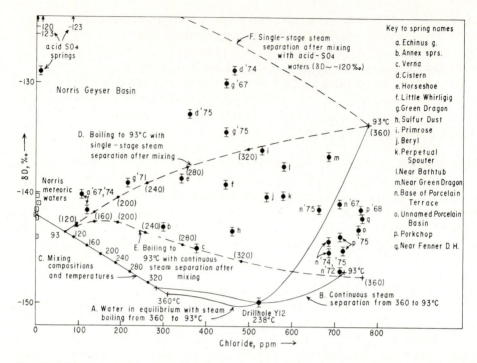

Fig. 5-6. Deuterium-chloride relations for meteoric and thermal waters in Norris Geyser Basin, Yellowstone National Park, Wyoming. Also shown are curves for: (A) water in equilibrium with steam boiling from 360° to 93°C, with water composition at 238°C equal to that from drill hole Y-12; (B) water boiling from 360° to 93°C, with continuous steam separation, and with water composition at 238°C equal to that from drill hole Y-12; (C) subsurface compositions and temperatures of mixtures of 360°C water (curve A) with typical Norris meteoric water at 5°C; (D) surface compositions of mixed waters of curve C after boiling to 93°C in equilibrium with steam (single-stage steam separation); (E) surface compositions of mixed waters of curve C after boiling to 93°C, with continuous steam separation; and (F) surface compositions of mixtures of 360°C thermal water with evaporated acid-sulfate waters after boiling to 93°C, with single-stage steam separation. Temperature before boiling of compositions along curves D and E indicated in parentheses. Estimated errors (±1 σ of 0.4‰) shown for each point. From Truesdell et al. (1977a).

deep thermal water), but steam formed at high temperatures that remains in contact with water until it separates at surface boiling temperatures produces a large deuterium enrichment of the remaining water (curve D). Variations in steam-separation mechanisms have caused certain hot-spring waters at Norris to vary widely in deuterium, with little change in chloride (springs $n-q$, Fig. 5-6). The wide ranges in chloride observed at Norris result from mixing of thermal water with low-Cl^- cold water. Because the diluting waters are higher in deuterium than the deep thermal water, both dilution and boiling increase the deuterium content of the hot-spring waters, and low-Cl^-, diluted thermal waters are similar in deuterium content to high-Cl⁻ waters that have lost large amounts of steam. Some hot-spring waters at Norris have been produced by non-equilibrium evaporation (acid-sulfate springs, Fig. 5-6) or are mixtures with these waters (springs a, d, and g).

The effects of boiling have not been considered in most isotopic studies of thermal systems; calculation of changes due to steam separation should permit more exact determination of the source of recharge and of the extent of oxygen isotope shift.

Steam and water equilibrium in vapor-dominated systems

Vapor-dominated geothermal systems differ from hot-water systems in that they have a limited permeability to liquid flow but a locally high permeability to vapor flow. Pressure reductions caused by drilling cause massive heat transfer from rock to immobilized liquid water, which vaporizes to form the steam produced from wells (Truesdell and White, 1973).

When wells in these systems are produced at low flow rates and high pressures, the isotopic composition of the steam differs from that of full-flow, low-pressure steam. This difference is close to that expected for liquid-water equilibrium at the temperature of the reservoir. Such an effect may be due to high-pressure steam originating as vapor and low-pressure steam originating as liquid, or may be due to partial condensation in the well bore. In either case the isotopic difference indicates the temperature of equilibration of steam and water in the reservoir or the well. Two examples are shown in Fig. 5-5; steam samples of low and high flows from two wells at The Geysers, California, showed in one case smaller (circles) and in the other case larger (crosses) isotopic differences than would be expected for equilibrium at the temperature of the reservoir. The differences in ^{18}O or deuterium taken separately would indicate discordant temperatures, but the ratios of the differences (circled cross in Fig. 5-5) agree exactly with the experimental data. Evidently, incomplete phase separation occurs in one well and multistage separation in the other.

^{18}O gradients, from more positive in the center to more negative at the margin, have been observed at Larderello (Panichi et al., 1974) and at The Geysers (Truesdell et al., 1977b). The isotopic gradient in the southeastern part of Larderello may be due in part to mixing of less oxygen-isotope-shifted recharge waters with more shifted deep waters, as indicated by the presence of tritium in the steam; however, at The Geysers and in the northeastern part of Larderello, where the steam is tritium free, Raleigh condensation of steam moving laterally from the center of the field toward the margin appears the more likely cause.

Non-equilibrium evaporation

Non-equilibrium surface evaporation of acid springs heated by the addition of steam causes large enrichments in both deuterium and ^{18}O but little increase in chloride unless some Cl^--containing water is also entering the spring. The increases in deuterium and ^{18}O were found by Craig (1963) to lie along lines with slopes near 3, rather than the equilibrium slope of 6 near 90°C (Fig. 5-5). Isotopic alteration of these steam-heated waters apparently occurs in part by exchange with atmospheric moisture, and the slope of the alteration line is influenced by the amount and composition of this moisture. Giggenbach (1978) found a $\delta D/\delta^{18}O$ slope of near 1.6 at El Tatio, Chile, where the air is exceptionally dry. These non-equilibrium evaporation lines may be used to calculate the isotopic composition of the deep thermal water; this is illustrated in Fig. 5-7 by unpublished data (A.H. Truesdell and N.L. Nehring, 1976) from Lassen Volcanic Park, California. At Lassen, isotopic compositions of fumaroles and acid hot springs within the park, and also of high-Cl^- alkaline boiling springs just south of the park boundary, reflect an oxygen isotope shift from local meteoric water (A). The isotopic composition of the major alkaline spring (Growler; Fig. 5-7, B) is apparently related to that of the highest temperature (113°C in 1975) fumarole in Bumpass Hell thermal area (Fig. 5-7, C) by equilibrium fractionation at 240°C, a temperature characteristic of vapor-dominated geothermal systems. (Open triangles indicate equilibrium steam compositions at 200° and 160°C.) When this steam mixes with local meteoric water (A), it produces (calcu-

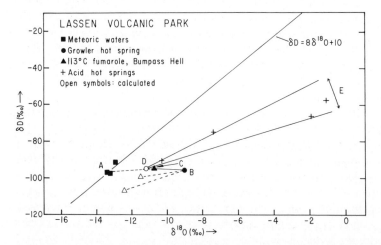

Fig. 5-7. ^{18}O and deuterium compositions of thermal waters and fumarolic steam in and near Lassen Volcanic National Park, California (A.H. Truesdell and N.L. Nehring, unpublished data, 1975). See text for explanation.

lated) boiling spring-water of composition (D), which evaporates along nonequilibrium lines (D-E) to produce acid-spring waters of the observed compositions (crosses). Fluids issuing from both the Bumpass fumarole and Growler spring appear to be cooled largely by conduction from 240°C to their surface temperatures because other cooling mechanisms (mixing, steam separation) would produce isotopic changes. The isolated position and limited flow of Growler spring indicate dominantly conductive cooling; cooling of Bumpass fumarole steam by this mechanism is less likely, but possible. If neither the fumarole nor the spring had been available for sampling, the composition of the deep water could have been approximated by taking the intersection of the acid-spring trend with the deuterium composition of meteoric water as the fumarole composition, and by assuming that the deep water was in equilibrium with fumarole steam at 240°C; this temperature is typical of vapor-dominated geothermal systems (James, 1968; White et al., 1971). The isotopic effects of non-equilibrium evaporation can also be seen in the composition of acid springs at Yellowstone, Wyoming, and The Geysers, California (Figs. 5-1 and 5-6; and other examples in Craig, 1963); at Hakone, Japan (Sakai and Matsubaya, 1977); and at El Tatio, Chile (Fig. 5-8, described in next section).

The isotopic hydrology of El Tatio, Chile

Most of the processes involved in the isotope hydrology of geothermal systems are illustrated by data from El Tatio, Chile. The chemical and isotopic composition of cold and hot waters of this system have been reported by Cusicanqui et al. (1976) and by Giggenbach (1978). El Tatio is situated at an altitude of 4250 m in the high Andes of Chile; over 200 hot springs exist, most at boiling temperatures (85°C at this altitude). Three types of hot-spring waters occur: (1) high-Cl^- spring waters with 4000—8000 ppm chloride; (2) moderate-Cl^-, high HCO_3^- spring waters; and (3) zero-Cl^-, high-SO_4^{2-} acid-spring waters. Drill holes of about 750 m depth discharge waters chemically similar to (1) but containing lower chloride. Each of these water types is distinguished by its isotopic composition, given in Fig. 5-8 (from Giggenbach, 1978).

Drill-hole waters lowest in deuterium and ^{18}O were considered to be samples of the deep thermal water (A). Giggenbach (1978) interprets the deep thermal water composition to have originated from oxygen isotope shift of meteoric water ($\delta D = -78\%_o$) infiltrating at high elevations east of the thermal area. He considers this water to boil with multistage steam separation, forming cooler waters enriched in chloride, deuterium and ^{18}O, and these waters then to mix with local meteoric water (B), forming the high- to moderate-Cl^- hot springs. An alternative interpretation of the data suggests that the mixing takes place before boiling and involves cold meteoric water of a composition similar to those analyzed (B'). The resulting mixed water iso-

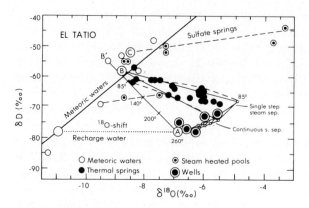

Fig. 5-8. ^{18}O and deuterium compositions of hot-spring and drill-hole thermal waters and meteoric waters of El Tatio, Chile, geothermal system. Data from Giggenbach (1978). See text for explanation.

tope compositions lie along line $(A\text{-}B')$ at the temperatures indicated, and boil after mixing to surface (85°C) compositions between the dotted lines. Cold meteoric water mixing in the subsurface may have a range in isotopic compositions similar to the range observed at the surface, which would account for the hot-spring compositions that lie outside the dotted lines. During mixing, the reaction of carbon dioxide dissolved in thermal water with rock minerals adds sodium and bicarbonate to produce the high-HCO_3^-, moderate-Cl^- waters.

Steam produced from boiling of the thermal waters will be correspondingly depleted in ^{18}O and D. Mixing of this steam with cold local meteoric water is considered to have produced an isotopic composition (C) before surface evaporation. Acid springs containing sulfate from the surface oxidation of hydrogen sulfide have an isotopic composition deriving from (C) by steady-state evaporation governed by the kinetic fractionation factor $\beta = \alpha \, (D/D')^{1/2}$, where α is the equilibrium vapor pressure ratio of H_2O to HDO or $H_2^{18}O$, and D/D' is the ratio of the corresponding diffusion coefficients. Some El Tatio meteoric water from higher altitude sources is much lower in deuterium and yields low-D sulfate springs (below B in Fig. 5-8).

The only possible meteoric recharge for the deep thermal water has a much lower deuterium content than any local meteoric water, and isotopic as well as geologic evidence suggest that recharge is from precipitation falling on higher mountains some 15 km to the east. During deep lateral circulation these waters are shifted by +4‰ in ^{18}O composition by high-temperature exchange with the country rock.

GEOTHERMOMETRY

Thermal fluids from springs, wells, and fumaroles may retain chemical or isotopic evidence of temperature conditions at depths below those of sampling. These temperature indicators or "geothermometers" represent the frozen-in equilibria of temperature-dependent reactions. In a single geothermal system several reservoirs may exist, differing in fluid residence time and generally increasing in temperature with depth. Deeper reservoirs are usually hotter but may not be reachable by drilling at reasonable cost. The presence of reservoirs at different temperatures can be indicated by geothermometer reactions that equilibrate at different rates. The rates of these reactions determine whether they will equilibrate in deep geothermal reservoirs and how rapidly they will reequilibrate in shallower reservoirs and during passage to the sampling point. Some isotopic reactions appear not to equilibrate in large geothermal reservoirs at temperatures of 200—300°C, and others equilibrate so rapidly that only the temperature of collection is indicated. Differing rates of reequilibration may often help in reconstructing the thermal history of a fluid (Hulston, 1977); at present, however, exact reconstruction is not possible owing to inadequate data on the rates of geothermometer

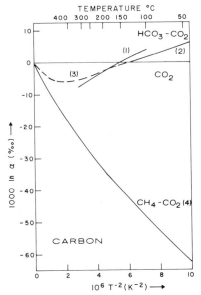

Fig. 5-9. Equilibrium fractionation of carbon isotopes between HCO_3^-, CO_2, and CH_4. Experimental data for CO_2-HCO_3^- from (1) Malinin et al. (1967) and (2) Mook et al. (1974), with proposed geothermometer curve (3) from Hulston (1978). The calculated curve for CO_2-CH_4 (4) is from Bottinga (1969).

TABLE 5-1

Geothermometer temperature ranges, equilibration rates and indicated reservoir temperatures * (in °C)

	Observed reservoir temperatures	Chalcedony saturation	Quartz saturation	Na-K	Na-K-Ca	Chemical mixing model	$\Delta^{18}O$ liq-vap
Useful temperature ranges and equilibration rates							
Useful temperature range (°C)		50—150	130—300	100—350	50—350	50—350	100—
Estimated $t_{1/2}$ at 250°C (years)		0.02	0.02	0.1	0.1	—	10⁻
Indicated temperatures in geothermal systems							
System:							
Seltjarnarnes, Iceland spring	119	115	140	68	109		
Raft River, Idaho shallow wells	147	110	137	90	139	145	
Ohtake, Japan wells	195		195—238	220	230		
Hveragerdi, Iceland springs, well G-8	220		174—212				
Wairakei, New Zealand well 44	248		248	255	259		
Broadlands, New Zealand springs well 8	273		203 to 278	311	218—302	270—306	
Salton Sea, Calif. well IID No. 2	300		230	354	308		
Yellowstone Park, Wyoming Upper Basin springs	—		to 210	~120	to 221	280	
Larderello, Italy steam wells	180—260						
The Geysers, California steam wells	240—250						241

* See Truesdell (1976) for more examples.

reactions. Extensive comparisons of temperatures indicated by isotopic and chemical geothermometers with those observed in drill holes were compiled by Truesdell (1976; see also Table 5-1).

Elements of low atomic mass that undergo changes in valence show large

	$\Delta^{18}O$ SO$_4$-H$_2$O	ΔD H$_2$-H$_2$O	ΔD H$_2$-CH$_4$	$\Delta^{13}C$ CO$_2$-CH$_4$	$\Delta^{13}C$ CO$_2$-HCO$_3^-$	$\Delta^{34}S$ H$_2$S-SO$_4$	References
-370	80—350+	50—300	50—300	300+	50—250	300+	
-5	1	10^{-3}	10^{-3}	$>10^3$	0.01	$>10^3$	
							Arnorsson (1974)
	142						McKenzie and Truesdell (1977)
	180—280						Sakai (1977)
		178—218					Arnason (1977b)
	303			360		373	Lyon and Hulston (1970), Lyon (1974a)
		275	265	385			Truesdell and Fournier (1976b), Lyon (1974a)
		220	255	380			H. Craig (personal communication, 1975), White (1968)
	260—314			244—380	~200	>350	Truesdell (1976)
-270	152—329			240—402			Panichi et al. (1977), Cortecci (1974)
	232			390			Craig (1953), Truesdell et al. (1977b)

isotopic fractionations in nature; of these the most thoroughly studied have been hydrogen, carbon, oxygen, and sulfur. Fractionation factors of geochemical interest for these elements have been compiled by Friedman and O'Neil (1977).

Carbon isotope geothermometers

Carbon isotopes were the first to be applied to the thermometry of geothermal fluids. Isotopic fractionations between methane, carbon dioxide, and dissolved bicarbonate ion are shown in Fig. 5-9; these factors were obtained by calculation from spectrographic data and by experiment, but require confirmation by more extensive experimental and field measurements. Hulston (1978) has suggested use of the dashed curve in Fig. 5-9 to reconcile discordant experimental HCO_3^-—CO_2 fractionation values.

Carbon dioxide—methane geothermometer

The ratios of $^{13}C/^{12}C$ in carbon dioxide and methane have been used as the basis of a geothermometer, assuming that the gases are in isotopic equilibrium through the exchange reaction:

$$CO_2 + 4\,H_2 \rightleftharpoons CH_4 + 2\,H_2O$$

Craig (1953) calculated the temperature dependence of ^{13}C isotope fractionation between carbon dioxide and methane and applied these calculations to isotopic measurements on gases from Yellowstone hot springs. The resulting temperatures were generally between 200 and 300°C, which he considered reasonable. Hulston and McCabe (1962a, b) in New Zealand measured $\Delta^{13}C(CO_2, CH_4)$ geothermometer temperatures of 250—440°C for volcanic fumarole gases, 235—260°C for Wairakei borehole waters and fumaroles, and 130—280°C for surface discharges of other thermal areas, using Craig's (1953) calibration curve. The Wairakei isotopic temperatures agreed with measured borehole temperatures. Similar agreement between indicated and measured temperatures of steam was found at Larderello, Italy, by G.C. Ferrara et al. (1963), who also used Craig's (1953) fractionation data. Bottinga (1969), however, recalculated the CO_2-CH_4 fractionation data and showed that Craig's calculations are in error. Geothermometric temperatures calculated using Bottinga's fractionation curve are from 75 to 100°C higher and do not agree with observed system temperatures. This has cast doubt on the significance of the indicated temperatures, and Gunter and Musgrave (1971) suggested that isotopic equilibrium is not established and that methane is formed by the thermal decomposition of organic matter, not by the reaction of hydrogen with carbon dioxide.

In a restudy of Larderello, Panichi et al. (1977) suggested that the relative temperature differences between wells are similar in both isotopic and physical measurements, and that isotopic temperatures are real but occur in deeper parts of the system. Similar suggestions have been made for Wairakei and Broadlands, New Zealand (Hulston, 1977). This hypothesis has been reinforced by the discovery of substantially higher temperatures in deep research drill holes at Larderello (310°C; R. Celati, oral communication,

1976) and at Broadlands (307°C; Macdonald, 1976), and by the agreement for New Zealand systems between carbon (CO_2, CH_4) and sulfur (H_2S, SO_4^{2-}) isotope temperatures. Confidence in carbon (CO_2, CH_4) isotope temperatures must await either experimental verification of the calculated frac-

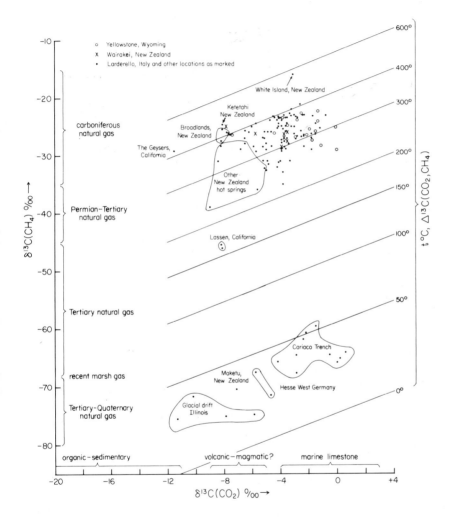

Fig. 5-10. Carbon isotope compositions of coexisting CH_4 and CO_2 from volcanic, geothermal, and sedimentary sources. Data sources: Lassen, The Geysers, Yellowstone — Craig (1953, 1963); additional Yellowstone — G.L. Lyon (written communication, 1971); New Zealand geothermal systems — Hulston and McCabe (1962a, b), Lyon and Hulston (1970), and Lyon (1974a); Larderello — Panichi et al. (1977); sedimentary gases — Wasserburg et al. (1963), Lyon (1974b), and Stahl (1974). Fractionation data from Bottinga (1969).

tionation data or corroboration of the indicated temperatures by deep drilling.

Agreement among several independent geothermometers is strong but not conclusive evidence. In studies of surface and near-surface thermal fluids at Yellowstone Park, Wyoming, temperatures of 244 to 390°C were indicated by $\Delta^{13}C(CO_2, CH_4)$ (recalculated from data of Craig, 1953, 1963; Gunter and Musgrave, 1971; and G.L. Lyon, written communication, 1971); with allowance for some reequilibration at lower temperatures, these values agree reasonably well with the temperatures indicated by enthalpy-chloride relations (340—360°C, Fournier et al., 1976; Truesdell and Fournier, 1976a) and by the sulfate-water geothermometer, using calculated corrections for steam separation and dilution (340—380°C; McKenzie and Truesdell, 1977). Deep drilling to confirm these indicated temperatures is unlikely at Yellowstone.

Available ^{13}C analyses of coexisting carbon dioxide and methane in geothermal gases, as well as some analyses of low-temperature gases with indicated isotopic temperatures, are shown in Fig. 5-10 (references in caption). Indicated and observed temperatures agree for very high temperatures (White Island, New Zealand) at which the reaction $CO_2 + 4\ H_2 = CH_4 + 2\ H_2O$ is rapid, and also at very low temperatures where bacteria are active and surface and source temperatures are nearly the same.

Carbon dioxide—dissolved bicarbonate geothermometer

Carbon dioxide dissolves in water to form aqueous CO_2, uncombined with water, and carbonic acid (H_2CO_3), in equilibrium with bicarbonate and carbonate ions (HCO_3^- and CO_3^{2-}). The equilibria between these species depend on temperature and pH and have been intensively studied. Carbon isotope fractionation between CO_2 and dissolved HCO_3^- has been studied experimentally at low temperatures by many workers, to 125°C by Mook et al. (1974), and to 286°C by Malinin et al. (1967). The high-temperature experimental data are not consistent and Hulston (1978) has suggested the dashed fractionation curve in Fig. 5-9 to reconcile the differences. In well-mixed systems the exchange of oxygen isotopes between aqueous CO_2 and H_2O is known to be rapid, and under these conditions the carbon isotope exchange between CO_2 and HCO_3^- is probably also rapid. In both equilibria the rate-controlling step is the exchange $CO_2(aq) = CO_2(g)$, and hence exchange time depends on the liquid-interface surface area available. Thus, if subsurface phase separation with separate conduits for liquid and vapor occurs, $CO_2(g)$-HCO_3^- exchange can be much slower.

This effect was demonstrated in the analyses of samples from Steamboat Springs, Nevada, and Yellowstone, Wyoming, which indicates that $CO_2(g)$ and $CO_2(aq)$ (mostly as HCO_3^-) in nearly neutral boiling springs are not in equilibrium at the spring temperature; furthermore, observed isotopic differences correspond to an equilibrium temperature of 160—210°C at Steamboat Springs and of 180—280°C at Yellowstone, using the fractionation data of Malinin et al. (1967). Extrapolation of the data of Mook et al. (1974) gives indicated temperatures of 140—190°C for Steamboat Springs and of 160—280°C for Yellowstone. Both sets of fractionation data indicate temperatures in reasonable agreement with those indicated by silica contents for near-surface reservoirs in which CO_2 reacts with rock minerals to form HCO_3^-. Contrary to expectations, carbon isotope equilibrium between CO_2 and HCO_3^- established during this reaction appears to freeze in and resist further exchange as fluids rise to the surface.

Hydrogen isotope geothermometers

At least five hydrogen-containing gases (H_2O, H_2S, H_2, CH_4, and NH_3) occur in major amounts in geothermal fluids, and four independent isotopic geothermometers based on hydrogen-deuterium fractionation appear possible. Fractionation factors for major hydrogen-containing gases have been calculated by Bottinga (1969) and by Richet et al. (1977); those for H_2-H_2S

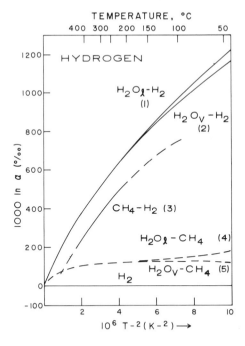

Fig. 5-11. Equilibrium fractionation of hydrogen isotopes between $H_2O(g)$, H_2, and CH_4. Data from (*1, 2*) Bottinga (1968b), and Richet et al. (1977), (*3*) Harmon Craig (oral communication, 1975), and (*4, 5*) Hulston (1978).

and H_2-H_2O agree well with the limited available experimental data. Agreement with experimental data for H_2-CH_4 is not as good (H. Craig, oral communication, 1975), and experimental data for ammonia is lacking. The D/H ratios for hydrogen sulfide and ammonia have not been tested as possible geothermometers, but limited measurements of D/H ratios for coexisting hydrogen, water, and methane in geothermal fluids have been made; the calculated and experimental deuterium fractionation factors for these gases are shown in Fig. 5-11.

For geothermometers involving water, isotopic fractionation due to boiling must be considered and the composition of water at the point of equilibration measured or calculated. In hot-water systems, the isotopic compositions and quantities of both steam and water from a separator can be measured and the composition of hot water feeding the drill hole calculated. If the fluid contains excess steam, the composition and quantity of this steam must be included in the calculations. Where only steam or water is sampled, the composition of reservoir fluid can be calculated on the basis of a model for the mechanism of steam separation, using the measured geothermal aquifer temperature.

Hydrogen-water geothermometer

Arnason (1977a, b) has made extensive measurements of the deuterium content of hydrogen and water from drill holes and hot springs of Iceland. He assumed that hydrogen equilibrates with liquid water and found excellent agreement with silica geothermometric temperatures and with measured downhole temperatures in most areas. At Reykjanes, however, the isotopic temperature is 70—90°C higher than other temperatures, and this was interpreted as due to contributions from a deeper, higher temperature reservoir. Using the equation discussed earlier, Arnason (1977a, b) corrected for changes in deuterium content due to single-stage steam separation during ascent to the surface.

Application of the hydrogen-water geothermometer to other systems appears to show variable degrees of reequilibration between the reservoir and the sampling point. Indicated temperatures include 115 and 122°C for Yellowstone fumaroles (Gunter and Musgrave, 1971), 265°C for Broadlands wells (Lyon, 1974a), and 140 and 220°C for Salton Sea wells (H. Craig, unpublished data, 1975). In each case the indicated temperature is similar to the collection temperature and, for Salton Sea and Yellowstone, 120—240°C lower than the measured or estimated aquifer temperature.

Methane-hydrogen geothermometer

Although isotopic exchange between methane and hydrogen in the region 200—500°C was achieved in the laboratory with the use of a catalyst (H. Craig, oral communication, 1975), in geothermal systems H_2-H_2O isotopic exchange is probably faster than H_2-CH_4. Fig. 5-11 shows that the equilibrium constants for these reactions are similar so it is not surprising that similar temperatures have been obtained: 70°C for Yellowstone fumaroles, 125 and 255°C for Salton Sea wells, and 275°C for Broadlands wells.

Isotopic exchange between methane and water is currently being explored as the basis for a geothermometer. However, no experimental measurements have been made, and the theoretical calculations contain uncertainties (Richet et al., 1977). The dashed curve in Fig. 5-11 shows an interim temperature scale suggested by Hulston (1978), which was selected from several possible theoretical scales as the one appearing to give temperatures between those of CH_4-CO_2 and H_2-H_2O equilibria for a number of unpublished New Zealand data (J.R. Hulston and G.L. Lyon). On this scale the Yellowstone Park data of Gunter and Musgrave (1971) give temperatures of 250—450°C (CH_4-CO_2 temperature, 380°C).

The remaining hydrogen-containing species, hydrogen sulfide and ammonia, are partially ionized in solution and would be expected to equilibrate too rapidly with water to be useful geothermometers, except possibly in systems containing only superheated steam between the reservoir and the surface.

Oxygen isotope geothermometers

As geothermometers, oxygen- and hydrogen-isotope systems have the basic disadvantage that a large reservoir of these elements is contained in

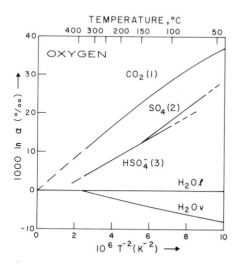

Fig. 5-12. Equilibrium fractionation of oxygen isotopes between CO_2, SO_4^{2-}, HSO_4^-, $H_2O(l)$ and $H_2O(g)$. Data from (*1*) Bottinga (1968a, b) and Bottinga and Craig (1968), (*2*) Lloyd (1968), and (*3*) Mizutani and Rafter (1969) and Mizutani (1972).

water itself, and boiling during ascent produces isotopic fractionation. ^{18}O is more highly fractionated than deuterium by steam separation occurring above 200°C, and is affected by reaction with rocks containing abundant oxygen but little hydrogen. In spite of these difficulties, several oxygen isotope geothermometers show great promise. Experimental fractionation data for oxygen isotope geothermometric systems are shown in Fig. 5-12.

Sulfate-water geothermometer

The best behaved isotope geothermometer for hot-water systems appears to be the fractionation of oxygen isotopes between sulfate (and bisulfate) ions and water. This exchange reaction is both sufficiently rapid to be at equilibrium in all geothermal reservoirs of economically significant size and temperature, and sufficiently slow to retain indication of temperatures above 300°C in certain hot-spring waters. Apparently reliable experimental data are available for oxygen isotope exchange between $HSO_4^- $-$ SO_4^{2-}$ and H_2O from 100 to 200°C and at 348°C (Lloyd, 1968; Mizutani and Rafter, 1969; Mizutani, 1972). In these experiments the acidity of the solution, and therefore the ratio of HSO_4^- to SO_4^{2-}, was varied. Most geothermal waters above 200°C are below or near pH 7 and contain more than 300 ppm Na; under these conditions almost all sulfate ion exists as HSO_4^- and $NaSO_4^-$, which probably exhibit similar isotopic behaviour.

Lloyd (1968) showed that the rate of exchange is strongly dependent on pH, with

$$\log t_{1/2} = 2500/T + b ,$$

where $t_{1/2}$ is the half-time of exchange in hours, T is the absolute temperature, and b is 0.28 at pH 9, —1.17 at pH 7, and —2.07 at pH 3.8. Most deep geothermal waters have pH values from 6 to 7, and the time for 90% exchange at 250°C is about 1 year, a short time compared to tritium ages (see below) of more than 30—50 years on reservoir fluids. Isotopic equilibrium between sulfate and water has been demonstrated for borehole waters from Wairakei, New Zealand (~270°C; Mizutani and Rafter, 1969; Kusakabe, 1974), Otake, Japan (~220°C; Mizutani, 1972), Ohnuma, Japan (220—260°C; Sakai, 1977), Raft River (~145°C; McKenzie and Truesdell, 1977) and Bruneau-Grandview, Idaho (~100°C; A.H. Truesdell, unpublished data) and for wet-steam wells at Larderello, Italy (~240°C; Cortecci, 1974). Because of their unusually rapid ascent to the surface, some Yellowstone thermal waters indicate a sulfate-water geothermometric temperature of about 360°C without having reequilibrated during ascent.

Application of the sulfate-water geothermometer requires measurement or calculation of the oxygen isotope composition of water at the point of equilibrium, as discussed above. If dilution is also involved, temperatures are more difficult to calculate, and mixing models (Fournier and Truesdell,

1974; Truesdell and Fournier, 1976b) must also be applied. Both mixing and steam-separation models were applied to sulfate-water geothermometric calculations by McKenzie and Truesdell (1977).

Carbon dioxide—water geothermometer

Oxygen isotope exchange between carbon dioxide and water vapor has been applied as a geothermometer with apparent success to indicate reservoir temperatures at Larderello, Italy (Panichi et al., 1977). This exchange probably takes place through the activated complex HCO_3^-, which exists only in the presence of liquid water; therefore, exchange would only occur in the reservoir, not in the rising steam column. Temperatures indicated by CO_2-$H_2O(g)$ fractionation data are equal to or higher than wellhead steam temperatures. The use of fractionation data for CO_2-$H_2O(g)$ may, however, be incorrect because oxygen isotope exchange occurs between carbon dioxide and liquid water in the reservoir, not between carbon dioxide and water vapor in the steam. The use of fractionation data for CO_2-$H_2O(l)$ eliminates much of the apparent difference between indicated and wellhead temperatures.

Water-steam geothermometer

Geothermometers that employ the fractionation of both hydrogen and oxygen isotopes between liquid water and steam are useful in those uncommon geothermal systems where both steam and water may be sampled separately and the isotopic differences compared with experimental data (Fig. 5-5; Bottinga and Craig, 1968; Bottinga, 1968a). Even in such systems, incomplete or multistage separation of steam and water may occur, as described above; thus the differences in ^{18}O or deuterium content of the two phases may be smaller or greater than the single-stage enrichment factors (Δ in Fig. 5-5), and divergent temperatures will be indicated by the two isotopes. However, provided these processes take place at a single temperature, the ratio of deuterium- to oxygen-enrichment factors ($\Delta D/\Delta^{18}O$ in Fig. 5-5) may indicate the correct temperature.

The use of this ratio is indicated in Fig. 5-5 for steam and water collected from a Weber cyclone separator at Roosevelt Hot Springs, Utah, and for reservoir steam at The Geysers, California, collected at different flow rates. At Roosevelt the differences in ^{18}O content between steam and water separated at 226°C indicated an isotopic temperature of 255°C, but ΔD and $\Delta D/\Delta^{18}O$ indicated 220 and 224°C. This disparity in isotopic temperatures suggested the presence of steam in the water line. At The Geysers, ΔD or $\Delta^{18}O$ alone indicated temperatures of 240 and 260°C, but the temperatures from $\Delta D/\Delta^{18}O$ (circled cross in Fig. 5-5) was 247°C, in exact agreement with observed well-bottom temperatures (Truesdell et al., 1977b). By similar arguments Giggenbach (1971) showed that Wairakei fumaroles produce steam that has separated from water at 114—122°C.

Although The Geysers is an extreme case in which only steam is produced because reservoir water is immobile and vaporizes completely in response to production-induced pressure decreases, this geothermometer could also be applied to other systems of limited permeability, such as at Broadlands, in which boiling in the reservoir occurs and bore-fluid enthalpy increases with flow rate.

Sulfur isotope geothermometers

Sulfur has multiple valence states and exhibits large isotopic fractionations, particularly between oxidized and reduced species. Calculated and experimentally measured fractionation factors between SO_2, H_2S, HS^-, HSO_4^-, and SO_4^{2-} are shown in Fig. 5-13 (references in caption). Sulfur dioxide has not been detected in deep geothermal fluids, although it occurs in volcanic gases. In laboratory studies the exchange between H_2S or HS^- or HSO_4^- and SO_4^{2-} has been found to be very fast under acid conditions and extremely slow under alkaline conditions. Thus the application of the sulfur isotope geo-

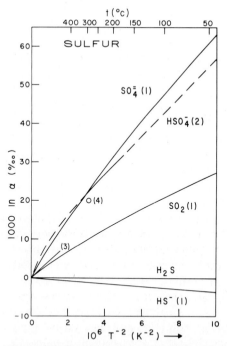

Fig. 5-13. Fractionation of sulfur isotopes between SO_4^{2-}, HSO_4^-, SO_2, H_2S, and HS^-. Data from (*1*) Sakai (1957, calculation), (*2*) Robinson (1973, experiment), and (*3*) Thode et al. (1971, experiment). Single experimental point (*4*) for SO_4^{2-}-H_2S is from Sakai and Dickson (1978).

thermometer to geothermal systems requires that an estimate of the pH of deep fluid be made before the equilibrium time can be calculated. In large reservoirs at 250—270°C such as Wairakei, New Zealand, where the deep pH is ~6, these sulfur species have remained well mixed without reestablishing equilibrium for times that must exceed hundreds of years. This behavior is similar to that of CO_2-CH_4 mixtures, and thus these equilibria are useful in indicating the temperature of a deeper fluid that is feeding the system.

Apparent chemical and isotopic disequilibrium at 260°C between H_2S and SO_4^{2-} (probably mostly as HSO_4^- in the reservoir) has been carefully documented at Wairakei by Rafter and Wilson (1963), Wilson (1966), and Kusakabe (1974). Each of these investigations showed that the Wairakei reservoir fluid contains approximately equal molal concentrations (0.3 mmol/kg) of H_2S ($\delta^{34}S$ = 4 ± 1‰) and of SO_4^{2-} ($\delta^{34}S$ = 23 ± 1‰). Using the calculated SO_4^{2-}-H_2S experimental factors of Robinson (1973), the indicated temperature is about 350°C, whereas the observed reservoir temperature is 260—270°C. In addition, equal molal concentrations of H_2S to SO_4^{2-} could only be in equilibrium under more oxidizing conditions than those indicated by gas and mineral equilibria. Giggenbach (1977) has suggested that the apparent chemical disequilibrium discussed by Kusakabe (1974) can be explained by the existence of sulfate ion pairs. If the 300°C experimental calibration of $NaSO_4^-$-H_2S fractionation by Sakai and Dickson (1978) is more applicable than the HSO_4^--H_2S calibration of Robinson (1973), then the indicated Wairakei temperature would be 315°C. Calculations based on these experiments and on those of Robinson (1978) suggest that the half-time of isotopic equilibrium for Wairakei conditions (temperature, pH, and sulfur concentration) should be about 30—600 years. It is tentatively suggested that the Wairakei fluid has equilibrated at about 350°C in a deeper reservoir and undergone partial reequilibration in the exploited reservoir, where it has remained for 50—500 years.

ISOTOPIC DATING OF GEOTHERMAL WATERS

Deuterium and ^{18}O analyses have shown that the water in most geothermal systems originates as local precipitation and circulates deeply to acquire heat. This process, though simple in principle, is undoubtedly very complex in detail; meteoric water descending to greater or lesser depths must mix with water of higher temperature and salinity ascending from deeper aquifers; only a part of the water circulates to maximum depth. Some evidence that these processes occur is given by the various temperatures indicated by the geothermometers of differing equilibration rates described above, and by studies of ore deposits where succeeding fluids differ in temperature and composition. A direct measure of the duration of circulation in aquifers at various levels would be very useful in determining the energy contained in a geothermal system; however, except for a few systems, established radioisotope clocks have been of little use.

Tritium in geothermal borehole fluids

Tritium is a radioactive isotope of hydrogen with a half-life of 12.3 years that is produced in the atmosphere by cosmic ray interactions. Its natural abundance is about 5—10 atoms of tritium per 10^{18} atoms of hydrogen (5—10 tritium units (TU)), but testing of thermonuclear bombs during 1954—1965 caused large increases of tritium to more than 1000 TU in the Northern and 50 TU in the Southern Hemisphere, so that tritium has been very useful in dating subsurface waters whose circulation times are less than ~20 years.

The circulation time of most deep geothermal waters is, however, much longer and their measured tritium content is usually less than the modern detection limit of about 0.1 TU. This level of tritium indicates only that the waters have been out of contact with the atmosphere for 60 years or more. Extensive measurements of tritium contained in borehole waters from Wairakei, New Zealand, were reported by Banwell (1963), Taylor et al. (1963), and Wilson (1963). The tritium content measured was uniformly low (0.04—0.4 TU for one collection) but varied with the collector and time of analysis. If near-surface tritium-containing waters were drawn down into the reservoir, the chloride and tritium content and possibly also the enthalpy of the reservoir fluid should be affected. However, no simple relation of tritium content to enthalpy or chloride content could be found by Wilson (1963), who suggested a radioactive source in the reservoir rock. Taylor et al. (1963) suggested that the experimental values represent the limits of the analytical method, and this seems the most reasonable explanation.

The tritium content of steam samples from The Geysers, California, collected away from areas of water injection has been analyzed by several laboratories and found to be near the analytical background. Steam from older wells in the central part of The Geysers contained 0.7—1.4 TU (C.B. Taylor, written communication to D.E. White, 1972), and steam from newer wells of the southeastern Geysers, 0.1—0.3 ± 0.1 TU (F.J. Pearson, Jr., written communication, 1977). Contamination may have occurred during collection or analysis (particularly in earlier analyses made during the time of high atmospheric tritium levels), and these values are close to the analytical background. In addition, tritium-containing water may be infiltrating the wells (especially older ones) from incompletely cased-off, near surface water entries; and so we must conclude that at The Geysers, as at Wairakei, there is little real evidence that the geothermal aquifer contains any water younger than 60 years.

Recent studies at Larderello, Italy, present a different picture. Unlike The Geysers, Larderello has been shown to have areas of infiltration south and east of the field in which permeable parts of the reservoir rock crop out. Wells close to these infiltration areas are usually unproductive because of flooding by surface water. Wells at somewhat greater distances from these areas produce steam that has been found to contain up to 40 TU, whereas steam from wells in the center of the field contains less than 0.5 TU (Celati et al., 1973). The pattern of tritium occurrence suggests that lateral recharge is taking place, but tritium contributions from entries above the reservoir should be considered because the wells containing tritium are among the oldest and shallowest in the field.

Tritium in hot springs

Since little tritium is contained in deep thermal waters, tritium found in hot springs may indicate the mixing of deep and shallow waters. This is of interest because warm springs of large flow can originate from mixing of hot and cold waters, or from conductive heating during deep circulation with little loss of heat during ascent. If only conductive heating is involved, drilling will produce water only slightly hotter than that issuing at the surface, and the thermal system has poor potential for electric power generation. If, on the other hand, the temperature of the spring results from mixing, drilling may produce high-temperature fluids of good potential for power production.

Tritium has been used in investigations of warm springs of the eastern U.S. by Hobba et al. (1978), and of warm springs of the Dead Sea Rift in Israel by Mazor et al. (1977). In both cases the highest temperature waters (40°C in Virginia, 60°C in Israel) contain no tritium, which suggests that mixing of deep high-temperature water with cold near-surface water is not likely. In both cases, temperatures indicated by chemical geothermometers are close to orifice temperatures and not consistent with subsurface mixing. It was concluded in both these investigations that drilling would produce waters only slightly hotter than those issuing at the surface.

Although reasonable for these studies, this interpretation must be applied cautiously because mixing may occur at deep levels where both hot and cold waters are old and lack tritium. At Yellowstone, Wyoming, extensive subsurface mixing of hot and cold waters is indicated by the chemical and isotopic composition of hot-spring waters (Truesdell and Fournier, 1976; Truesdell et al., 1977a), but in general only the more dilute waters contain tritium (Begemann and Libby, 1957; Pearson and Truesdell, 1978).

^{14}C in geothermal fluids

^{14}C has a much longer half-life (5800 years) than tritium and could be much more useful in dating geothermal fluids were it not for the extreme dilution of introduced radiocarbon by old, ^{14}C-free carbon dioxide in the fluid and carbonate contained in the reservoir. In these systems, small concentrations of carbon dioxide in recharge waters are diluted by much larger quantities of radiocarbon-free carbon dioxide introduced from thermal metamorphism of limestone (or possibly from magma). More important, much of this metamorphic carbon dioxide is deposited in the reservoir as hydrothermal calcite, which is available for isotopic exchange with ^{14}C-containing carbon dioxide. Isotopic exchange between calcite and dissolved carbon dioxide occurs rapidly at temperatures above 200°C, and the introduced ^{14}C mixes with this large reservoir of old carbon as well.

In view of these processes it is not surprising that deep geothermal fluids have been found to be very low in ^{14}C. Borehole fluids from The Geysers, Steamboat Springs, and Salton Sea were reported by Craig (1963) to contain less than 0.5 percent modern carbon (pmc); carbon dioxide associated with the high-Cl^- springs of Yellowstone contains less than 0.2 pmc (F.J. Pearson, Jr., written communication, 1976) and water from a deep Wairakei borehole contains 0.5 ± 0.1 pmc which was suggested to indicate that it might be as old as 40,000 years (Fergusson and Knox, 1959). This age must be considered as an upper limit because the processes described above would deplete ^{14}C and increase the apparent age.

^{14}C measurements of surface thermal fluids are, like tritium measurements, useful in demonstrating dilution. As a result of dilution, low-Cl^- Yellowstone spring waters contain much larger quantities of ^{14}C than do high-Cl^- waters (F.J. Pearson, Jr., written communication, 1976). This difference probably reflects not only ^{14}C in the young, cold diluting water but also less removal of ^{14}C by dilution and exchange in the lower temperature, near-surface aquifers in which dilution takes place.

In warm-spring systems at Hammat Gader, Israel (Mazor et al., 1973), where no entry of metamorphic carbon dioxide and less rapid isotopic exchange occur, ^{14}C has been used to demonstrate the mixing of an old, high-temperature component (0 pmc, ~70°C) with a young, cold component; and, at the Arkansas hot springs, to calculate the age (~5000 years) of waters heated (to 62°C) by deep circulation (Pearson et al., 1973).

Future prospects for dating geothermal fluids

The application of more exotic radioisotopes to geothermal systems may be justified by the great economic importance of being able to estimate geothermal energy reserves. In particular it may be worthwhile to investigate cosmic-ray-produced ^{39}Ar, which has a half-life of 269 years and is not affected by reactions other than boiling in hydrologic systems (Oeschger et al., 1974). The input function of ^{39}Ar is, however, poorly understood, and its collection, separation, and analysis are very difficult and expensive. Even if the age of a geothermal water could be determined, it would be difficult to relate this age to the volume of the reservoir because it is unlikely that these reservoirs correspond to any simple hydrologic model. Nevertheless, even imperfect estimates may be of value for certain reservoirs.

ORIGIN OF CHEMICAL CONSTITUENTS

Although it has been shown by isotopic analysis that all or nearly all the water of geothermal fluids is of meteoric origin, a possible magmatic or juvenile origin for salts and gases dissolved in the fluids remains controversial. Elements once considered diagnostic of magmatic origin (Cl, B, Li) have been shown to be leachable in significant quantities from volcanic and sedimentary reservoir rocks of geothermal systems (Ellis and Mahon, 1964). Since volcanic rocks are considered present (at depth if not at the surface) in all high-temperature systems and must themselves be of magmatic origin, the question becomes whether these rocks constitute an intermediate reservoir for the dissolved constituents of geothermal waters, or whether the waters receive these constituents directly.

Extensive comparisons of the isotopic compositions of constituents of geothermal fluids and of the content and ratios of dissolved inert gases with those of possible sources have been made in efforts to solve this problem. These comparisons are seldom entirely diagnostic for isotopes because low temperature surface fractionation processes produce wide ranges in isotopic compositions including possible magmatic values, and because isotopic fractionation during crystallization and leaching is so very slight that it is probably impossible to distinguish components leached from volcanic rocks from those contributed directly from magma.

Carbon isotope origins in geothermal fluids

The isotopic compositions of carbon in gaseous and dissolved carbon dioxide from Steamboat Springs, Yellowstone, The Geysers, and Larderello were reported by Craig (1963), and from the Salton Sea by Lang (1959). Measurements of ^{13}C contents of carbon dioxide and methane in Larderello steam were reported by G.C. Ferrara et al. (1963) and by Panichi et al. (1977); the origin of carbon dioxide in Italian geothermal systems in relation to its isotopic composition was discussed by Panichi and Tongiorgi (1976). The carbon isotope compositions of carbon dioxide and methane from Wairakei and other New Zealand geothermal areas were reported by Hulston and McCabe (1962b) and Lyon (1974a). Where $\delta^{13}C$ for both carbon dioxide and methane were reported, these results are shown in Fig. 5-10. The isotopic compositions of carbon dioxide from Yellowstone ($\delta^{13}C = -1$ to $-5‰$), Monte Amiata (-1 to $-2‰$), Larderello (-1 to $-5‰$), the Salton Sea ($-3.5‰$), and Wairakei (-3.0 to $-4.5‰$) fall within the range of $\delta^{13}C$ for marine limestone ($+3$ to $-4‰$; Craig, 1953), which possibly indicates a carbon dioxide origin from either magmatic assimilation or high-temperature leaching of limestone.

Other systems have lighter carbon isotopes. Craig (1963) found δ^{13}C-(CO_2) values of −8.5 to −9.5‰ at Steamboat Springs, −8 to −9‰ at Lassen, and −11 to −12‰ at The Geysers. Lyon (1974a) reported δ^{13}C(CO_2) of about −8‰ for Broadlands and −7 to −8.5‰ for other New Zealand geothermal areas. Although these values are in the range calculated for juvenile or deep-seated carbon and found in carbonatites and carbon-bearing basalt (Craig, 1953), they could as well indicate marine organic carbon (−7 to −17‰; Craig, 1963) or a mixture of limestone with terrestrial organic carbon (to −30‰).

The isotopic composition of methane is probably the result of high-temperature equilibration with carbon dioxide, which is the major carbon-containing gas in geothermal systems. However, despite evidence of chemical equilibrium of CO_2, CH_4, H_2, and H_2O at Wairakei and Larderello (Hulston and McCabe, 1962a, b; Truesdell and Nehring, 1978), the apparent isotope disequilibrium of these gases and the relatively restricted range in δ^{13}C from −22 to −35‰ for most methane samples (Fig. 5-10) allows the possibility that in some systems these gases have separate origins (Gunter and Musgrave, 1971).

Other isotopic origines

Other isotopes in geothermal systems have not received as much attention. Doe et al. (1966) found that lead and strontium in the Salton Sea geothermal brine is leached from sedimentary reservoir rocks. Extensive measurements of δ^{34}S on H_2S and SO_4^{2-} of Wairakei have been reported by Rafter and Wilson (1963), Wilson (1966), and Kusakabe (1974). Wairakei borehole waters contain equal molal amounts of H_2S and SO_4^{2-}, with δ^{34}S values of +4 and +23 ± 1‰ respectively. Steiner and Rafter (1966) suggested that the dissolved SO_4^{2-} has a seawater origin, whereas the H_2S comes from upper crustal magma. Kusakabe (1974) suggested that source sulfur in the Wairakei geothermal system originates from sulfate in greywacke of Jurassic age, a period when seawater sulfate is considered to have had a δ^{34}S value of +17‰. However, recent sulfur isotopic analyses by Giggenbach (1977) of greywackes of the Taupo volcanic zone, New Zealand, indicated average δ^{34}S values of +6.5‰, which does not support this proposal. Giggenbach favored a sulfur origin from magmatic sulfur dioxide, whose δ^{34}S, after addition of sedimentary sulfur from rocks, shifts from ~0 to +4 to +5‰ and which disproportionates on contact with surface water into H_2S and SO_4^{2-}. The δ^{34}S composition of the total sulfur in the Wairakei system is close to this value and the small amount of sulfur in the fluids is isotopically heavier (~+15‰) due to continued deposition in the geothermal system of sulfide of lighter isotopic composition.

Inert gases in geothermal fluids

Non-reactive gases dissolved in geothermal fluids provide tracers for solute contributions from the atmosphere, from the radioactive decay of rock components, and possibly from mantle outgassing.

The common isotopes of neon, argon, krypton, xenon, and (in most cases) nitrogen dissolved in thermal waters originate almost entirely from the atmospheric gas content of infiltrating recharge waters (Mazor and Wasserberg, 1965; Mazor, 1976). In thermal waters that have not boiled in the subsurface (i.e., that have behaved as a closed system to inert gases), the contents of these gases depend only on the temperature of equilibration of the water with the atmosphere at the point of recharge. If boiling has occurred, the remaining water is depleted in these gases, with the pattern of depletion depending on gas solubilities and on the temperature and extent of boiling. High-temperature solubilities of these gases have been measured by Potter and Clynne (1978) and may be useful in determining the temperature and mechanism of subsurface boiling (Potter et al., 1977).

The mixture of completely degassed deep thermal water with air-saturated cold water may produce waters having atmospheric noble gas ratios that apparently indicate closed-system behavior but with greatly reduced gas conconcentrations. This mixed-water pattern is found in many of the Yellowstone samples reported by Mazor and Fournier (1973). Yellowstone gases separated from water at high temperatures retain the noble gas pattern of the original water except for neon enrichment (Mazor, 1976) probably because above 250°C neon is significantly less soluble in water than the other atmospheric noble gases.

Certain isotopes of noble gases (^{4}He, ^{40}Ar, ^{136}Xe, and ^{222}Rn) are produced by radioactive decay of uranium, thorium, and ^{40}K in rocks). These isotopes are indicators of the content of radioactive elements in a geothermal system and of the water/rock ratio; the ratio of radiogenic to atmospheric argon may indicate residence time within the system. The average content of radiogenic argon in Larderello steam analyzed by G. Ferrara et al. (1963) decreased from 23 to 12% from 1951 to 1963, which suggests that atmospheric argon is introduced with increased recharge due to pressure drawdown. No excess radiogenic argon was found in thermal fluids from New Zealand (Hulston and McCabe, 1962a, b), or in most Yellowstone thermal waters (Mazor and Fournier, 1973).

Because of its short half-life of 3.8 days, ^{222}Rn is being tested as a tracer in geothermal studies of the reservoirs of vapor-dominated systems (Stoker and Kruger, 1976). Radium, from which radon is produced by radioactive decay, is not soluble in steam, and so radon concentration must be related to the areas of rock and water surface in contact with steam during the last days before it reaches the surface. Kruger et al. (1977) found that, when the flow of a Geysers well was decreased from 90 to 45 tons/hour, the radon concentration decreased from an average of 16 to 8 nCi/kg. This may indicate that during passage to the well the steam flows through large fractures having a low surface/volume ratio for times similar to the half-life of radon. Studies at Larderello, Italy (D'Amore, 1975), have shown that the ratio of radon to total dry gas is greatest at the margins of the field. These relatively large radon/gas ratios may indicate heavily produced zones containing steam boiling from gas-depleted water into large, dried volumes where radon emanates from the rock surfaces. Radon in vapor-dominated geothermal reservoirs is a promising tool but requires further study.

Radon originates from ^{238}U by way of ^{226}Ra, which is chemically similar to calcium and may be expected to coprecipitate with calcite in places where carbon dioxide is lost at the surface or in zones of subsurface boiling. Because of this, radon in hot-water systems is associated with silicic rocks of relatively high uranium content and with travertine-depositing springs. No systematic data are available on the distribution of radon in hot-water reservoirs or on its concentration in borehole fluids as a function of flow.

^{3}He is not formed by the radioactive disintegration of rock material (although relatively small amounts are formed by tritium decay) and occurs in minor amounts relative to ^{4}He in the atmosphere (^{3}He/^{4}He ~ 1.4×10^{-6}). ^{3}He is found in greater relative abundance in geothermal systems and in ocean waters near spreading centers, where it is suggested to originate from the outgassing of primordial mantle. Craig (H. Craig, oral communication, 1975; Craig and Lupton, 1976) has found ^{3}He/^{4}He ratios at Lassen, the Salton Sea, and Kilauea that range from 3 to 16 times that of the atmosphere. Russian workers have found high ^{3}He/^{4}He ratios for geothermal fluids of Iceland (Kononov and Polak, 1976) and of Kamchatka, U.S.S.R. (Gutsalo, 1976). The occurrence of anomalous ^{3}He/^{4}He ratios in the volcanic rocks of Iceland and East Africa (Mamryn et al., 1974) and in oceanic basalt (Craig and Lupton, 1976) suggests that the observed ^{3}He/^{4}He ratios of geothermal fluids could as well result from rock leaching as by direct contributions from the mantle.

SOLID PHASE STUDIES

This review has emphasized isotope studies of geothermal fluids because these fluids are often available before drilling and thus are useful in exploration, and because the isotopic composition of fluids can provide information about deep processes taking place below drilled depths in modern systems. The isotopes contained in reservoir rock minerals provide complementary information about the shallower parts of modern geothermal systems and in some cases permit the history of a system to be reconstructed.

Although the deeper parts of ancient geothermal systems in which ore mineralization has occurred have been the subject of many isotope studies (reviewed in White, 1974), investigations of minerals in cores and cuttings from modern geothermal aquifers are fewer. Clayton et al. (1968) showed that although quartz does not equilibrate, calcite from a Salton Sea (Imperial Valley, California) geothermal well is in equilibrium with modern reservoir water from 150 to 340°C and suggested that the water/rock volume ratio of this system is about 1 : 1. Kendall (1976), who studied cuttings from other wells of the same field, found that isotopic equilibrium exists only in zones of high permeability. Quartz-water isotopic equilibrium was found only at temperatures above 290°C, in zones where high permeability permits fluid access; calcite equilibrates above 100°C in permeable zones. Coplen (1976), in a study of the lower temperature Mesa field (also in the Imperial Valley) demonstrated wide variations in the isotopic equilibration of water with calcite, little equilibration in shale and poorly sorted sandstones, and greater equilibration in the most permeable, well-sorted sandstone. The present reservoir water in the system consists of evaporated Colorado River water, but calcite veins in the system were deposited from an isotopically different water. Like the high-temperature Salton Sea system, the water/rock volume ratio of the Mesa field is 1 : 1, which implies that flushing has occurred at least five times.

Volcanic aquifer rocks of the Taupo volcanic zone, New Zealand, were originally more reactive chemically than clastic rocks of the Imperial Valley, California. In these systems isotopic equilibration of fluid with quartz and layer silicates at temperatures above 230°C was found at Broadlands (Eslinger and Savin, 1973), and with quartz at 250°C at Wairakei (Clayton and Steiner, 1975). From this work the water/rock ratio at Wairakei was calculated to be 4.3 : 1, but this ratio has been revised to 2 : 1 by Stewart (1978) from new isotopic analyses of Wairakei thermal waters. These water/rock ratios can only apply to the rock volume drilled, but alteration in geothermal systems should extent to maximum depth of water circulation. If the amount of isotopic alteration of these deep rocks differs from that of rock in the drilled reservoir, the calculated water/rock ratio will be significantly affected.

Oxygen isotope exchange of minerals appears to be largely related to mineral reactions that result in recrystallization. Existing crystals of quartz and most other rock-forming minerals may preserve their original $\delta^{18}O$ values for long periods of time. Blattner (1975) has emphasised that different generations of crystals provide information on different stages of development of geothermal systems. Giletti et al. (1978) have made calculations of the times requires for diffusional feldspar-water oxygen isotope exchange which suggest that this process may be significant at geothermal temperatures only for grains smaller than 100 μm. A large change in the $\delta^{18}O$ of a subsurface water may be caused by only a small fraction of reservoir rock reacting or recrystallising.

Isotope studies of hydrothermal minerals may indicate complex histories for geothermal systems. In a comprehensive study of cuttings from The Geysers, California, Lambert (1976) found evidence for an early (Jurassic or Cretaceous), 170—320°C geothermal system involving ocean water trapped in sediment with little convective flow, followed much later (Pliocene-Holocene) by a second geothermal system containing large quantities of convecting meteoric water flowing along fractures. This second hot-water geothermal system underwent heating from 170 to 320°C before cooling to 250°C, when the present vapor-dominated geothermal system was formed. This complex history is in marked contrast to that of the hot-water system at Broadlands, where fluid inclusions supply no evidence of changes in salinity or temperature over the lifetime of the system (Browne et al., 1976).

SUMMARY

Isotopes provide a key to our modern understanding of geothermal systems. Although questions about the source of some salts and gases and about deep connection between the magmatic heat source and convecting geothermal water remain unanswered, much progress has been made. Water in geothermal systems has been shown to be of local meteoric or oceanic origin, and no detectable contribution from magmatic water has been found. Dissolved carbon and sulfur show isotopic evidence of possible recycling in surface processes, but the data also permit a magmatic origin. Most dissolved noble gases derive from the atmosphere or from radioactive decay, but some may represent mantle outgassing either directly or by intermediate incorporation into volcanic rocks.

The hydrology of geothermal waters can be best studied isotopically because processes occurring in these systems produce large isotopic changes. In fact, some mechanisms of subsurface and surface boiling can only be followed through isotopic changes because chemical changes are not sufficiently diagnostic. More than half the apparent oxygen isotope shifts of some high-temperature thermal waters may result from boiling and steam separation. Isotopes can also indicate the source and amount of diluting waters that commonly mix with ascending thermal waters in the near surface. However, radioactive isotopes have found little application in the hydrologic study of thermal systems except at Larderello, Italy, where rapid recharge is occurring in response to exploitation.

Subsurface temperatures at various levels in a geothermal system may be indicated by chemical and isotopic geothermometers that equilibrate at different rates. The isotopic fractionations of oxygen, carbon and sulfur suggest that temperatures exceeding 350°C exist below drilled depths within several geothermal systems.

Finally, although little discussed here, the isotope study of minerals in geothermal reservoirs may indicate the history of these systems: how much water they have contained, where this water came from, and what temperatures prevailed in the past. Studies of exhumed fossil geothermal systems, many of which contain ore minerals, may help us to understand the deep interactions of magma with hot water.

ACKNOWLEDGEMENTS

We wish to thank our colleagues at the U.S. Geological Survey and at the Institute of Nuclear Sciences and Chemistry Division of the N.Z. Department of Scientific and Industrial Research for their stimulating discussions of isotope chemistry of geothermal systems and for the use of their unpublished data. In particular, we wish to thank Donald White, Philip Bethke and Robert Rye of the U.S.G.S. and Brian Robinson, Michael Stewart, Graeme Lyon and Werner Giggenbach of the D.S.I.R. for reviews of the manuscript. We also wish to thank George Havach and Polly Bennett, U.S.G.S., for careful and perceptive editing and preparation of the manuscript.

APPENDIX — METHODS OF COLLECTION AND ANALYSIS

^2H and ^{18}O in thermal waters

It is important in sampling hot springs to avoid surface influences (evaporation, oxidation) by choosing springs with large flows issuing from small orifices, and by collecting samples as close to the vent as possible. Collection from a hot-water geothermal well is best accomplished with a production cyclone separator or a portable minicyclone separator if the well is flowing, or with a downhole sampler if it is shut in (Ellis et al., 1968; Klyen, 1973; Nehring and Truesdell, 1978). Because of fractionation during phase separation, conditions of separation (pressure, temperature, flow rate) and enthalpy of total discharge should be known exactly, and cooling should be adequate to prevent loss of water vapor.

^{14}C and ^3H in thermal waters

There is very little tritium or ^{14}C in thermal waters, and careful collection is required to avoid contamination from the atmosphere. During the period of relatively high atmospheric tritium in 1970, it was apparently impossible to collect uncontaminated samples of Yellowstone hot-spring waters, but this problem did not recur in 1975 (F.J. Pearson, Jr., unpublished data, 1976). ^{14}C in water may be collected as $SrCO_3$ by treatment of a large volume of the water with $SrCl_2$ and carbonate-free NaOH (Pearson and Bodden, 1975). However, it has been found that ^{14}C may enter spring water by atmospheric exchange during convective circulation in pools, while gas collected from the same spring was found to be ^{14}C free (F.J. Pearson, Jr., unpublished data, 1976). The use of $CdCl_2$ as a flocculent permits the simultaneous collection as CdS of trace amounts of dissolved hydrogen sulfide for isotope study.

^{13}C in dissolved CO_2

Geothermal waters from springs and wells are highly supersaturated in CO_2 relative to the atmosphere and should be preserved for chemical and carbon isotope analysis by immediately treating the water with concentrated ammonia saturated with $SrCl_2$, or by collection in a gas-tight bottle for subsequent gas evolution. With high-SO_4^{2-} waters, $SrSO_4$ is precipitated with $SrCO_3$, and CO_2 yield must be measured to determine total CO_2 content. Problems of analytical reproducibility with $SrCO_3$ may be due to incomplete homogenization before portions are taken for analysis.

^{34}S and ^{18}O in dissolved SO_4^{2-}

If sulfate is abundant and hydrogen sulfide is absent, no special collection procedures are necessary; however, some thermal waters contain low concentrations of sulfate and high concentrations of hydrogen sulfide that may be oxidized to sulfate by sulfur-oxidizing bacteria in spring pools or in bottles after collection. Sulfate may be extracted by field ion exchange, samples may be preserved from bacterial action with formaldehyde, or the hydrogen sulfide may be fixed with Zn^{2+} or Cd^{2+}. Surface- and near-surface-produced sulfate may be avoided in part by selecting springs with minimum SO_4^{2-}/Cl^- ratios. Carbon dioxide for mass spectrometric analysis is usually produced from dissolved sulfate by carbon reduction of precipitated $BaSO_4$ at ~1000°C (Rafter, 1967; Mizutani, 1971; Sakai and Krouse, 1971). Thermal waters usually require ion exchange to concentrate and purify the sulfate before precipitation (Nehring et al., 1977), but high-silica bore waters present problems. Sulfur dioxide for sulfur isotopic analysis of dissolved sulfate is produced by oxidation with CuO of AgS precipitated from a solution of BaS remaining after carbon reduction of $BaSO_4$ (Rafter, 1957), or by heating a mixture of $BaSO_4$ and silica glass to a temperature near 1800°C and converting any released SO_3 to SO_2 over hot Cu (Bailey and Smith, 1972).

^{34}S in dissolved H_2S

The collection of dissolved hydrogen sulfide has proved difficult. Sweeping with nitrogen into $AgNO_3$ solution has worked erratically (Rafter and Wilson, 1963; Kusakabe, 1974). Precipitation with Zn^{2+} and filtration, or direct precipitation with $CdCl_2$ from large volumes of water in connection with separation of dissolved carbon dioxide, has been somewhat more successful, but recovery of less than 1 ppm of hydrogen sulfide remains unreliable (F.J. Pearson, Jr., R.O. Rye, and the authors, unpublished data, 1977). The precipitated ZnS or CdS may be treated with acid to release hydrogen sulfide gas for absorption in $AgNO_3$ solution, or it may be oxidized with H_2O_2 or Br to SO_4^{2-} for precipitation as $BaSO_4$.

^2H and ^{18}O in $H_2O(g)$ and ^{18}O in CO_2

Steam from whatever source must be completely condensed for isotopic analysis. Collection of the steam (and gas) in a cooled, evacuated bottle, or the use of a water- or air-cooled condenser with the exit maintained at temperatures below 30°C, is necessary. Steam of very high CO_2 content may undergo low-temperature exchange of oxygen isotopes unless this is prevented by direct freezing from the vapor phase. Similarly, collection of carbon dioxide in steam for ^{18}O analysis requires immediate freezing to prevent isotope exchange (Panichi et al., 1977).

^{13}C and ^2H in CO_2, CH_4, and H_2

Carbon dioxide, hydrogen sulfide, and residual gases are conveniently collected in an evacuated bottle containing NaOH solution. This procedure permits large quantities of gas to be transported in small containers, prevents the oxidation of H_2S, and assists the later separation of carbon dioxide and methane for isotopic analysis. Carbon dioxide can be precipitated with $SrCl_2$ after lowering the pH to 10 to prevent precipitation of $Sr(OH)_2$. Hydrogen sulfide is best oxidized to SO_4^{2-} with H_2O_2 and precipitated as $BaSO_4$. If hydrogen is not present, methane in the residual gas can be oxidized to CO_2 and H_2O over CuO at 800°C. Methane and hydrogen can be separated in a molecular sieve column at −78°C and separately oxidized and trapped out of the carrier gas. Alternately, the hydrogen can be oxidized with CuO at 400°C, followed by oxidation of methane at 800°C (G.L. Lyon, written communication, 1971).

^{14}C in CO_2 and ^{34}S in H_2S

Ammonium hydroxide obtained free of carbon dioxide effectively absorbs CO_2 and H_2S from geothermal gases. The CO_2 and H_2S may be evolved with acid and the H_2S separated by bubbling through CdOAc solution (F.J. Pearson, Jr., written communication, 1976). ^{14}C collection by this method is preferred to collection from hot-spring water because evolved gases are seldom contaminated with atmospheric carbon dioxide. Volcanic gases contain other sulfur compounds, such as SO_2 and S, that may be separated by special methods for isotope study (Giggenbach, 1976).

Noble gases

Radon collection and analysis in geothermal fluids have been described by D'Amore (1975) and Stoker and Kruger (1976). Collection and analysis of other noble gases was described by Craig and Lupton (1976) and by Mazor (1976).

REFERENCES

Allen, E.T. and Day, A.L., 1935. Hot springs of the Yellowstone National Park. *Carnegie Inst. Washington Publ.*, 466: 525 pp.

Araña, V. and Panichi, C., 1974. Isotopic composition of steam samples from Lanzarote, Canary Islands. *Geothermics*, 3: 142—145.

Arnason, B., 1977a. The hydrogen and water isotope thermometer applied to geothermal areas in Iceland. *Geothermics*, 5: 75—80.

Arnason, B., 1977b. Hydrothermal systems in Iceland traced by deuterium. *Geothermics*, 5: 125—151.

Arnórsson, S., 1974. The composition of thermal fluids in Iceland and geological features related to the thermal activity. In: L. Kristjansson (Editor), *Geodynamics of Iceland and the North Atlantic Area*. Reidel, Dordrecht, pp. 307—323.

Bailey, S.A. and Smith, J.W., 1972. Improved method for the preparation of sulfur dioxide from barium sulfate for isotope ratio studies. *Anal. Chem.*, 44: 1542—1543.

Baldi, P., Ferrara, G.C. and Panichi, C., 1976. Geothermal research in western Campania (southern Italy): Chemical and isotopic studies of thermal fluids in the Campi Flegrei. In: *Proceedings, Second United Nations Symposium on the Development and Use of Geothermal Resources, San Francisco, Calif., 1975*, U.S. Government Printing Office, Washington, D.C., pp. 687—697.

Banwell, C.J., 1963. Oxygen and hydrogen isotopes in New Zealand thermal areas. In:

E. Tongiorgi (Editor), *Nuclear Geology on Geothermal Areas, Spoleto, 1963*. Consiglio Nazionale delle Ricerche, Laboratorio di Geologia Nucleare, Pisa, pp. 95—138.

Begemann, F. and Libby, W.F., 1957. Continental water balance, ground water inventory and storage items, surface ocean mixing rates and world-wide water circulation patterns from cosmic-ray and bomb tritium. *Geochim. Cosmochim. Acta*, 12: 277—296.

Blattner, P., 1975. Oxygen isotopic composition of fissure-grown quartz, adularia and calcite from Broadlands geothermal field, New Zealand. *Am. J. Science*, 275: 785—800.

Bottinga, Y., 1968a. Isotope Fractionation in the System Calcite-Graphite-Carbon Dioxide-Methane-Hydrogen-Water. Ph.D. Thesis, University of California, San Diego, Calif., 126 pp. (unpublished).

Bottinga, Y., 1968b. Calculation of fractionation factors for carbon and oxygen isotopic exchange in the system calcite-carbon dioxide-water. *J. Phys. Chem.*, 72: 800—808.

Bottinga, Y., 1969. Calculated fractionation factors for carbon and hydrogen isotope exchange in the system calcite-carbon dioxide-graphite-methane-hydrogen-water vapor. *Geochim. Cosmochim. Acta*, 33: 49—64.

Bottinga, Y. and Craig, H., 1968. High temperature liquid-vapor fractionation factors for $H_2O-HDO-H_2{}^{18}O$. *Trans. Am. Geophys. Union*, 49: 356—357 (abstract).

Bradley, E., Brown, R.M., Gonfiantini, R., Payne, B.R., Przewlocki, K., Sauzay, G., Yen, C.K. and Yurtsever, Y., 1972. Nuclear techniques in ground-water hydrology. In: *Ground-water Studies, an International Guide for Research and Practice*. UNESCO, Paris, Chapter 10: 38 pp.

Browne, P.R.L., Roedder, E. and Wodzicki, A., 1976. Comparison of past and present geothermal waters from a study of fluid inclusions. In: J. Cadek and T. Paces (Editors), *Proceedings, 1st International Symposium on Water-Rock Interactions, Prague, Czechoslovakia, 1974*. Geological Survey, Praha, pp. 140—149.

Celati, R., Noto, P., Panichi, C., Squarci, P. and Taffi, L., 1973. Interactions between the steam reservoir and surrounding aquifers in the Larderello Geothermal Field. *Geothermics*, 2: 174—185.

Clayton, R.N. and Steiner, A., 1975. Oxygen isotope studies of the geothermal system at Wairakei, New Zealand. *Geochim. Cosmochim. Acta*, 39: 1179—1186.

Clayton, R.N., Friedman, I., Graff, D.L., Mayeda, T.K., Meents, W.F. and Shimp, N.F., 1966. The origin of saline formation waters, 1. Isotopic composition. *J. Geophys. Res.*, 71: 3869—3882.

Clayton, R.N., Muffler, L.J.P. and White, D.E., 1968. Oxygen isotope study of calcite and silicates of the River Ranch No. 1 well, Salton Sea geothermal field, California. *Am. J. Sci.*, 266: 968—979.

Coplen, T.B., 1976. Cooperative geochemical resource assessment of the Mesa geothermal system. *Univ. Calif., Riverside, Rep.*, IGPP-UCR-76-1: 97 pp.

Cortecci, G., 1974. Oxygen isotopic ratios of sulfate ions-water pairs as a possible geothermometer. *Geothermics*, 3: 60—64.

Craig, H., 1953. The geochemistry of the stable carbon isotopes. *Geochim. Cosmochim. Acta*, 3: 53—92.

Craig, H., 1961. Isotopic variations in meteoric waters. *Science*, 133: 1702.

Craig, H., 1963. The isotopic geochemistry of water and carbon in geothermal areas. In: E. Tongiorgi (Editor), *Nuclear Geology on Geothermal Areas, Spoleto, 1963*. Consiglio Nazionale delle Ricerche, Laboratorio di Geologia Nucleare, Pisa, pp. 17—53.

Craig, H., 1966. Isotopic composition and origin of the Red Sea and Salton Sea brines. *Science*, 154: 1544—1547.

Craig, H. and Lupton, J.E., 1976. Primordial neon, helium, and hydrogen in oceanic basalts. *Earth Planet. Sci. Lett.*, 31: 369—385.

Craig, H., Boato, G. and White, D.E., 1956. Isotopic geochemistry of thermal waters. *Natl. Acad. Sci. Natl. Res. Counc., Publ.*, 400: 29—38.

Cusicanqui, H., Mahon, W.A.J., and Ellis, A.J., 1976. The geochemistry of the El Tatio geothermal field, Northern Chile. In: *Proceedings, Second United Nations Symposium on the Development and Use of Geothermal Resources, San Francisco, Calif., 1975, 1.* U.S. Government Printing Office, Washington, D.C., pp. 703—712.

D'Amore, F., 1975. Radon-222 survey in Larderello geothermal field, Italy, 1. *Geothermics*, 4: 96—108.

Doe, B.R., Hedge, C.E. and White, D.E., 1966. Preliminary investigation of the source of lead and strontium in deep geothermal brines underlying the Salton Sea geothermal area. *Econ. Geol.*, 61: 462—483.

Ellis, A.J. and Mahon, W.A.J., 1964. Natural hydrothermal systems and experimental hotwater/rock interactions. *Geochim. Cosmochim. Acta*, 28: 1323—1357.

Ellis, A.J., Mahon, W.A.J. and Ritchie, J.A., 1968. Methods of collection and analysis of geothermal fluids. *N.Z. Dep. Sci. Ind. Res., Rep.*, CD2103: 51 pp.

Eslinger, E.V., and Savin, S.M., 1973. Mineralogy and oxygen isotope geochemistry of the hydrothermally altered rocks of the Ohaki-Broadlands, New Zealand geothermal area. *Am. J. Sci.*, 273: 240—267.

Fergusson, G.J. and Knox, F.B., 1959. The possibilities of natural radiocarbon as a ground water tracer in thermal areas. *N.Z. J. Sci.*, 2: 431—441.

Ferrara, G., Gonfiantini, R. and Pistoia, P., 1963. Isotopic composition of argon from steam jets of Tuscany (Italy). In: E. Tongiorgi (Editor), *Nuclear Geology on Geothermal Areas, Spoleto, 1963.* Consiglio Nazionale dell Ricerche, Laboratorio di Geologia Nucleare, Pisa, pp. 267—275.

Ferrara, G.C., Ferrara, G. and Gonfiantini, R., 1963. Carbon isotopic composition of carbon dioxide and methane from steam jets of Tuscany (Italy). In: E. Tongiorgi (Editor), *Nuclear Geology on Geothermal Areas, Spoleto, 1963).* Consiglio Nazionale delle Ricerche, Laboratorio di Geologia Nucleare, Pisa, pp. 277—284.

Fournier, R.O. and Truesdell, A.H., 1974. Geochemical indicators of subsurface temperature, 2. Estimation of temperature and fraction of hot water mixed with cold water. *J. Res. U.S. Geol. Surv.*, 2: 263—269.

Fournier, R.O., White, D.E. and Truesdell, A.H., 1976. Convective heat flow in Yellowstone National Park. In: *Proceedings, Second United Nations Symposium on the Development and Use of Geothermal Resources, San Francisco, Calif., 1975, 1.* U.S. Government Printing Office, Washington, D.C., pp. 731—739.

Friedman, I. and O'Neil, J.R., 1977. Compilation of stable isotope fractionation factors of geochemical interest. In: M. Fleischer (Editor), *Data of Geochemistry. U.S. Geol. Surv. Prof. Paper*, 440-KK: 12 pp., 49 figs.

Giggenbach, W.F., 1971. Isotopic composition of waters of the Broadlands geothermal field (New Zealand). *N.Z. J. Sci.*, 14: 959-970.

Giggenbach, W.F., 1976. A simple method for the collection and analysis of volcanic gas samples. *Bull. Volcanol.*, 39: 132—145.

Giggenbach, W.F., 1977. The isotopic composition of sulfur in sedimentary rocks bordering the Taupo volcano zone (New Zealand). In: *Geochemistry, 1976. N.Z. Dep. Sci. Ind. Res., Bull.*, 218: 57—64.

Giggenbach, W.F., 1978. The isotopic composition of waters from the El Tatio geothermal field, northern Chile. *Geochim. Cosmochim. Acta*, 42: 979—988.

Giletti, B.J., Semet, M.P. and Yund, R.A., 1978. Studies on diffusion, III. Oxygen in feldspars: an ion microprobe determination.

Gonfiantini, R. (Editor), 1977. The application of nuclear techniques to geothermal studies; report of the Advisory Group Meeting, Pisa (Italy), 1975. *Geothermics*, 5: 1—184.

Gunter, B.D. and Musgrave, B.C., 1971. New evidence on the origin of methane in hydrothermal gases. *Geochim. Cosmochim. Acta*, 35: 113—118.

Gupta, M.L., Saxena, V.K. and Sukhija, B.S., 1976. An analysis of the hot spring activity of the Manikaran area, Himachal Pradesh, India, by geochemical studies and tritium concentration of spring waters. In: *Proceedings, Second United Nations Symposium on the Development and Use of Geothermal Resources, San Francisco, Calif., 1975, 1*. Government Printing Office, Washington, D.C., pp. 741—744.

Gutsalo, L.K., 1976. Helium isotopic geochemistry in thermal waters of the Kuril Islands and Kamchatka (U.S.S.R.). In: *Proceedings, Second United Nations Symposium on the Development and Use of Geothermal Resources, San Francisco, Calif., 1975, 1*. U.S. Government Printing Office, Washington, D.C., pp. 745—749.

Hobba, W.A., Fisher, D.W., Pearson, F.J., Jr. and Chemerys, J.C., 1978. Hydrology and geochemistry of thermal springs of the Appalachians, II. Geochemistry. *U.S. Geol. Surv. Paper*, 1044-E (in press).

Hulston, J.R., 1977. Isotope work applied to geothermal systems at the Institute of Nuclear Sciences, New Zealand. *Geothermics*, 5: 89—96.

Hulston, J.R., 1978. Interim temperature scales for the methane-water and carbon dioxide-bicarbonate isotopic equilibria. *N.Z. Inst. Nucl. Sci., Rep., INS-R-249:* 7 pp., 2 figs., 5 tables.

Hulston, J.R. and McCabe, W.J., 1962a. Mass spectrometer measurements in the thermal areas of New Zealand, 1. Carbon dioxide and residual gas analyses. *Geochim. Cosmochim. Acta*, 26: 383—397.

Hulston, J.R. and McCabe, W.J., 1962b. Mass spectrometer measurements in the thermal areas of New Zealand, 2. Carbon isotopic ratios. *Geochim. Cosmochim. Acta*, 26: 399—410.

James, R., 1968. Wairakei and Larderello: Geothermal power systems compared *N.Z. J. Sci.*, 11: 706—719.

Kendall, C., 1976. *Petrology and Stable Isotope Geochemistry of Three Wells in the Buttes Area of the Salton Sea Geothermal Field, California, U.S.A.* M.Sc. Thesis, University of California, Riverside, Calif., 211 pp. (unpublished).

Kharaka, Y.K., Berry, F.A.F. and Friedman, I., 1973. Isotopic composition of oil-field brines from Kettleman North Dome, California, and their geologic implications. *Geochim. Cosmochim. Acta*, 37: 1899—1908.

Klyen, L.E., 1973. A vessel for collecting subsurface water samples from geothermal drillholes. *Geothermics*, 2: 57—60.

Kononov, V.I. and Polak, B.G., 1976. Indicators of abyssal heat recharge of recent hydrothermal phenomena. In: *Proceedings, Second United Nations Symposium on the Development and Use of Geothermal Resources, San Francisco, Calif., 1975, 1*. U.S. Government Printing Office, Washington, D.C., pp. 767—773.

Kruger, P., Stoker, A.K. and Umaña, A., 1977. Radon in geothermal reservoir engineering. *Geothermics*, 5: 13—20.

Kusakabe, M., 1974. Sulphur isotopic variations in nature, 10. Oxygen and sulfur isotope study of Wairakei (New Zealand) geothermal well discharges. *N.Z. J. Sci.*, 17: 183—191.

Lambert, S.J., 1976. *Stable Isotope Studies of Some Active Geothermal Areas*. Ph.D. Thesis, California Institute of Technology, Pasadena, Calif., 362 pp. (unpublished; University Microfilms 76-6533).

Lang, W.B., 1959. The origin of some natural carbon dioxide gases. *J. Geophys. Res.*, 64: 127—131.

Lloyd, R.M., 1968. Oxygen isotope behavior in the sulfate-water system. *J. Geophys. Res.*, 73: 6099—6110.

Lyon, G.L., 1974a. Geothermal gases. In: I.R. Kaplan (Editor), *Natural Gases in Marine Sediments*. Plenum, New York, N.Y., pp. 141—150.

Lyon, G.L., 1974b. Isotopic analysis of gas from the Cariaco Trench sediments. In: I.R. Kaplan (Editor), *Natural Gases in Marine Sediments*. Plenum, New York, N.Y., pp. 91—97.

Lyon, G.L. and Hulston, J.R., 1970. Recent carbon isotope and residual gas measurements in relation to geothermal temperatures. *N.Z. Inst. Nucl. Sci., Rep.*, INS-R-79: 11 pp.

Macdonald, W.J.P., 1976. The useful heat contained in the Broadlands geothermal field (New Zealand). In: *Proceedings, Second United Nations Symposium on the Development and Use of Geothermal Resources, San Francisco, Calif., 1975*, 2. U.S. Government Printing Office, Washington, D.C., pp. 1113—1119.

Malinin, S.D., Kropotova, O.I. and Grinenko, V.A., 1967. Experimental determination of carbon isotope exchange constants in the CO_2(gas) $\rightleftharpoons$ HCO_3^-(solution) system under hydrothermal conditions. *Geokhimiia 1967*, 8: 927—935 (in Russian, with English abstract).

Mamryn, B.A., Gerasimovskiy, V.I. and Khabarin, L.V., 1974. Helium isotopes in the rift rocks of east Africa and Iceland. *Geochem. Int.*, 11: 488—502.

Mariner, R.H. and Willey, L.M., 1976. Geochemistry of thermal waters in Long Valley, Mono County, California. *J. Geophys. Res.*, 81: 792—800.

Matsubaya, O., Sakai, H., Kusachi, I. and Setake, H., 1973. Hydrogen and oxygen isotopic ratios and major element chemistry of Japanese thermal water systems. *Geochem. J. (Japan)*, 7: 123—151.

Mazor, E., 1976. Atmospheric and radiogenic noble gases in thermal waters: Their potential application to prospecting and steam production studies. In: *Proceedings, Second United Nations Symposium on the Development and Use of Geothermal Resources, San Francisco, Calif., 1975*, 1. U.S. Government Printing Office, Washington, D.C., pp. 793-802.

Mazor, E. and Wasserburg, G.J., 1965. Helium, neon, argon, krypton and xenon in gas emanations from Yellowstone and Lassen volcanic National Parks. *Geochim. Cosmochim. Acta*, 29: 443—454.

Mazor, E. and Fournier, R.O., 1973. More on noble gases in Yellowstone National Park hot waters. *Geochim. Cosmochim. Acta*, 37: 515—525.

Mazor, E., Kaufman, A. and Carmi, I., 1973. Hammat Gader (Israel): geochemistry of a mixed thermal spring complex. *J. Hydrol.*, 18: 289—303.

Mazor, E., Levitte, D., Truesdell, A.H., Healy, J., Gat, J. and Nissenbaum, A., 1977. Mixing models and geothermometers applied to the warm (up to 60°C) springs of the Jordan rift valley, Israel. *Weizmann Inst. (Israel), Int. Rep.*, 32 pp.

McKenzie, W.F. and Truesdell, A.H., 1977. Geothermal reservoir temperatures estimated from the oxygen isotope compositions of dissolved sulfate and water from hot springs and shallow drillholes. *Geothermics*, 5: 51—62.

Mizutani, Y., 1971. An improvement in the carbon-reduction method for the oxygen isoisotopic analysis of sulphates. *Geochem. J. (Japan)*, 5: 69—77.

Mizutani, Y., 1972. Isotopic composition and underground temperature of the Otake geothermal water, Kyushu, Japan. *Geochem. J. (Japan)*, 6: 67—73.

Mizutani, Y. and Rafter, T.A., 1969. Oxygen isotopic composition of sulphates, 3. Oxygen isotopic fractionation in the bisulfate ion-water system. *N.Z. J. Sci.*, 12: 54—59.

Mook, W.G., Bommerson, J.C. and Staverman, W.H., 1974. Carbon isotope fractionation between dissolved bicarbonate and gaseous carbon dioxide. *Earth Planet. Sci. Lett.*, 22: 169—176.

Nehring, N.L. and Truesdell, A.H., 1978. Collection of chemical, isotope, and gas samples from geothermal wells. *Proceedings, Second Workshop on Sampling and Analysis of Geothermal Effluents, Las Vegas, Nev., 1977. Environ. Prot. Agency Rep.*, EPA-600/7-78-121: 130—140.

Nehring, N.L., Bowen, P.A. and Truesdell, A.H., 1977. Techniques for the conversion to carbon dioxide of oxygen from dissolved sulfate in thermal waters. *Geothermics*, 5: 63—66.

Oeschger, H., Gugelmann, A., Loosli, H., Schotterer, U., Siegenthaler, U. and Wiest, W., 1974. ^{39}Ar dating of groundwater. In: *Isotope Techniques in Groundwater Hydrology*, 2. IAEA, Vienna, pp. 179—190.
Ohmoto, H. and Rye, R.O., 1974. Hydrogen and oxygen isotopic compositions of fluid inclusions in the Kuroko deposits, Japan. *Econ. Geol.*, 69: 947—953.
Panichi, C. and Tongiorgi, E., 1976. Carbon isotopic composition of CO_2 from springs, fumaroles, mofettes, and travertines of central and southern Italy: A preliminary prospection method of a geothermal area. In: *Proceedings, Second United Nations Symposium on the Development and Use of Geothermal Resources, San Francisco, Calif., 1975, 1.* U.S. Government Printing Office, Washington, D.C., pp. 815—825.
Panichi, C. and Gonfiantini, R., 1978. Environmental isotopes in geothermal studies. In: *Proceedings, Symposia Internacional Sobre Energia Geotermica en America Latina, Guatemala, 1976.* Instituto Italo-Latino Americano, Roma, pp. 29—70.
Panichi, C., Celati, R., Noto, P., Squarci, P., Taffi, L. and Tongiorgi, E., 1974. Oxygen and hydrogen isotope studies of the Larderello (Italy) geothermal system. In: *Isotope Techniques in Groundwater Hydrology*, 2. IAEA, Vienna, pp. 3—28.
Panichi, C., Ferrara, G.C. and Gonfiantini, R., 1977. Isotope thermometry in the Larderello (Italy) geothermal field. *Geothermics*, 5: 81—88.
Pearson, F.J., Jr. and Bodden, M., 1975. U.S. Geological Survey, Water Resources Division, radiocarbon measurements, I. *Radiocarbon*, 17: 135—148.
Pearson, F.J., Jr. and Truesdell, A.H., 1978. Tritium in the waters of Yellowstone National Park (Wyoming). In: R.E. Zartman (Editor), *Short Papers of the Fourth International Conference of Geochronology, Cosmochronology, and Isotope Geology, 1978. U.S. Geol. Surv., Open-File Rep.*, 78—701: 327—329.
Pearson, F.J., Jr., Bedinger, M.S. and Jones, B.F., 1973. Carbon-14 ages of water from the Arkansas hot springs. In: *Proceedings of the 8th International Conference on Radiocarbon Dating, Wellington, New Zealand, 1972, 1.* pp. 330—341.
Potter, R.W., II and Clynne, M.A., 1978. The solubility of the noble gases He, Ne, Ar, Kr and Xe in water up to the critical point. *J. Solution Chem.*, 7: 837—844.
Potter, R.W., II, Mazor, E. and Clynne, M.A., 1977. Noble gas partition coefficients applied to the conditions of geothermal steam formation. *Geol. Soc. Am. Abstr. Progr.*, 9(7): 1132—1133 (abstract).
Rafter, T.A., 1957. Sulfur isotopic variations in nature, 1. The preparation of sulfur dioxide for mass spectrometer examination. *N.Z. J. Sci. Technol., Sect. B*, 38: 849—857.
Rafter, T.A., 1967. Oxygen isotopic composition of sulphates, 1. A method for the extraction of oxygen and its quantitative conversion to carbon dioxide for isotope ratio measurements. *N.Z. J. Sci.*, 10(2): 493—510.
Rafter, T.A. and Wilson, S.H., 1963. The examination of sulphur isotopic ratios in the geothermal and volcanic environment. In: E. Tongiorgi (Editor), *Nuclear Geology on Geothermal Areas, Spoleto, 1963.* Consiglio Nazionale delle Ricerche, Laboratorio di Geologia Nucleare, Pisa, pp. 139—172.
Richet, P., Bottinga, Y. and Javoy, M., 1977. A review of hydrogen, carbon, nitrogen, oxygen, sulfur, and chlorine stable isotope fractionation among gaseous molecules. *Annu. Rev. Earth Planet. Sci.*, 5: 65—110.
Rightmire, C.T., Young, H.W. and Whitehead, R.L., 1976. Geothermal investigations in Idaho, IV. Isotopic and geochemical analyses of water from the Bruneau—Grand View and Weiser areas, southwest Idaho. *Idaho Dep. Water Admin., Water Inf. Bull.*, 30: 28 pp.
Robinson, B.W., 1973. Sulphur isotope equilibrium during sulphur hydrolysis at high temperatures. *Earth Planet. Sci. Lett.*, 18: 443—450.
Robinson, B.W., 1978. Isotope equilibria between sulphur solute species at high temperatures: In: *Stable Isotopes in the Earth Sciences. N. Z. Dep. Sci. Ind. Res. Bull.*, 220: 203—206.

Sakai, H., 1957. Fractionation of sulphur isotopes in nature. *Geochim. Cosmochim. Acta*, 12: 150—169.

Sakai, H., 1977. Sulfate-water isotope thermometry applied to geothermal systems. *Geothermics*, 5: 67—74.

Sakai, H. and Krouse, H.R., 1971. Elimination of memory effects in $^{18}O/^{16}O$ determinations in sulphates. *Earth Planet. Sci. Lett.*, 11: 369—373.

Sakai, H. and Matsubaya, O., 1977. Stable isotope studies of Japanese geothermal systems. *Geothermics*, 5: 97—124.

Sakai, H. and Dickson, F.W., 1978. Experimental determination of the rate and equilibrium fractionation factors of sulfur isotope exchange between sulfate and sulfide in slightly acid solutions at 300°C and 1000 bars. *Earth Planet. Sci. Lett.*, 39: 151—161.

Stahl, W., 1974. Carbon isotope fractionations in natural gases. *Nature*, 251: 134—135.

Steiner, A. and Rafter, T.A., 1966. Sulfur isotopes in pyrite, pyrrhotite, alunite and anhydrite from steam wells in the Taupo volcanic zone, New Zealand. *Econ. Geol.*, 61: 1115—1129.

Stewart, M.K., 1978. Stable isotopes in waters of the Wairakei geothermal area, New Zealand. In: *Stable Isotopes in the Earth Sciences. N.Z. Dep. Sci. Ind. Res. Bull.*, 220: 113—119.

Stoker, A.K. and Kruger, P., 1976. Radon in geothermal reservoirs. In: *Proceedings, Second United Nations Symposium on the Development and Use of Geothermal Resources, San Francisco, Calif., 1975, 3*. U.S. Government Printing Office, Washington, D.C., pp. 1797—1803.

Taylor, C.B., Polach, H.A. and Rafter, T.A., 1963. A review of the work of the Tritium Laboratory, Institute of Nuclear Sciences — New Zealand. In: E. Tongiorgi (Editor), *Nuclear Geology on Geothermal Areas, Spoleto 1963*. Consiglio Nazionale delle Ricerche, Laboratorio di Geologia Nucleare, Pisa, pp. 185—233.

Taylor, H.P., Jr., 1974. The application of oxygen and hydrogen isotope studies to problems of hydrothermal alteration and ore deposition. *Econ. Geol.*, 69: 843—883.

Taylor, H.P., Jr., 1977. Water/rock interactions and the origin of H_2O in granitic batholiths. *J. Geol. Soc. (London)*, 133: 509—558.

Thode, H.G., Cragg, C.B., Hulston, J.R. and Rees, C.E., 1971. Sulfur isotope exchange between sulfur dioxide and hydrogen sulfide. *Geochim. Cosmochim. Acta*, 35: 35—45.

Truesdell, A.H., 1976. Summary of section III — geochemical techniques in exploration. In: *Proceedings, Second United Nations Symposium on the Development and Use of Geothermal Resources, San Francisco, Calif., 1975, 1*. U.S. Government Printing Office, Washington, D.C., pp. iii—xiii.

Truesdell, A.H. and White, D.E., 1973. Production of superheated steam from vapor-dominated geothermal reservoirs. *Geothermics*, 2: 154—173.

Truesdell, A.H. and Fournier, R.O., 1976a. Conditions in the deeper parts of the hot spring systems of Yellowstone National Park, Wyoming. *U.S. Geol. Surv., Open-File Rep.*, 76-428: 22 pp.

Truesdell, A.H. and Fournier, R.O., 1976b. Calculation of deep temperatures in geothermal systems from the chemistry of boiling spring waters of mixed origin. In: *Proceedings, Second United Nations Symposium on the Development and Use of Geothermal Resources, San Francisco, Calif., 1975, 1*. U.S. Government Printing Office, Washington, D.C., pp. 837—844.

Truesdell, A.H. and Nehring, N.L., 1978. Cases and water isotopes in a geochemical section across the Larderello, Italy geothermal field. *Pure Appl. Geophys.*, 117: 276—289.

Truesdell, A.H., Nathenson, M. and Rye, R.O., 1977a. The effects of subsurface boiling and dilution on the isotopic compositions of Yellowstone thermal waters. *J. Geophys. Res.*, 82: 3694—3703.

Truesdell, A.H., Nehring, N.L. and Frye, G.A., 1977b. Steam production at The Geysers, California, comes from liquid water near the well bottoms. *Geol. Soc. Am. Abstr. Progr.*, 9(7): 1206 (abstract).

Wasserburg, G.J., Mazor, E. and Zartman, R.E., 1963. Isotopic and chemical composition of some terrestrial natural gases. In: J. Geiss and E.D. Goldberg (Editors), *Earth Science and Meteoritics*. North-Holland, Amsterdam, pp. 219—240.

White, D.E., 1957. Thermal waters of volcanic origin. *Geol. Soc. Am. Bull.*, 68: 1637—1658.

White, D.E., 1968. Environments of generation of some base-metal ore deposits. *Econ. Geol.*, 63: 301—335.

White, D.E., 1974. Diverse origins of hydrothermal ore fluids. *Econ. Geol.*, 69: 954—973.

White, D.E., Muffler, L.J.P. and Truesdell, A.H., 1971. Vapor-dominated hydrothermal systems compared with hot-water systems. *Econ. Geol.*, 66: 75—97.

White, D.E., Barnes, I. and O'Neil, J.R., 1973. Thermal and mineral waters of nonmeteoric origin, California Coast Ranges. *Geol. Soc. Am. Bull.*, 84: 547—560.

White, D.E., Fournier, R.O., Muffler, L.J.P. and Truesdell, A.H., 1975. Physical results of research drilling in thermal areas at Yellowstone National Park, Wyoming. *U.S. Geol. Surv., Prof. Paper*, 892: 70 pp.

Wilson, S.H., 1963. Tritium determinations on bore waters in the light of chloride-enthalpy relations. In: E. Tongiorgi (Editor), *Nuclear Geology on Geothermal Areas, Spoleto 1963*. Consiglio Nazionale delle Ricerche Laboratorio di Geologia Nucleare, Pisa, pp. 173—184.

Wilson, S.H., 1966. Sulphur isotope ratios in relation to volcanological and geothermal problems. *Bull. Volcanol.*, 29: 671—690.

Chapter 6

SULPHUR AND OXYGEN ISOTOPES IN AQUEOUS SULPHUR COMPOUNDS

F.J. PEARSON, Jr. and C.T. RIGHTMIRE

INTRODUCTION

Dissolved sulphur compounds are common in natural waters. Under oxidizing conditions sulphate (SO_4^{2-}) is the dominant sulphur species, while under reducing conditions sulphide as H_2S or HS^- prevails. The ratios of the stable isotopes of sulphur in these compounds frequently suggest the source of the dissolved species and/or provide information about the reactions and geochemical processes affecting them. The stable oxygen isotope ratios of sulphate provide information about the source of sulphate species, and about the geochemical environment of the water mass in which the sulphate is found. Here we will discuss the distribution and fractionation of sulphur and oxygen isotopes in sulphur species in natural waters. We will touch on some aspects of the geochemistry of sulphur and discuss some examples of the uses of sulphur and oxygen isotopes in groundwater studies.

ISOTOPE FRACTIONATION

Sulphur has four stable isotopes, ^{32}S, ^{33}S, ^{34}S, and ^{36}S. These have natural abundances of 95.0, 0.76, 4.22, and 0.014%. There are several other short-lived, radioactive isotopes of sulphur which have been made artificially. The longest lived of these, ^{35}S, has a half-life of 87 days and is frequently used in sulphur tracer experiments. There are three stable isotopes of oxygen — ^{16}O, ^{17}O, and ^{18}O, with relative natural abundances of 99.759, 0.037, and 0.204%. Normally only $^{34}S/^{32}S$ and $^{18}O/^{16}O$ ratios are measured. Very little additional information would be gained in hydrologic investigations by measuring the ratios of the other isotope pairs. Hoefs (1973) discusses measurement techniques and the stable isotope geochemistry of both sulphur and oxygen. Garlick (1969) has reviewed the geochemistry and measurement techniques for oxygen isotopes.

The ratios of the stable isotopes of an element in reacting compounds normally are different in each compound. Fractionation of isotopes between the compounds occurs because a chemical bond involving a heavy isotope has lower vibrational frequency than the equivalent bond with the light isotope. The bond with the heavy isotope therefore is stronger than that with the light isotope. If vibrational frequencies of a molecule are known, as they are for many simple molecules, the amount of isotope fractionation which may take place can be calculated using statistical mechanical methods (Urey, 1947; Bigeleisen and Mayer, 1947; Richet et al., 1977). The reaction:

$$H_2{}^{34}S + H^{32}S^- \rightleftharpoons H_2{}^{32}S + H^{34}S^- \tag{1}$$

is an isotope exchange reaction with equilibrium constant:

$$K = \frac{[H_2{}^{32}S][H^{34}S^-]}{[H_2{}^{34}S][H^{32}S^-]} \tag{2}$$

The isotope fractionation factor, α, for this reaction is:

$$\alpha_{HS^- - H_2S} = \frac{({}^{34}S/{}^{32}S)_{HS^-}}{({}^{34}S/{}^{32}S)_{H_2S}} \tag{3}$$

which equals the equilibrium constant, K, of equation (2).

Because bonds with ^{34}S are stronger than those with ^{32}S, the chemical properties of the species containing the different isotopes are not the same, and so K and α do not generally equal one. In fact, HS^- is depleted in ^{34}S relative to H_2S with which it is in equilibrium at 25°C by 0.38% and α in equation (3) equals 0.9962.

Stable isotope compositions are reported as δ-values in parts per thousand (termed per mil and noted as ‰) enrichment (or depletion if negative) relative to an agreed upon standard. Therefore:

$$\delta^{34}S \text{ (in ‰)} = \frac{({}^{34}S/{}^{32}S)_{sample} - ({}^{34}S/{}^{32}S)_{standard}}{({}^{34}S/{}^{32}S)_{standard}} \times 1000 \tag{4}$$

Likewise, $\delta^{18}O$ values are reported from $^{18}O/^{16}O$ ratio measurements. The standard for sulphur isotope measurements is troilite (FeS) from the Canyon Diablo iron meteorite (abbreviated CD); the oxygen standard is Vienna Standard Mean Ocean Water (V-SMOW) (Hoefs, 1973). One of the primary reasons for reporting isotopic compositions as differential measurements is that mass spectrometers measure differences between the ratios of two gaseous samples, rather than the absolute ratio of a sample. This technique allows a precision of 0.1‰ or better to be attained.

The theoretical temperature dependence of isotopic equilibrium constants is discussed by Urey (1947), Bigeleisen and Mayer (1947), and Richet et al. (1977). At low temperatures, generally much below room temperature, the natural logarithm of α (ln α) is inversely proportional to absolute temperature, T. At higher temperatures ln α is inversely proportional to T^2. The definition of low or high temperature depends upon the vibrational frequencies of the molecules involved. The pressure effect upon α is negligible (Clayton et al., 1974).

The relation between the fractionation factor, α (equation 3), and δ-values (equation 4) is:

$$\alpha_{A-B} = \frac{1 + \frac{\delta_A}{1000}}{1 + \frac{\delta_B}{1000}} = \frac{1000 + \delta_A}{1000 + \delta_B} \tag{5}$$

Frequently, the "permil fractionation" Δ, is used, where:

$$\Delta_{A-B} = \delta_A - \delta_B \tag{6}$$

Because $\ln(1 + X) \simeq X$ when X is small, Δ is nearly identical to $10^3 \ln \alpha$. In the HS$^-$-H$_2$S example above, for example, where $\alpha = 0.9962$, $10^3 \ln \alpha = -3.8$.

Fractionation factor variations with temperature are usually expressed or graphed as $10^3 \ln \alpha$ ($\simeq \Delta$) against $1/T$ or $1/T^2$ (see Fig. 6-1 and equation (10), below).

Kinetic isotope fractionation may occur in addition to equilibrium fractionation because of reaction rate differences between isotopes of different masses. Kinetic effects in the sulphur system are discussed elsewhere in this volume.

Rates of approach to sulphur and oxygen isotopic equilibrium

In an early use of radioactive ^{35}S as a tracer, Voge (1939) studied the reaction:

$$^{35}SO_4^{2-} + HS^- \rightarrow H^{35}S^- + SO_4^{2-} \tag{7}$$

He could detect no radioactivity in the HS$^-$ after 36 hours of reaction at 100°C. More recently, Robinson (1973) investigated the similar reaction:

$$H^{34}SO_4^- + H_2{}^{32}S \rightleftharpoons H^{32}SO_4^- + H_2{}^{34}S \tag{8}$$

between 200 and 320°C. His data fit an expression characteristic of a first-order reaction:

$$\ln(1-f) = -kt \tag{9}$$

where f = fraction exchanged at time t, and k = first-order exchange rate constant. When discussing first-order reactions, it is convenient to use the reaction half-life, $t_{1/2}$, which is simply $\ln 0.5/-k$. The expression:

$$\log t_{1/2} \text{ (hours)} = 3.06(10^3/T) - 4.85 \tag{10}$$

where the temperature, T, is in Kelvin, can be derived from Robinson's data.

Lloyd (1968) measured rates of approach to equilibrium in the reaction:

$$\tfrac{1}{4}(S^{18}O_4^{2-}) + H_2{}^{16}O \rightleftharpoons \tfrac{1}{4}(S^{16}O_4^{2-}) + H_2{}^{18}O \tag{11}$$

between 25 and 448°C. He found that while this rate, like that of all reactions, increases with temperature, it also increases with decreasing pH so that it is considerably more rapid in acid than in alkaline solutions. The expression:

$$\log t_{1/2} \text{ (hours)} = 2.15(10^3/T) + 0.44 \text{ pH} - 3.09 \tag{12}$$

can be derived from Lloyd's data. Lloyd's rates at pH = 7 are some two orders of magnitude slower than rates reported earlier by Teis (1956) in solutions Teis termed "neutral", but for which no measured pH values are given.

The sulphur in the sulphate molecule is surrounded by four oxygens, and it therefore seems unlikely that sulphate sulphur could exchange with sulphide in aqueous solution if sulphate oxygen were not also exchanging with water oxygen of the solution. Thus, if the work of Robinson (1973) and Lloyd (1968) are consistent, Robinson's sulphur exchange rates should be no faster than Lloyd's oxygen exchange rates.

To compare the two sets of data, the pH's of Robinson's experiments are needed. Although they were not measured, the experimental conditions were such that the solutions had pH values of no more than two, and were perhaps as low as zero. At 200°C, Robinson's data show a sulphur exchange half-time of 40 hours. From equation (12) from Lloyd's oxygen data, the same exchange half-time at 200°C would occur at pH $\approx$ 0.4, within the probable range of Robinson's experiments.

Thus, it seems reasonable to expect that sulphate-sulphide exchange rates at lower temperatures and higher pH values than have been measured will be no more rapid than sulphate-oxygen exchange rates. Lloyd's (1968) data at 25°C and pH = 7 show that the half-time of the latter reaction is about 1900 years, and at 100°C and the same pH, it is about 65 years. The minimum rate detectable by Voge's (1939) experiments had a half-time of 16 days, far below this rate.

Sulphur isotope fractionation

Slow reaction rates make it difficult to experimentally determine equilibrium fractionation factors between sulphur species at low temperatures. Thus, it has been necessary to use the statistical mechanical techniques of Bigeleisen and Mayer (1947) and of Urey (1947) to calculate the fractionation factors at low temperatures. Such calculations have been made by Tudge and Thode (1950), Sakai (1957, 1968), Thode et al. (1971), and Richet et al. (1977). A critical review and compilation of sulphur fractionation factors to temperatures as low as 50°C is given by Ohmoto and Rye (1979). Fig. 6-1 is a graph of fractionation factors in the sulphur system plotted against $1/T^2$. What little experimental data exist are also shown on this graph. Although not of importance in aqueous systems, fractionation involving gaseous SO_2 is shown to suggest the size of errors which may be present in the calculated fractionation factors due to imperfect knowledge of molecular vibrational frequencies. The two SO_2 curves are based on two sets of vibrational spectra for the H_2S and SO_2 molecules. As the figure shows, at 25°C they differ by 6‰. In addition, neither of them agrees particularly well with the high-temperature experimental data. This suggests that at low temperatures, at least, calculations using sulphide-sulphate fractionation factors more precise than several permil are unwarranted.

Because oxygen isotope exchange between sulphate and water is also slow, one might deduce that bond breakage in the sulphate molecule limits the rate of isotope exchange, and thus that H_2S-HS^- exchange need not

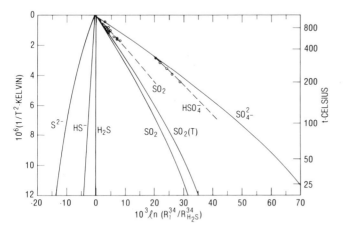

Fig. 6-1. Sulphur isotope fractionation factors, relative to H_2S(gas) = H_2S(aqueous). Solid lines are theoretical and, except $SO_{2(g)}(T)$ are from calculations of Sakai (1968). Experimental data are shown as circles and extrapolations from them by dashed lines. HSO_4^--H_2S data are from Robinson (1973), and the SO_2-H_2S data and the theoretical curve $SO_{2(g)}(T)$ are from Thode et al. (1971).

be slow. No experimental data exist to test this hypothesis, although Shavan, Galas, Lees, and Zhook (1974, cited by Friedman and O'Neil, 1977) report that at 20°C dissolved H_2S is depleted in ^{34}S by 1.6‰ relative to gaseous H_2S.

In spite of the slow approach to sulphur isotope equilibrium at earth's surface temperatures, significant differences in the $^{34}S/^{32}S$ ratios of coexisting oxidized and reduced sulphur compounds do exist. These are due to kinetic fractionation accompanying inorganic or bacterial reduction or oxidation of sulphur compounds. Harrison and Thode (1957) studied the chemical reduction of sulphate and found that $^{32}SO_4^{2-}$ was reduced some 2.2% faster than $^{34}SO_4^{2-}$ so that:

$$\Delta_{(SO_4^{2-}-H_2S)} \text{ (kinetic)} = 22‰$$

at temperatures between 18 and 50°C. This is considerably lower than the equilibrium fractionation, $\approx 70‰$ at 25°C (Fig. 6-1). There have been many studies of isotope fractionation during bacterial sulphate reduction which Goldhaber and Kaplan (1974) and Krouse (Chapter 11) have reviewed at length. For the purposes of this review, it is necessary only to mention that the amount of fractionation found varies from nearly zero to almost 50‰ and depends variously on the rate of reduction and the particular electron donor used in the experiment.

Kinetic fractionation during sulphide oxidation by bacteria is in the opposite direction from equilibrium fractionation and from kinetic fractionation during sulphate reduction by bacteria in that the product SO_4^{2-} is depleted in ^{34}S. Kaplan and Rittenberg (1964) found ^{34}S depletions of as much as $-18‰$ in SO_4^{2-} bacterially produced in the laboratory, though most of their results show very little fractionation during oxidation. Nissenbaum and Rafter (1967) report gypsum depleted by as much as $-14‰$ below the pyrite from which the gypsum was formed. In both the laboratory and field studies cited, there was less than a few permil difference between the sulphide and elemental sulphur formed from it.

Oxygen isotope fractionation

The fractionation factor for the exchange of oxygen between sulphate and water (reaction (11)) was calculated by Urey (1947) along with fractionation factors for reactions between a number of other simple molecules. Experimental work in the system has been done by Teis (1956) and by Lloyd (1967, 1968) who measured the fractionation factors between 25 and 575°C and by Mizutani and Rafter (1969a) who worked between 110 and 200°C. The fractionation factors calculated by Urey differ from those found by experiment by more than 10‰ at lower temperatures. Agreement between the two sets of experiments, however, is rather good. The greatest difference between them is about 2‰ at 110°C. Truesdell and Hulston (Chapter 5, this volume) discuss fractionation factors between sulphate and water in detail and present a graph of the variation of these fractionation factors with temperature. In addition, Friedman and O'Neil (1977) show the several fractionation factors graphically.

Lloyd (1967, 1968) has also measured the fractionation factor between oxygen in the mineral anhydrite ($CaSO_4$) and water from which it precipitates. The fractionation factor has the form:

$$10^3 \ln \alpha_{(anhydrite-water)} = \frac{3.878 \times 10^6}{T^2} - 3.4 \qquad (13)$$

Friedman and O'Neil (1977, fig. 15) present unpublished data of B.W. Robinson and M. Kusakabe on fractionation in the barite ($BaSO_4$) water system. Fontes (1965) has investigated the fractionation between solution and the water of hydration in gypsum ($CaSO_4 \cdot 2\,H_2O$). He finds that the water of hydration is about 3—4‰ enriched in ^{18}O over the oxygen in the water with which the gypsum is in equilibrium.

Kinetic fractionation affects the oxygen isotopes during sulphate reduction or sulphide oxidation as it does the sulphur isotopes (Chapter 11, this volume). Lloyd (1967, 1968) and Mizutani and Rafter (1969b, 1973) examined kinetic effects and found that during bacterial sulphate reduction, ^{18}O is enriched in the residual sulphate because it forms the stronger oxygen bond. The fractionation involved is 4.5‰.

Lloyd (1968) also examined oxygen isotope fractionation during the oxidation of sulphide to sulphate. He found that sulphide oxidation by dissolved oxygen proceeds in two steps. First, sulphite (SO_3^{2-}) is produced by the reaction:

$$S^{2-} + H_2O + O_2 \rightarrow SO_3^{2-} \qquad (14)$$

during which 2 parts of molecular oxygen are consumed to 1 part water oxygen. Molecular ^{16}O is consumed in preference to molecular ^{18}O, with a kinetic fractionation of 8.7‰. The second reaction step appears to be sulphite oxidation by molecular oxygen by the reaction:

$$SO_3^{2-} + \tfrac{1}{2} O_2 \rightarrow SO_4^{2-} \qquad (15)$$

Junge and Ryan (1957) pointed out that while reaction (15) proceeds much more slowly than reaction (14) in pure solutions, it is catalyzed by trace amounts of transition metals. Lloyd was unable to measure fractionation in reaction (15) because exchange between sulphite oxygen and water occurred too rapidly — some 10^5 times faster than sulphate-water exchange.

According to reactions (14) and (15), molecular oxygen and water oxygen are used in a 3 : 1 ratio during sulphide oxidation to sulphate. However, because of the rapid exchange of the intermediate sulphite with water, more than 1/3 of the oxygen in the product sulphate appears to come from water oxygen. For example, Schwarcz and Cortecci (1974) found that when an aqueous suspension of pyrite is oxidized by bubbling air through it, about half the oxygen in the sulphate is from the water.

GEOCHEMISTRY AND ISOTOPE DISTRIBUTION OF AQUEOUS SULPHUR COMPOUNDS

Geochemistry

Sulphur commonly exists in one of three oxidation states in natural systems. In oxidizing environments, the +6 state with sulphur as the sulphate (SO_4^{2-}) ion predominates. Under reducing conditions sulphur exists in the −2 valence state in the form of bisulphide (HS^-) in waters with pH values less than 7, and in the form of hydrogen sulphide (H_2S) in more acid waters. Sulphur can also occur in solution as elemental sulphur (S^0), but the concentration of this species is generally insignificant at low temperature and normally it is not measured in water analysis. Other sulphur valence states may also be represented by such species as sulphite (SO_3^{2-}) and thiosulphate ($S_2O_3^{2-}$), but these, generally, have vanishing small concentrations or only transitory existence during reactions. They may, however, be important in such processes as pyrite oxidation and play a large role in some types of uranium mineralization processes (Granger and Warren, 1969). The equilibrium distribution of aqueous sulphur species under various pH and redox conditions is discussed by Garrels and Christ (1965), Stumm and Morgan (1970), and Thorstenson (1970). Goldhaber (1974) and Goldhaber and Kaplan (1974) present a detailed discussion of the sulphur cycle, directed primarily at seawater and marine sediment phenomena, but applicable in many respects to terrestrial waters as well.

The common sulphur-bearing minerals gypsum ($CaSO_4 \cdot 2 H_2O$) and anhydrite ($CaSO_4$) contain sulphur in the +6 valence state, while the common sulphide ore minerals have sulphur in the −2 valence state. The most common reduced sulphur-containing mineral, pyrite, contains sulphur in the form S_2^{2-} where, however, it does not have a formal valence of −1, but rather is sulphur in the zero and −2 valence states (Berner, 1970).

Sulphate found in groundwaters can have several sources. Most comes from the solution of evaporite sulphate minerals such as anhydrite or gypsum in the aquifer. Such solution can raise the sulphate concentration to as much as 2000 mg/l. Sulphate may also be derived from the oxidation of sulphide minerals, and the resulting concentrations may be high where such minerals are exposed to attack. A few milligrams per litre of sulphate occur in atmospheric precipitation. This sulphate is derived from oceanic aerosols, from the solution of gaseous sulphur oxides, and particularly in arid regions, from wind-blown mineral sulphate dust from soils and playas. Gaseous oxides are from such natural sources as volcanoes or from the oxidation of H_2S produced in natural reducing environments such as bogs, or they may come from man-made sources such as the burning of fossil fuels.

Sulphide, which in normal groundwaters is found in the species H_2S and HS^-, occurs only under reducing conditions and normally at concentrations of only a few milligrams per litre. Because of the extremely low solubility of sulphide minerals, only a small amount of sulphide in groundwater can be derived from their solution. More commonly sulphide is thought to form within the aquifer system itself by reduction of sulphate ions in the water. Also, decomposition of organic material in the aquifer may release sulphide to a groundwater system. These two processes probably account for the high (>200 mg/l) sulphide concentrations found in some oilfield brines.

As Fig. 6-1 shows, the amount of equilibrium sulphur isotope fractionation which may occur during a geochemical reaction depends on the particular sulphur species involved. For example, in acid solutions, reduced sulphur is present as H_2S, while in alkaline solutions, HS^- is dominant. They have equal concentrations at pH $\simeq$ 7. (The stability field of S^{2-} is poorly known, but certainly it is not an important solution species at pH values below 12—14; Goldhaber, 1974). Because of fractionation between H_2S and HS^-, equilibrium fractionation factors during sulphate reduction will vary with pH. At high temperatures ($\geq$ 200—300°C) where equilibrium rather than kinetic fractionation may be expected to control sulphur isotope distribution, solution composition considerations are important and have been discussed by Ohmoto (1972).

Isotope distribution

Oceanic sulphate and evaporite minerals. The $^{34}S/^{32}S$ ratios in various natural sulphur compounds vary widely. As noted earlier, both equilibrium and kinetic fractionation tend to enrich ^{34}S in the oxidized sulphate species and ^{32}S in more reduced species. Thus sulphate minerals and oceanic sulphate are enriched in ^{34}S relative to coexisting reduced sulphur compounds. Ore minerals, organic sulphide, and the sulphur in coal and oil tend to contain less ^{34}S than do sulphates. The distribution of sulphur isotopes in a number of natural sulphur-containing materials is shown by Krouse (this volume), and is also discussed by Goldhaber (1974), Holser and Kaplan (1966), and Ohmoto and Rye (1979).

Many measurements of the ^{34}S content of modern oceanic sulphate and of the sulphate of evaporite minerals of various geologic ages have been reported. While the $\delta^{34}S$ value of sulphate throughout the modern ocean is constant at +20‰, $\delta^{34}S$ values of evaporites, which presumably represent the sulphate in the ocean when they were deposited, vary with age. Summaries of analyses of evaporite sulphate and interpretations thereof have been given by Nielsen (1965), Holser and Kaplan (1966), and by Holser (1977) among others.

These authors, as well as Krouse (this volume), show diagrams of the changes in δ^{34}S with time from as low as 10‰ in late Permian/early Triassic evaporites to nearly 35‰ in late Proterozoic/early Cambrian deposits.

The ^{34}S content of oceanic sulphate at any given time has been interpreted as a balance among several competing processes (Nielsen, 1965; Holser and Kaplan, 1966; Rees, 1970; Holser, 1977). Most sulphate enters the ocean through rivers. As will be discussed in more detail below, rivers contain sulphate derived from solution of older evaporite minerals and sulphate formed from the oxidation of sulphide minerals. Sulphate leaves the ocean as sulphate in evaporite minerals; the δ^{34}S of precipitating evaporite minerals is very close to that of the ocean. Sulphate may also be removed from the ocean after reduction to sulphide. Both kinetic and equilibrium fractionation during sulphate reduction tend to deplete the precipitating sulphide in ^{34}S. Goldhaber and Kaplan (1974, 1975) discuss the sulphate-sulphide fractionation observed in marine sediments. The inclusion of ^{32}S-enriched sulphide minerals in sediments tends to leave oceanic sulphate enriched in ^{34}S. The δ^{34}S value of oceanic sulphate at any time is thus a result of competing rates among the processes of sulphate introduction to the ocean and sulphide removal from the ocean. Holser (1977) has pointed out that the sulphate isotope changes can be extremely rapid and suggests that they may represent catastrophic changes in oceanic circulation and behavior.

A number of measurements have shown that the ^{18}O content of sulphate throughout the modern ocean is constant, with a δ^{18}O of +9.6‰. Longinelli and Craig (1967) suggested that this value might represent isotopic equilibrium between the oxygen of sulphate and of ocean water (reaction 11), based on extrapolation to low temperature of Teis' (1956) exchange rate data. However, Lloyd's (1968) work shows that this suggestion of equilibrium is incorrect for two reasons. First, Lloyd's exchange rates, from more carefully controlled and longer-time experiments than those by earlier workers, are too slow for oxygen isotope equilibrium to be attained during the lifetime of an average sulphate molecule in the ocean. Second, Lloyd's (1967, 1968) and Mizutani and Rafter's (1969a) fractionation factor data predict that sulphate at equilibrium with seawater (δ^{18}O = 0‰) should have a δ^{18}O value of about +38‰.

To account for the discrepancy between the predicted and observed ^{18}O content of oceanic sulphate, Lloyd (1968) proposed a steady-state model based on his measured kinetic fractionation factors (above). The oxygen incorporated in sulphate during sulphide oxidation in seawater is 1/3 from atmospheric oxygen dissolved in the water (with δ^{18}O of +23‰, Kroopnick and Craig, 1972) and 2/3 from the water itself. With a kinetic fractionation of −8.7‰ during the consumption of atmospheric oxygen, the sulphate produced would have a δ^{18}O of +4.6‰. The source of oceanic sulphide is largely the bacterial reduction of sulphate, which, Lloyd found experimentally, kinetically enriches the residual sulphate in ^{18}O by 4.5‰. With this process at steady state, the predicted $\delta^{18}O$ of oceanic sulphate would be +9.1‰, close to the observed +9.6‰.

Cortecci and Longinelli (1971, 1973) have found that the ^{18}O content of sulphate in the shells of living marine organisms is close to that of the sulphate dissolved in the water in which they grow. They find, however, that fossil shells exhibit a wide range of sulphate oxygen isotopic compositions which they interpret as a result of post-depositional change. Thus the ^{18}O content of sulphate from shells gives no information on past ocean sulphate $\delta^{18}O$ values. Holser (1977), on the other hand, reports that there is little variation in the ^{18}O content of the sulphate in evaporite minerals with time.

Sulphur compounds in terrestrial waters. The terrestrial hydrologic cycle begins with atmospheric precipitation (rainfall or snow). Near the oceans rainfall contains dissolved material from sea spray with ratios of dissolved ions near those of normal seawater. Inland, the ion ratios change and, in particular, the sulphate to chloride ratio increases over its value in normal seawater (Fisher, 1968; Pearson and Fisher, 1971). This excess of sulphate inland clearly implies the existence of terrestrial sulphate sources.

Early isotope studies of sulphate in atmospheric precipitation (Östlund, 1959, Jensen and Nakai, 1961) found $\delta^{34}S$ values in the limited range of +3.2 to +8.2‰ in non-industrial regions, but with values as high as +15.6‰ in industrial areas, the high values corresponding to the ^{34}S contents of coal burned locally (Jensen and Nakai, 1961; Nakai and Jensen, 1967).

These studies did not consider the chemistry of the precipitation. However, Mizutani and Rafter (1969c) in a study of rainfall in the vicinity of Gracefield, New Zealand, and Cortecci and Longinelli (1970) in a study around Pisa, Italy, found linear relationships between the $\delta^{34}S$ of rainfall sulphate and the amount of excess sulphate (that sulphate present above sulphate which would be present from sea spray alone). Where the excess sulphate was equal to zero, that is when only seawater sulphate was present, they found $\delta^{34}S$ sulphate ratio values of +20 to +22‰. In the New Zealand study, Mizutani and Rafter found that the isotopic composition of the excess sulphate itself was in the range of −2 to −4‰. Krouse (Chapter 11) describes these studies in more detail. Other sources of excess sulphate in rainfall, particularly in arid regions, may be the dispersion of mineral sulphates by wind from evaporite deposits at the surface or from sulphate minerals which have precipitated from terrestrial surface water in playa or saline environments, or in soils.

The oxygen isotopic composition of rainfall sulphate has also been studied. Rafter and Mizutani (1967) and Mizutani and Rafter (1969c) found a $\delta^{18}O$ range between +8 and +14‰ with a mean of +10.2‰, close to the $\delta^{18}O$ value of seawater sulphate: 9.6‰. They found no correlation between the amount of excess sulphate in their samples and its oxygen isotopic composition. Cortecci and Longinelli (1970) found a larger $\delta^{18}O$ variation in their rainwater sulphate (from +5 to +17‰), and also noted a seasonal variation in $\delta^{18}O(SO_4^{2-})$, parallel with seasonal $\delta^{18}O$ changes in the rainwater itself. From the $\delta^{34}S$ values of their samples (−3 to +3‰), it appears that most of their rainfall sulphate is not oceanic, but is produced by oxidation of a reduced sulphur compound such as SO_2 from fossil fuel burning. Cortecci (1973) suggests that SO_2 oxidation to SO_4^{2-} involves oxygen from both water vapor and atmospheric O_2 and that the $\delta^{18}O$ of rainfall SO_4^{2-} can be written as:

$$\delta^{18}O(SO_4^{2-}) = \frac{2\delta^{18}O(SO_2) + \delta^{18}O(O_2) + \delta^{18}O(H_2O)}{4} \tag{16}$$

The sulphur isotopic composition of sulphate in streams and rivers reflects the isotopic composition of the sulphate sources. The sulphate in some rivers — those with relatively low sulphate concentrations — can be shown by sulphate mass balance studies (Fisher, 1968; Pearson and Fisher, 1971) to have been derived entirely from the sulphate in the rain falling on the basin. Other potential sources of sulphate to river water include the solution of sulphate minerals, disseminated or as evaporite deposits, in the basin feeding the river, the oxidation of sulphide minerals again in the rocks of the basin, and soil sulphate and oxidized soil organic sulphur as well.

Isotopically the sources represent two groups: relatively heavy (^{34}S-enriched) sulphate derived from evaporite minerals and relatively light (^{34}S-depleted) sulphate derived from reduced sulphur sources such as sulphide minerals or soil organic sulphur. The actual isotopic composition of a sulphate source will vary with the age of the evaporites in the basin. It could be as light as 10‰, from Permo-Triassic evaporites to as heavy as 35‰, from late Proterozoic evaporites (Chapter 11, this volume). The isotopic composition of sulphate derived from reduced sulphur will depend on the composition of the reduced sulphur itself, the size of the reduced sulphur reservoir and on the type of process by which the sulphur is oxidized.

An immediate source of sulphate to rivers is likely to be sulphate and sulphur oxidized from reduced sulphur in the soil. Lowe et al. (1971) have investigated the isotopic composition of sulphate from soils in western Canada. They find $\delta^{34}S$ values from as high as +5, equivalent to the sulphate of rain in the area, to as low as −28‰. There appears to be no systematic difference between the $\delta^{34}S$ values of water-soluble sulphate present in the soil and organic sulphur derived from other soil fractions. $\delta^{34}S$ values of pyrite and other sulphide minerals disseminated widely in such rock as black shale vary widely (Chapter 11, this volume), but they tend to be concentrated at values below +10‰, the lower limit of evaporitic sulphate (Holser and Kaplan, 1966, fig. 1). In addition to the fact that sulphide minerals tend to be isotopically lighter than sulphate minerals, the sulphate produced by oxidation of sulphide minerals may be even more depleted in ^{34}S by the oxidation process itself. This has been demonstrated in laboratory studies by Kaplan and Rittenberg (1964) and has been found to occur naturally in the oxidation of pyrite (Nissenbaum and Rafter, 1967). Thus, although there is a wide range of δ^{34}S values which can be expected to be found in stream sulphate waters, there is a tendency for sulphate derived from evaporite deposits to be isotopically more enriched in ^{34}S than sulphate derived from the oxidation of organic sulphur or sulphide minerals.

There have been a number of studies of the isotopic composition of sulphate in river systems. Longinelli and Cortecci (1970) measured both the sulphur and oxygen isotopic content of sulphate in two rivers in Tuscany. They noted that both ^{34}S and ^{18}O were progressively enriched in sulphate from the head waters toward the outlet of the river systems. In the upper reaches of both rivers the sulphate concentration was less than 10 mg/l and the δ^{34}S values ranged from -3 in one system to $-12‰$ in the other. δ^{18}O values of the sulphate were from 0 to $+4‰$. Near the points of discharge of of rivers to the ocean the sulphate concentration was considerably higher and the δ^{34}S values were in the range of $+3$ to $+12‰$. This is still more depleted than normal oceanic sulphate which is about $+20‰$. The ^{18}O content of the sulphate, however, was in the range expected of oxygen in oceanic sulphate being in the range of $+6$ to $+11‰$. Longinelli and Cortecci were unable to correlate the changes they found in these rivers with the geology of the regions.

Hitchon and Krouse (1972) report a number of analyses from rivers and streams in the McKenzie River system of northwestern Canada. They found a range of δ^{34}S values for sulphate from -20 to $+20‰$. They noted a rough but clear relationship between the isotopic composition of the river sulphate and the geology of the basin. In particular, in basins draining Paleozoic rocks known to contain evaporites at depth, all the δ^{34}S values were positive. On the other hand, streams draining basins underlain by marine Cretaceous rock had negative δ^{34}S values. In these basins it is probable that the sulphate is derived from the oxidation of pyrite or from groundwater sulphate of unspecified origin. Finally, Chukhrov et al. (1975) present measurements made on a number of rivers throughout the U.S.S.R. They too found systematic relations between the geology of the basin and the δ^{34}S values in the sulphate of the rivers draining them. For example, the upper reaches of the Kuma River receives water from Cretaceous and Jurassic rocks containing a certain amount of evaporitic sulphate minerals. δ^{34}S values of the sulphate in this portion of the river range from $+8$ to $+14‰$. Further downstream, this river begins to receive sulphate presumably derived from pyrite in the basin and the δ^{34}S of the river sulphate decreases to $-5‰$.

The sulphur isotopic composition of lake water depends on that of the rivers or groundwater feeding the lake. It also depends on geochemical processes within the lake. An important process in many lakes is the reduction of sulphate to sulphide in the sediments on the lake's bottom or in the reducing bottom waters of stratified lakes. If the isotopic composition of the sulphur entering the lake is known, measurements of the isotopic composition of sulphate dissolved in the lake can be used as an indication of whether or not sulphate reduction is occurring and may even give some suggestion as to the amount of that reduction. Deevey and Nakai (1962) and Deevey et al. (1963) describe sulphur isotope studies on two lakes. In Linsley Pond, Connecticut, which is thoroughly mixed except during the late summer of each year, they found that while the $\delta^{34}S$ of the dissolved sulphate was usually about +7‰, during the autumn, toward the end of the lake's period of stratification, the bottom water sulphate concentration decreased and its $\delta^{34}S$ increased to +13‰. Sulphate extracted from bottom mud had a $\delta^{34}S$ of −7‰. The ^{34}S enrichment in the dissolved sulphate coupled with its concentration decrease was interpreted as a result of sulphate reduction, with ^{34}S-depleted sulphide being formed in the bottom muds. The measured $\delta^{34}S$ values suggested a fractionation during reduction of about 20‰.

There have also been a number of analyses of sulphur isotope ratios in the bottom waters of permanently stratified, meromictic, lakes. Deevey et al. (1963), and Matrosov et al. (1975) have found significant ^{34}S enrichment in the sulphate in the permanently stratified, reducing bottom layers of such lakes, coupled with ^{34}S-depleted reduced sulphur in the bottom mud and dissolved in the waters of these lakes. The process they find seems to exhibit all the features noted in laboratory bacterial sulphate reduction experiments. Fractionations between sulphate and sulphide as high as 55‰ have been observed. In addition, Matrosov et al. note that there tends to be less fractionation at lower sulphate concentration. There is considerable similarity between the behavior of sulphur compounds and sulphur isotopes in lake sediments and their behavior in oceanic sediments. A number of studies of the sulphur cycle in oceanic sediments are summarized by Goldhaber and Kaplan (1974, 1975).

The sources of sulphur compounds in groundwater are essentially the same as those to stream and river water. Groundwater may contain sulphate derived from rainfall, from the solution of evaporite minerals, or from the oxidation of sulphide minerals or organic sulphur. The range of conditions under which groundwater exists is much wider than the range of rivers, and therefore there are more chances that the isotopic composition of groundwater sulphur compounds may be changed by the operation of various geochemical processes. For example, the temperature range over which groundwaters are found may be from nearly zero to the critical point of water, a much wider range than rivers ever attain. At high temperatures, as in geothermal areas (Chapter 5, this volume) or in regions where hydrothermal ore deposits are being formed (Ohmoto, 1972), isotopic equilibrium may be attained between coexisting sulphate and sulphide species. Further, most surface waters, except the bottom portion of stratified lakes, are under oxidizing conditions so that the stable sulphur species present is sulphate. Groundwaters, like the lower portions of stratified lakes, may be reducing enough that sulphide can occur. Fractionation between coexisting sulphate and sulphide in reducing groundwaters is commonly in the range observed in laboratory bacterial experiments. However, in at least one case the residence time of the groundwater has apparently been long enough that fractionation between sulphate and sulphide approaches equilibrium at a low temperature.

FIELD STUDIES OF GROUNDWATER SYSTEMS

As examples of the use of sulphur and oxygen isotopes of dissolved sulphur compounds in groundwater investigations, we will describe three studies. One of these descriptions — that of the springs of Central Italy — is based on papers by Zuppi et al. (1974) and by Fontes and Zuppi (1976). The other two are based on work of Rightmire et al. (1974) and on unpublished data of R.O. Rye, F.J. Pearson, Jr., William Back, C.T. Rightmire, and B.B. Hanshaw.

Sulphide-bearing springs of Central Italy

The axis of the central Italian peninsula consists of Mesozoic limestones and dolostones, which are covered by Cenozoic sedimentary deposits in the Adriatic and Mediterranean coastal regions. In the western part of the region, in the vicinity of Rome, Quaternary volcanics have penetrated the sedimentary deposits and produced great thicknesses of detrital deposits and tufa. These rocks have regions of low permeability due to the presence of clay minerals resulting from the weathering of volcanic silicate minerals.

The region has been folded and deeply faulted. The main direction of faulting is north-northwest to south-southeast but east-west faults and thrusts are also present. Karst has developed on the carbonates in the central portion of the peninsula which, with fractures associated with the faulting, provide good conditions for infiltration. Deeper circulation is linked to the faults in the region and so the pattern of groundwater flow is neither simple nor obvious. Because of the faulting one cannot assume that surface water sheds correspond to groundwater basins.

Zuppi et al. (1974) and Fontes and Zuppi (1976) describe the isotope hydrology of springs issuing from the Mesozoic carbonates near the center of the peninsula at Antrodoco, Monte Sant'Angelo, and Cotilia, and from a carbonate outlier on the Adriatic coastal plain at Acquasanta. They also examined a spring at Tivoli, where the carbonates dip below the Quaternary sediments at the edge of the Mediterranean coastal plain, and one at

Lavinio which emerges from these sediments near the coast southeast of Rome. Among other stable and radio-isotope measurements they determined the ^{18}O contents of the waters and of the sulphate dissolved in them and the ^{34}S contents of both sulphate and sulphide.

All the springs but that at Acquasanta have relatively constant sulphate contents and temperatures. Dissolved sulphate contents range from 130 ± 10 mg/l at Monte Sant'Angelo to 790 ± 80 mg/l at Tivoli, and spring temperatures range from 13 ± 1°C at Monte Sant'Angelo to 23 ± 1°C at Tivoli. At Acquasanta, sulphate ranges from 300 to 750 mg/l and temperature from 24 to 35°C. The springs also contain dissolved sulphide, but its concentration is not reported.

The object of the isotope study was, in part, to determine the origin of the sulphur compounds in the springs, and from this to infer something of the pattern of groundwater circulation in the region. The geology suggests two possible sources of sulphur compounds: (1) sulphate from solution of gypsum or anhydrite, which is known to occur in Triassic sediments underlying the carbonates of the axis of the peninsula; and (2) reduced sulphur (as sulphide or elemental sulphur) associated with volcanic rocks.

Fig. 6-2 shows the ^{18}O content of the dissolved sulphate relative to that of the water from the several springs. With the exception of the spring at Lavinio there is a much wider variation in the ^{18}O content of the sulphate than that of the water. This variation suggests that there may well be two sources of sulphate to these waters, one of which, with ^{18}O values in the

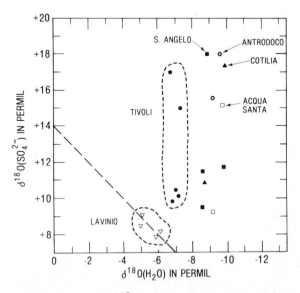

Fig. 6-2. Graph of $\delta^{18}O$ values of water and sulphate, central Italian springs (data from Zuppi et al., 1974).

range of +9 to +10‰, could be ocean-derived sulphate. It also demonstrates that oxygen isotope equilibrium between the water and the sulphate does not occur for if it did there would be a much closer relationship between the ^{18}O contents of the sulphate and the water.

There is some correlation between the ^{18}O contents of the sulphate and the water from the spring at Lavinio. The straight line in Fig. 6-2 is drawn through points from Lavinio and represents isotopic equilibrium between sulphate and water oxygen with a fractionation between them of +14‰. This fractionation corresponds to equilibrium at a temperature of about 135°C (Lloyd, 1968). It is of interest that the silica content of the spring at Lavinio ranges from 88 to 96 mg/l, which corresponds to quartz saturation at temperatures between 130 and 135°C (Truesdell, 1976, table 2). Although the surface temperature of the Lavinio spring is constantly within two degrees of 19°C the agreement between the calculated quartz and sulphate oxygen isotopic temperatures suggests that at depth water of 130 to 135°C is present in the system.

Fig. 6-3 shows the ^{34}S content of the sulphate and sulphide co-existing in the springs. This graph also suggests two sources of sulphur to the system. A heavy sulphur component with $\delta^{34}S$ of +15‰ or greater which appears in some of the samples from the springs at Tivoli and Antrodoco could well be marine in origin, either from modern oceanic sulphate as the ^{18}O suggests or from the solution of evaporite sulphate. The fact that the sulphate concentrations in these waters are relatively high is evidence that there is solution of the evaporitic sulphate known in the region.

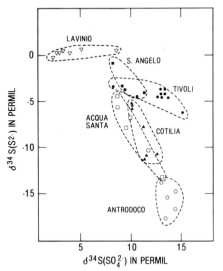

Fig. 6-3. Graph of $\delta^{34}S$ values of sulphate and sulphide, central Italian springs (after Zuppi et al., 1974, fig. 4).

The second source of sulphur in the system is exemplified by samples from the Lavinio spring in which both the sulphate and sulphide $\delta^{34}S$ values approach 0‰. Sulphur with this $\delta^{34}S$ is common in volcanic terrains (Schoen and Rye, 1970) and its presence in the region is quite likely.

The variation of $\delta^{34}S$ values from several of the springs is seasonal. In some of the springs, Acquasanta especially, there is very good correlation between the ^{34}S content of the sulphate and that of the sulphide. In Acquasanta there is also good correlation between the dissolved sulphate concentration and its $\delta^{34}S$ value, with low-sulphate waters having a higher ^{34}S content than higher-sulphate waters from Acquasanta. The correlations in Fig. 6-3 cannot be related to the attainment of sulphur isotope equilibrium in the system however, for they are in the wrong direction for equilibrium fractionation. Certainly however, chemical oxidation of elemental sulphur plays a part in the pattern shown in Fig. 6-3 in that, for Acquasanta at least, higher sulphate concentrations associated with less ^{34}S imply that additional sulphate is being derived from some relatively ^{34}S-poor source such as volcanic sulphur.

As mentioned, both the silica concentration and the ^{18}O contents of the dissolved sulphate and the water in the spring at Lavinio suggest a temperature at depth of at least 130°C. The pH of the Lavinio spring is about 3. According to equation (12), the half-time for attaining sulphate-water oxygen isotope equilibrium at these pH and temperature conditions is about 150 days. This time should also be sufficient to attain sulphur isotope equilibrium between dissolved sulphate and sulphide species. At 130°C, though, equilibrium fractionation between sulphate and H_2S, the dominant sulphide species in acid waters, is at least 35 to 40‰ (Fig. 6-1). Yet, the largest measured $\delta^{34}S$ difference between sulphate and sulphide at Lavinio is less than 10‰.

The relatively low ^{34}S content of the dissolved sulphate suggests that it results from oxidation of, presumably, volcanic sulphur with $\delta^{34}S$ around zero — the oxidation being accompanied by little fractionation. If this oxidation occurred during recharge to the system and the sulphate-bearing, but still oxidizing, water were then heated, the sulphur and oxygen isotopic composition of the sulphate would be accounted for. The sulphide in the spring, under this hypothesis, would result from near-discharge, kinetic reduction of a small amount of the sulphate. It is also possible that sulphide oxidation occurs throughout the flow system. In this case, the correlation between the ^{18}O contents of the water and sulphate (Fig. 6-2) would be merely an artifact of the kinetics of the oxidation process, and would have no temperature significance.

From this work, Zuppi et al. (1974), and Fontes and Zuppi (1976) conclude that:

(1) There are two sources of dissolved sulphur compounds to these springs: a ^{34}S-enriched source, probably evaporite sulphate from Triassic

sediments at depth, and a ^{34}S-depleted source, the reduced sulphur associated with the volcanic rocks.

(2) The range of variation in the isotopic compositions of the various springs reflects the circulation in the groundwater system feeding them. Antrodoco, for example, contains the most ^{34}S-enriched sulphate derived from evaporites at depth. Lavinio, on the other hand, derives sulphur from a volcanic source alone. The other springs have variable sulphur isotopic compositions which result from changing mixtures of waters with both volcanic and evaporitic sulphur.

Floridan aquifer

The principal artesian aquifer of Florida (the Floridan aquifer) is of Eocene age and is one of the most extensive limestone aquifers in the United States. In central Florida the maximum thickness of the fresh water in the aquifer is approximately 650 m. The area of study in Florida extends from the southern tip of the mainland to slightly north of Ocala. The area is tropical to subtropical and the primary climatic controls are its relatively low latitude and its proximity to the Atlantic Ocean, the Gulf of Mexico and numerous inland lakes. Mean annual rainfall ranges from 1300 to 1500 mm. The potentiometric high is in the center of the peninsula, southeast of Ocala. Water moves outward in all directions from the high and discharges near the coast. Rightmire et al. (1974) describe the hydrology of the region and discuss the sulphur isotope geochemistry of the aquifer.

The distribution of sulphate in the Floridan aquifer is essentially the same as that of the other constituents in the water, with the lowest concentrations, less than 100 mg/l, in the recharge area and in the area of highest potentiometric surface and a gradual sulphate increase downgradient to values greater than 500 mg/l.

A map of the δ^{34}S values of the sulphate ion does not show a systematic correlation with flow pattern, except that the lowest values are in the area of major recharge. Therefore, in order to identify the origin of sulphate ion in the groundwater, ratios of sulphate to chloride were plotted against sulphate concentration for the groundwater analyses, and ocean water (Fig. 6-4). Samples 2 through 5 have sulphate to chloride ratios of 0.2—0.3 and contain less than 5 mg/l of sulphate in solution. These samples were all taken from the area of high recharge south of Ocala and are characteristic of recharge to the system as a whole. Samples 36, 39, 40, and 41 are all near the coast or salt water embayments, have sulphate concentrations of 500 mg/l or more, and sulphate to chloride ratios approaching 0.1. They fall on a line defining a mixing trend between recharge water and ocean water. Another group of waters, samples 9, 10, 14, 16, 18, and 21 show increasing sulphate to chloride ratios with increasing sulphate concentration. Solution of sedimentary gypsum in waters of the recharge type would produce waters

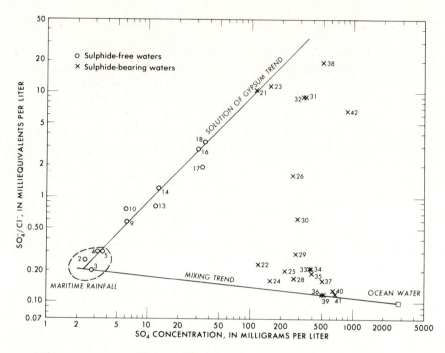

Fig. 6-4. Graph of sulphate to chloride ratio and sulphate concentration, Floridan aquifer (Rightmire et al., 1974, fig. 2).

such as these. The remainder of the waters shown in Fig. 6-4 are mixtures of sulphate derived from both solution of gypsum and from water chemically similar to ocean water.

A plot of the $\delta^{34}S$ values of the dissolved sulphate against sulphate concentrations (Fig. 6-5) substantiates these conclusions. The $\delta^{34}S$ values of the samples range from +8.1 to +29.6‰. The waters of low sulphate concentration in the recharge area have $\delta^{34}S$ values from +8 to +15‰, and sulphate to chloride ratios about twice those of seawater (Fig. 6-4). As discussed above, these are values typical of rainfall with some excess sulphate.

Samples with sulphate concentrations exceeding 100 mg/l have $\delta^{34}S$ values more positive than 20.7 with the majority being around +24‰. The $\delta^{34}S$ value of oceanic sulphate is +20‰, and that of 5 samples of evaporite sulphate minerals from the Floridan aquifer ranged from +19 to +22‰. The line in Fig. 6-5 is calculated for the addition of sulphate of $\delta^{34}S = +22‰$ to a recharge water with sulphate of $\delta^{34}S = +10‰$ and concentration of 3 mg/l.

Although the $\delta^{34}S$ values of many of the high-sulphate waters are consistent with a modern oceanic or Eocene evaporite mineral source, some sulphide-bearing waters have $\delta^{34}S$ values which are too positive. Sulphur isotope fractionation during sulphide production may result in the enrichment

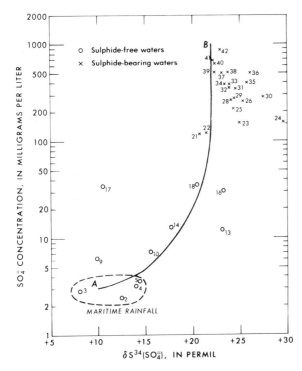

Fig. 6-5. Graph of $\delta^{34}S$ value and concentration of sulphate, Floridan aquifer (Rightmire et al., 1974, fig. 3).

of ^{34}S in the residual sulphate. Production of sulphide, by reduction of sulphate dissolved in these waters, could, therefore, result in a higher $\delta^{34}S$ value in the residual sulphate in solution and could have brought about the ^{34}S enrichment in these samples.

The sulphide content of water in the Floridan aquifer ranges from zero in the recharge area to 4.6 mg/l. The geographic distribution of dissolved sulphide concentrations is less regular than that of dissolved sulphate, but the $\delta^{34}S$ values of the dissolved sulphide (Fig. 6-6), are systematically distributed. Positive $\delta^{34}S$ values are in the upgradient area and they gradually become more negative downgradient to an extreme value of $-42‰$. These values are within the range of the $\delta^{34}S$ found in sulphide minerals within sedimentary rocks. If sulphide minerals have precipitated from water in the Floridan aquifer, the $\delta^{34}S$ in the minerals would be similar to the values shown for the water analysis.

The difference between the $\delta^{34}S$ values of sulphate and sulphide ($\Delta^{34} = \delta^{34}S(SO_4^{2-}) - \delta^{34}S(H_2S)$) are shown in Fig. 6-7. The distribution of δ^{34} values is similar to the pattern of groundwater flow in the aquifer. The lowest Δ^{34} values ($<58‰$) are present where sulphide occurs nearest the recharge area,

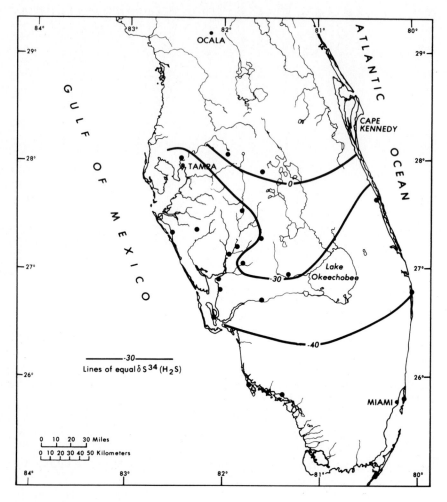

Fig. 6-6. Map of $\delta^{34}S$ values of H_2S in water of the Floridan aquifer.

while the highest (66‰) is at the extreme southern tip of the peninsula. The Δ^{34} values seem to be directly related to water residence time in the aquifer system, in that the highest Δ^{34} values are found in those regions of the aquifer most distant from its recharge area.

The most remarkable feature of the Δ^{34} values in the Floridan aquifer is how large they are. Water in this aquifer has pH values around 7.5, so that the dominant sulphide species is HS^-; its temperature is from 25 to 28°C. According to Sakai's (1968) calculations (Fig. 6-1), equilibrium fractionation between SO_4^{2-} and HS^- at 28°C is about 74‰. Ohmoto and Rye (1979) suggest that the fractionation is nearer to 65 to 66‰. Despite this

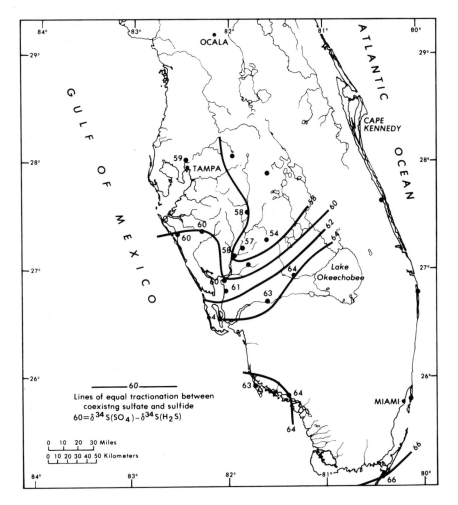

Fig. 6-7. Map of Δ^{34} ($= \delta^{34}S(SO_4^{2-}) - \delta^{34}S(H_2S)$) of water of the Floridan aquifer.

uncertainty it appears that the observed fractionation in the Floridan aquifer closely approaches and may actually represent isotopic equilibrium.

Because of the low rates of approach to sulphate-sulphide isotopic equilibria at low temperatures, it has long been assumed that observed low-temperature fractionation results from kinetic isotope effects accompanying bacterial sulphate reduction. Rye and Ohmoto (1974) point out, for example, that isotopic disequilibrium between reduced and oxidized sulphur species is not uncommon in hydrothermal solutions below 300°C.

Most laboratory bacterial sulphate reduction experiments are run at relatively high rates and produce sulphide some 30‰ different from the starting sulphate. There is experimental evidence (Kaplan and Rittenberg, 1964) that the fractionation increases as the rate of sulphate reduction decreases, and field evidence (Goldhaber and Kaplan, 1975) shows that equilibrium fractionation is approached at very low reduction rates. The low sulphide concentrations in Floridan groundwater and the scarcity of reduced sulphide minerals in the aquifer itself suggest that the rate of sulphate reduction there is low and accounts for the approach to sulphur isotope equilibrium.

From the chemistry and sulphur isotope ratios, the following conclusions about the geochemistry of sulphur in the Floridan aquifer can be drawn:

(1) The dissolved sulphate in the recharge area is rainfall derived.

(2) The increase in sulphate away from the recharge area results from solution of evaporite minerals in the aquifer itself, plus some mixing with seawater near coastal discharge points.

(3) Dissolved sulphide results from the reduction of sulphate. As a result of this reduction, the remaining sulphate is slightly enriched in ^{34}S.

(4) The rate of sulphate reduction is low enough that the differences between $\delta^{34}SO_4^{2-}$ and $\delta^{34}H_2S$ approach those expected at isotopic equilibrium.

Edwards aquifer, central Texas

The Edwards and associated limestones form the aquifer which is the chief source of water to the city of San Antonio and vicinity. The north and northwest parts of the San Antonio area are in the Edwards plateau region, while the south and southeast parts are on the Gulf coastal plain. These two physiographic areas are separated by the Balcones fault zone. The base flow of the streams that drain the Edwards plateau is spring flow from the water table aquifer in the plateau. This base flow and part of the flood flow are lost by seepage into the outcrop of the Edwards aquifer in the Balcones fault zone. Recharge to the aquifer is partly by this seepage and partly by direct infiltration of precipitation on the outcrop. The unconfined part of the aquifer nearest the fault zone and the downgradient, confined section adjacent to the south, contain an oxidizing calcium bicarbonate water with a total dissolved solids content of 250—300 mg/l. The general flow path in this part of the Edwards aquifer, which is used for water supply, is to the east and northeast with natural discharge at large springs. Downgradient, the character of the water changes abruptly. It becomes strongly reducing, has a high sulphate content and contains considerable quantities of hydrogen sulphide. At depth to the east and south, the Edwards contains petroleum and natural gas associated with brines containing over 100,000 mg/l chloride and more than 200 mg/l H_2S.

Rightmire et al. (1974) describe the sulphur isotope geochemistry of the Edwards. Chemical and isotopic data on Edwards waters are given by Pearson and Rettman (1976).

The composition of the water in the oxidizing, upgradient part of the aquifer varies little from place to place. The sulphate concentration there is from 7 to 31 mg/l, but has no regular pattern of spatial variation. Downgradient, in the reducing part of the aquifer, the dissolved solids content increases to over 4000 mg/l in the most southeasterly wells sampled. The dissolved sulphate and sulphide contents increase in the same direction. Water from the most southeasterly wells is saturated with respect to gypsum and contains nearly 2000 mg/l sulphate and 60 mg/l H_2S.

Fig. 6-8 displays the relationship between the sulphate to chloride ratio and the sulphate concentration of Edwards samples. The reducing waters are not all of the same chemical type, but have varying sulphate to chloride (Fig. 6-8) and calcium + magnesium to sodium + potassium ratios. Waters to the east of San Antonio have relatively higher sodium and chloride contents than those to the south. The downgradient increases of sodium and chloride in the reducing area result from increasing admixtures with the brines which occur in the Edwards at depth. The cause of the directional variability in the sulphate to chloride and calcium to sodium ratios is not known. An hypothesis to explain the variability is that to the east, the rate of gypsum solution relative to the rate of brine admixing is lower than it is to the south.

The variation of the dissolved sulphate concentration with its $\delta^{34}S$ value is shown in Fig. 6-9. This figure shows that the $\delta^{34}S$ values in the high-

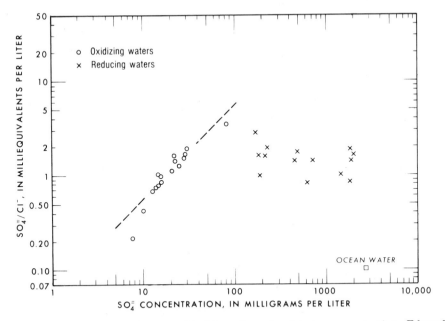

Fig. 6-8. Graph of sulphate to chloride ratio and sulphate concentration, Edwards aquifer (after Rightmire et al., 1974, fig. 3).

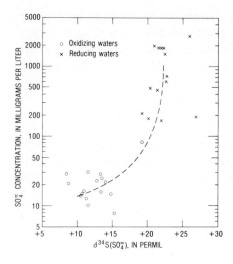

Fig. 6-9. Graph of $\delta^{34}S$ and concentration of sulphate, Edwards aquifer (after Rightmire et al., 1974, fig. 6).

sulphate, reducing waters are around +20 to +22 ‰, in agreement with the view that this sulphate is derived from the solution of evaporite gypsum. Unlike the Floridan aquifer (Fig. 6-5) the Edwards has few samples significantly more enriched in ^{34}S than +22‰, although the one brine sampled, to the south of San Antonio, has $\delta^{34}SO_4$ of +25.9‰. Enrichment of the Floridan sulphate was attributed to its being residual from sulphide reduction. By that reasoning, it would appear that relatively less reduction is occurring in the Edwards in spite of the much higher sulphide content, except possibly in the brines.

The variation in the sulphate to chloride ratio to the sulphate concentration in oxidizing waters from the Edwards (Fig. 6-8) resembles that of the sulphide-free waters of Florida (Fig. 6-4) in that both have some source of sulphate unaccompanied by chloride. In Florida, this was interpreted as solution of gypsum in the aquifer, a contention supported by the $\delta^{34}S$ values of the sulphate (Fig. 6-5). In the oxidizing Edwards, on the other hand (Fig. 6-9), there appears to be little correlation between sulphate concentration and $\delta^{34}S$ values, which suggests, the ratio plot (Fig. 6-8) not withstanding, that the solution of gypsum in the aquifer is not the primary source of the observed sulphate.

A possible interpretation of the $\delta^{34}S$ values of sulphate in the oxidizing waters in the Edwards is as follows: atmospheric precipitation in the region of Texas from which recharge to the Edwards is derived often occurs in brief, but rather violent, storms which are accompanied by high winds and dust. When the winds in the area of study are from the west they have had the opportunity to gather Permian evaporite dust. This dust settling or washed to the ground could then be a source of sulphate in the recharging water. Surface runoff would quickly dissolve sulphate dust fallout and thereby greatly increase the dissolved sulphate concentrations in the recharge waters.

The chloride in the rainfall is likely to be fairly constant as a result of oceanic aerosols in the precipitation but because of blowing dust sulphate concentration may vary greatly from storm to storm, depending upon the amount of dust associated with the storm. Thus, it is possible that waters entering the Edwards could have a widely varying sulphate to chloride ratio coupled with a relatively constant $\delta^{34}S$.

This hypothesis would explain the relatively constant chloride concentrations of the waters in the oxidizing Edwards. It would also account for the lack of systematic variation of the sulphate concentration within the aquifer; such lack resulting from the aquifer's many recharge points with each point deriving its water from a region with varying climatic, dust, and precipitation conditions.

The concentration and ^{34}S content of sulphide in the reducing part of the Edwards are also strikingly different from their values in the sulphide-bearing section of the Floridan aquifer. In Florida, sulphide concentrations (as H_2S) are less than 5 mg/l and have an irregular spatial distribution. In the Edwards, sulphide concentrations increase regularly downgradient in the reducing section of the aquifer and reach levels of 60 mg/l. $\delta^{34}S$ values of Edwards sulphide range from -10 to $-27‰$, and Δ^{34} values from 31 to 50‰. In Florida, $\delta^{34}S$ values are from $+7$ to $-42‰$ and Δ^{34} values reach 66‰. Thus, in contrast to Florida, the sulphide sulphur isotope geochemistry appears to be a result of more rapid sulphide production, and also to mixing between several sulphide sources.

The lack of isotopic equilibrium and the pattern of increasing sulphide with distance from the recharge area suggests that some of the sulphide in the wells sampled did not originate and is not being formed near the sampling points. Geochemical modeling of the Edwards system, now in progress, suggests that about 40% of the sulphide found is migrating into the system, presumably from the higher sulphide brines further downdip. The known extremely high mobility of hydrogen sulphide in water supports this hypothesis. The decrease in sulphide toward the oxidizing portion of the area may be due to a diffusion controlled concentration gradient with progressive oxidation of sulphide in the region near the upper limit of the reducing zone.

The chemistry of sulphur in the Edwards can be summarized as follows:

(1) The $\delta^{34}S$ content of sulphate in the oxidizing Edwards suggests that much of the sulphate is not derived from solution of evaporite minerals from the aquifer. Instead, this sulphate may be primarily recharge derived.

(2) The ^{34}S content of sulphate in the reducing part of the aquifer indicates that the high sulphate concentrations here do result from evaporite mineral solution. The absence of samples with ^{34}S contents higher than

expected from evaporites suggests that little sulphate reduction is occurring.

(3) The relatively constant ^{34}S contents of sulphide, the regular downgradient increase in sulphide concentrations, and the distribution of Δ^{34} values suggest that the sulphide in the reducing part of the Edwards is not entirely being formed where it is found. This is supported by the assertion in (2) that little sulphate reduction is occurring.

SUMMARY

(1) Ratios of stable sulphur (^{34}S/^{32}S) and oxygen (^{18}O/^{16}O) isotopes of sulphur compounds in natural waters may give information about the sources of the compounds and (or) geochemical processes and environments affecting them.

(2) Rates of approach to isotopic equilibrium between sulphate and water vary directly with temperature and inversely with solution pH, but are slow under earth's surface conditions. At 25°C and pH = 7, the half-time of the equilibrium exchange reaction is about 1900 years.

(3) Rates of approach to sulphur isotope equilibrium between sulphate and sulphide are also slow at earth's surface conditions. They are probably at least as slow as the rates of approach to oxygen isotope equilibrium between sulphate and water.

(4) Equilibrium fractionation factors between sulphate and water oxygen have been measured to temperatures as low as 25°C. Such factors for sulphate and sulphide sulphur have not been measured at low temperatures, but are available from statistical mechanical calculations.

(5) Kinetic fractionation effects bring about most of the observed differences in the sulphur isotope ratios of sulphate and sulphide coexisting at earth's surface temperatures.

(6) Sulphur occurs in the +6 oxidation state as sulphate (SO_4^{2-}) in oxidizing environments, as elemental sulphur (S^0) and in the —2 state as sulphide (H_2S or HS^-) in reducing environments. Common evaporite minerals include sulphate sulphur while many ore minerals are sulphides.

(7) ^{34}S tends to be enriched in sulphate and depleted in sulphide compounds. Modern oceanic sulphate and evaporite minerals have δ^{34}S values of +20‰. The δ^{34}S values of older evaporites vary with their age.

(8) The δ^{18}O value of modern oceanic sulphate is +9.6‰. Ocean water and sulphate are not in oxygen isotopic equilibrium, but at a steady state with respect to kinetic fractionation factors.

(9) Sulphate in surface waters may come from and have δ^{34}S values characteristic of sulphate in atmospheric dust and dissolved in precipitation, solution of soil sulphate and evaporite minerals, and from oxidation of sulphide minerals and soil organic sulphur. The ^{34}S content of sulphate in lakes may be increased by the preferential incorporation of ^{32}S in sulphide which may form in reducing bottom waters and sediments.

(10) Sulphur compounds in groundwaters have the same sources as those in surface waters. However, the isotopic composition of groundwater sulphur compounds may be affected by the operation of more geochemical processes than commonly occur in surface waters.

(11) Sulphur and oxygen isotope studies of sulphur compounds in three groundwater systems illustrate the use of these isotopes in determining the sources of sulphur compounds and the geochemical environment of the groundwater systems.

REFERENCES

Berner, R.A., 1970. Sedimentary pyrite formation. *Am. J. Sci.*, 268: 1—23.
Bigeleisen, J. and Mayer, M.G., 1947. Calculation of equilibrium constants for isotopic exchange reactions. *J. Chem. Phys.*, 15: 261—267.
Chukhrov, F.V., Churikov, V.S., Yermilova, L.P. and Nosik, L.P., 1975. On the variation of sulfur isotopic composition in some natural waters. *Geochem. Int.* 12(2): 20—33 (translation from *Geokhimiya*, 1975, 3: 343—356).
Clayton, R.N., Goldsmith, J.R., Kavel, K.J., Mayeda, T.K. and Newton, R.C., 1974. Limits on the effect of pressure on isotopic fractionation. *Geochim. Cosmochim. Acta*, 39: 1197—1201.
Cortecci, G., 1973. Oxygen-isotope variations in sulfate ions in the water of some Italian lakes. *Geochim. Cosmochim. Acta*, 37: 1531—1542.
Cortecci, G. and Longinelli, A., 1970. Isotopic composition of sulfate in rainwater, Pisa, Italy. *Earth Planet. Sci. Lett.*, 8: 36—40.
Cortecci, G. and Longinelli, A., 1971. $^{18}O/^{16}O$ ratios in sulfate from living marine organisms. *Earth Planet. Sci. Lett.*, 11: 273—276.
Cortecci, G. and Longinelli, A., 1973. $^{18}O/^{16}O$ ratios in sulfate from fossil shells. *Earth Planet. Sci. Lett.*, 19: 410—412.
Deevey, E.S. and Nakai, N., 1962. Fractionation of sulfur isotopes in lake waters. In: M.L. Jensen, *Symposium on Biogeochemistry of Sulfur Isotopes*. Yale University, New Haven, Conn., pp. 169—178.
Deevey, E.S., Nakai, N. and Stuiver, M., 1963. Fractionation of sulfur and carbon isotopes in a meromictic lake. *Science*, 139: 407—408.
Fisher, D.W., 1968. Annual variations in chemical composition of atmospheric precipitation, eastern North Carolina and southeastern Virginia. *U.S. Geol. Surv. Water-Supply Paper*, 1535-M: 1—21.
Fontes, J.Ch., 1965. Fractionnement isotopique dans l'eau de cristallisation du sulfate de calcium. *Geol. Rundsch.*, 55: 172—178.
Fontes, J.Ch. and Zuppi, C.M., 1976. Isotopes and water chemistry in sulphide-bearing springs of central Italy. In: *Interpretation of Environmental Isotope and Hydrochemical Data in Ground Water Hydrology*. International Atomic Energy Agency, Vienna, pp. 143—158.
Friedman, I. and O'Neil, J.R., 1977. Compilation of stable isotope fractionation factors of geochemical interest. In: M. Fleischer (Editor), *Data of Geochemistry. U.S. Geol. Surv. Prof. Paper*, 440-KK: 1—12 (6th ed.).
Garlick, G.D., 1969. The stable isotopes of oxygen. In: K.J. Wedepohl (Editor), *Handbook of Geochemistry*. Springer-Verlag, New York, N.Y., Sect. 8-B, pp. 1—26.
Garrels, R.M. and Christ, C.L., 1965. *Solutions, Minerals and Equilibria*. Harper and Row, New York, N.Y., 450 pp.

Goldhaber, M.B., 1974. *Equilibrium and Dynamic Aspects of the Marine Geochemistry of Sulfur*. Ph.D. Dissertation, University of California, Los Angeles, Calif., 399 pp.

Goldhaber, M.B. and Kaplan, I.R., 1974. The sulfur cycle. In: E.D. Goldberg (Editor), *The Sea, 5. Marine Chemistry*. John Wiley and Sons, New York, N.Y., pp. 569—655.

Goldhaber, M.B. and Kaplan, I.R., 1975. Controls and consequences of sulfate reduction rates in recent marine sediments. *Soil Sci.*, 119: 42—54.

Granger, H.C. and Warren, C.G., 1969. Unstable sulfur compounds and the origin of roll-type uranium deposits. *Econ. Geol.*, 64: 160—171.

Harrison, A.G. and Thode, H.G., 1957. The kinetic isotope effect in the chemical reduction of sulfate. *Trans. Faraday Soc.*, 53: 1648—1651.

Hitchon, B. and Krouse, H.R., 1972. Hydrogeochemistry of the surface waters of the Mackenzie River drainage basin, Canada, III. Stable isotopes of oxygen, carbon and sulfur. *Geochim. Cosmochim. Acta*, 36: 1337—1357.

Hoefs, J., 1973. *Stable Isotope Geochemistry*. Springer-Verlag, New York, N.Y., 140 pp.

Holser, W.T., 1977. Catastrophic chemical events in the history of the ocean. *Nature*, 267: 403—408.

Holser, W.T. and Kaplan, I.R., 1966. Isotope geochemistry of sedimentary sulfates. *Chem. Geol.*, 1: 93—135.

Jensen, M.L. and Nakai, N., 1961. Sources and isotopic composition of atmospheric sulfur. *Science*, 134: 2102—2104.

Junge, C.E. and Ryan, T., 1957. The oxidation of sulfur dioxide in dilute solutions. *Q. J. R. Meteorol. Soc.*, 15: 46—55.

Kaplan, I.R. and Rittenberg, S.C., 1964. Microbiological fractionation of sulphur isotopes. *J. Gen. Microbiol.*, 34: 195—212.

Kroopnick, P.M. and Craig, H., 1972. Atmospheric oxygen: Isotopic composition and solubility fractionation. *Science*, 175: 54—55.

Lloyd, R.M., 1967. Oxygen-18 composition of oceanic sulfate. *Science*, 156: 1228—1231.

Lloyd, R.M., 1968. Oxygen isotope behavior in the sulfate-water system. *J. Geophys. Res.*, 73: 6099—6110.

Longinelli, A. and Cortecci, G., 1970. Isotopic abundance of oxygen and sulfur in sulfate ions from river water. *Earth Planet. Sci. Lett.*, 7: 376—380.

Longinelli, A. and Craig, H., 1967. Oxygen-18 variations in sulfate ions in sea water and saline lakes. *Science*, 156: 56—59.

Lowe, L.F., Sasaki, A. and Krouse, H.R., 1971. Variations of sulfur-34: sulfur-32 ratios in soil fractions in western Canada. *Can. J. Soil Sci.*, 51: 129—131.

Matrosov, A.G., Chebotarev, Ye.N., Kudryavtseva, A.J., Zyukun, A.M. and Ivanov, M.V., 1975. Sulfur isotope composition in freshwater lakes containing H_2S. *Geochem. Int.*, 1975: 217—221 (translation from *Geokhimiya*, 1975: 943—947).

Mizutani, Y. and Rafter, T.A., 1969a. Oxygen isotopic composition of sulfates, 3. Oxygen isotopic fractionation in the bisulfate-water system. *N.Z. J. Sci.*, 12: 54—59.

Mizutani, Y. and Rafter, T.A., 1969b. Oxygen isotopic composition of sulphates, 4. Bacterial fractionation of oxygen isotopes in the reduction of sulphate and the oxidation of sulphur. *N.Z. J. Sci.*, 12: 60—68.

Mizutani, Y. and Rafter, T.A., 1969c. Oxygen isotopic composition of sulphates, 5. Isotopic composition of sulphate in rain water, Gracefield, New Zealand. *N.Z. J. Sci.*, 12: 69—80.

Mizutani, Y. and Rafter, T.A., 1973. Isotope behavior of sulphate oxygen isotopes in the bacterial reduction of sulphate. *Geochem. J.*, 6: 183—191.

Nakai, N. and Jensen, M.L., 1967. Sources of atmospheric sulfur compounds. *Geochem. J.*, 1: 199—210.

Nielsen, H., 1965. Schwefelisotope im marinen Kreislauf und das $\delta^{34}S$ der früheren Meere. *Geol. Rundsch.*, 55: 160—172.

Nissenbaum, A. and Rafter, T.A., 1967. Sulfur isotopes in altered pyrite concretions from Israel. *J. Sediment. Petrol.*, 37: 961—962.

Ohmoto, H., 1972. Systematics of sulfur and carbon isotopes in hydrothermal ore deposits. *Econ. Geol.*, 67: 551—578.

Ohmoto, H., and Rye, R.O., 1979. Isotopes of sulfur and carbon. In: H.L. Barnes (Editor), *Geochemistry of Hydrothermal Ore Deposits*. Holt, Rinehart and Winston, New York, N.Y., 2nd ed. (in press).

Östlund, G., 1959. Isotopic composition of sulfur in precipitation and sea-water. *Tellus*, 11: 478—480.

Pearson, F.J., Jr. and Fisher, D.W., 1971. Chemical composition of atmospheric precipitation in the northeastern United States. *U.S. Geol. Surv., Water-Supply Paper*, 1335-P: 1—23.

Pearson, F.J., Jr. and Rettman, P.L., 1976. Geochemical and isotopic analyses of waters associated with the Edwards limestone aquifer, central Texas. *Edwards Underground Water Distr. Rep.*, 35 pp.

Rafter, T.A. and Mizutani, Y., 1967. Oxygen isotopic composition of sulphates, 2. Preliminary results on oxygen isotopic variation in sulphates and the relationship to their environment and to their $\delta^{34}S$ values. *N.Z. J. Sci.*, 10: 816—840.

Rees, C.E., 1970. The sulphur isotope balance of the ocean: an improved model. *Earth Planet. Sci. Lett.*, 7: 366—370.

Richet, P., Bottinga, Y. and Javoy, M., 1977. A review of hydrogen, carbon, nitrogen, oxygen, sulphur, and chlorine stable isotope fractionation among gaseous molecules. *Annu. Rev. Earth Planet. Sci.*, 5: 65—110.

Rightmire, C.T., Pearson, F.J., Jr., Back, W., Rye, R.O. and Hanshaw, B.B., 1974. Distribution of sulfur isotopes of sulfates in ground waters from the principal artesian aquifer of Florida and the Edwards aquifer of Texas, United States of America. In: *Isotope Techniques in Ground Water Hydrology, 2. Vol. II*. IAEA, Vienna, pp. 191—207.

Robinson, B.W., 1973. Sulphur isotope equilibrium during sulphur hydrolysis at high temperatures. *Earth Planet. Sci. Lett.*, 18: 443—450.

Rye, R.O. and Ohmoto, H., 1974. Sulfur and carbon isotopes and ore genesis: a review. *Econ. Geol.*, 69: 826—842.

Sakai, H., 1957. Fractionation of sulfur isotopes in nature. *Geochim. Cosmochim. Acta*, 12: 150—169.

Sakai, H., 1968. Isotopic properties of sulfur compounds in hydrothermal processes. *Geochem. J.*, 2: 29—49.

Schoen, R. and Rye, R.O., 1970. Sulfur isotope distribution in solfataras, Yellowstone National Park. *Science*, 170: 1082—1084.

Schwarcz, H.P. and Cortecci, G., 1974. Isotopic analyses of spring and stream water sulfate from the Italian Alps and Apennines. *Chem. Geol.*, 13: 285—294.

Stumm, W. and Morgan, J.J., 1970. *Aquatic Chemistry*. Wiley-Interscience, New York, N.Y., 583 pp.

Teis, R.V., 1956. Isotopic composition of oxygen in natural sulfates. *Geochem. (USSR), Eng. Trans.*, pp. 257—263.

Thode, H.G., Cragg, C.B., Hulston, J.R. and Rees, C.E., 1971. Sulphur isotope exchange between sulphur dioxide and hydrogen sulphide. *Geochim. Cosmochim. Acta*, 35: 35—45.

Thorstenson, D.C., 1970. Equilibrium distribution of small organic molecules in natural waters. *Geochim. Cosmochim. Acta*, 34: 745—770.

Truesdell, A.H., 1976. Geochemical techniques in exploration, summary of section III. In: *Proceedings of the 2nd United Nations Symposium on the Development and Use of Geothermal Resources, San Francisco, Calif., 1975, 1*. U.S. Government Printing Office, Washington, D.C., pp. liii—lxiii.

Tudge, A.P. and Thode, H.G., 1950. Thermodynamic properties of isotopic compounds of sulphur. *Can. J. Res., Sect. B*, 28: 567—578.

Urey, H.C., 1947. The thermodynamic properties of isotopic substances. *J. Chem. Soc. (London)*, pp. 562—581.

Voge, H.H., 1939. Exchange reactions with radiosulphur. *J. Am. Chem. Soc.*, 61: 1032—1035.

Zuppi, G.M., Fontes, J.Ch. and Letolle, R., 1974. Isotopes du milieu et circulations d'eaux sulfurees dans le Latium. In: *Isotope Techniques in Groundwater Hydrology, 1*. IAEA, Vienna, pp. 341—361.

Chapter 7

URANIUM DISEQUILIBRIUM IN HYDROLOGIC STUDIES

J.K. OSMOND

INTRODUCTION

Under conditions present at the earth's surface, uranium tends to be a mobile element, and natural waters usually carry 0.1—10.0 µg/l. The presence of uranium in the environment is of special interest because of its radioactivity and that of its daughter products, which in turn makes it easy to detect and measure at low concentrations.

The principal factor influencing uranium geochemistry is its multiple valency: +4 under reducing conditions, and +6 in oxidizing environments. In the +6 state uranium is soluble, particularly if there is enough CO_2 in the system to allow the formation of the uranyl carbonate complex ions (Hostetler and Garrels, 1962). In the +4 state, it forms insoluble precipitates, such as UO_2.

The great range of concentration of uranium in groundwaters, as shown in Fig. 7-1, is largely due to this change in solubility at an Eh of about −0.1 to −0.2 (Hostetler and Garrels, 1962) and a pH of about 6 (Dement'yev and Syromyatnikov, 1968). Fig. 7-1 also shows an even more interesting aspect of uranium geochemistry, namely, the very extreme range of its isotopic composition, as expressed by the $^{234}U/^{238}U$ ratio. Whereas isotopic ratio variations of stable isotopes such as $^{18}O/^{16}O$ may be expressed in parts per thousand of relative difference with respect to a reference standard, those of $^{234}U/^{238}U$ can be measured in absolute ratios. These variations are not the result of mass fractionation effects, as in the case of light isotopes such as hydrogen and oxygen; nor are they the result of radiogenic accumulations of stable isotopes, as in the case of lead or argon.

^{234}U is a member of the ^{238}U radioactivity decay chain (lower part of Fig. 7-2). Inasmuch as each ^{238}U atom must experience the same series of events, the flux of atoms passing through each step should be the same, even though there will be fewer atoms at a given time at a short-lived isotopic stage. A useful analogy is that of a stretch of river channel with deep pools and shallow rapids where, as the old saying goes, "still water runs deep". Thus, in the absence of istopic fractionation effects, ^{234}U should have the same

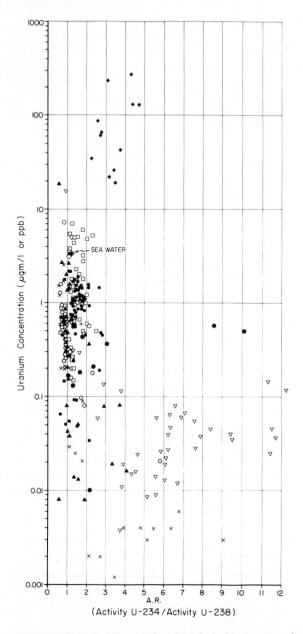

Fig. 7-1. Scatter diagram showing the wide ranges of variation of uranium concentration and $^{234}U/^{238}U$ activity ratio (A.R.) in groundwaters. The group of samples with high concentration (above 10 ppb) are from the vicinity or ore-bearing sandstone aquifers. The low-concentration, high-A.R. samples are from the reducing zones in deep aquifers. Most groundwater samples are seen to cluster between 0.1 and 10 ppb and A.R. in the range 0.7—2.0; this is also the region on the diagram where most surface waters would plot. This diagram is taken from Osmond and Cowart (1976); the various symbols refer to sources of data as assembled in their review.

alpha activity as its parent, and therefore a constant mass ratio:

$A_{234} = A_{238}$ (at radioactive equilibrium) \hfill (1)

$A_{234} = (N^{234})(\lambda_{234})$ \hfill (2)

$(N^{234})(\lambda_{234}) = (N^{238})(\lambda_{238})$ \hfill (3)

$N^{234}/N^{238} = \lambda_{238}/\lambda_{234} = T_{234}/T_{238}$ \hfill (4)

where N = number of atoms present; A = alpha activity (disintegrations per minute); λ = decay constant (reciprocal years); and T = half life (years), = to $0.693/\lambda$.

Because the half-life of ^{234}U, 250,000 years, is much shorter than that of ^{238}U, 4.5 billion years, equation (4) shows that the mass ratio of ^{234}U to ^{238}U is small (1:18,000) and ^{234}U can be neglected in computing uranium concentrations in minerals or waters. (The same can be said of ^{235}U, which seldom varies in abundance relative to ^{238}U, and will not be discussed here.)

The natural disequilibrium of $^{234}U/^{238}U$ may be measured either by mass spectrometer, and the results compared with the ratio calculated by equation (4); or by alpha spectrometry, and the results compared with equation (1) (as in Fig. 7-1). The latter is the more common procedure (see Appendix).

Isotopic variations of uranium in natural waters have been extensively investigated during the past decade as documented in the general reviews by Cherdyntsev (1969) and Osmond and Cowart (1976), the latter with 190 references. Confusion regarding the mechanism of fractionation, and uncertainty regarding the significance of regional variations, has tended to hold back progress in applications to hydrologic problems. This condition of tentativeness, we hope, is rapidly improving.

ISOTOPIC FRACTIONATION OF ^{234}U

The unexpected variation in alpha activity ratios (A.R.) of uranium in natural waters was first noted by Cherdyntsev et al. (1955) and other Russian investigators (Baranov et al., 1958; Starik et al., 1958; Chalov, 1959; Syromyatnikov, 1960); and later elaborated on by Thurber (1962), Rosholt et al. (1963) and others (Sakanoue and Hashimoto, 1964; Ku, 1965).

The cause of disequilibrium has been somewhat in dispute, but clearly depends in some way on the radiogenic origin of ^{234}U. The process which appears to explain most convincingly the natural variations and laboratory experiments is one in which ^{234}Th, the intervening daughter, escapes from mineral surfaces by alpha recoil.

"Recoil" is the term given to the sudden reactive movement of daughter isotope, through a distance of a few hundred angstrom units, when an alpha particle is expelled in the opposite direction. Gamma decay, resulting in no significant mass loss, produces no recoil; beta decay although involving enough energy to break chemical bonds, also does not cause any significant recoil. Early thinking about the causes of uranium isotope fractionation usually considered recoil as a mechanism which caused crystal lattice damage, but it was not until 1971 that Kigoshi suggested that recoil of the daughter isotope out of grain surfaces might be the dominant process. We shall refer here to "recoil transfer" as this direct fractionation mechanism in order to distinguish it from other "recoil" effects, which only served to place the daughter isotope in sites more vulnerable to leaching.

The recoil transfer model of uranium series disequilibrium is shown schematically in Fig. 7-2. Here the importance of an interface between high and low uranium concentration phases is emphasized. Such a model would appear to explain the data of Fig. 7-1, wherein high A.R.'s are found in very low concentration waters, and also in those traversing high-concentration orebodies.

An alternative possible fractionation process is one in which crystal lattice destruction due to recoil, and/or decay-related micro-environmental oxidation, renders the ^{234}U daughter more vulnerable to leaching than its parent ^{238}U. This view was the more popular among researchers, until Kigoshi (1971) observed that ^{234}Th atoms tend to escape from uraniferous zircon grains in water, even when no observable leaching occurs. Still more convincing evidence has recently been noted by J.N. Rosholt (in a semi-offical U.S. Geological Survey report, "Cross Sections, 1975"). He observes that in a very high uranium concentration slime from a processing plant the uranium in the water phase tends to develop a *low* A.R.; and in the suspended, lower concentration, solid particles, it tends to develop a *high* A.R. This is exactly the expected outcome according to the recoil transfer model depicted in Fig. 7-2 (if liquid and solid phases are reversed in terms of relative concentration) but appears to be inconsistent with models based on selective leaching.

The recoil transfer model would readily explain the natural fractionation between tetravalent and hexavalent uranium observed in natural phosphate nodules (Chalov and Merkulova, 1966; Kolodny and Kaplan, 1970) provided the reduced phases of the nodules have the higher uranium concentration (a point which appears to be plausible, but does not appear to have been checked).

It is very likely that such processes as selective leaching, oxidation, and chemical fractionation of ^{234}Th ($T = 25$ days) do occur, and contribute to some extent to the natural variation in ^{234}U/^{238}U, but recoil across phase boundaries appears to be the dominant process (Osmond and Cowart, 1976). (For an example of the continuing debate on this subject, see Yaron et al., 1976.) Inasmuch as recoil transfer involves a readily understood, and

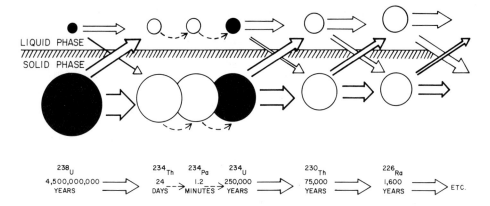

Fig. 7-2. The uranium decay series, and diagrammatic depiction of the alpha-recoil transfer model disequilibrium.

At the bottom of the figure is the decay scheme of the ^{238}U series through the first five daughters (after eight further short-lived events, stable ^{206}Pb results). The large arrows indicate alpha-decay events, the dotted arrows indicate beta-decay events.

The emission of an alpha particle, of several million electron-volt energy (but not of a beta particle) results in a recoil displacement of the daughter nucleus of a few hundred angstroms in solids and liquids. When this occurs near a solid-liquid phase boundary, as for example at the surface of a mineral grain in an aquifer, the transfer of the daughter atom from the solid phase to the liquid phase, or vice versa, is likely (upper part of diagram). Minerals usually have a higher uranium concentration than do pore waters. As a result of recoil displacement probabilities, the relative activities of successive daughters in the solid (near the interface) are expected to decrease, whereas the relative activities in the liquid should increase. Sizes of atoms and arrows represent relative activity. Thus, if leaching effects are minimal, we should, and do, find low alpha activity ratios (A.R.'s) between ^{234}U and ^{238}U (shaded spheres) in fine-grained rocks and sediments, and correspondingly high A.R.'s in groundwater. This is what is referred to in the text as the "recoil transfer" model of uranium isotopic disequilibrium.

A similar model probably explains the observed variations in isotopic composition of ^{232}Th series isotopes (^{228}Th/^{232}Th and ^{224}Ra/^{228}Ra) and of ^{235}U series isotopes (^{227}Th/^{231}Th). However, inasmuch as ^{235}U and ^{238}U are both parents only, their isotopic ratio in nature is generally constant.

to some extent quantifiable physical relationship, patterns of uranium isotope variations in nature are more significant clues to pore water and aquifer interaction process than would be the case if they were the result of subtle chemical leaching conditions.

If uranium in a natural initially closed system, such as unweathered rock, becomes separated into two phases, such as (1) dissolved in pore water and (2) in residual grains, the total uranium in the system will remain in equilibrium, even though its parts become isotopically fractionated. That is, if uranium in pore water develops a high A.R., the residual uranium in the insoluble minerals must have a low A.R., and the degree of disequilibrium in uranium in the two phases will be related to the amount of uranium in each phase:

$$M_s Y_s + M_w Y_w = M_t Y_t \tag{5}$$

where M_s, M_w and M_t are the amount of uranium in solid phase, liquid phase, and total system respectively ($M_s + M_w = M_t$), and Y_s, Y_w and Y_t are the corresponding activity ratios.

Rearranging terms, and letting $Y_t = 1.00$ (the total system is in equilibrium):

$$M_s/M_w = (Y_w - 1)/(1 - Y_s) \tag{6}$$

and:

$$Y_w = 1 + M_s/M_w(1 - Y_s) \tag{7}$$

These equations may be applied to the system soil/pore water, and also to deeper groundwater aquifers, if the conditions of a closed system can be approximated. It also may be applied to the earth's surficial zones as a whole. The average A.R. of the world's dissolved uranium, which is dominated by that in seawater with A.R. of 1.14 (Thurber, 1962; Koide and Goldberg, 1965; Umemoto, 1965) is balanced by the average A.R. of the residual uranium in weathered rocks, soils, and sediments. If the latter is taken to be about 0.92 (Rosholt et al. 1966; Scott, 1968, Sacket and Cook, 1969) then according to equation (6) the amount of uranium in weathered rock, soils and sediments must be about twice that dissolved in the sea.

In the case of aquifers in which very little leaching is occurring, Y_s depends on recoil distance relative to grain size and M_s/M_w depends on the phase concentrations and porosity. Applying equation (7), Y_w then reduces to a function primarily determined by the surface area to mass concentration ratio. A complicating variable is the decay of ^{234}U which affects the rate of development and/or destruction of "depleted rinds" on grain surfaces and "excess" terms in pore water. Most of the applications of uranium isotopic disequilibria to hydrologic problems depend on the relationships of equations (6) and (7). However, the field of study is somewhat underdeveloped, so far, in terms of quantification.

Whatever the model, it is clear that the principal locus of development of uranium isotopic disequilibrium is in soils and weathered rock. The only systematic study of uranium and thorium isotopes in soils is that of Rosholt et al. (1966). Fig. 7-3 is a diagrammatic simplification of their results, which shows that (a) uranium-thorium fractionation is more pronounced than $^{234}U/^{238}U$, but that both are appreciable; and (b) the ratio $^{234}U/^{238}U$ decreases progressively during the development of the C horizon, is maintained at the

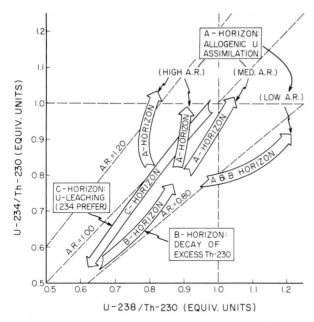

Fig. 7-3. Development of uranium series isotopic disequilibrium in soils (from Osmond and Cowart, 1976; highly modified from Rosholt et al., 1966). Starting with unweathered rock in isotopic equilibrium ($^{234}U/^{230}Th$, $^{238}U/^{230}Th$, and $^{234}U/^{238}U$ activity ratios are all equal to 1.00), weathering produces progressively greater U/Th fractionation as the C horizon develops, followed by a trend toward U/Th re-equilibration in the B horizon. This may be due in part to decay of excess ^{230}Th, in part to accumulation of secondary uranium, and in part to the fact that only very resistant minerals, such as quartz, zircon, etc., survive through the B horizon development. The $^{234}U/^{238}U$ A.R.'s (dashed diagonals) also diminish through the C horizon, but persist through development of the B horizon as well. Uranium isotopic disequilibrium is not produced by chemical fractionation, as is U/Th, but rather by alpha-recoil-related process, as indicated in the text, and in Fig. 7-2. The various trend lines in the A horizon reflect the accumulation of organically bound uranium, which may have a high A.R., a low A.R., or a value near equilibrium, depending on the climate and source of surface waters.

level of about 0.85 in the B horizon, and varies in the A horizon in accordance with local organic and climatic conditions.

From Fig. 7-3 and equation (7), we conclude that the A.R. of uranium in soil water and run-off water is a function of the maturity of soil development in a region. Where chemical weathering is dominant the ratio of uranium A.R. in regional run-off should be low to moderate. Conversely, in regions of arid climate or high relief, we expect immature soils, high M_s/M_w and consequently a high A.R. in regional run-off. A review of reported data by Osmond and Cowart (1976) seems to confirm this hypothesis.

In northern and central Florida, where the uranium in many samples has been analyzed, A.R.'s less than equilibrium are common (Osmond et al.,

1968; 1974; Kaufman et al., 1969). This is due, according to Osmond and Cowart (1976), to a regional accumulation of secondary residual uranium, held back by reducing barriers in the aquifer, in a fashion similar to that of a supergene enrichment process. In a swampy area like that of North Florida, we consider that nearly all of the uranium is dissolved out of the weathered rock and soil zone. This would tend to produce a regional run-off and surface aquifer uranium A.R. value close to unity. However, a major part of this uranium is immobilized by reducing conditions and held back, presumably on organic matter and/or clays. Water which passes through the reduced zone to greater depths will be characterized by low uranium concentrations and high A.R.'s (due to the recoil transfer process) like that of many other soil water systems. The volume of recharge water is great in this karstic system even though its uranium concentration is low, and an appreciable amount of low-A.R. uranium is left behind (the requirement of over-all equilibrium). Some of this low-A.R. uranium is oxidized and dissolved to escape in run-off water.

There is virtually no physical weathering in this region, and the mass balance equation can be applied, with run-off uranium and uranium in recharging water as the two components. If the A.R. value of run-off water is 0.85, and the corresponding value for water in the deep aquifer is 2.5, one may calculate that the mass ratio of uranium in run-off to that in recharge is about 10 (equation (6)). If the corresponding concentration values are 0.6 ppb and 0.03 ppb, respectively, then the water volume ratio is 1:2; that is twice as much water is infiltrating through the aquifer as is escaping as run-off. This is only a rough analysis. It is a useful exercise, however, because the standard water balance approach requires not only the monitoring of rainfall and run-off, but also an estimate of evapotranspiration rates in the region.

MIXING STUDIES: CONTINENTAL WATERS

In the previous examples of uranium "balance" calculations, equation (6) was applied by assuming initial, or total system, equilibrium, and calculating the proportions of uranium passing through the geochemical cycle via different pathways. Conversely, the A.R.'s of two-component parts of a system may be used to deduce their relative proportions as they combine, to form a product mixture. If the A.R.'s of uranium in all the component waters are determined, along with uranium concentration values, the combining proportions of up to three components may be calculated.

This has been the approach of the Florida State group in its studies of groundwater circulation patterns and surface water mixing proportions in Florida (Osmond et al., 1968; 1974). In effect, we are using dissolved uranium isotopes as "fingerprints" of natural water masses, and applying isotope dilution equations to solve for water volume proportions. The results are usually plotted on mixing diagrams, which have some of the properties of phase diagrams, where mixing components and resulting products are found along straight lines or within triangles.

There are several sets of equations by which this may be done, but each involves the determination of two parameters. One is ^{238}U concentration, in activity units or in concentration units (in the latter case, simple total uranium suffices). The other is either ^{234}U concentration (in activity units or "equivalent" concentration units); or some measure of the ratio of ^{234}U to ^{238}U. This may take the form of activity ratio (A.R.); or "excess", equal to $C(A.R. - 1)$, where C = concentration in equivalent concentration units. These equations and plots have been described in Kigoshi (1973), Osmond et al. (1974) and in Osmond and Cowart (1976).

The excess ^{234}U vs. uranium concentration type of plot is used for several of the figures of this review, rather than ^{234}U/^{234}U activity ratio. This is done because it tends to spread out the points on the Y-axis, and permits a better distinction among samples with A.R.'s close to unity. It should be emphasized, however, that any of the "correct" plotting methods will show mixing, leaching, dilution, and decay relationships as straight lines.

Osmond et al. (1974) studied the uranium isotopes in two major springs in central Florida, as well as in possible groundwater sources in the surrounding areas. They were able to estimate the proportions of the various sources to the springs which were consistent with other hydrologic data, such as the potentiometric surface and aquifer permeability data. However, they point out that there is inherent ambiguity to some extent if more than three sources are involved.

A very intensive application of this approach is Briel's analysis of the source waters of a river flowing through karstic terrain in North Florida (Briel, 1976). He noted that a plot of all surface and subsurface waters in the area formed a triangle, whose apices could be thought of as end-member waters (see Fig. 7-4). The upstream reaches of the river were dominated by a low concentration water with an excess of ^{234}U relative to ^{238}U. This type of water (A) is the normal surface run-off component of the area. In the middle reaches of the river, the influence of an underground source type increases. This source type (B) is characteristic of some of the springs of the area, and has a relatively high uranium concentration and a mild deficiency of ^{234}U. In the downstream reaches, another underground source and spring type (C) becomes important; it carries less uranium with a relatively greater deficiency of ^{234}U.

To visualize how such components isotopically affect the total river system, note (Fig. 7-4) how in each case the discharge of Ichetucknee springs (sample 31, type C end-member), "pulls" the river isotopic character from one plotted point (sample 26) to another plotted point (sample 36), along a straight-line mixing curve.

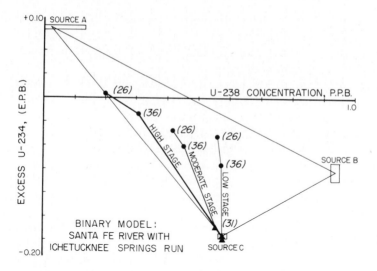

Fig. 7-4. Diagram showing how uranium isotope variations can be used to deduce water mixing proportions (from Briel, 1976). The abcissa, ^{238}U concentration in parts per billions, is equivalent to total uranium, inasmuch as ^{234}U in natural uranium makes up about 1 part in 10^4 by weight. The ordinate, "excess", is a measure of the ^{234}U present, as a proportion of the amount that would be in equilibrium with the ^{238}U, expressed as equivalent parts per billion. Samples plotting above the horizontal scale line have an A.R. greater than 1.00; those below have an A.R. less than 1.00.

Each trio of points represents the mixing system of Ichetucknee springs run (sample 31) as it joins the Santa Fe River (Florida, U.S.A.) sampled just upstream from the confluence (sample 26), to form the downstream waters of the Santa Fe (sample 36). The Ichetucknee springs are escaping groundwater, and produce a generally steady supply of water of stable uranium concentration and isotopic composition. The Santa Fe River varies in discharge stages with the seasons, and its isotopic character varies also, e.g., at low stages the ^{238}U concentration increases, and the relative amount of ^{234}U decreases (i.e., the proportion of "source B" is greater). The position of plotted points of samples from site 36 (the resultant) shows that the volume contribution of the springs is greater at low river stage. By using mixing equations, the volume proportions of 26 and 31 to make up 36 can be calculated; and also the relative amounts of the three regional sources (A, B, and C) to any river sample can be determined. Source C is Ichetucknee springs-type groundwater (sample 31); source B is another groundwater type (an end member whose isotopic character is inferred) which characterizes the region further upstream; source A is representative of run-off water near the headwaters of the stream.

As indicated in Fig. 7-4, the relative volumes of components varies with river stage, and the positions of plotted points reflect these changes in an easily visualized fashion. Deviations from this pattern were noted by Briel, and these are in themselves instructive. For example, in Fig. 7-4, at high river discharge stage, the geographic relationships of points 26, 31, and 36 are the same as at other stages, yet a significant component of source B-like water is suggested, because the plotted position of 36 is not on a straight line joining 26 and 31.

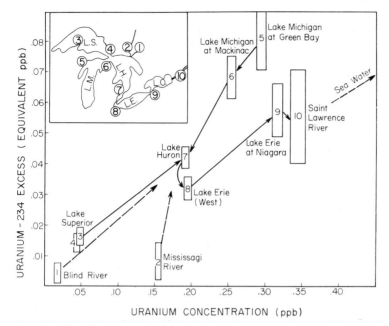

Fig. 7-5. The Great Lakes of North America cannot be considered to be parts of a closed system, but they do exhibit certain aspects of mixing systems. In this case, all samples have a positive excess of ^{234}U. Lake Superior water and Lake Michigan water types appear to be mixing to form Lake Huron water and Upper Lake Erie water. However, at the Niagara outlet from Erie, the water has more uranium and more excess ^{234}U, perhaps due to mixing with an unknown source, or by evaporation. Inasmuch as it is not a cause of isotopic fractionation for heavy elements, evaporation would cause both 234 excess and 238 concentration to increase in the same proportions, and thus form a radial vector away from the origin, as is observed. Another possibility is the mixing of upper Lake Erie water with more concentrated water with a similar A.R., like seawater (but if this were the case the salinity would increase also). Perhaps re-solution of uranium adsorbed on bottom sediments is occurring. In any event, mixing with another source of uranium with any other A.R. would be revealed by non-radial displacement. For example, pollution by uranium with A.R. near 1.0 would produce a displacement vector from 8 to 9 that would be horizontal.

Water passing through Lake Ontario, on the other hand, appears not to have changed much from that at Niagara (Osmond and Cowart, 1976).

Although this analysis of the Santa Fe River system is consistent with flow data monitored by the U.S. Geological Survey, Briel calculated other tributary mixing volumes by the uranium isotopic method which were not consistent. He cautions that in collecting river water samples one has to be careful to go far enough downstream from a confluence of sources to allow adequate mixing.

If analysis of 150 samples over a two-year period is necessary to study a small river system extending over about 100 km (above), one can be only tentative in judging the significance of the data reported by Osmond and

Cowart (1976) which attempts to characterize the vast Great Lakes system of North America with only 10 essentially synchronously collected samples (Fig. 7-5). Isotopically, Lake Huron (and the western basin of Lake Erie) can be considered to be a mixture of "Lake Superior-like" and "Lake Michigan-like" waters; but eastern Lake Erie, and Lake Ontario, have higher uranium concentrations. This pattern follows very closely that of total dissolved solids, but whether the increase across Lake Erie is due to evaporation, pollution, or other causes, is in dispute (Weiler and Chawla, 1969). That the increase in uranium does not involve a change in activity ratio (excess ^{234}U increases proportionately to ^{238}U in Fig. 7-5) argues against some unusual source of man-caused pollution.

MIXING AND URANIUM BALANCES: MARINE WATERS

On a global scale, there have been a number of attempts to compute the residence time of uranium in the seas by regarding the uranium in seawater as a mixture of "new" uranium coming from in-flow of rivers, with their low concentration and high A.R., and of "old" uranium at a high concentration in seawater which has been there long enough for the excess A.R. to have decayed significantly (Baht and Krishnaswamy, 1969; Sackett et al., 1973; Ku et al., 1974; Osmond and Cowart, 1977). Such attempts have been unsatisfactory thus far, in part because of an observation by Ku (1965) that a significant proportion of the excess ^{234}U in seawater may be derived by diffusion from sediments.

With respect to balance studies, and to studies of mixing of surface waters, Spalding and Sackett (1972) have raised the question whether uranium pollution has disrupted the natural hydrospheric system. Phosphate-based fertilizer carries about 100 ppm of uranium, which is readily leached and presumably mobilized in the run-off water and groundwater. However, preliminary studies by Kaufman (1974, and in progress) suggest that the effect may not be severe. For example, he appears to have a reasonable uranium isotopic balance (equation (6)) for the Mississippi River system based on natural fractionation assumptions only. The resulting calculation of flux ratio for the Mississippi River drainage area (a ratio of sediment carried to dissolved uranium, M_s/M_w, of about 4:1) agrees roughly with independent estimates of uranium concentration and sediment load studies.

A number of researchers, including Ku (1965) and Sacket and Cook (1969) have pointed out that the estimated flux of uranium reaching the sea is not balanced by the estimated removal rates into normal sediments, e.g., carbonates, red clays, etc. Anoxic basins on the shelf or slope, where the uranium content is known to be high (Veeh, 1967) may or may not be extensive enough to effect a balance. We are left with the possibility that the contemporary concentration of uranium in rivers is anomalously high (and steady-state estimates of uranium residence time in the seas accordingly suspect), or that a significant fraction of uranium dissolved in rivers is lost by adsorption onto sediments in estuaries and bays. It is possible that anoxic bays and fjords (Kolodny and Kaplan, 1973) may be sites of uranium removal even if they have no river in-puts, i.e., the source of the uranium is seawater itself. Martin et al. (1978) report on a study of uranium isotopes in waters and sediments of a river estuary in which co-precipitation with phosphate appears to be occurring. In this case, they conclude, uranium is behaving non-conservatively under oxidizing conditions.

Fig. 7-6 shows how the study of dissolved uranium isotopes might illuminate this and related subjects. The Indian River estuary along the east coast of Florida is not associated with any major rivers; it is a parallic bay, separated from the sea except at a few points, by continuous offshore islands. The two principal water sources to the estuary are seawater and groundwater. Uranium isotopic plots of estuary samples should form a

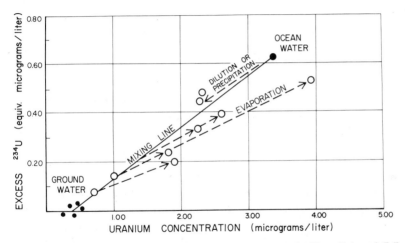

Fig. 7-6. The same kind of mixing diagram as shown in Figs. 7-4 and 7-5 can be used to evaluate the water and uranium mass balances in an estuary. The Indian River estuary along Florida's east coast has no major river inputs; its waters have two main sources: seawater and groundwater. (Several canals are feeding water into the estuary, but their isotopic character is that of groundwater.) The position of estuary samples along the line joining seawater and groundwater points indicates the relative volumes of the two sources. Samples plotting to the right of this line give evidence of uranium enrichment due to water evaporation, i.e., sample points move radially away from the origin. Samples plotting to the left of this line give evidence of uranium loss due to sediment scavenging, i.e., sample points move radially toward the origin. Radial vectors are based on the fact that neither evaporation nor co-precipitation will affect the isotopic ratio (radials on this type of plot are equal-A.R. lines). In this example, the very low concentration of uranium in the groundwater has produced a mixing line which is almost radial. The distinction between "mixing" and "evaporation/precipitation" variations, and thus the usefulness of the study, would have been enhanced if the two trend lines were at more of an angle.

straight-line mixing curve, each point falling along this line according to the proportions of the two sources. As shown in Fig. 7-6, deviations from this straight line can be interpreted, providing other sources are ruled out, as being due either to water evaporation, causing an increase in uranium concentration along a family of lines extending radially away from the origin; or rainfall or uranium precipitation, causing a decrease in uranium concentration along radials toward the origin.

AQUIFER INTERACTIONS

The use of dissolved uranium isotopes as mixing tracers depends on their conservative nature in near-surface waters. The uranium concentration and isotopic fingerprints are usually established in soil water (where initial acidity tends to be neutralized), and subsequent flow overland or in the shallow aquifer does not significantly alter them. This is a realistic approximation under oxidizing, high-CO_2 conditions. But at some depth, which may be quite shallow, if the aquifer contains reducing agents, the soluble +6 uranium complex breaks down and the precipitation of +4 uranium occurs. At this point, not only does non-conservation of dissolved uranium become a factor, but isotopic fractionation effects due to recoil transfer are amplified.

The study of the isotopic character of water in these reducing aquifers cannot be applied to mixing problems; however, several significant water-mineral interactions become amenable to study.

Fig. 7-7 shows two examples of how the uranium concentration and isotopic ratio change across reducing barriers. The precipitation of uranium is nearly quantitative and the concentration in the water decreases by 2 or 3 orders of magnitude, but the small amount remaining in solution is readily detected, and its A.R. measured. The increase in A.R. across such barriers is in many cases dramatic. A.R. values of 5, 10, 15, or more have been reported (Kronfeld and Adams, 1974; Osmond and Cowart, 1976). Cowart and Osmond (1974, 1977) believe that the alpha-recoil transfer model of uranium isotopic fractionation best explains this, and a similar view is held by Kronfeld et al. (1975). The argument here is best presented in terms of equation [7]. At these accumulation zones, where precipitation rather than leaching is occurring, much of the uranium is present as secondary coating on grain surfaces, and the M_s/M_w (ratio of concentration in solid vs. liquid phases) is very high. Furthermore, these coatings do not consist of primary crystals, where lattice defects due to recoil damage would be a factor in selective leaching.

An obvious application of this pattern is in the exploration for orebodies beneath the water table, those which may still be in the process of formation (Cowart and Osmond, 1977). The traditional approach to hydrogeochemical

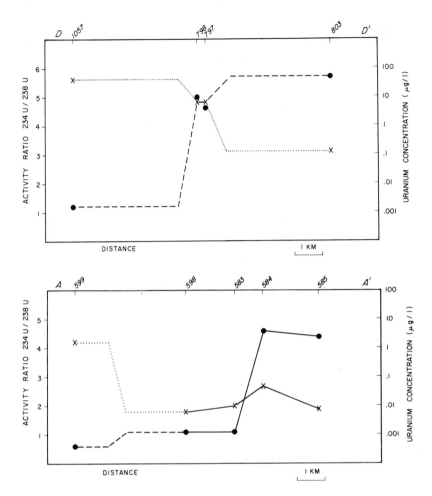

Fig. 7-7. Diagrams showing the effects of a reducing zone, and uranium accumulation, in a sandstone aquifer on uranium isotopic composition (Cowart and Osmond, 1978). As groundwater moves past a reducing barrier, from left to right on the diagrams, the uranium concentration (×'s) decreases dramatically, as indicated by the log scale of concentration (right). Concomitantly, the A.R. increases from values near unity or below, to 5 or 6 (circles and left-hand scale). The upper diagram is based on water samples near a known orebody, located at numbers 797 and 798. Lower diagram is from a normal sandstone aquifer, in which the reducing barrier has been identified, but in which an accumulation of uranium is only inferred.

prospecting, searching for high concentration uranium haloes, works only for those deposits near the surface where oxidizing waters are now dispersing previous accumulations. Non-dispersed orebodies will be indicated by *low* uranium concentrations, and by high A.R.'s in flow-through water.

Reducing barriers do not seem to be present in several other well-studied aquifers for which extensive uranium isotopic data are available e.g., those investigated by Wakshal and Yaron (1974), Kigoshi (1973), and by Alekseev et al. (1973).

Wakshal and Yaron (1974) studied several aquifers in Israel and noted that each appeared to be characterized by distinct isotopic ratios of their dissolved uranium. They suggested that this is the result of differing aquifer histories. Osmond and Cowart (1976) believe that the characteristic isotopic ratios in such aquifers without reducing barriers is due principally to climatic and topographic factors as they affect fractionation of uranium in the weathered zone.

Alternative possible explanation for the changing isotopic character of dissolved uranium in deep aquifers are that isotopic re-equilibration occurs, or that the down-dip aquifer water may be mixing with other waters due to faulting or breaching of confining beds. Some rather deep oil field brines have been analyzed by Kronfeld et al. (1975), and the very high A.R.'s they

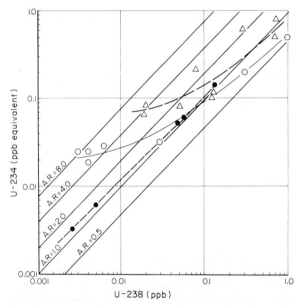

Fig. 7-8. The variations in uranium isotopic composition in three aquifer systems, as shown on log ^{234}U vs. log ^{238}U scales (Osmond and Cowart, 1976). Equal A.R. values plot as parallel diagonals. The triangles are samples from a karstic area in North Florida, characterized by moderate uranium concentrations and moderate variations in A.R. The open circles are samples from a sandstone aquifer in Texas, displaying extreme variations in concentration and A.R. (see lower diagram, Fig. 7-7). The solid circles are samples from the Yellowstone Park, U.S.A., geothermal region. The latter contain uranium with A.R.'s near equilibrium, even at very low concentrations, which probably reflects isotopic equilibration in geothermal systems.

found imply that *normal* geothermal gradients are not enough to counteract recoil-related isotopic fractionation effects at that depth.

Abnormal geothermal conditions may be another matter. A number of water samples from high-temperature geothermal systems have been analyzed by Cowart and Osmond (1976) and indications are that the high temperatures do in fact cause equilibration of dissolved and precipitated uranium isotopes (see Fig. 7-8). The unfractionated character is not due to the absence of a reducing barrier; Wollenberg (1975) presents strong evidence that precipitation of uranium does occur in geothermal systems.

The study of uranium series isotopes in geothermal waters is only beginning, but there appear to be at least two promising applications: (1) as evidence of high-temperature conditions at depth (re-equilibration), and (2) as an indicator of possible mixing with cooler surface water.

Aside from ore and energy considerations, the study of deep-aquifer uranium characteristics is of environmental interest in other respects.

A deep saline aquifer in Florida has been used increasingly as a liquid

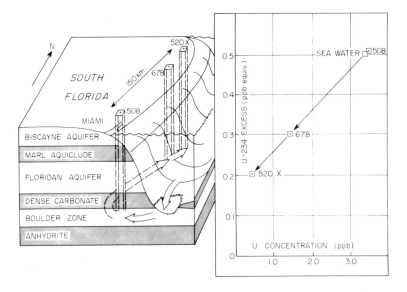

Fig. 7-9. Uranium isotopic evidence for convective circulation of seawater into the deep groundwater system of South Florida (Osmond and Cowart, 1977; based on data from Cowart et al., 1978). Kohout (1965) has proposed such a flow pattern, based on temperature profiles of wells drilled through the "boulder zone", which is actually only a cavernous limestone aquifer, at a depth of some 1000 m. The hypothetical intake site is in the bottom of the Florida Strait off Miami. Only three "boulder zone" water samples have been analyzed for uranium isotopes, but their plotted evolution appears to confirm the hypothesized flow pattern, as shown. Samples from inland wells have not yet been obtained. The "boulder zone" is being used at some localities as a liquid waste injection horizon. The wisdom of this practice will obviously depend on what is learned about the water circulation pattern.

waste injection horizon. This practice is based on the premise that the deep saline aquifer is hydraulically isolated from the extensive Floridan aquifer above, a major source of potable water. In the view of Kohout (1965), however, the water in this deep aquifer (called the "boulder zone" because its cavernous nature affects the drilling bit as if it were in gravel) is saline because it is part of a geothermally driven convective system, and the source of the water is the ocean, by way of "sink holes" in the bottom of Florida Strait. Isotopic analysis of dissolved uranium in the few samples available from this horizon are consistent with Kohout's hypothesis (Cowart et al., 1978) (see Fig. 7-9). If so, the waste water that is injected may not be sequestered, as originally hoped, but may in fact flow inland where danger of contamination of underground water supplies becomes more serious.

If uranium isotopic fractionation effects are in part a function of pore water to surface area ratios, it would seem to be a promising line of study of earthquake dilatency phenomena. That is, dilatant formation of crack surfaces and the "pumping" of pore water might favor the production of high A.R.'s in the groundwater as a precursor effect. (The increase in radon in groundwater is well known as a precursor phenomenon (Spiridonov and Tyminskiy, 1971). That this increase in radon might be due to a recoil transfer process across phase boundaries has been suggested by Osmond and Cowart (1976).) During the Tashkent series of earthquakes in 1966 some high uranium A.R.'s in groundwater were observed by Spiridonov and Tyminskiy (1971). Later, the same waters showed lower A.R.'s. Unfortunately, "base-line" analyses were not available to show whether, as hypothesized, the high A.R.'s were gradually developed prior to the earthquake.

AGE DATING

The half-life of ^{234}U, 250,000 years, is such as to make it attractive to the study of Pleistocene sediments (Ku, 1965), coral reefs (Veeh, 1966), playa lakes (Thurber, 1965), and cave deposits (Van and Lalou, 1969a; G.M. Thompson et al., 1975; P. Thompson et al., 1975). Further discussion of the subjects is outside the scope of this review.

Worth special mention, however, is the attempt to date groundwater itself. Kigoshi (1973) and Kronfeld et al. (1975) speculate that *increasing* A.R.'s should develop in groundwaters (the result of the recoil transfer mechanism) as a function of residence time in the aquifer, and Kigoshi proposed equations which could yield the "age" of the underground water. Alekseev et al. (1973) and Kronfeld and Adams (1974) observed *decreasing* A.R. values as water moves through well-defined aquifer flow paths. They proposed, independently, that this is a time-related function, due to decay of unsupported excess ^{234}U. This process also yields an easily calculated theoretical age for the far down-dip water.

Cowart and Osmond (1974), however, noted both a sharp *increase* and then a gradual *decrease* in the A.R. of underground water in the Carrizo aquifer of Texas. They note that the rapid increase in A.R., though probably related to recoil transfer, is associated with a synchronous *decrease* in uranium concentration, which violates the conditions required for the application of Kigoshi's formulae. Furthermore, in this aquifer, ^{14}C dating of the water shows that circulation is too fast (about 1.5 m/year) for decay of ^{234}U to be apparent down-dip. They suggest varying positions of the reducing barrier with time is the cause of the varying A.R.'s. Considering all the possible water-mineral interactions possible, the dating of "fossil" waters by uranium isotopes must be regarded at this time as speculative.

SUMMARY

Uranium can be used to study geochemical problems in the same way as other ionic species, but with the added dimension that isotopic variation gives. The parameter of ^{234}U/^{238}U activity ratio, or ^{234}U excess, is similar to other elemental isotopic measures, except for some significant advantages: (a) it is easy to measure by alpha radiation spectrometry, (b) the degree of variation is very great, (c) the concentration of uranium, by weight, in a sample is essentially independent of its isotopic variation, because the *number* of ^{234}U atoms is small, and (d) the mechanism of fractionation is much more interesting, geochemically, than if it were due to a simple mass effect.

The application of uranium isotopes to environmental studies can be grouped in four categories.

(1) The use of the conservative properties of dissolved uranium isotopes as fingerprints of water masses: to compute mixing volumes, and to identify water sources. The tracing of pollution sources is a variation of this approach.

(2) The use of dissolved uranium isotopes as non-conservative indicators of water-mineral interactions in aquifers. We have mentioned the location of orebodies, and the study of geothermal systems, and also more normal aquifers as indicators of paleohydrologic conditions. The potential for use of groundwater uranium in earthquake prediction is also a possibility, but at this time unevaluated.

(3) The use of uranium isotopes in rocks, soils, sediments, and natural waters as mass balance parameters. Efforts so far have been only tentative, but the potential for elucidation of soil formation processes and routes of sediment movement in the external parts of the rock cycle is great.

(4) The use of the ^{234}U/^{238}U ratio as an age dating parameter. Although firm ages, based on uranium only, have been few in number, the possibility of qualitative contributions to age estimates of sediments and cave deposits

is promising; as to the ages of underground water itself the question is still open.

ACKNOWLEDGEMENTS

The advice and criticism of Dr. J.B. Cowart of Florida State University is gratefully acknowledged. Marjorie Knapp and Allen Brown assisted with the illustrations and Louise Cox did the typing. Funds from the following sources were involved in part of the research and study reported here: U.S. National Science Foundation grant No. GA-41474; U.S. Geological Survey grant No. 14-08-001-G-262; and Office of Water Resources and Technology grant No. C-6066. Some of the Information summarized was collected while the author was on a NATO Senior Post-Doctoral Fellowship in Europe.

APPENDIX — ANALYTICAL TECHNIQUES FOR ^{234}U AND ^{238}U ANALYSIS

Although mass analysis to obtain ^{234}U/^{238}U isotopic ratios is possible and sometimes done (Bir, 1966; Rosholt and Nobles, 1969; P. Thompson et al., 1975), the more common procedure, requiring less expensive equipment, is to measure the alpha activity ratio of ^{234}U to ^{238}U. The two techniques produce approximately equivalent results relative to sensitivity and uncertainty limits; only the alpha pulse height analysis procedure will be described here.

A typical alpha spectrum, from a natural water sample, is shown in Fig. 7-10. Background alpha count rates are very low, less than a few counts per day per isotope usually, and interfering thorium isotopes are easily removed and not often abundant in waters. We would like to have at least a microgram of uranium on the counting planchet; with 30% counting geometry, this will produce about 0.2 counts per minute in the ^{238}U peak. Thus, with 3 or 4 days of counting, uncertainty limits of 3% can be achieved in individual isotope count rates, and about 4—5% in isotope abundance and isotope ratio determinations. The most frequently utilized synthetic yield tracer (spike) for determining concentration is ^{232}U.

Sample collection. If a natural water carries 0.1 µg/l, then 20 litres of water must be collected to achieve the desired count rates, given that chemical yields run about 50%. Many surface waters and some groundwaters have higher concentrations than this, and smaller samples can be collected and/or shorter counting times used. Many groundwaters and some surface waters are even lower in uranium concentration, and larger samples are required, combined with even longer counting times. It is obviously helpful if the approximate concentration can be estimated, not only so as to collect sufficient sample, but also to add the right amount of spike (too little adds unnecessarily to the counting times; too much adds to the ^{234}U peak background).

Because spike calibration is an important consideration, most investigators, if possible, spike and acidify to a pH of about 1 in the field, and transport the necessary sized samples to the laboratory. However, some researchers report that quantitative extraction of uranium in the field is possible, thus eliminating the need to transport large-volume samples. The preferred technique is by ion exchange (Bhat and Krishnaswami, 1969), or by activated charcoal (Van and Lalou, 1969b).

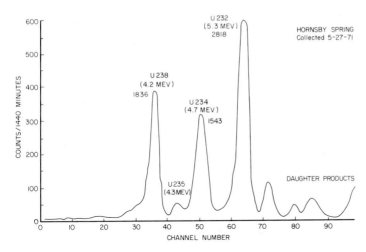

Fig. 7-10. Alpha energy spectrum of uranium from a Florida spring water sample (from Briel, 1976). The three naturally occurring isotopes ^{238}U, ^{235}U, and ^{234}U are identifiable, along with the ^{232}U isotopic spike, which was added at the time of sample collection. In this sample, the ^{234}U/^{238}U alpha activity ratio is 1543:1836, or 0.84. The ^{232}U spike activity added was 11.1 disintegrations per hour per liter, which is equivalent in activity to 0.25 µg/l of ^{238}U, so the uranium concentration is 1836:2818 × 0.25 = 0.16 µg/l. At these count rates, the background alpha activity can be neglected.

Laboratory separation. Co-precipitation, ion exchange, and solvent extraction can be used alternatively or in combination to effect separation of uranium from water and interfering elements. A useful summary and comparison of techniques has been reported by Veselsky (1974). Co-precipitation can be done with Fe(OH)$_3$ (Moore, 1967; Lal et al., 1964; Kronfeld and Adams, 1974; Briel, 1976) or with aluminum phosphate (Edwards, 1968). Various ion exchange procedures are described by Korkisch and Godl (1974), Tataru et al. (1971), Dement'yev (1967), and Miyake et al. (1966). Solvent extraction techniques are reported by Van and Lalou (1969b), Scott (1968), and Kigoshi (1973).

The final step is preparation of a low mass counting planchet, usually on stainless steel or platinum. The two standard procedures are evaporation of solvent (Scott, 1968; Van and Lalou, 1969b) or electrodeposition (Mitchel, 1960; Puphal and Olsen, 1972). In recent years, solid-state detectors in conjunction with multichannel pulse height analyzers have been used almost exclusively for the alpha-energy measurements.

REFERENCES

Alekseev, F.A., Gottikh, R.P., Zverev, V.L., Spiridov, A.I. and Grumbkov, A.P., 1973. The utilization of the isotope ratio ^{234}U/^{238}U in hydrologic research. *IAWA Panel on the Application of Uranium Isotope Disequilibrium in Hydrology, Vienna, March 1973* (unpublished manuscript).

Baht, S.F. and Krishnaswami, S., 1969. Isotopes of uranium and radium in Indian rivers. *Proc. Indian Acad. Sci., Sect. A*, 70: 1—17.

Baranov, V.L., Surkov, Yr.A. and Vilenskii, V.D., 1958. Isotope shifts in natural uranium compounds. *Geochemistry (USSR)*, 1958: 591—599.

Bir, R., 1966. Mesure de l'abondance naturelle de ^{234}U. In: W.L. Mead (Editor), *Advances in Mass Spectrometry. 3*. Institute of Petroleum, London, pp. 569—570.

Briel, L.I., 1976. *An Investigation of the $^{234}U/^{238}U$ Disequilibrium in the Natural Waters of the Santa Fe River Basin of North-Central Florida*. Ph.D. Dissertation, Florida State University, Tallahassee, Fla.,

Chalov, P.I., 1959. The $^{234}U/^{238}U$ ratio jn some secondary minerals. *Geochemistry*, 2: 203—3 210 (translated from *Geokhimiya*, 1959: 165).

Chalov, P.I. and Merkulova, K.I., 1966. Comparative oxidation rates of U-234 and U-238 atoms in certain minerals. *Dokl. Akad. Nauk SSSR*, 167: 146—148.

Cherdyntsev, V.V., 1969. *Uranium-234*. Israel Program for Scientific Translations, Jerusalem (translation of *Uran-234*, Atomizdat, Moskva, 1969) 230 pp.

Cherdyntsev, V.V., Chalov, P.I. and Khaidarov, G.Z., 1955. *Transactions of the Third Session of the Committee for the Determination of Absolute Ages of Geological Formations*. Akademii Nauk SSSR, Moskva, 175 pp.

Cowart, J.B. and Osmond, J.K., 1974. ^{234}U and ^{238}U in the Carrizo Sandstone aquifer of South Texas. In: *Isotope Techniques in Groundwater Hydrology, 2*. IAEA, Vienna, pp. 131—149.

Cowart, J.B. and Osmond, J.K., 1976. Dissolved uranium series nuclides in geothermal waters. *Geol. Soc. Am., Abstr. Progr.*, November.

Cowart, J.B. and Osmond, J.K., 1977. Uranium isotopes in groundwater: their use in prospecting for sandstone-type uranium deposits. *J. Geochem. Explor.*, 8: 365—380.

Cowart, J.B., Kaufman, M.I. and Osmond, J.K., 1978. Uranium-isotope variations in groundwaters of the Floridan Aquifer and Boulder Zone of South Florida. *J. Hydrol.*, 36: 161—172.

Dement'yev, V.S., 1967. Absorption of uranium (VI) from natural waters and carbonate-containing solutions by cationites. *Sov. Radiochem.* 9: 156—158.

Dement'yev, V.S. and Syromyatnikov, N.G., 1968. Conditions of formation of a sorption barrier to the migration of uranium in an oxidizing environment. *Geokhimiya*, 4: 459—465.

Edwards, K.W., 1968. Isotopic analysis of uranium in natural waters by alpha spectrometry. *U.S. Geol. Surv., Water-Supply Paper*, 1696-F: 26 pp.

Hostetler, P.B. and Garrels, R.M., 1962. Transportation and precipitation of uranium and vanadium at low temperature, with special reference to sandstone-type uranium deposits, Econ. Geol., 57: 137—167.

Kaufman, M.I., 1974. River mass balances as inferred from uranium isotope disequilibrium. *Geol. Soc. Am., Abstr. Progr.*, 6.

Kaufman, M.I., Rydell, H.S. and Osmond, J.K., 1969. $^{234}U/^{238}U$ disequilibrium as an aid to hydrologic study of the Floridan Aquifer. *J. Hydrol.*, 9: 374—386.

Kunihiko Kogoshi, 1971. Alpha-recoil thorium-234: dissolution into water and the uranium-234/uranium-238 disequilibrium in nature. *Science*, 173: 47—48.

Kigoshi, K., 1973. Uranium-238/234 disequilibrium and age of underground water. *IAEA Panel on Application of Uranium Isotope Disequilibrium in Hydrology*, Vienna, March 1973 (unpublished manuscript).

Kohout, F.A., 1965. A hypothesis concerning cyclic flow of salt water related to geothermal heating in the Floridan Aquifer. *N.Y. Acad. Sci., Ser. 2*, 28: 249—271.

Koide, M. and Goldberg, E.D., 1965. $^{234}U/^{238}U$ Ratios in Sea Water. In: M. Sears (Editor), *Progress in Oceanography, 3*. Pergamon Press, London, pp. 173—177.

Kolodny, Y. and Kaplan, I.R., 1970. Uranium isotopes in sea-floor phosphorites, *Geochim. Cosmochim. Acta*, 34: 3—24.

Kolodny, Y. and Kaplan, I.R., 1973. Deposition of uranium in the sediment and interstitial water of an anoxic fjord. In: *Proceedings of Symposium on Hydrogeochemistry and Biogeochemistry, Tokyo, 1970, 1*. Clarke, Washington, D.C., p. 418.

Korkisch, J. and Godl, L., 1974. Determination of uranium in natural waters after anion, exchange separation. Anal. Chim. Acta. 71: 113—121.

Kronfeld, J. and Adams, J.A.S., 1974. Hydrologic investigations of the ground waters of central Texas using U-234/U-238 disequilibrium. J. Hydrol., 22: 77—88.

Kronfeld, J., Gradztan, E., Muller, H.W., Radin, J., Yaniv, A. and Zach, R., 1975. Excess ^{234}U: an aging effect in confined water. Earth Planet. Sci. Lett., 27: 342—345.

Ku, T.-L., 1965. An evaluation of the ^{234}U/^{238}U method as a tool for dating pelagic sediments. J. Geophys. Res., 70: 3457—3474.

Ku, T.-L., Knauss, K.G. and Mathieu, G.G., 1974. Uranium in open ocean: concentration and isotopic composition (ABS). Trans. Am. Geophys. Union, 4: 314.

Lal, D., Arnold, J.R. and Somayajulu, B.L.K., 1964. A method for the extraction of trace elements from sea water. Geochim. Cosmochim. Acta, 28: 1111—1117.

Martin, J.-M., Nijampurkar, V. and Salvadori, F., 1979. Uranium and thorium isotope behavior in estuarine systems. In: E.D. Goldberg (Editor), Biogeochemistry of Estuarine Systems (in press).

Mitchell, R.F., 1960. Electrodeposition of actinide elements at tracer concentrations. Anal. Chem., 32: 326—328.

Miyake, Y., Sugimura, Y. and Uchida, T., 1966. Ratio ^{234}U/^{238}U and the uranium concentration in seawater in the western North Pacific. J. Geophys. Res., 71: 3083—3087.

Moore, W.S., 1967. Amazon and Mississippi river concentrations of uranium, thorium, and radium isotopes. Earth Planet. Sci. Lett., 2: 231—234.

Osmond, J.K. and Cowart, J.B., 1976. Theory and uses of natural uranium isotopic variations in hydrology. At. Energy Rev., 14: 621—680.

Osmond, J.K., Rydell, H.S. and Kaufman, M.I., 1968. Uranium disequilibrium in groundwater: an isotope dilution approach in hydrologic investigations. Science, 162: 997—999.

Osmond, J.K., Kaufman, M.I. and Cowart, J.B., 1974. Mixing volume calculations, sources, and aging trend of Floridan Aquifer water by uranium isotopic methods. Geochim. Cosmochim. Acta, 38: 1083—1100.

Puphal, K. and Olsen, D.R., 1972. Electrodeposition of alpha-emitting nuclides from a mixed oxalate-chloride electrolyte. Anal. Chem., 44: 284—289.

Rosholt, J.N. and Noble, D.C., 1969. Loss of uranium from crystallized silicic volcanic rocks. Earth Planet. Sci. Lett., 6: 268—270.

Rosholt, J.N., Shields, W.R. and Garner, E.L., 1963. Isotopic fractionation of uranium in sandstone. Science, 139: 224—226.

Rosholt, J.N., Doe, B.R. and Tatsumoto, M., 1966. Evolution of the isotopic composition of uranium and thorium in soil profiles. Geol. Soc. Am. Bull., 77: 987—1003.

Sackett, W.M. and Cook, G., 1969. Uranium geochemistry of the Gulf of Mexico. Trans. Gulf Coast Assoc. Geol. Soc., 19: 233—238.

Sackett, W.M., Mo, T., Spalding, R.F. and Exner, M.E., 1973. A re-evaluation of the marine geochemistry of uranium. Radioactive Contamination of the Marine Environment. IAEA, Vienna, pp. 757—769.

Sakanoue, M. and Hashimoto, T., 1964. A study of ^{234}U/^{238}U ratio in natural samples. Nippon Kagoku Zasshi, 85: 622—627.

Scott, M.R., 1968. Thorium and uranium concentrations and isotope ratios in river sediments, Earth Planet. Sci. Lett., 4: 245—252.

Spalding, R.F. and Sackett, W.M., 1972. Uranium in runoff from the Gulf of Mexico distributive province: anomalous concentrations. Science, 175: 629—631.

Spiridonov, A.I. and Tyminskiy, V.G., 1971. On changes in the ^{234}U/^{238}U isotope ratio in underground water after the Tashkent earthquake of 1966-1967. Izv., Acad. Sci. USSR, Phys. Solid Earth, 3: 214—216.

Starik, I.E., Starik, F.E. and Mikhailov, B.A., 1958. Shifts of isotopic ratios in natural materials. Geochemisty, 5: 587—590.

Syromyatnikov, N.F., 1960. Interphase isotopic exchange of ^{234}U and ^{238}U, Geochemistry. 3: 320—327.

Tataru, S., David, G. and Filip, G., 1971. Absorption of uranium by ion-exchange resins in the presence of SO_4 anions. Rev. Roum. Chim., 16: 625—630.

Thompson, G.M., Lumsden, D.N., Walker, R.L. and Carter, J.A., 1975. Uranium series dating of stalagmites from Blanchard Springs Cavern, U.S.A., Geochim. Cosmochim. Acta, 39: 1211—1218.

Thompson, P., Ford, D.C. and Schwartz, H.P., 1975. ^{234}U/^{238}U ratios in limestone cave seepage waters and speleothems from West Virginia. Geochim. Cosmochim. Acta, 39: 661—670.

Thurber, D.L., 1962. Anomalous ^{234}U/^{238}U in nature. J. Geophys. Res., 67: 4518—4520.

Thurber, D., 1965. The concentrations of some natural radio-elements in the waters of the Great Basin. Bull. Volcanol., 28: 195—201.

Umemoto, S., 1965. ^{234}U/^{238}U in seawater from the Kuroshio region. J. Geophys. Res., 70: 5326—5327.

Van, N.H. and Lalou, C., 1969a. Comportement géochimique des isotopes des familles de l'uranium et du thorium dans les concrétionnements de grottes: application à la datation des stalagmites. C. R. Acad. Sci. Paris, Ser. D, 269: 560—563.

Van, N.H. and Lalou, C., 1969b. Determination simultanes du rapport isotopique ^{234}U/^{238}U et de la teneur in uranium 238 dans les eaux naturelles. Radiochim. Acta, 12: 156—160.

Veeh, H.H., 1966. ^{230}Th/^{238}U ages of pleistocene high sea level stands. J. Geophys. Res., 71: 3379—3386.

Veeh, H.H., 1967. Deposition of uranium from the ocean, Earth Planet. Sci. Lett., 3: 145—150.

Veselsky, J., 1974. An improved method for the determination of the ratio ^{234}U/^{238}U in natural waters, Radiochim. Acta, 21: 151—154.

Wakshal, E. and Yaron, F., 1974. ^{234}U-^{238}U disequilibrium in waters of the Judea group (Cenomanian-Turonian) aquifer in Galilee, northern Israel. In: Isotope Techniques in Groundwater Hydrology, 2. IAEA, Vienna, pp. 151—175.

Weiler, R.R. and Chawla, V.K., 1969. Dissolved mineral quality of Great Lakes waters. Proceedings, 12th Conference on Great Lakes Research. University of Michigan, Ann Arbor, Mich., pp. 801—818.

Wollenberg, H.A., 1975. Radioactivity of geothermal systems. In: Second U.N. Symposium on Development and Use of Geothermal Resources, San Francisco, Calif., 1975, 3. U.S. Government Printing Office, Washington, D.C., p. 95.

Yaron, F., Kronfeld, J., Gradsztan, E., Radin, J., Yaniv, A., Zach, R., Starinsky, A., Kolodny, Y. and Katz, A., 1976. Excess ^{234}U: an aging effect in confined waters: comments and replies. Earth Planet. Sci. Lett., 33: 176—182.

Chapter 8

OXYGEN AND HYDROGEN ISOTOPE EFFECTS IN LOW-TEMPERATURE MINERAL-WATER INTERACTIONS

SAMUEL M. SAVIN

INTRODUCTION

The isotopic composition of a mineral is a function of the temperature and of the isotopic composition of the (usually aqueous) environment with which it has interacted. Where isotopic equilibrium between the mineral and the environment has not been achieved, the isotopic composition of the mineral also depends on the nature and extent of interaction with the environment. Therefore, in many instances the isotopic composition of a mineral or a suite of co-existing minerals can yield information about some aspect of its history. In some cases this may be information about the environment in which the mineral was formed or was obviously altered, and in other cases simply the environment in which it existed for some time.

Exchange of oxygen or hydrogen isotopes between a mineral and water must affect the isotopic composition of both phases. In most surface and near-surface environments the amount of water is much larger than the amount of exchanging solid phase, and the isotopic composition of the water is therefore not measurably altered on a local scale. (On a worldwide scale, however, over geologically long times, the isotopic composition of the hydrosphere must be measurably affected by such processes.) Where the amount of water is comparable to or less than the amount of rock with which it interacts, the isotopic composition of the water may be substantially altered. This could be important in some hydrologic investigations using stable isotope techniques.

This chapter deals with isotopic exchange between minerals and water at temperatures normally considered "sedimentary" or "diagenetic". Emphasis is on the following groups of minerals: silica minerals, aluminosilicates (primarily clay minerals), hydroxides, oxides, phosphates, sulfates and carbonates. It may be useful to keep in mind the following generalization, which will be supported by the data presented. *At sedimentary temperatures isotopic exchange between most minerals and water is slow except when*

Contribution No. 121, Department of Earth Sciences, Case Western Reserve University.

chemical or mineralogical reaction involving the two phases occurs. At higher temperatures minerals are more susceptible to isotopic exchange without obvious chemical alteration than they are at low temperatures. The temperature at which such exchange becomes appreciable differs for different minerals.

ISOTOPIC FRACTIONATIONS BETWEEN MINERALS AND WATER

Equilibrium fractionation processes

Equilibrium isotopic fractionations are frequently expressed as fractionation factors, α:

$\alpha = R_a/R_b$

where a and b denote two phases in equilibrium and R is an isotope ratio: D/H, $^{13}C/^{12}C$, $^{18}O/^{16}O$, etc. (There is some ambiguity in the literature, as α is used by some authors to express the magnitude of isotopic fractionation regardless of whether equilibrium is inferred.)

Oxygen isotope fractionations. Equilibrium oxygen isotope fractionations between minerals and water have been estimated using a number of techniques including: isotopic exchange experiments; growth of minerals by precipitation from solution or by reaction between a solution and an existing solid; theoretical calculations; and comparison of the $\delta^{18}O$ values of naturally occurring phases with those of the solutions with which they are assumed to have formed. Estimates of fractionation factors at low temperatures obtained in different ways may differ. While for many purposes the values given here are sufficiently accurate, revised values for many fractionation factors will certainly be published in the future.

Oxygen isotope fractionation factors typically vary with temperature over large temperature ranges according to the relationship:

$1000 \ln \alpha = AT^{-2} + B$

where T is in °K. To the extent that this relationship holds, fractionation factors can be estimated by linear extrapolation outside the temperature range for which experimental data are available. Where data are available over a limited temperature range, isotopic fractionations are sometimes expressed simply as linear or quadratic functions of temperature (°C).

Estimated oxygen isotope fractionation factors of a number of minerals commonly involved in low-temperature processes are given in Tables 8-1, 8-2, and 8-3. The discrepancies among fractionations between the same

Mineral-water oxygen isotope fractionation factors expressed in the form: $1000 \ln \alpha = AT^{-2} + B$

Mineral	Temperature range (°C)	Method	$10^6 A$	B	Source
Quartz	200—500	synthesis experiments; incomplete exchange experiments	3.38	−3.40	Clayton et al. (1972)
	500—750	complete exchange experiments	2.51	−1.96	Clayton et al. (1972)
Muscovite	400—650	exchange experiments; synthesis experiments	2.38	−3.89	O'Neil and Taylor (1969)
Illite	160—270	naturally occurring quartz-illite pairs; Clayton et al. quartz-water fractionation (extrapolated)	2.43	−4.82	Eslinger and Savin (1973)
Smectite *	0—270	naturally occurring low-temperature smectite; Eslinger and Savin illite-water fractionation (extrapolated)	2.67	−4.82	Yeh and Savin (1977)
Mixed-layer illite/smectite *	0—270	linear interpolation between Eslinger and Savin illite-water fractionation and Yeh and Savin smectite-water fractionation	2.43 + 0.24 × (% expandable layers)	−4.82	Yeh and Savin (1977)
Serpentine	not specified	combination of estimates from naturally occurring mineral pairs and laboratory studies; approximate	0.6	−5.4	Wenner and Taylor (1971)
Calcite	0—700	equilibration experiments; direct precipitation	2.78	−3.39	O'Neil et al. (1969)
Dolomite	300—510	partial exchange experiments	3.20	−2.00	Northrop and Clayton (1966)
	350—400	mineral-CO_2 exchange experiments combined with calcite-water fractionation (O'Neil et al., 1969)	3.34	−2.94	O'Neil and Epstein (1966)
Proto-dolomite	25— 80	direct precipitation	2.62 **	+2.2	Fritz and Smith (1970)

* Fractionations for the aluminosilicate portion of the mineral after removal of interlayer water.
** Calculated by Garlick (1974) from data of Fritz and Smith (1970).

TABLE 8-2

Mineral-water oxygen isotope fractionation factors expressed in the form: $t(°C) = a + b(\delta_m - \delta_w) + c(\delta_m - \delta_w)^2$

Mineral	Temperature range (°C)	Method	a	b	c	Source
Shell phosphate	2–28	naturally occurring marine organisms	109.2 *	−4.3	—	Longinelli and Nuti (1973a) corrected by Friedman and O'Neil (1977)
Biogenic silica (dehydroxylated)	3–30	naturally occurring biogenic silica	169 *	−4.1	—	Labeyrie (1974)
Calcite	7–30	calcite biogenically precipitated under controlled or natural conditions	16.4 **	−4.2	0.13	Epstein (in preparation) from Epstein et al. (1953)

* δ_w and δ_m are the isotopic compositions of water and mineral relative to SMOW.
** δ_w in the carbonate equation is defined as the $\delta^{18}O$ value of CO_2 equilibrated with the water at 25°C. δ_m is the isotopic composition of the CO_2 extracted from the carbonate using 100% H_3PO_4 at 25°C. Another frequently used version of this equation has been modified by Craig (1965) from that given in Epstein et al. (1953). In Craig's version, a has a value of 16.9.

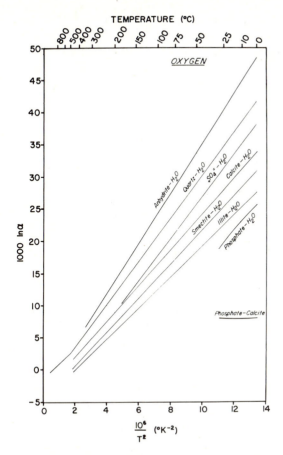

Fig. 8-1. Oxygen isotope fractionations plotted as a function of temperature. Sources: anhydrite-H_2O and SO_4^{2-}-H_2O, Lloyd (1968); quartz-H_2O, Clayton et al. (1972); calcite-H_2O, O'Neil et al. (1969); smectite-H_2O, Yeh and Savin (1977); illite-H_2O, Eslinger and Savin (1973); phosphate-H_2O, Longinelli and Nuti (1973a).

phases, as estimated in different studies, are indicative of the uncertainties with which many of these values are known.

Oxygen isotopic fractionation curves for several minerals are plotted in Fig. 8-1. Additional useful isotopic fractionations have been compiled by Friedman and O'Neil (1977).

Possible effect of grain size on isotopic fractionation factors. At 200°C the oxygen isotope fractionation between CO_2 and the surface layer of calcite crystals is 5 or 6‰ greater than the fractionation between CO_2 and the bulk of the crystals (Hamza and Broecker, 1974). An alternate way to think of this is that the surface layer of the crystal is depleted in ^{18}O by 5 or 6‰ rela-

TABLE 8-3

Mineral-water oxygen and hydrogen isotope fractionations estimated at single temperatures from compositions of naturally occurring materials

Mineral	Estimated temperature (°C)	Type of occurrence	α^{oxygen}	$\alpha^{hydrogen}$	Source
Quartz	0 * (?)	euhedral crystals dredged from ocean floor	1.036		Garlick (1969)
	0 *	cherty layers in sepiolite dredged from sea floor	1.036		Savin (1973)
	0	chert dredged from sea floor	1.039		Knauth and Epstein (1975)
Phillipsite	0	phillipsite from 3 deep sea cores	1.034		Savin and Epstein (1970b)
Glauconite	6	glauconite dredged from Blake Plateau	1.026	0.93	Savin and Epstein (1970a)
Illite	22	laboratory experiments (partial equilibration)	1.024		James and Baker (1976)

Mineral	Temp	Description	Value	Value2	Reference
Smectite	0—15	marine bentonites; authigenic marine smectites	1.027	0.94	Savin and Epstein (1970a)
	"weathering conditions"	soils from continental U.S.A., Hawaii, Central America	1.027	0.970	Lawrence and Taylor (1971)
	0	authigenic smectite from sea floor	1.0308		Yeh and Savin (1976)
Kaolinite	17	Southeastern U.S.A. kaolinite deposits	1.027	0.97	Savin and Epstein (1970a)
Sepiolite	0	Mid-Atlantic Ridge dredge haul; approximate value	1.033(?)		Savin (1973)
Deweylite	15—30	supergene serpentine-like material, Cedar Hill, Pennsylvania	1.0185	0.946	Wenner (1971)
Gibbsite	"weathering conditions"	gibbsite in soils	1.018	0.985	Lawrence and Taylor (1971)
Magnetite (maghemite?)	9	chiton tooth	1.0056		O'Neil and Clayton (1964)
Manganese nodule (mineralogy unspecified)	0	ocean bottom	1.010—1.016	0.923	Savin and Epstein (1970b) Dymond et al. (1973)

* Temperatures estimated in this work—modified from or not given in original paper.

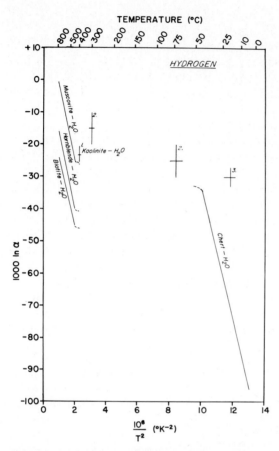

Fig. 8-2. Hydrogen isotope fractionations between minerals and water plotted as a function of temperature. Crosses are kaolinite-water fractionations estimated at isolated temperatures by: *1* = Suzuoki and Epstein (1976); *2* = Sheppard et al. (1969); *3* = Savin and Epstein (1970a). The chert-H_2O curve is from Knauth (1973). Muscovite-H_2O, hornblende-H_2O and biotite-H_2O curves are from Suzuoki and Epstein (1976).

tive to the bulk of the crystal at 200°C. This presumably reflects anisotropic forces acting upon the atoms of the outer few molecular layers of the crystals. This is generally of little significance in geochemical studies because surface layers ordinarily contain a negligible fraction of the oxygen of crystals. The possibility has not been investigated, however, that among the finest naturally occurring crystals, such as the fine fractions of clay mineral suites and carbonates (with their high surface/volume ratios), this surface effect could significantly affect the isotopic composition of the mineral.

Hydrogen isotope fractionations. Hydrogen isotope fractionations between clay minerals and water at sedimentary temperatures, as estimated by several authors, are listed in Table 8-3. The results of systematic series of experimental measurements of hornblende-water, biotite-water and muscovite-water hydrogen isotope fractionations made by Suzuoki and Epstein (1976) are shown in Fig. 8-2. Also shown are hydrogen isotope fractionations between water and the hydroxyl groups of chert at sedimentary temperatures estimated by Knauth (1973) and hydrogen isotope fractionations between kaolinite and water at isolated temperatures estimated by several authors.

The single most important variable affecting the mineral-water hydrogen isotope fractionation factor is the nature of the atom to which the OH group is bonded (Suzuoki and Epstein, 1976). At elevated temperatures the tendency to concentrate deuterium is: Al-OH > Mg-OH > Fe-OH. The scanty data available suggest that this ranking also exists at low temperatures.

Knauth and Epstein (1975) reported that water extracted from marine opal by heating was depleted in deuterium relative to seawater by 56 to 87‰. They concluded from the similarity between this fractionation and the fractionations observed for clay minerals that the water extracted from the chert had been present in the mineral as hydroxyl groups.

Water of crystallization in H_2O-bearing minerals. Oxygen and hydrogen isotope fractionations occur between the water of crystallization in hydrated

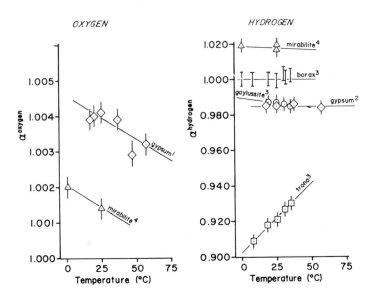

Fig. 8-3. Oxygen and hydrogen isotopic fractionations between water of crystallization in several minerals and the solutions from which the minerals crystallized. Sources of data: *1* = Fontes (1965); *2* = Fontes and Gonfiantini (1967); *3* = Matsuo et al. (1972); *4* = Stewart (1974).

salts and the solutions from which they formed (Baertschi, 1953; Barrer and Denny, 1964; Fontes, 1965; Fontes and Gonfiantini, 1967; Matsuo et al., 1972; Stewart, 1974; and others). Fractionation factors are plotted as functions of temperature in Fig. 8-3. For most of the minerals shown there is very little change of the fractionation factor with temperature, and the isotopic composition of the water therefore depends almost entirely on that of the solution in which the salt crystallized.

Where water of crystallization occupies crystallographically distinct sites isotopic fractionations may occur among the sites. These fractionations can be measured in cases where it is possible to dehydrate the crystal in a series of discrete steps, each increment of water lost corresponding to water in a different position. Heinzinger and Götz (1975) showed that the fractionation of oxygen isotopes among different sites of $CuSO_4 \cdot 5\ H_2O$ varied sufficiently with temperature to be used to estimate the temperature of crystallization.

At room temperature over periods of months, water, once incorporated into crystals, is remarkably resistant to subsequent isotopic exchange with aqueous solutions, at least in the case of $Na_2SO_4 \cdot H_2O$ and $SrCl_2 \cdot 6\ H_2O$ (Usitalo, 1958).

Time-dependent processes

Kinetic isotope effects. Mineral phases, and in some instances dissolved species, may be formed and continue to exist for long periods of time out of equilibrium with their environments. A disequilibrium isotopic distribution which results from the kinetics of a reaction is referred to as exhibiting a *kinetic effect.*

Kinetic effects may be produced by differences in the rates of chemical reaction (or physical process) of isotopically substituted compounds. For example, dissociation reactions of isotopically light-substituted chemical bonds are more rapid than those of isotopically heavy-substituted bonds. Also, species substituted with a light isotope should diffuse more quickly than species of the same molecule substituted with a heavy isotope. The rates of many other reactions may also depend on the isotopic substitution of reacting species.

Once a phase is formed out of isotopic equilibrium with its environment, slow rates of isotopic exchange may prevent the attainment of isotopic equilibrium. McCrea (1950) showed that when $CaCO_3$ was precipitated slowly from aqueous solution it formed in isotopic equilibrium with the solution. However, when it precipitated rapidly it was out of isotopic equilibrium. Apparently, the carbonate ions accreting on the growing crystal have an isotopic composition similar to that of the dissolved HCO_3^-. When the crystal grows rapidly, each layer of carbonate ions becomes covered over by the next layer before it has time to equilibrate with the solution. However, when the crystal grows slowly continuous equilibration between carbonate ions and solution can occur.

Many organisms form minerals which are not in isotopic equilibrium with their environments. Such disequilibrium is sometimes referred to as a *vital effect* or a *biological isotope effect*. There is no reason to believe that such effects are produced by factors other than the equilibrium and kinetic effects discussed above. However, a long series of complex chemical reactions may be involved in the formation of a mineral by an organism. Each step in the series will have associated with it an equilibrium or disequilibrium fractionation, which contributes to the total fractionation between the mineral and the growth environment. Few attempts have been made to deconvolve these total fractionations into the fractionations of the component steps.

Experimental studies of the rate of isotopic exchange between clay minerals and water. Before discussing the isotopic compositions of naturally occurring samples it is useful to consider the results of several experimental studies of the kinetics of isotopic exchange in clay-water and related systems. Essentially complete isotopic exchange between atmospheric water vapor and the interlayer water of expandable clay minerals occurs in a matter of hours or days (Savin, 1967; Lawrence, 1970).

Exchange between water and the oxygen and hydrogen of clay mineral structures is generally slow. For example exchange of oxygen and/or hydrogen isotopes between kaolinite and water at room temperature proceeds to the extent of 2% or less (in some cases substantially less) over times ranging up to months (McAuliffe et al., 1947; Roy and Roy, 1957; Halevy, 1964). An exception is halloysite, and expandable 1 : 1-type layer silicate (i.e. one hydroxyl-bearing octahedral sheet and one tetrahedral sheet per structural unit). Halloysite exchanges hydrogen isotopes with water quite rapidly, presumably because water in the interlayer sites is in intimate contact with the hydroxyl hydrogen atoms of the mineral (Lawrence, 1970). Kaolinite, a non-expandable analogue of halloysite, exchanges rapidly when it is artificially expanded with hydrazine (Ledoux and White, 1964).

Oxygen isotope exchange between illites (or mixed-layer illite/smectites) and water is very slow at room temperature but is greatly accelerated when accompanied by interlayer cation exchange through the agency of sodium-organic compounds (James and Baker, 1976). Increased accessibility of water to the aluminosilicate framework during ion exchange may play a role in the greater rate of exchange, in a manner analogous to that noted for the system muscovite-water (O'Neil and Taylor, 1969). However, the low rate of exchange between smectite and water suggests that a factor other than simple ion exchange, perhaps related to the presence of the organic compound, has caused the rapid isotopic exchange in the illite-Na-organic-water system.

In the temperature range 100—350°C, the rates of oxygen and hydrogen isotopic exchange of illite, smectite, and kaolinite with water increase with increasing temperature (O'Neil and Kharaka, 1976). At temperatures in the range of 100°C significant hydrogen isotope exchange and barely detectable oxygen exchange between water and both 1 : 1- and 2 : 1-type clays can be observed over periods of weeks to two years. Exchange between water and expandable clay minerals occurs more rapidly than between water and non-expandable clays. At temperatures of 350°C over periods of a year or more, massive hydrogen exchange and substantial oxygen exchange can occur between clay minerals and water. Much more oxygen exchange occurs between kaolinite and water than between illite and water. This is probably because the hydroxyl groups of kaolinite are exposed at the surface of each aluminosilicate sheet while in illite the hydroxyl groups are well within the interior of each sheet.

ISOTOPE EFFECTS DURING WEATHERING AND SOIL FORMATION

Clays and related minerals formed as weathering products approach isotopic equilibrium with the environments in which they are formed. Chemically and mineralogically unaltered parent materials remain isotopically unaltered in the weathering environment (Taylor and Epstein, 1964; Savin and Epstein, 1970a; Lawrence, 1970; Lawrence and Taylor, 1971, 1972).

δD and $\delta^{18}O$ values of meteoric water lie very close to a straight line described by the equation $\delta D = 8\, \delta^{18}O + 10$ (Craig, 1961). δD and $\delta^{18}O$ values of a series of clay minerals of similar chemical composition which have equilibrated with different meteoric waters at the same temperature must also lie along a straight line. The equation of this line is:

$$\delta D = 8 \frac{\alpha^{hy}_{clay-H_2O}}{\alpha^{ox}_{clay-H_2O}} \delta^{18}O + 1000 \left(8 \frac{\alpha^{hy}_{clay-H_2O}}{\alpha^{ox}_{clay-H_2O}} - 6.99\alpha^{hy}_{clay-H_2O} - 1\right)$$

(Savin and Epstein, 1970a). For each clay mineral there exists a family of lines on a graph of δD vs. $\delta^{18}O$ — one line for each temperature of equilibration. This is illustrated schematically in Fig. 8-4.

Isotopic compositions of kaolinites and gibbsites (and amorphous hydroxides) of low temperature origin are plotted in Fig. 8-5. Also shown are the "kaolinite line" of Savin and Epstein (1970a) and the "gibbsite line" of Lawrence and Taylor (1971). The data points cluster quite closely about the lines, indicating that for the samples analyzed, the following conditions prevailed:

(1) The samples either formed in isotopic equilibrium with meteoric water or subsequent to their formation underwent isotopic exchange and equilibration with meteoric waters.

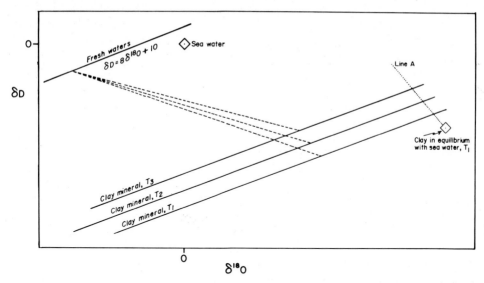

Fig. 8-4. Schematic representation of $^{18}O/^{16}O$ and D/H systematics in clay minerals (or other minerals such as chert which contain both oxygen and hydrogen). At temperature T_1 the isotopic compositions of a suite of clays of similar chemistry and mineralogy and in equilibrium with different fresh waters lie along line "clay mineral, T_1". Similarly for other temperatures, $T_3 > T_2 > T_1$. Dashed lines connect composition of an arbitrarily chosen fresh water to the clay in equilibrium with it at different temperatures. Clays in equilibrium with seawater at a range of temperatures plot along line A.

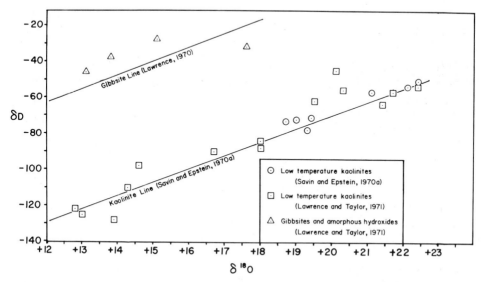

Fig. 8-5. Isotopic composition of kaolinites and gibbsites formed at earth's surface conditions.

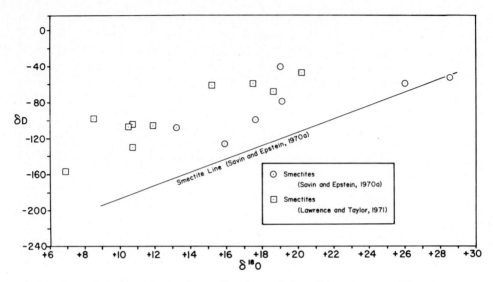

Fig. 8-6. Isotopic compositions of smectites formed at earth's surface conditions.

(2) The samples equilibrated with meteoric waters at a fairly narrowly restricted range of temperatures. The restricted temperature range is reasonable since nearly all the samples were from the continental United States or Hawaii.

The isotopic compositions of Tertiary and Quaternary smectite-rich soils and other smectites of low-temperature origin, taken from Savin and Epstein (1970a) and Lawrence (1970) are plotted in Fig. 8-6. While these data show a rough correlation between δD and $\delta^{18}O$, there is not a well-defined clustering about a single line as was the case for kaolinite and gibbsite. The "smectite line" of Savin and Epstein (1970a) shown in this figure was calculated using fractionation factors estimated from the compositions of authigenic marine smectites, undoubtedly formed at colder temperatures than most of the other smectites analyzed.

The scatter of analytical data on the figure probably reflects two factors: the large range in temperature over which smectites may commonly form by weathering (0° to perhaps 35°C); and the large range of chemical compositions of smectites formed during weathering. Additional scatter may be introduced by the greater rate of isotopic exchange of smectite than of kaolinite, and perhaps by the different rates of exchange of hydrogen isotopes and oxygen isotopes as inferred from experiments at high temperatures (O'Neil and Kharaka, 1976). However, analyses of detrital ocean sediments (Yeh and Savin, 1976; Yeh and Epstein, 1978) suggest that at sedimentary temperatures rates of oxygen and hydrogen isotope exchange of clays may be similar.

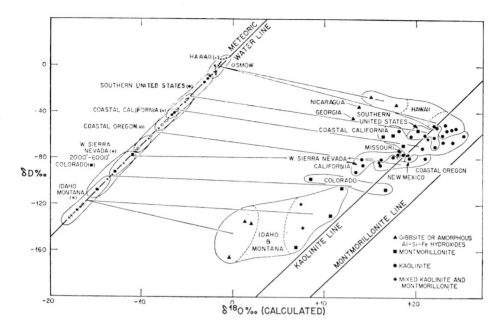

Fig. 8-7. Isotopic compositions of clay minerals and hydroxides from Quaternary soils of the United States and the compositions of meteoric waters from related localities (from Lawrence and Taylor, 1971). Reprinted by permission.

In Fig. 8-7 are plotted the isotopic compositions of the clay minerals and hydroxides from Quaternary soils analyzed by Lawrence and Taylor (1971). Complexities are introduced into the interpretation of the isotopic compositions of soils because of the complex mineralogy of the weathering products and the variable ratios of weathered material to unweathered parent rock. In spite of these complexities several conclusions can be drawn from this isotopic data.

(1) The close relationship between isotopic compositions of the soils and those of meteoric waters in the regions from which the soils were sampled indicates that the isotopic composition of the ambient water, and not the isotopic composition of the parent rock, is the primary factor in determining the isotopic composition of the weathering product in the soil. (In fact, there are only a few cases reported in the literature in which the isotopic composition of a precursor solid phase has been demonstrated to be inherited by a daughter mineral following mineral transformation in the presence of water. O'Neil and Kharaka (1976) reported that when kaolinite reacted in the presence of water at 350°C to form pyrophyllite + diaspore, isotopic exchange was incomplete and the pyrophyllite inherited unexchanged oxygen from the kaolinite precursor. In that case, structural similarities between the two minerals might permit the units of kaolinite lattice to be incorporated in toto into pyrophyllite.)

(2) During the formation of the weathering products isotopic equilibrium with the ambient water is closely approached.

(3) Mineralogically unaltered parent materials such as quartz retain the isotopic compositions they had in the parent rock.

In addition to the weathering products discussed above which formed on crystalline rocks, Lawrence and Taylor (1972) analyzed three soil profiles which formed on shales. They found that weathering had very little effect on the hydrogen or oxygen isotopic composition of the shales until it proceeded to so great an extent that the clay minerals of the shale underwent obvious mineralogical alteration.

ISOTOPIC STUDIES OF MARINE SEDIMENTATION, HALMYROLYSIS, AUTHIGENESIS AND EARLY DIAGENESIS

Detrital vs. authigenic clays

Isotopic analyses of a number of ocean sediments, mostly of Quaternary age, from the world's major oceanic basins indicate that in most instances the clays have not reached oxygen isotopic equilibrium with the bathyal marine environment in thousands or even hundreds of thousands of years. Most clay minerals in the ocean basins are detrital, having been derived from the continents. Oxygen isotope ratios of this detritus remain essentially unchanged during marine sedimentation over long periods of time. They therefore reflect the conditions of formation of the detritus as weathering or diagenetic minerals, and have potential use as indicators of provenance. The detrital components also retain the K-Ar ages (Hurley et al., 1963) and $^{87}Sr/^{86}Sr$ ratios (Dasch, 1969) acquired at their continental sources.

By measuring progressively finer sized particles of detrital ocean sediments Yeh and Savin (1976) were able to show that a *small amount* of oxygen isotope exchange can occur between detrital clays and seawater. Estimated percentages of exchange of different size fractions of three ocean sediments are shown in Fig. 8-8. Only for the finest sizes, which constitute a small percentage of the sample, were substantial amounts of isotopic exchange noted. However, if the amount of chemical interaction between minerals and seawater is comparable to the amount of oxygen isotopic exchange, this may be sufficient to play a role in regulating the chemical composition of seawater.

Yeh and Epstein (1978) have presented evidence that detrital clay minerals undergo hydrogen isotope exchange with seawater to roughly the same extent as they do oxygen isotope exchange.

In some instances smectite in marine sediments is clearly authigenic, having been formed by the submarine weathering (halmyrolysis) of volcanic ejecta. Evidence for the authigenic nature of such smectites includes high mineralogic purity, the presence of associated volcanic materials, and occur-

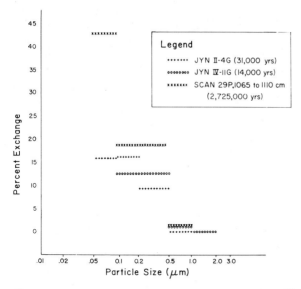

Fig. 8-8. Percent exchange vs. particle size for three North Pacific core samples (from Yeh and Savin, 1976). Values of percent exchange for SCAN 29P are maximum values. Actual values may be lower. Reproduced by permission.

rence in regions of submarine volcanism. From comparison of oxygen isotopic compositions and mineralogic compositions of a suite of primarily detrital deep-sea sediments, Savin and Epstein (1970b) estimated that the average $\delta^{18}O$ value of detrital smectite in ocean sediments was approximately $17 \pm 2‰$. This is in marked contrast to the $\delta^{18}O$ values of clearly authigenic marine smectites (+26 to +29‰: Savin and Epstein, 1970a; 30.8‰: Yeh and Savin, 1976). The greater ^{18}O content of the authigenic phase reflects both the greater ^{18}O content of the water and the colder temperatures prevalent on the ocean bottom than in the continental source areas of most detritus. The marked contrast in $\delta^{18}O$ values permits authigenic phases to be distinguished from detrital ones where no other chemical or mineralogical differences exist.

An instance in which the occurrence of an authigenic clay mineral phase was demonstrated using oxygen isotopic compositions has been described by Eslinger and Savin (1976). In this case, fine-grained quartz (0.1—0.5 μm) and clay (finer than 0.1 μm) from Deep Sea Drilling Project Site 323 had $\delta^{18}O$ values typical for detrital components: quartz, +9 to +19‰; clay, +16 to +25‰. Within this range however, quartz and clay $\delta^{18}O$ values varied sympathetically. In addition, the clay mineral fractions of those samples with the highest quartz $\delta^{18}O$ values contained the greatest proportions of mixed-layer illite/smectite, and the illite/smectite mixed-layer clay contained the greatest proportion of smectite (i.e., expandable) layers. At this drilling site the sediments were interpreted to be primarily detrital. However, many samples contained some proportion of authigenic quartz and clay, presumably formed by submarine alteration of volcanic material.

Weathering of submarine volcanic products, both glassy and crystalline, takes place on the sea floor. Garlick and Dymond (1970) measured $\delta^{18}O$ values of Tertiary and volcanic glass shards and found that with increasing age the shards progressively acquired more water (without undergoing devitrification) and became enriched in ^{18}O. The correlation between $\delta^{18}O$ and water content is only fair. Since $\delta^{18}O$ values become much greater than those of seawater, the glass, in addition to acquiring water must be undergoing isotopic exchange with the water, with a silicate-water oxygen isotope fractionation factor similar to that of other silicates. Surprisingly, potassium-argon ages of the shards remain largely unaltered by this hydration and oxygen isotope exchange (Garlick and Dymond, 1970).

$^{18}O/^{16}O$ ratios of fresh and weathered basalts dredged from the ocean floor are plotted as a function of water content (H_2O^+) in Fig. 8-9 (Garlick and Dymond, 1970; Muehlenbachs, 1971; Pineau et al., 1976). The H_2O^+ content of the basalts should increase as their clay and hydroxide contents increase.

The correlation between $\delta^{18}O$ and water content indicates that the basalts are mixtures of fresh rock and low-temperature weathering products. How-

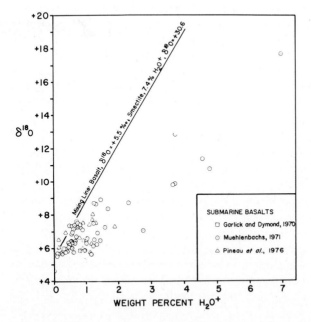

Fig. 8-9. $\delta^{18}O$ vs. content of H_2O^+ in a large number of submarine basalts.

ever, most analyses plot below a mixing line for two components: fresh basalt and smectite formed at 0°C. The weathering products are probably primarily smectite formed at low temperatures but often substantially warmer than the 0°C of the modern sea floor, and secondarily, low-temperature oxide and hydroxide phases. (Oxides and hydroxides in isotope equilibrium with water at low temperatures have markedly lower $\delta^{18}O$ values than do silicates in equilibrium with the same water at the same temperature.)

While $^{18}O/^{16}O$ ratios frequently serve, by themselves, to distinguish detrital ocean sediments from those which are largely authigenic, there are some instances in which isotope ratios do not provide a clearcut distinction. When authigenic silicates ($\delta^{18}O$ values between +26 and +34‰) are mixed with large proportions of iron and manganese oxides ($\delta^{18}O$ values between +10 and +16‰; Dymond et al., 1973) the mixtures can have $\delta^{18}O$ values similar to those of typical detrital sediments (+15 to +20‰). The nature of such sediments can be clarified, however, by consideration of isotopic compositions in conjunction with chemical and mineralogic compositions.

Dymond et al. (1973) examined metalliferous sediments of the East Pacific Rise in an attempt to determine the conditions under which they

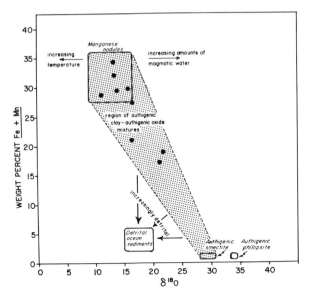

Fig. 8-10. $\delta^{18}O$ vs. Fe + Mn content of metalliferous sediments from the East Pacific Rise (solid circles) (from Dymond et al., 1973). Mixtures of authigenic clays and authigenic oxides formed at low temperatures in seawater would lie within the shaded area. Arrows in the upper part of the diagram show the effects expected for increasing amounts of magmatic water and increasing temperature in the environment of authigenic mineral formation. Arrows in the lower part of the diagram show the effects expected as the fractions of the sediments which are detrital increase.

formed. When $\delta^{18}O$ values of the metalliferous sediments are plotted as a function of the iron + manganese contents (Fig. 8-10), their compositions fall in the range expected for mixtures of authigenic silicates and oxides formed at low temperatures. It is conceivable, although unlikely, that the $\delta^{18}O$ values reflect formation at somewhat elevated temperatures in the presence of water containing an appreciable component of magmatic water — the temperature and water $\delta^{18}O$ effects just cancelling one another.

Detrital vs. authigenic quartz

Most silt-size and finer-grained quartz in ocean sediments is detrital in origin (Rex and Goldberg, 1958). This is apparent from distributional data and is confirmed by oxygen isotope studies of the quartz (Rex et al., 1969; Clayton et al., 1972; and others). There is no evidence that even the finest size fractions of quartz undergo oxygen isotopic exchange with seawater. The isotopic composition of quartz therefore reflects the provenance of the mineral. In the North Pacific, silt-size quartz has a $\delta^{18}O$ value between +16.5 and +19‰. This range of values must reflect mixtures of quartz of high- and low-temperature origins homogenized through sedimentary recycling on the continents. Quartz in the South Pacific has lower $\delta^{18}O$ values (+12 to 16‰) indicating a smaller component of quartz of low-temperature origin (Clayton et al., 1972).

Siliceous sediments

Siliceous debris of organic origin comprises an important component of ocean sediments but has been subject to very little oxygen or hydrogen isotopic study. Mopper and Garlick (1971) analyzed radiolarian tests from tropical and high-latitude Quaternary piston cores and concluded that the tests underwent oxygen isotopic re-equilibration with bottom water subsequent to deposition. They did not remove the water from their opaline samples prior to analysis, and attributed lower-than-normal analytical reproducibility to the irreproducible loss of water during their analytical procedures.

Labeyrie (1974) analyzed numerous samples of sponge spicules and diatom frustules, as well as the water in which they grew. Analysis of the opaline silica was preceded by treatment with NaOCl to remove organic matter, and by dehydration at temperatures above 600°C. Analytical reproducibility was good. The isotopic fractionation measured between water and biogenically precipitated opaline silica analyzed in this fashion is similar to the extrapolated quartz-water isotopic fractionation of Clayton et al. (1972) (Table 8-1). The potential for using silica-water oxygen isotopic fractionations as the basis for paleoclimatic reconstructions of Pleistocene time is therefore good.

Knauth (1973) made a detailed study of the hydrogen isotope geochemistry of diatomite from the marine Miocene Monterey Formation. Although samples were collected from uplifted sediments, the material was extremely hydrous, amorphous silica which probably was chemically and mineralogically similar to newly-deposited marine diatomaceous remains. Some of the hydrogen of the opal is readily exchangeable at 25°C while other hydrogen was resistant to isotopic exchange on a laboratory time scale. By proper choice of outgassing temperatures, it is possible to achieve a fairly clean separation of hydrogen in the two types of sites, so that the non-exchangeable fraction can be isotopically analyzed. It is not yet known whether the "non-exchangeable" hydrogen is indeed resistant to isotopic exchange over thousands or millions of years. If it is, however, its composition should provide paleoenvironmental information.

Marine sulfates

The oxygen isotope geochemistry of marine sulfates is complex and not completely understood. A general outline of factors important in determining the $^{18}O/^{16}O$ ratios of marine sulfates follows.

(1) The $\delta^{18}O$ value of dissolved SO_4^{2-} in seawater varies within a narrow range from +9.3 to 10.0‰ (Lloyd, 1967; Longinelli and Craig, 1967). Lloyd's (1968) experiments indicate that oxygen isotopic exchange between seawater and SO_4^{2-} is slow and the rate is dependent upon pH. In ocean water at 4°C, equilibration would require 2.5×10^5 years (Lloyd, 1968). The residence time of sulfate ion in seawater (i.e., oceanic SO_4^{2-} reservoir size ÷ annual river input) is about 10^7 years, and Longinelli and Craig (1967) therefore argued that the dissolved SO_4^{2-} was in isotopic equilibrium with seawater. However, the results of exchange experiments by Lloyd (1968) and Mizutani and Rafter (1969a), when extrapolated to low temperatures (Fig. 8-1), indicate that at equilibrium with deep ocean water, SO_4^{2-} should have a $\delta^{18}O$ value of 35 to 39‰. Lloyd suggested that this disequilibrium resulted from rapid (relative to the residence time) cycling of sulfur through reduced (S^{2-}) and oxidized (SO_4^{2-}) compounds, in part by biological processes which have attendant non-equilibrium isotopic fractionations.

(2) Lloyd (1968) and Mizutani and Rafter (1969b, 1973) investigated the bacterially-induced non-equilibrium oxygen isotopic fractionations which occur during oxidation or reduction of sulfur compounds. When SO_4^{2-} is reduced bacterially to S^{2-} an oxygen isotope fractionation occurs such that the sulfate remaining in solution generally becomes progressively enriched in ^{18}O as the fraction of the initial SO_4^{2-} pool which remains decreases (Lloyd, 1968; Mizutani and Rafter, 1969b). However, because of the formation and back-reaction of complex intermediate species which can exchange isotopically with water, the isotopic composition of the SO_4^{2-} remaining during bac-

terial reduction is also affected by the isotopic composition of the water in which the reaction occurs (Mizutani and Rafter, 1973). In a closed or semi-closed system such as a sediment—pore water system, therefore, the isotopic composition of the sulfate remaining in solution while bacterial reduction is occurring is not a unique function of the temperature and the amount of SO_4^{2-} remaining. Under normal ocean sediment conditions, however, $\delta^{18}O$ values of dissolved SO_4^{2-} become progressively greater as reduction of SO_4^{2-} to S^{2-} proceeds (Mizutani and Rafter, 1973).

When S^{2-} is inorganically oxidized to SO_3^{2-} and SO_4^{2-} in the presence of an oxygen atmosphere the isotopic composition of the sulfate is a function both of that of the water and of the O_2 (Lloyd, 1968). Additionally, the intermediate SO_3^{2-} undergoes fairly rapid oxygen isotope exchange with water. When native sulfur is bacterially oxidized to SO_4^{2-}, the $\delta^{18}O$ of the SO_4^{2-} reflects primarily the $\delta^{18}O$ value of the water, and there is little isotopic fractionation in this process (Mizutani and Rafter, 1969b).

(3) Isotopic equilibrium is not achieved between solid sulfate phases and water during slow inorganic precipitation of the solid phase in the laboratory. The $\delta^{18}O$ value of dissolved SO_4^{2-} is probably the most important factor in determining the $\delta^{18}O$ value of the solid. Extrapolation of Lloyd's (1968) experimental results to 25°C indicates that the equilibrium anhydrite-dissolved SO_4^{2-} oxygen isotope fractionation is in the vicinity of 6‰. Anhydrite and gypsum probably show similar isotopic behavior. Crystallization of gypsum in the laboratory by slow evaporation of a saturated solution at room temperature occurred with a gypsum-dissolved SO_4^{2-} fractionation of approximately 2‰ (Lloyd, 1968). It is not certain at this point whether the discrepancy between these two values reflects isotopic disequilibrium during precipitation, errors in the estimated fractionations introduced during the extrapolation, or both. It seems, however, that the $\delta^{18}O$ value of a sulfate mineral precipitated from a solution at low temperature should reflect, but not exactly equal, the $\delta^{18}O$ of the dissolved SO_4^{2-}.

Oxygen isotopic analyses of marine barites are given by Church (1970) and by Cortecci and Longinelli (1972) and are shown in Fig. 8-11. Only those samples composed of large (greater than 5 μm) barite crystals have $\delta^{18}O$ values similar to or slightly greater than dissolved seawater SO_4^{2-}. The isotopic data are therefore consistent with inorganic crystallization of the large crystals from seawater. All of the large crystals analyzed are associated with manganese nodules or manganiferous sediments, as such large crystals most frequently are (Church, 1970).

$\delta^{18}O$ values of three fine-grained (finer than 5 μm) disseminated barites are depleted in ^{18}O by 1 to 5‰ relative to dissolved seawater sulfate (Church, 1970). SO_4^{2-} in one co-existing pore water sample was depleted in ^{18}O by about 1.3‰ relative to seawater sulfate. One manner in which such ^{18}O-depleted sulfates could be generated is through the oxidation of S^{2-} to SO_4^{2-} as described by Mizutani and Rafter (1969b).

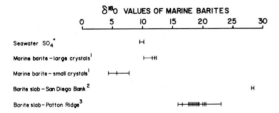

Fig. 8-11. Oxygen isotopic compositions of seawater SO_4^{2-} and marine barites. Sources: 1 = Church (1970); 2 = Friedman (quoted by Church, 1970); 3 = Cortecci and Longinelli (1972).

The barite slabs, all from the sea floor off southern California, are greatly enriched in ^{18}O relative to the other marine barites and seawater sulfates. Isotopically similar Cretaceous barite concretions were reported by Rafter and Mizutani (1967). These and the southern California slabs are also greatly enriched in ^{34}S. It is most likely that biological reduction processes caused the dissolved SO_4^{2-} from which these barites formed to become highly fractionated with respect to the isotopes of both sulfur and oxygen. Because of the different isotopic effects produced by oxidation and reduction of sulfur-containing species, the isotopic compositions of these barite slabs cannot be uniquely interpreted at this time. However, it is clear, on the basis of isotopic evidence, that the barite slabs are formed by a process different from the fine-grained or coarse-grained disseminated marine barite.

The isotope geochemistry of gypsum and anhydrite is discussed in the section on evaporites.

Marine phosphates

Phosphate occurs in marine sediments both as phosphorite nodules and in an ill-defined form extractable in small amounts from biogenically precipitated $CaCO_3$. Early work on the isotope geochemistry of phosphate was done by Tudge (1960). In recent years $^{18}O/^{16}O$ measurements of marine phosphates have been reported by Longinelli (1965, 1966), Longinelli and Nuti (1968a, 1968b, 1973a, 1973b), and others. These studies show that consistent, temperature-dependent, and probably equilibrium oxygen isotopic fractionations occur between PO_4^{2-} from apatite (fish teeth, etc.) and water. Fractionations of similar magnitude exist between PO_4^{2-} extracted from biogenically precipitated $CaCO_3$ and water. Variations with temperature of the PO_4^{2-}-water and $CaCO_3$-water oxygen isotope fractionations are very similar, so there is essentially no variation with temperature of the PO_4^{2-}-$CaCO_3$ fractionation (Fig. 8-1).

Most published oxygen isotope analyses of phosphorite PO_4^{2-} and PO_4^{2-} from $CaCO_3$ have been made on Tertiary and older material.

Marine carbonates

There is an enormous literature on the oxygen and carbon isotope geochemistry of marine carbonates. Most calcium carbonate in the deep sea is biogenically precipitated in the upper portions of the water column by foraminifera and coccolithophorids. While there is a possibility that small departures from oxygen isotope equilibrium with the seawater do occur (Shackleton et al., 1973), planktic foraminifera do deposit their tests in close approach to equilibrium (Emiliani, 1954; Savin and Douglas, 1973). $\delta^{18}O$ values of the tests may therefore be used to calculate temperatures of growth when $\delta^{18}O$ of the water in which the carbonate grew can be estimated. Carbon isotope ratios of foraminiferal calcite will not be discussed here except to note that they are not in isotopic equilibrium with, nor do they simply reflect, the $^{13}C/^{12}C$ ratio of the dissolved bicarbonate in which they grow (Vincent and Shackleton, 1975; Williams, 1976; Douglas and Savin, 1978). The complex factors which control the $^{13}C/^{12}C$ ratios of these tests are not completely understood.

Coccoliths in marine sediments have been analyzed by Douglas and Savin (1975), Margolis et al. (1975) and Anderson and Cole (1975). Isotopic analyses of coccoliths grown in the laboratory under controlled conditions have been done by Dudley (1976) and Lindroth et al. (1980). In some instances coccoliths have $\delta^{18}O$ values which appear to reflect isotopic equilibrium with the natural or laboratory growth medium while in other instances they do not. In all instances temperature is one of the important variables in determining the $\delta^{18}O$ values. The general usefulness of coccoliths in isotopic paleotemperature studies is still somewhat uncertain although they can yield climatic information in ocean cores in which foraminifera are not preserved. Coccolith samples used in paleotemperature studies must be chosen with care. Coccoliths frequently become encrusted with carbonate overgrowths which formed diagenetically and which have $^{18}O/^{16}O$ ratios reflecting the diagenetic environment (Douglas and Savin, 1975).

Nearshore marine carbonate sediments are composed primarily of calcium carbonate of biogenic origin. The isotopic composition of the biogenic component is determined by the temperature and isotopic composition of water at the time of precipitation and by the equilibrium or non-equilibrium nature of formation of the carbonate. (In some instances, the sediments may also contain, of course, reworked carbonate from older sediments or rocks.) Early diagenesis may affect the isotopic composition, mineralogy, chemistry and fabric of carbonate sediments in a number of ways. Many studies of the isotopic effects of early diagenesis have been published, and a few representative ones are reviewed here.

Gross (1964) studied Pleistocene limestones from Bermuda which had undergone diagenesis and lithification. Common obvious mineralogical changes include the solution of aragonitic components and the formation of calcite cement. The cement is depleted in both ^{18}O and ^{13}C relative to the biogenic primary constituents. The depletion in ^{18}O reflects the ^{18}O-depleted meteoric water from which the calcite was precipitated. The depletion in ^{13}C reflects the derivation of dissolved HCO_3^- in part from ^{13}C-depleted soil gases.

In several parts of the world dolomite has been observed to form in Recent marine sediments. Isotopic, chemical, petrographic, and field studies indicate that there is undoubtedly more than one way in which dolomite can form. Degens and Epstein (1964) isotopically analyzed coexisting dolomite and calcite from Recent sediments of the Bahamas, Florida Bay, and South Australia and found little difference between the $\delta^{18}O$ values of calcite and dolomite in each sample. Because extrapolation of the results of laboratory work at high temperatures (O'Neil and Epstein, 1966; Northrop and Clayton, 1966) and measurement of naturally occurring samples formed at high temperatures (Engel et al., 1958; Clayton and Epstein, 1958) suggests that at surface temperatures the fractionation between the two phases should be 4—7‰, Degens and Epstein argued that the dolomite in their samples formed by substitution of magnesium for calcium in the calcite lattice in a solid-state reaction without any concomitant isotopic exchange. Other authors have interpreted isotopic data as supporting other mechanisms of dolomitization. Clayton et al. (1968a) concluded that in Deep Springs Lake, a saline lake in California, the $\delta^{18}O$ of dolomite reflected its formation in isotopic equilibrium with the lake water rather than through an intermediate, isotopically unexchanged calcite phase. A re-examination of sediments from Coorong Lagoon, South Australia, by Clayton et al. (1968b), an area originally studied by Degens and Epstein (1964), indicated that dolomite there is commonly enriched in ^{18}O and depleted in ^{13}C relative to co-existing calcite by 2‰ or more. This led Clayton et al. to conclude that the dolomite and calcite formed under different conditions (perhaps at different times of the year) and that the dolomite did not form by replacement of calcite without isotopic exchange.

Careful study of Pleistocene reef carbonates from Jamaica led Land (1970, 1973a, 1973b) and Land and Epstein (1970) to conclude that Recent diagenesis of the sediments occurred much more intensively in the phreatic zone (below the water table) than in the vadose zone (zone of aeration). The $\delta^{13}C$ and $\delta^{18}O$ values of the diagenetic products indicated the influence of meteoric water and dissolved soil-derived CO_2 in the alteration process. This was true even for the dolomite, indicating that highly saline solutions are not necessary for the formation of this mineral. The dolomite was interpreted to have formed in regions where the fresh, meteoric water mixed with ocean water (Land, 1973a).

Allan and Matthews (1977) have presented isotopic data for limestones from Barbados that indicate major diagenetic alteration in the vadose zone. They cautioned that isotopic studies of carbonates which emphasize surface samples might yield results which were not representative of the diagenetic history of the limestones studied and that studies based on core samples would yield a more realistic picture of diagenesis of a limestone terrane such as that of Barbados.

A sequence of early marine diagenetic processes of a Mississippian carbonate sediment, including cementation and partial dolomitization was inferred by Choquette (1968) from careful isotopic and petrographic study.

EVAPORITE FORMATION

Isotopic studies of minerals from evaporite deposits have been concentrated on the carbonate phases. The role of oxygen isotope ratios in understanding the mechanism of formation of dolomite in high-salinity environments has already been discussed. The limited number of measurements of non-carbonate minerals from evaporite environments indicates the potential usefulness of the isotopic approach in determining the conditions under which evaporite minerals formed and the origin of the water and/or dissolved constituents in the evaporite basin.

Interpretation of isotopic compositions of evaporite minerals may be complicated by the change in isotopic composition of the water in the basin resulting from evaporation. This enrichment changes in a complex manner and may approach 8‰ or more in ^{18}O when a large fraction of the water in a closed basin has evaporated (Fontes, 1966; Fontes and Gonfiantini, 1970; Lloyd, 1966).

$\delta^{18}O$ values of gypsum and anhydrite in evaporite deposits should reflect primarily the source of the SO_4^{2-} (marine vs. continental) as discussed prev-

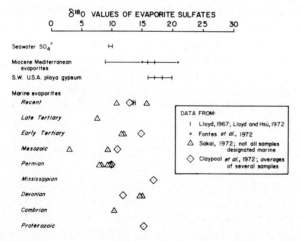

Fig. 8-12. Oxygen isotopic compositions of SO_4^{2-} in marine evaporites.

iously. Secondarily, processes in the evaporite depositional area may affect the isotopic compositions of the minerals. $\delta^{18}O$ values of a large number of gypsum and anhydrite samples from evaporites are shown in Fig. 8-12. Most samples are marine. The data of Claypool et al. (1972) indicate that $\delta^{18}O$ of marine evaporites may vary with time within relatively narrow limits. Claypool et al., however, presented average values only, not analytical data or sample descriptions. From the data available it appears that $\delta^{18}O$ values of anhydrite and gypsum from marine evaporite sequences of a single age generally fall within a fairly narrow range. Marine evaporite sulfates usually have $^{18}O/^{16}O$ ratios similar to or slightly greater than modern dissolved seawater SO_4^{2-}.

Sulfate minerals from Mediterranean Miocene marine evaporite sequences drilled by the Deep Sea Drilling Project have a greater range of $\delta^{18}O$ values than do most marine evaporite deposits (Lloyd and Hsü, 1972; Fontes et al., 1972). This suggests that the sulfate of these deposits is derived from a variety of isotopically different sources, with a large component derived from continental runoff. Studies of this type may prove useful in understanding the genesis of other geologically complex evaporite deposits.

$\delta^{18}O$ and δD values of the waters from which evaporites have precipitated can be inferred from the $\delta^{18}O$ and δD values of the water of crystallization in the minerals. The water of hydration of gypsum is enriched, relative to the solution from which it crystallizes, by about 4‰ in ^{18}O and is depleted in deuterium by about 15‰ (Gonfiantini and Fontes, 1963; Fontes, 1965; Fontes and Nielsen, 1966). These fractionations do not vary measurably with temperature. Therefore, to the extent that the water of crystallization remains unexchanged during diagenesis, it should directly reflect the water from which the evaporite minerals crystallized. This has been shown to be the case for gypsum precipitating naturally from evaporating seawater (Fontes, 1966). The water of crystallization in gypsum from the Mediterranean evaporites discussed above, however, has clearly undergone post-depositional isotopic exchange with normal seawater and does not reflect the isotopic compositions of the highly ^{18}O- and deuterium-enriched water from which it is undoubtedly precipitated.

The mineral mirabilite ($Na_2SO_4 \cdot 10\ H_2O$) forms at low temperature when seawater becomes sufficiently concentrated, by freezing and/or evaporation. The isotopic compositions of the water of crystallization of several mirabilite samples from Antarctica (Bowser et al., 1970) show a broad spread of values ($\delta^{18}O$ between -7 and $-38‰$). When interpreted in the light of the isotopic fractionation factors of Stewart (1974) the $\delta^{18}O$ and δD values of the water of crystallization indicate precipitation from mixtures of meteoric and seawaters (Fig. 8-13). $\delta^{18}O$ values of the sulfate itself indicate in most instances derivation of the SO_4^{2-} from seawater. In a few instances it appears that bacterial reduction and re-oxidation caused the isotopic composition of the sulfate to be altered.

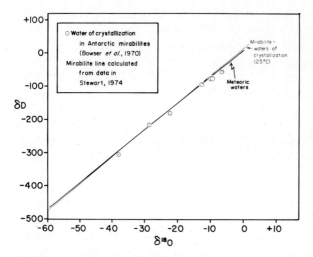

Fig. 8-13. Oxygen and hydrogen isotopic compositions of waters of crystallization of mirabilite samples from Antarctica (from Bowser et al., 1970). The fine line is the line along which lie waters of crystallization of mirabilite in equilibrium with meteoric waters at 25°C. An equivalent 0°C line is shifted only slightly from the 25°C line.

Another example of the use of oxygen isotope measurements to determine the nature of a mineralogic reaction and the source of water involved in the reaction is provided by the study of chert formation in saline East African lakes by O'Neil and Hay (1973). In these lakes, unusual sodium silicates such as magadiite ($NaSi_7O_{13}(OH)_3 \cdot H_2O$) are converted to chert. In an attempt to determine whether this conversion occurred through the interaction of sodium silicates with fairly fresh groundwater or with highly saline, highly evaporated water, O'Neil and Hay measured $\delta^{18}O$ values of a number of cherts from the lakes of the region. A rough correlation was observed between $\delta^{18}O$ values of cherts and the salinities at the time of chert formation as estimated from mineralogical considerations. Additionally, some of the cherts from high salinity environments are among the most ^{18}O-rich cherts analyzed (approximately +44‰). These are even more ^{18}O-rich than most deep-sea cherts formed at substantially lower temperatures. It therefore seems that cherts in these lakes must have formed in the presence of waters strongly enriched in ^{18}O by evaporation.

LATER DIAGENETIC PROCESSES

Diagenesis of cherts

With increasing age, temperature and depth of burial, the biogenic silica component of siliceous ocean sediments frequently becomes altered to chert. The series of stages commonly followed is: opal-A → opal-CT →

microcrystalline quartz. Materials composed primarily of microcrystalline quartz are often classified as cherts, while those which are principally opal-CT are frequently referred to as porcellanites. Studies of oxygen and hydrogen isotope variations in siliceous sediments which have been diagenetically altered to different stages have provided insight into the mechanisms of some of the diagenetic processes involved. They have also indicated ways in which isotopic compositions of cherts can be used as paleoenvironmental indicators.

Knauth and Epstein (1975) and Kolodny and Epstein (1976) have analyzed coexisting porcellanite and chert from a large number of Deep Sea Drilling Project (DSDP) ocean sediment cores from Jurassic to Eocene in age. The $\delta^{18}O$ values of porcellanites are reasonable ones for the formation of that phase at sea bottom temperature. Porcellanites are invariably enriched in ^{18}O by 0.5—3‰ relative to coexisting cherts. This probably reflects the formation of chert after burial to perhaps 100—300 m below the sea floor at somewhat warmer temperatures and in the presence of pore waters slightly depleted in ^{18}O relative to bottom water (see pp. 319—320). It renders the interpretation of $\delta^{18}O$ values of cherts more difficult than those of porcellanites because of the greater uncertainty in the isotopic composition of the pore water at the greater depths in the sediment at which chert is formed.

Murata et al. (1977) studied the $^{18}O/^{16}O$ ratios of diagenetic silica minerals in two sequences of the Monterey Formation, California. In these rocks the diagenetic modifications of the silica minerals discussed above are observed as abrupt transitions, with increasing depth of burial, from opal-A to opal-CT and from opal-CT to microcrystalline quartz. Additionally, within the opal-CT zone there is a progressive increase with depth in the degree of ordering as interpreted from X-ray diffraction spacings. Abrupt changes in isotopic composition occur at the two sharply defined mineralogical boundaries (Fig. 8-14), and $\delta^{18}O$ values are consistent with formation of the diagenetic phases in equilibrium with interstitial waters at temperatures inferred from heat flow measurements and stratigraphic thicknesses (49°C for the opal-A → opal-CT transition and 83°C for the opal-CT → quartz transition). On the contrary, the progressive ordering with depth observed in the opal-CT zone is accompanied by no oxygen isotopic change. The results may therefore be taken to indicate that the mineralogic transitions involve pore waters (presumably as an agent for solution and redeposition) while the ordering of cristobalite occurs in the solid state.

The hydrogen extractable from siliceous minerals is present in more than one site (Knauth, 1973). Hydrogen extracted (as water) from opal-A, opal-CT and microcrystalline quartz at room temperature is readily exchangeable with the atmosphere while that extracted at higher temperatures is much less exchangeable. Knauth (1973) showed that essentially all the hydrogen in opal-A is exchangeable at 100°C and therefore samples subject to percolation by heated groundwaters have probably lost their initial isotopic compositions. Hydrogen extracted (as water) at high temperatures from opal-CT

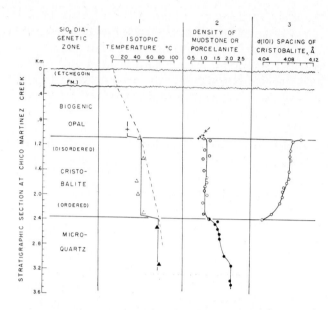

Fig. 8-14. Isotopic temperature, density, and cristobalite $d(101)$ spacing of chert from a Monterey Shale section, Chico Martinez Creek, California (from Murata et al., 1977). Isotopic temperatures were calculated from $\delta^{18}O$ values of SiO_2, an assumed $\delta^{18}O$ of water of 0.0‰ and the quartz-water fractionation curve of Clayton et al. (1972). The dashed line through the isotopic temperature plot is a geothermal gradient estimated from heat flow calculations. Reproduced by permission.

and microcrystalline chert is less readily exchangeable and is probably derived from hydroxyl groups (Knauth, 1973; Knauth and Epstein, 1975).

The δD and $\delta^{18}O$ values acquired at the time of formation of microcrystalline quartz are frequently preserved. They thus can be used as indicators of the environment of formation. Isotopic compositions of Paleozoic and Mesozoic chert nodules and beds in sediments indicate that microcrystalline quartz forms in hydrogen and oxygen isotope equilibrium with its environment (Knauth and Epstein, 1975). Using a treatment analogous to that described earlier for clay minerals, the isotopic compositions of cherts formed at the same temperature in the presence of seawater or different meteoric waters lie along a straight line with slope of approximately 8 on a δD vs. $\delta^{18}O$ plot. As the temperature changes the position of the line shifts (Fig. 8-4). Envelopes drawn around the isotopic compositions of cherts of different ages analyzed by Knauth (1973) and Knauth and Epstein (1976) are shown in Fig. 8-15. These are elongate, their long axes lying roughly parallel to the meteoric water line, and have been interpreted by Knauth and Epstein as reflecting changes in temperature at the time of chert formation.

According to this interpretation, points plotting anywhere but at the high δ-value end of the envelope drawn about the data for samples of a given age

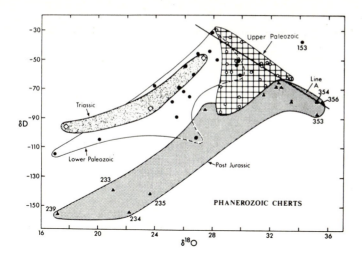

Fig. 8-15. Isotopic compositions of Phanerozoic cherts (from Knauth and Epstein, 1976). Envelopes enclose data of a given age range. Note elongation of most envelopes, generally parallel to meteoric water line. Line A represents isotopic compositions of chert in equilibrium with modern seawater over a range of temperatures (see Fig. 8-4). Reproduced by permission.

must have formed in the presence of meteoric waters rather than seawater. Some of the samples which are clearly marine on the basis of fossil evidence have isotopic compositions indicative of interaction with fresh water. Knauth and Epstein have interpreted these cherts as having been formed under the influence of meteoric waters, but at a time close to the time of sedimentation. Thus the temperature at the time of diagenetic chert formation was similar to the temperature at the time of sediment deposition.

Knauth's results for Paleozoic and Mesozoic cherts, when interpreted in this way, imply that the fractionation factor $\alpha^{hy}_{SiO_2-H_2O}$ approaches unity with increasing temperature (i.e., the depletion of the solid phase in deuterium relative to water becomes more marked as temperatures become cooler). There is an apparent conflict between this conclusion and the results obtained for DSDP cherts and porcellanites of Tertiary and Mesozoic age by Kolodny and Epstein (1976). Kolodny and Epstein found a wide spread in $\delta^{18}O$ values (+27 to +42‰) but very little change in δD values (−78 to −95‰). If the hydrogen isotope fractionation factor were temperature dependent, one would normally expect a change in δD values correlating with the change in $\delta^{18}O$ values. It is possible, as pointed out by Kolodny and Epstein, that the temperature range involved in the formation of the DSDP cherts was too small to produce a clear cut effect on δD values. Any temperature effect might be masked by variations in the deuterium content of sediment pore waters such as those observed by Friedman and Hardcastle (1973).

$\delta^{18}O$ values of cherts and carbonates have been used as the basis of argu-

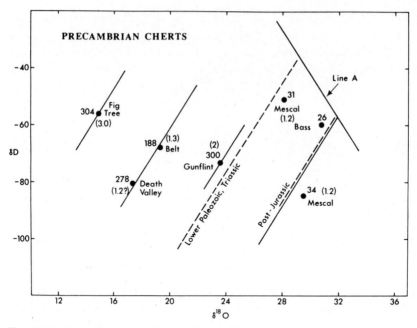

Fig. 8-16. Isotopic compositions of Precambrian cherts. Numbers in parentheses are ages (billions of years). Other numbers are sample numbers of Knauth and Epstein (1976). Reproduced from Knauth and Epstein by permission.

ments concerning the constancy of the $\delta^{18}O$ value of the oceans throughout geologic time (Weber, 1965; Perry, 1967). Large numbers of analyses of these materials have indicated a tendency toward lower $\delta^{18}O$ values with increasing age. One interpretation of those data, and the one favored by Weber (1965) and Perry (1967), is that the oceans contained less ^{18}O in the past (Paleozoic and Precambrian) than at present. An alternative explanation is that post-depositional exchange of the minerals with meteoric waters, combined perhaps with warmer temperatures during parts of the Precambrian, caused much of the depletion in ^{18}O with increasing age. Analyses of Precambrian cherts by Knauth and Epstein (1976) are shown in Fig. 8-16. Knauth and Epstein have argued that these data are not consistent with a progressive change in the $\delta^{18}O$ value of the hydrosphere over the past 2 billion years (b.y.). This is especially indicated by the enrichment in ^{18}O of the Gunflint Chert (2 b.y.) relative to the Belt (1.3 b.y.) and Death Valley (1.2 b.y.) cherts.

Burial diagenesis of shales

When detrital clay minerals undergo burial in thick sedimentary sequences, such that their temperatures become elevated significantly above those at the earth's surface, they commonly undergo chemical and mineralogic reactions. Perhaps the most ubiquitous reaction observed under burial conditions is the conversion of detrital smectite and mixed-layer illite/smectite of high expandability (large percentage of smectite layers) to illite/smectite of low expandability and ultimately to illite. This transformation commonly begins at temperatures of 60—80°C (Perry and Hower, 1970; Hower et al., 1976). The reactions which occur can be summarized as:

$K^+ + Al^{3+}$ + smectite → illite + quartz (± chlorite)

The conversion of smectite layers to illite layers in the mixed-layer clay involves the gain of Al^{3+} in, and the loss of Si^{4+} from, the tetrahedrally coordinated sites. Such a major chemical change would be expected to involve the breaking of oxygen bonds, permitting isotopic exchange of oxygen between the clay and the pore fluid.

Yeh and Savin (1977) isotopically analyzed quartz and clay minerals of several separated fine size fractions (in the range 10 μm and finer), as well as whole rock and total carbonate, from three deep wells penetrating Tertiary and late Cretaceous sediments in the Gulf of Mexico region. In addition to

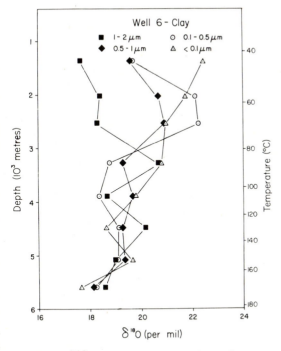

Fig. 8-17. $\delta^{18}O$ values of separated size fractions of clay minerals from a Gulf Coast well (CWRU Gulf Coast Well No. 6) plotted as a function of depth (from Yeh and Savin, 1977).

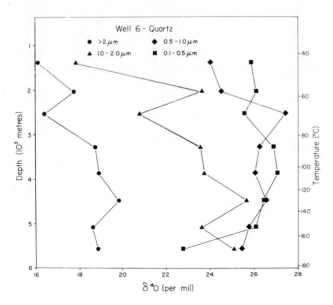

Fig. 8-18. $\delta^{18}O$ values of separated size fractions of quartz from a Gulf Coast well (CWRU Gulf Coast Well No. 6) plotted as a function of depth (from Yeh and Savin, 1977).

following the isotopic variations of the separated phases they attempted to compute "isotopic temperatures" of diagenesis from the $\delta^{18}O$ values of coexisting phases. $\delta^{18}O$ values of quartz and clay from the well sampled in most detail (CWRU Gulf Coast Well No. 6) are shown in Figs. 8-17 and 8-18. The conclusions of this study are summarized below.

(1) The detrital clay minerals of the sediment are not in isotopic equilibrium with one another when deposited in the sedimentary basin. This is apparent from the large range of $\delta^{18}O$ values of different size fractions from single samples in the upper portions of the well. This range is too great to reflect differing mineralogic compositions of different size fractions, all in isotopic equilibrium with one another.

(2) With increasing temperature and depth of burial isotopic exchange between the clay minerals and pore fluids begins. Finer size fractions begin exchanging at shallower depths than coarser ones. By the time the temperature has reached 85°C (burial depth 3300 m), the range of isotopic variation within a single sample has been greatly reduced. At this depth also, a significant amount of smectite has been converted to illite.

(3) The different size fractions of quartz in single samples are also out of internal isotopic equilibrium. Unlike the clay minerals, however, the approach to isotopic equilibrium with increasing depth of burial does not proceed very far. Quartz does not seem to undergo isotopic exchange with the pore fluids at temperatures attained here (170°C maximum). However,

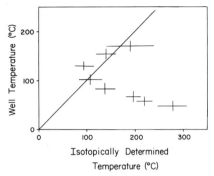

Fig. 8-19. Temperatures calculated from oxygen isotope fractionations between coexisting quartz (0-1—0.5 μm) and clay (0.1 μm) from CWRU Gulf Coast Well No. 6 plotted versus measured well temperatures (from Yeh and Savin, 1977). If the isotopically determined and measured well temperatures were concordant they would plot along the 45° line through the origin.

the results of hydrofluoric acid etching experiments indicate that detrital quartz cores of the grains are mantled by secondarily precipitated quartz rims. These rims presumably form from the Si^{4+} lost from the clay mineral lattice during the conversion of smectite layers to illite.

(4) Isotopic temperatures were calculated from the fractionation between fine quartz, fine clay, and calcite, using the appropriate geothermometric equations calculated from the mineral-water fractionations in Table 8-1. Isotopic temperatures calculated from the $\delta^{18}O$ values of the finest fractions of coexisting quartz and clay separable from the rocks (Fig. 8-19) approach measured well temperatures as measured temperatures increase. At measured temperatures of 100°C and above, the agreement between isotopic and measured temperatures is quite good. Isotopic temperatures calculated using $\delta^{18}O$ values of calcite were unreasonable and indicate that calcite continued to undergo isotopic exchange with pore fluids after the silicates had ceased to undergo isotopic exchange. Quartz-clay isotopic temperatures from two other wells similarly approached measured temperatures as measured temperatures increased. They did not, however, equal measured temperatures even where the measured temperatures reached 120°C. This probably reflects the presence of some ^{18}O-depleted detrital quartz in the samples analyzed and indicates that further work is necessary before oxygen isotope geothermometry can be routinely applied to determine the maximum temperatures of shale diagenesis.

SERPENTINIZATION OF ULTRAMAFIC ROCKS

Serpentinized ultramafic rocks are found in a variety of geologic environments, including mid-ocean ridges and other tectonically active submarine

TABLE 8-4

Estimated temperatures of isotopic equilibration of serpentines (from Wenner and Taylor, 1971)

Type	Isotopic temperature (°C)
Continental lizardite-chrisotile	85—115
Oceanic lizardite	130
Oceanic chrisotile	185
Oceanic antigorite	235
Continental antigorite	220—460

sites, ophiolite complexes, and alpine ultramafic bodies. While serpentinization of many ultramafics has been considered to be a high temperature process, much evidence, including that from stable isotope analyses indicates that serpentines form over a variety of temperatures, some fairly low.

Wenner and Taylor (1971) derived a serpentine-magnetite oxygen isotope geothermometer. This was based largely on isotopic compositions of naturally occurring samples since adequate data from laboratory calibration studies were not available. The temperatures of serpentinization they obtained using this geothermometer (Table 8-4) must therefore be considered only approximate. In addition to the temperatures listed in Table 8-4, deweylite, a serpentine-like material, appears on the basis of isotopic and field evidence to form under surface or near-surface conditions.

In a study of oceanic serpentines and serpentines from continental ophiolite complexes, Wenner and Taylor (1973) concluded that the oceanic serpentines formed through the interaction of ultramafic rocks with water largely of marine origin but with a possible admixture of some water of magmatic origin. The serpentines from the continental ophiolite complexes were isotopically distinct from the oceanic serpentines, and appeared to have formed in the presence of water largely of meteoric origin. Serpentinization therefore occurred subsequently to uplift of the ophiolite complex. In a very detailed study of the Franciscan formation, California, Magaritz and Taylor (1976) similarly concluded that most serpentinization occurred in this way. Magaritz and Taylor (1974) examined the serpentines of the Troodos ophiolite complex, Cyprus, and proposed that here, too, serpentinization most probably occurred at about 100°C in a non-oceanic environment. In this case, serpentinization was inferred to have occurred in the presence of water strongly enriched in ^{18}O, perhaps as a result of the intense evaporation of water that occurred in the Mediterranean during Miocene time.

EFFECT OF MINERAL-WATER INTERACTION ON THE ISOTOPIC COMPOSITION OF PORE WATER

When minerals undergo isotopic exchange with water, the isotopic composition of both the mineral and the water are affected. When the amount of the mineral is small relative to the amount of water with which it is interacting, the change in isotopic composition of the water may be negligible. However, when the amount of water is similar to or smaller than that of the mineral the isotopic composition of the water may be markedly affected. This situation can exist in the case of pore waters in relatively impermeable rocks or in sedimentary reservoirs with restricted circulation.

$\delta^{18}O$ and δD analyses have been made on pore waters squeezed from a number of DSDP cores (Lawrence, 1973, 1974; Lawrence et al., 1975a;

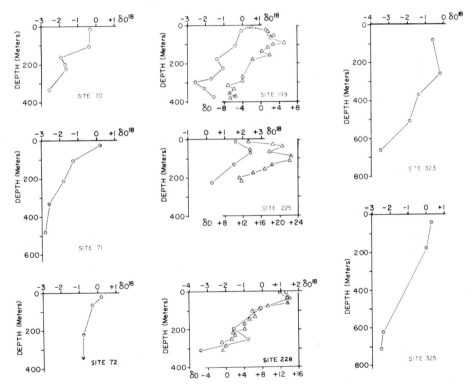

Fig. 8-20. Oxygen (circles) and hydrogen (triangles) isotopic compositions of pore waters squeezed from a number of Deep Sea Drilling Project sediments plotted as a function of depth below the sediment water interface. Sources: Sites 70, 71, 72 (Equatorial Pacific), Lawrence et al. (1975b); Site 149 (Venezuela Basin), oxygen from Lawrence (1973), hydrogen from Friedman and Hardcastle (1973); Sites 225, 228 (Red Sea), oxygen from Lawrence (1974), hydrogen from Friedman and Hardcastle (1974); Sites 323, 325 (Antarctic), Lawrence et al. (1975a).

Friedman and Hardcastle, 1973, 1974). Samples were from the South Atlantic, Equatorial Pacific and Red Sea. Sediments ranged in age from Pleistocene to Eocene and depths of burial were as great as 400 m. Typical data are shown in Fig. 8-20. In most instances there is a depletion in both ^{18}O and deuterium with increasing depth of burial. Some of this depletion may reflect changes in the isotopic composition of pore water initially trapped in the sediment due to waxing and waning of the ice sheets, or, in the case of the Red Sea sediments, to influx of fresh water. $\delta^{18}O$ variations may also reflect low temperature alteration or exchange between minerals and water. Lawrence et al. (1975b) have argued convincingly that the ^{18}O depletion in pore water from the lower portions of their Equatorial Pacific (predominantly carbonate) cores is the result of upward diffusion of water depleted in ^{18}O during low-temperature interaction with basalts of Layer II. Such interaction, they estimate, over geologically long periods could affect the isotopic composition of the oceans. In other instances ^{18}O variations in pore waters may have been caused by reactions in the sediment column — principally the alteration of volcanic ash to smectite. It is unlikely that decreases in the deuterium content of pore waters have been caused by isotopic exchange between minerals and water. All known hydrogen-bearing marine silicates are depleted in deuterium relative to the waters with which they have exchanged. The work of Suzuoki and Epstein (1976) suggests that this is also true at higher temperatures. Therefore pore water would undergo enrichment rather than depletion of deuterium as a result of such exchange. δD values of interstitial waters in marine sediments are much more likely than $\delta^{18}O$ values to reflect original pore water isotopic composition.

When pore waters exchange isotopically with the enclosing rocks in most geological environments, $\delta^{18}O$ values of the waters are much more likely than δD values to be affected (although this depends in part on the initial isotopic compositions of the two phases). This reflects the much greater amount of oxygen than hydrogen present in all common rock-forming minerals. Commonly, δD values of oilfield formation waters are similar to those of meteoric waters in the local groundwater recharge area, while ^{18}O values are higher than those of the local meteoric waters. Examples are the Michigan Basin and the U.S. Gulf Coast (Clayton et al., 1966). Similar isotopic behavior has been reported for the waters of hot springs (see Chapter 10).

An additional factor affecting the D/H and $^{18}O/^{16}O$ ratios of formation waters may be isotopic fractionation produced by ultrafiltration of waters through shales. Coplen and Hanshaw (1973) demonstrated that fractionations could be produced in this way under laboratory conditions. Isotopic variations in waters from several basins have been attributed, at least in part to fractionations produced by this mechanism (Graf et al., 1965; Hitchon and Friedman, 1969; Kharaka et al., 1973).

ACKNOWLEDGEMENTS

I am grateful to the faculty, staff and students of the Department of Oceanography, University of Hawaii, for making my vist, during which this chapter was written, stimulating and thoroughly enjoyable. Special thanks go to Peter Kroopnick and Stanley Margolis. The manuscript was improved by the comments of Peter Kroopnick, Philip Choquette and the editors of this book. Financial support was provided by Case Western Reserve University and by the National Science Foundation under grants DES 75-20431 and OCE 76-01457.

REFERENCES

Allan, J.R. and Matthews, R.K., 1977. Carbon and oxygen isotopes as diagenetic and stratigraphic tools: surface and subsurface data, Barbados, West Indies. *Geology*, 5: 16—20.
Anderson, T.F. and Cole, S.A., 1975. The stable isotope geochemistry of marine coccoliths: a preliminary comparison with planktonic foraminifera. *J. Foraminiferal Res.*, 5: 188—192.
Baertschi, P., 1953. Über die relativen Unterschiede im $H_2^{18}O$-Gehalt natürlicher Wässer. *Helv. Chim. Acta*, 36: 1352—1369.
Barrer, R.M. and Denny, A.F., 1964. Water in hydrates, I. Fractionation of hydrogen isotopes by crystallization of salt hydrates. *J. Chem. Soc.*, 4677—4684.
Bowser, C.J., Rafter, T.A. and Black, R.F., 1970. Geochemical evidence for the origin of mirabilite deposits near Hobbs Glacier, Victoria Land, Antarctica. *Mineral. Soc. Am., Spec. Paper*, 3: 261—272.
Choquette, P.M., 1968. Marine diagenesis of shallow marine lime-mud sediments: insights from ^{18}O and ^{13}C data. *Science*, 161: 1130—1132.
Church, T., 1970. *Marine Barite*. Ph.D. Thesis, University of California, San Diego, Calif., 100 pp.
Claypool, G.E., Holser, W.T., Kaplan, I.R., Sakai, H., and Zak, I., 1972. Sulfur and oxygen isotope geochemistry of evaporite sulfates. *Geol. Soc. Am. Abstr. Progr.*, 4: 473.
Clayton, R.N. and Epstein, S., 1958. The relationship between $^{18}O/^{16}O$ ratios in coexisting quartz, carbonate and iron oxides from various geological deposits. *J. Geol.*, 66: 352—371.
Clayton, R.N., Friedman, I., Graf, D.L., Mayeda, T.K., Meents, W.F. and Shimp, N.F., 1966. The origin of saline formation waters, I. Isotopic composition. *J. Geophys. Res.*, 71: 3869—3882.
Clayton, R.N., Jones, B.F. and Berner, R., 1968a. Isotope studies of dolomite formation under sedimentary conditions. *Geochim. Cosmochim. Acta*, 32: 415—432.
Clayton, R.N., Skinner, H.C.W., Berner, R.A. and Rubinson, M., 1968b. Isotopic composition of recent South Australian lagoonal carbonates. *Geochim. Cosmochim. Acta*, 32: 983—988.
Clayton, R.N., O'Neil, J.R. and Mayeda, T.K., 1972a. Oxygen isotope exchange between quartz and water. *J. Geophys. Res.*, 77: 3057—3067.
Clayton, R.N., Rex, R.W., Syers, J.K. and Jackson, M.L., 1972b. Oxygen isotope abundance in quartz from Pacific pelagic sediments. *J. Geophys. Res.*, 77: 3907—3913.
Coplen, T.B. and Hanshaw, B.B., 1973. Ultrafiltration by a compacted clay membrane, I.

Oxygen and hydrogen isotopic fractionation. *Geochim. Cosmochim. Acta*, 37: 2295—2310.

Cortecci, G. and Longinelli, A., 1972. Oxygen-isotope variations in a barite slab from the sea bottom of southern California. *Chem. Geol.*, 9: 113—117.

Craig, H., 1961. Isotopic variations in meteoric waters. *Science*, 133: 1702—1703.

Dasch, E.J., 1969. Strontium isotopes in weathering profiles, deep sea sediments and sedimentary rocks. *Geochim. Cosmochim. Acta*, 33: 1521—1552.

Degens, E.T. and Epstein, S., 1964. Oxygen and carbon isotope ratios in coexisting calcites and dolomites from recent and ancient sediments. *Geochim. Cosmochim. Acta*, 28: 23—44.

Douglas, R.G. and Savin, S.M., 1975. Oxygen and carbon isotope analyses of Tertiary and Cretaceous microfossils from Shatsky Rise and other sites in the North Pacific Ocean. In: R.L. Larson, R. Moberly et al., *Initial Reports of the Deep Sea Drilling Project, 32*, U.S. Government Printing Office, Washington, D.C., pp. 509—520.

Douglas, R.G. and Savin, S.M., 1978. Depth stratification in Tertiary and Cretaceous foraminifera based on oxygen isotope ratios. *Mar. Micropaleontol.*, 3: 175—196.

Dudley, W.C., 1976. *Paleooceanographic Application of Oxygen Isotopic Analyses of Calcareous Nannoplankton Grown in Culture*. Ph.D. Thesis, University of Hawaii, Honolulu, Hawaii, 168 pp.

Dymond, J., Corliss, J.B., Heath, G.R., Field, C.W., Dasch, E.J. and Veeh, H.H., 1973. Origin of metalliferous sediments from the Pacific Ocean. *Geol. Soc. Am. Bull.*, 84: 3355—3372.

Emiliani, C., 1954. Depth habitats of some species of pelagic foraminifera as indicated by oxygen isotope ratios. *Am. J. Sci.*, 252: 149—158.

Engel, A.E.J., Clayton, R.N. and Epstein, S., 1958. Variations in isotopic composition of oxygen and carbon in Leadville limestone (Mississippian, Colorado) and in its hydrothermal and metamorphic phases. *J. Geol.*, 66: 374—393.

Epstein, S., Buchsbaum, R., Lowenstam, H.A. and Urey, H.C. 1953. Revised carbonate water isotopic temperature scale. *Geol. Soc. Am. Bull.*, 64: 1315—1326.

Eslinger, E.V. and Savin, S.M., 1973. Mineralogy and oxygen isotope geochemistry of the hydrothermally altered rocks of the Ohaki-Broadlands, New Zealand geothermal area. *Am. J. Sci.*, 273: 240—267.

Eslinger, E.V. and Savin, S.M., 1976. Mineralogy and $^{18}O/^{16}O$ ratios of fine-grained quartz and clay from Site 323. In: C.D. Hollister, C.C. Craddock et al., *Initial Reports of the Deep Sea Drilling Project, 35*. U.S. Government Printing Office, Washington, D.C., pp. 489—496.

Fontes, J.C., 1965. Fractionnement isotopique dans l'eau de cristallisation du sulfate de calcium. *Geol. Rundsch.*, 55: 172—178.

Fontes, J.C., 1966. Intérêt en géologie d'une étude isotopique de l'évaporation. Cas de l'eau de mer. *C.R. Acad. Sci. Paris, Sér. D*, 263: 1950—1953.

Fontes, J.C. and Gonfiantini, R., 1967. Fractionnement isotopique de l'hydrogène dans l'eau de cristallisation du gypse. *C.R. Acad. Sci. Paris, Sér. D*, 265: 4—6.

Fontes, J.C. and Gonfiantini, R., 1970. Comportement isotopique au cours de l'évaporation de deux bassins Sahariens. *Earth Planet. Sci. Lett.*, 3: 258—266.

Fontes, J.C. and Nielsen, H., 1966. Isotopes de l'oxygène et du soufre dans le gypse parisien. *C.R. Acad. Sci. Paris, Sér. D*, 262: 2685—2687.

Fontes, J.C., Létolle, R. and Nesteroff, W.D., 1972. Les forages DSDP en Méditerranée (Leg 13): reconnaissance isotopique. In: D.J. Stanley (Editor), *The Mediterranean Sea: A Natural Laboratory*. Dowden, Hutchinson and Ross, Stroudsburg, Pa., pp. 671—680.

Friedman, I. and Hardcastle, K., 1973. Interstitial water studies, Leg 15 — isotopic composition of water. In: B.C. Heezen, I.D. MacGregor et al., *Initial Reports of the Deep Sea Drilling Project, 20*. U.S. Government Printing Office, Washington, D.C., pp. 901—903.

Friedman, I. and Hardcastle, K., 1974. Deuterium in interstitial waters from Red Sea Cores. In: R.B. Whitmarsh, O.E. Weser et al., *Initial Reports of the Deep Sea Drilling Project, 23*. U.S. Government Printing Office, Washington, D.C., pp. 969—970.

Friedman, I. and O'Neil, J.R., 1977. Compilation of stable isotope fractionation factors of geochemical interest. In: M. Fleischer (Editor), *Data of Geochemistry, U.S. Geol. Surv., Prof. Paper*, 440-KK (6th ed.).

Fritz, P. and Smith, D.G.W., 1970. The isotopic composition of secondary dolomites. *Geochim. Cosmochim. Acta*, 34: 1161—1173.

Garlick, G.D., 1969. The stable isotopes of oxygen. In: K.H. Wedepohl (Editor), *Handbook of Geochemistry, 2*. Springer-Verlag, New York, N.Y., Part 1, Chapter 8B.

Garlick, G.D., 1974. The stable isotopes of oxygen, carbon and hydrogen in the marine environment. In: E.D. Goldberg (Editor), *The Sea*, 5. Wiley, New York, N.Y., pp. 393—425.

Garlick, G.D. and Dymond, J.R., 1970. Oxygen isotope exchange between volcanic material and ocean water. *Geol. Soc. Am. Bull.*, 81: 2137—2142.

Gonfiantini, R. and Fontes, J.C., 1963. Oxygen isotopic fractionation in the water of crystallization of gypsum. *Nature*, 200: 644—646.

Graf, D.L., Friedman, I. and Meents, W.F., 1965. Origin of saline formation waters. II. Isotopic fractionation by shale micropore systems. *Ill. Geol. Surv., Circ.*, 393: 32 pp.

Gross, M.G., 1964. Variations in the $^{18}O/^{16}O$ and $^{13}C/^{12}C$ ratios of diagenetically altered limestones in the Bermuda Islands. *J. Geol.*, 72: 170—194.

Halevy, E., 1964. The exchangeability of hydroxyl groups in kaolinite. *Geochim. Cosmochim. Acta*, 28: 1139—1145.

Hamza, M.S. and Broecker, W.S., 1974. Surface effect on the isotopic fractionation between CO_2 and some carbonate minerals. *Geochim. Cosmochim. Acta*, 38: 669—682.

Heinzinger, K. and Götz, D., 1975. Method to determine crystallization temperatures of hydrated crystals by intracrystalline oxygen isotope effects. *Earth Planet. Sci. Lett.*, 27: 219—220.

Hitchon, B. and Friedman, I., 1969. Geochemistry and origin of formation waters in the Western Canada sedimentary basin, I. Stable isotopes of hydrogen and oxygen. *Geochim. Cosmochim. Acta*, 33: 1321—1349.

Hower, J., Eslinger, E.V., Hower, M. and Perry, E.A., 1976. The mechanism of burial metamorphism of argillaceous sediments, 1. Mineralogical and chemical evidence. *Geol. Soc. Am. Bull.*, 87: 725—737.

Hurley, P.M., Heezen, B.C., Pinson, W.H. and Fairbairn, H.W., 1963. K-Ar age values in pelagic sediments of the North Atlantic. *Geochim. Cosmochim. Acta*, 27: 393—399.

James, A.T. and Baker, D.R., 1976. Oxygen isotope exchange between illite and water at 22°C. *Geochim. Cosmochim. Acta*, 40: 235—239.

Kharaka, Y.K., Berry, A.F. and Friedman, I., 1973. Isotopic composition of oil-field brines from Kettleman North Dome, California and their geologic implications. *Geochim. Cosmochim. Acta*, 37: 1899—1908.

Knauth, L.P., 1973. *Oxygen and Hydrogen Isotope Ratios in Cherts and Related Rocks*. Ph.D. Thesis, California Institute of Technology, Pasadena, Calif., 369 pp.

Knauth, L.P. and Epstein, S., 1975. Hydrogen and oxygen isotope ratios in silica from the JOIDES Deep Sea Drilling Project. *Earth Planet. Sci. Lett.*, 25: 1—10.

Knauth, L.P. and Epstein, S., 1976. Hydrogen and oxygen isotope ratios in nodular and bedded cherts. *Geochim. Cosmochim. Acta*, 40: 1095—1108.

Kolodny, Y. and Epstein, S., 1976. Stable isotope geochemistry of deep sea cherts. *Geochim. Cosmochim. Acta*, 40: 1195—1209.

Labeyrie, L., 1974. New approach to surface seawater paleotemperatures using $^{18}O/^{16}O$ ratios in silica of diatom frustules. *Nature*, 248: 40—42.

Land, L.S., 1970. Phreatic versus vadose meteoric diagenesis of limestones: evidence from a fossil water table. *Sedimentology*, 14: 175—185.

Land, L.S., 1973a. Holocene meteoric dolomitization of Pleistocene limestones, North Jamaica. *Sedimentology*, 20: 411—424.

Land, L.S., 1973b. Contemporaneous dolomitization of middle Pleistocene reefs by meteoric water, North Jamaica. *Bull. Mar. Sci.*, 23: 64—92.

Land, L.S. and Epstein, S., 1970. Late Pleistocene diagenesis and dolomitization, North Jamaica. *Sedimentology*, 14: 187—200.

Lawrence, J.R., 1970. $^{18}O/^{16}O$ *and D/H Ratios of Soils, Weathering Zones, and Clay Deposits.* Ph.D. Thesis, California Institute of Technology Pasadena, Calif., 263 pp.

Lawrence, J.R., 1973. Interstitial water studies, Leg 15 — oxygen and carbon isotope variations in water, carbonates and silicates from the Venezuela Basin (Site 149) and the Aves Rise (Site 148). In: B.C. Heezen, I.D. McGregor et al., *Initial Reports of the Deep Sea Drilling Project, 20.* U.S. Government Printing Office, Washington, D.C., pp. 891—899.

Lawrence, J.R., 1974. Stable oxygen and carbon isotope variations in the pore waters, carbonates and silicates, Sites 225 and 228, Red Sea. In: R.B. Whitmarsh, D.E. Weser et al., *Initial Reports of the Deep Sea Drilling Project, 23.* U.S. Government Printing Office, Washington, D.C., pp. 939—942.

Lawrence, J.R. and Taylor, H.P., Jr., 1971. Deuterium and oxygen-18 correlation: clay minerals and hydroxides in Quaternary soils compared to meteoric waters. *Geochim. Cosmochim. Acta*, 35: 993—1003.

Lawrence, J.R. and Taylor, H.P., Jr., 1972. Hydrogen and oxygen isotope systematics in weathering profiles. *Geochim. Cosmochim. Acta*, 36: 1377—1393.

Lawrence, J.R., Gieskes, J. and Anderson, T.F., 1975a. Oxygen isotope material balance calculations, Leg 35. In: C.D. Hollister, C.C. Craddock et al., *Initial Reports of the Deep Sea Drilling Project, 35.* U.S. Government Printing Office, Washington, D.C. pp. 507—512.

Lawrence, J.R., Gieskes, J.M. and Broecker, W.S., 1975b. Oxygen isotope and cation composition of DSDP pore waters and the alteration of Layer II basalts. *Earth Planet. Sci. Lett.*, 27: 1—10.

Ledoux, R.L. and White, J.L., 1964. Infrared studies of OH groups in expanded kaolinite. *Science*, 143: 244—246.

Lindroth, K.J., Miller, L.G., Durazzi, J.T. McIntyre, A. and Van Donk, J., 1980. Coccoliths as isotopic temperature indicators: a preliminary investigation (in press).

Lloyd, R.M., 1966. Oxygen isotope enrichment of sea water by evaporation. *Geochim. Cosmochim. Acta*, 30: 801—814.

Lloyd, R.M., 1967. ^{18}O composition of oceanic sulfate. *Science*, 156: 1228—1231.

Lloyd, R.M., 1968. Oxygen isotope behavior in the sulfate-water system. *J. Geophys. Res.*, 73: 6099—6110.

Lloyd, R.M. and Hsü, K.J., 1972. Stable isotope investigations of sediments from the DSDP III cruise to South Atlantic. *Sedimentology*, 19: 45—58.

Longinelli, A., 1965. Oxygen isotopic composition of orthophosphate from shells of living marine organisms. *Nature*, 207: 716—719.

Longinelli, A., 1966. Ratios of $^{18}O/^{16}O$ in phosphate and carbonate from living and fossil marine organisms. *Nature*, 211: 923—927.

Longinelli, A. and Craig, H., 1967. Oxygen-18 variations in sulfate ions in sea-water and saline lakes. *Science*, 156: 56—59.

Longinelli, A. and Nuti, S., 1968a. Oxygen-isotope ratios in phosphate from fossil marine organisms. *Science*, 160: 879—882.

Longinelli, A. and Nuti, S., 1968b. Oxygen isotopic composition of phosphorites from marine formations. *Earth Planet. Sci. Lett.*, 5: 13—16.

Longinelli, A. and Nuti, S., 1973a. Revised phosphate-water isotopic temperature scale. *Earth Planet. Sci. Lett.*, 19: 373—376.
Longinelli, A. and Nuti, S., 1973b. Oxygen isotope measurements of phosphate from fish teeth and bones. *Earth Planet. Sci. Lett.*, 20: 337—340.
Magaritz, M. and Taylor, H.P., Jr., 1974. Oxygen and hydrogen isotope studies of serpentinization in the Troodos ophiolite complex, Cyprus. *Earth Planet. Sci. Lett.*, 23: 8—14.
Magaritz, M. and Taylor, H.P., Jr., 1976. Oxygen, hydrogen and carbon isotope studies of the Franciscan formation, Coast Ranges, California. *Geochim. Cosmochim. Acta*, 40: 215—234.
Margolis, S.V., Kroopnick, P.M., Goodney, D.E., Dudley, W.C. and Mahoney, M.E., 1975. Oxygen and carbon isotopes from calcareous nannofossils as paleooceanographic indicators. *Science*, 189: 555—557.
Matsuo, S., Friedman, I. and Smith, G.I., 1972. Studies of Quaternary saline lakes, I. Hydrogen isotope fractionation in saline minerals. *Geochim. Cosmochim. Acta*, 36: 427—436.
McAuliffe, C.D., Hall, W.S., Dean, L.A. and Hendricks, S.B., 1947. Exchange reactions between phosphates and soils: hydroxylic surfaces of soil minerals. *Soil Sci. Soc. Am., Proc.*, 12: 119—123.
McCrea, J.M., 1950. On the isotopic chemistry of carbonates and a paleotemperature scale. *J. Chem. Phys.*, 18: 849—857.
Mizutani, Y. and Rafter, T.A., 1969a. Oxygen isotopic composition of sulphates, III. Oxygen isotopic fractionation in the bisulphate ion-water system. *N. Z. J. Sci.*, 12: 54—59.
Mizutani, Y. and Rafter, T.A., 1969b. Oxygen isotopic composition of sulphates, IV. Bacterial fractionation of oxygen isotopes in the reduction of sulphate and in the oxidation of sulphur. *N. Z. J. Sci.*, 12: 60—68.
Mizutani, Y. and Rafter, T.A., 1973. Isotopic behavior of sulphate oxygen in the bacterial reduction of sulphate. *Geochem. J.*, 6: 183—192.
Mopper, K. and Garlick, G.D., 1971. Oxygen isotope fractionation between biogenic silica and ocean water. *Geochim. Cosmochim. Acta*, 35: 1185—1187.
Muehlenbachs, K., 1971. *Oxygen Isotope Studies of Rocks from Mid-Ocean Ridges*. Ph.D. Thesis, University of Chicago, Chicago, Ill., 124 pp.
Murata, K.J., Friedman, I. and Gleason, J.D., 1977. Oxygen isotope relations between diagenetic silica minerals in Monterey Shale, Temblor Range, California. *Am. J. Sci.*, 277: 259—272.
Northrop, D.A. and Clayton, R.N., 1966. Oxygen-isotope fractionations in systems containing dolomite. *J. Geol.*, 74: 174—196.
O'Neil, J.R. and Clayton, R.N., 1964. Oxygen isotope geothermometry. In: H. Craig, S.L. Miller and G.J. Wasserburg (Editors), *Isotopic and Cosmic Chemistry*. North-Holland, Amsterdam, pp. 157—168.
O'Neil, J.R. and Epstein, S., 1966. Oxygen isotope fractionation in the system dolomite-calcite-carbon dioxide. *Science*, 152: 198—201.
O'Neil, J.R. and Hay, R.L., 1973. $^{18}O/^{16}O$ ratios in cherts associated with the saline lake deposits of East Africa. *Earth Planet. Sci. Lett.*, 19: 257—266.
O'Neil, J.R. and Kharaka, Y.F., 1976. Hydrogen and oxygen isotope exchange reactions between clay minerals and water. *Geochim. Cosmochim. Acta*, 40: 241—246.
O'Neil, J.R. and Taylor, H.P., Jr., 1969. Oxygen isotope equilibrium between muscovite and water. *J. Geophys. Res.*, 74: 6012—6022.
O'Neil, J.R., Clayton, R.N. and Mayeda, T.K., 1969. Oxygen isotope fractionation in divalent metal carbonates. *J. Chem. Phys.*, 51: 5547—5558.
Perry, E.A. and Hower, J., 1970. Burial diagenesis in Gulf Coast pelitic sediments. *Clays Clay Miner.*, 18: 165—177.

Perry, E.C., Jr., 1967. The oxygen isotope chemistry of ancient cherts. *Earth Planet. Sci. Lett.*, 3: 62—66.

Pineau, F., Javoy, M., Hawkins, J.W. and Craig, H., 1976. Oxygen isotope variations in marginal basin and ocean ridge basalts. *Earth Planet. Sci. Lett.*, 28: 299—307.

Rafter, T.A. and Mizutani, Y., 1967. Oxygen isotopic composition of sulfates, II. Preliminary results on oxygen isotopic variation in sulphates and the relationship to their environment and to their $\delta^{34}S$ values. *N.Z. J. Sci.*, 10: 816—840.

Rex, R.W. and Goldberg, E.D., 1958. Quartz contents of pelagic sediments of the Pacific Ocean. *Tellus*, 10: 153—159.

Rex, R.W., Syers, J.K., Jackson, M.L. and Clayton, R.N., 1969. Eolian origin of quartz in soils of Hawaiian Islands and in Pacific pelagic sediments. *Science*, 163: 277—279.

Roy, D.M. and Roy, R., 1957. Hydrogen-deuterium exchange in clays and problems in the assignment of infra-red frequencies in the hydroxyl region. *Geochim. Cosmochim. Acta*, 11: 72—85.

Savin, S.M., 1967. *Oxygen and Hydrogen Isotope Ratios in Sedimentary Rocks and Minerals*. Ph.D. Thesis. California Institute of Technology, Pasadena, Calif., 220 pp.

Savin, S.M., 1973. Oxygen and hydrogen isotope studies of minerals in ocean sediments. In: *Proceedings of Symposium on Hydrochemistry and Biochemistry, Tokyo, September, 1970*. Clarke Co., Washington, D.C., pp. 372—391.

Savin, S.M. and Douglas, R.G., 1973. Stable isotope and magnesium geochemistry of Recent planktonic foraminifera from the South Pacific. *Geol. Soc. Am. Bull.*, 84: 2327—2342.

Savin, S.M. and Epstein, S., 1970a. The oxygen and hydrogen isotope geochemistry of clay minerals. *Geochim. Cosmochim. Acta*, 34: 25—42.

Savin, S.M. and Epstein, S., 1970b. The oxygen and hydrogen isotope geochemistry of ocean sediments and shales. *Geochim. Cosmochim. Acta*, 34: 43—63.

Shackleton, N.J., Wiseman, J.D.H. and Buckley, H.A., 1973. Non-equilibrium isotopic fractionation between seawater and planktonic foraminiferal tests. *Nature*, 242: 177—179.

Sheppard, S.M.E., Nielson, R.L. and Taylor, H.P., Jr., 1969. Oxygen and hydrogen isotope ratios of clay minerals from porphyry copper deposits. *Econ. Geol.*, 64: 755—777.

Stewart, M.K., 1974. Hydrogen and oxygen isotope fractionation during crystallization of mirabilite and ice. *Geochim. Cosmochim. Acta*, 38: 167—172.

Suzuoki, T. and Epstein, S., 1976. Hydrogen isotope fractionation between OH-bearing minerals. *Geochim. Cosmochim. Acta*, 40: 1229—1240.

Taylor, H.P. and Epstein, S., 1964. Comparison of oxygen isotope analysis of tektites, soils, and impactite glasses. In: H. Craig, S.L. Miller and G.J. Wasserburg (Editors), *Isotopic and Cosmic Chemistry*. North-Holland, Amsterdam, pp. 181—199.

Tudge, A.P., 1960. A method of analysis of oxygen isotopes in orthophosphate — its use in the measurement of paleotemperatures. *Geochim. Cosmochim. Acta*, 18: 81—93.

Usitalo, E., 1958. Die Hydratisierung von Salzen in Mischungen von leichtem und schwerem Wasser. *Suomen Kemistilehti*, 31-B: 362—366.

Vincent, E. and Shackleton, N.J., 1975. Oxygen and carbon isotopic composition of Recent planktonic foraminifera from the southwest Indian Ocean. *Geol. Soc. Am., Abstr. Progr.*, 7: 1308.

Weber, J.N., 1965. Changes in the oxygen isotopic composition of sea water during the Phanerozoic evolution of the oceans. *Geol. Soc. Am., Spec. Paper*, 82: 218—219 (abstract).

Wenner, D.B., 1970. *Hydrogen and Oxygen Isotopic Studies of Serpentinization of Ultramafic Rocks*. Ph.D. Thesis, California Institute of Technology, Pasadena, Calif., 368 pp.

Wenner, D.B. and Taylor, H.P., Jr., 1971. Temperatures of serpentinization of ultramafic rocks based on $^{18}O/^{16}O$ fractionation between coexisting serpentine and magnetite. *Contrib. Mineral. Petrol.*, 32: 165—185.

Wenner, D.B. and Taylor, H.P., Jr., 1973. Oxygen and hydrogen isotope studies of the serpentinization of ultramafic rocks in oceanic environments and continental ophiolite complexes. *Am. J. Sci.*, 273: 207—239.

Williams, D.F., 1976. *Planktonic Foraminiferal Paleoecology in Deep-Sea Sediments of the Indian Ocean.* Ph.D. Thesis. University of Rhode Island, Kingston, R.I., 282 pp.

Yeh, H.W. and Epstein, S., 1978. Hydrogen isotope exchange between clay minerals and sea water. *Geochim. Cosmochim. Acta*, 42: 140—143.

Yeh, H.W. and Savin, S.M., 1976. The extent of oxygen isotope exchange between clay minerals and sea water. *Geochim. Cosmochim. Acta*, 40: 743—748.

Yeh, H.W. and Savin, S.M., 1977. The mechanism of burial metamorphism of argillaceous sediments, 3. Oxygen isotopic evidence. *Geol. Soc. Am. Bull.*, 88: 1321—1330.

Chapter 9

THE ISOTOPIC COMPOSITION OF REDUCED ORGANIC CARBON

PETER DEINES

INTRODUCTION

One of the most important processes affecting changes in the carbon isotopic composition in the geochemical cycle is the abstraction of carbon from the carbon dioxide reservoir of the atmosphere and surface waters by photosynthetic fixation in the form of complex organic molecules. In general such carbon shows a high degree of enrichment in the light isotope compared to its source. The products of this process are either subject to rapid decay or are incorporated into sedimentary rocks and occur there as solids, liquids or gases, either in dispersed form or concentrated, forming in the latter case often mineable deposits of the fossil fuels, coal, petroleum or natural gas. The isotopic composition of the carbon fixed in this manner is governed by the process of photosynthesis, as well as any subsequent changes occurring during accumulation, diagenesis, katagenesis, and metamorphism, as well as migration processes. In this chapter the isotopic fractionation during photosynthesis and the isotopic composition of the resulting products will be reviewed with the aid of a collection of some 8000 analyses compiled from literature. All reported analyses have been converted to a common scale of reference and are given as permil deviations from the PDB standard. The literature review and manuscript were completed in the spring of 1976.

PHOTOSYNTHESIS AND THE CARBON ISOTOPIC COMPOSITION OF PLANTS

Photosynthesis

Carbon fixation in photosynthesis may proceed by two pathways which differ by the number of carbon atoms in the first formed intermediate compounds (Bassham, 1971). The two pathways are known as the Calvin-Benson (C_3 or non-Kranz) and the Hatch-Slack (C_4 or Kranz) cycle respectively. It has been found that the carbon isotopic composition of the plant material formed is highly correlated with the type of photosynthetic cycle followed by the organism.

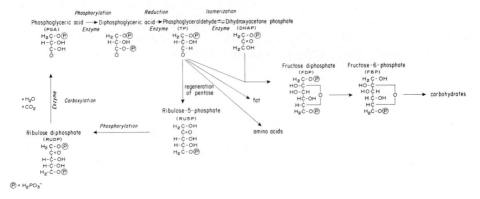

Fig. 9-1. Condensed version of the Calvin cycle.

A condensed version of the Calvin cycle is given in Fig. 9-1. The major reactions are: (1) the carboxylation of RUDP to form PGA; (2) the reduction of PGA to TP; and (3) the regeneration of RUDP via RU5P. TP may be condensed to give sugar monophosphates and eventually carbohydrates including starch, sugar and cellulose. Alternatively TP can be converted to glycerophosphate for fat synthesis or transformed to various amino acids, fatty acids, and other molecules.

In the Hatch-Slack cycle the CO_2 taken up by the plants appears first in four carbon acids, in particular malic and aspartic acid, and subsequently in PGA and other intermediates of the Calvin cycle. Plants following this photosynthetic cycle have chloroplasts in the mesophyll cells near the sur-

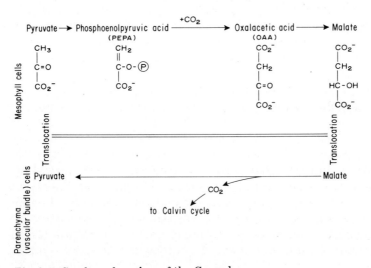

Fig. 9-2. Condensed version of the C_4 cycle.

face of the leaf that are different from the chloroplasts in the parenchyma vascular bundle cells deeper inside the leaves. Fig. 9-2 gives a condensed version of the C_4 cycle. In the chloroplasts of the mesophyl pyruvate is converted to PEPA. PEPA undergoes carboxylation to form OAA, which is then reduced to give malic acid, or is converted to aspartic acid. Malic acid and aspartic acid are translocated to the chloroplasts of the vascular bundle cells where the compounds are decarboxylated to form pyruvate. This compound is translocated back to the chloroplasts of the mesophyll cells where it enters the cycle anew, while the liberated CO_2 is used in Calvin cycle reactions.

Photosynthesis of C_3 type plants takes place via the enzyme ribulose-1,5-diphosphate carboxylase (RUDPC), while photosynthesis of C_4-type plants takes place via the enzyme phosphoenolpyruvate carboxylase (PEPC). In plants with crassulacean acid metabolism (CAM) both carboxylases are present. These plants alone are capable of net CO_2 fixation via the primary carboxylase reactions of C_3 and C_4 plants. Dark CO_2 fixation in CAM plants is functionally equivalent to photosynthesis in C_4 plants while CO_2 fixation in the light follows the C_3 pathway (Osmond, 1975).

The carbon isotopic composition of plants

Nier and Gulbransen (1939) and Murphey and Nier (1941) observed that the carbon isotopic composition of living organic matter and carbonate showed systematic differences, the organic matter being depleted in ^{13}C. Wickman (1952) and Craig (1953a, 1954a, b) studied the carbon isotopic composition of plants in more detail and related the depletion of ^{13}C in the plant material compared to the source CO_2 to the process of photosynthesis. The early studies focused on the difference in isotopic composition between terrestrial and marine plants and attributed the observed variation in ^{13}C content to a difference in the isotopic composition of the source carbon. Terrestrial plants were thought to show a relatively narrow $\delta^{13}C$ range around $-25^0/_{00}$, the very few heavier values were interpreted to arise from special local environmental conditions.

Although several authors (Wickman, 1952; Craig, 1953a; Stuiver and Deevey, 1962; Hall, 1967; Tauber, 1967) had measured plant carbon isotopic compositions more than $10^0/_{00}$ heavier than $-25^0/_{00}$, it was not realized until a study of Bender (1968) that consistent isotopic composition differences exist between certain plant types. Bender (1968) found that corn and other tropical grasses showed a systematic ^{13}C enrichment and pointed out that the tropical grasses follow the C_4 photosynthetic pathway. Smith and Epstein (1971) observed that relative ^{13}C enrichment in plants is not an uncommon phenomenon, and classified plants on the basis of their carbon isotopic composition into two groups with low and high $\delta^{13}C$ values, respectively. A compilation of over 1000 published $\delta^{13}C$ analyses of total plants, wood, leaves and seeds (Fig. 9-3A) demonstrates this separation. Tregunna

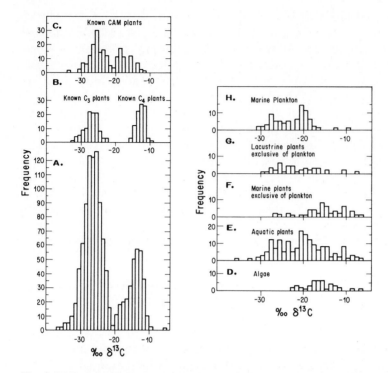

Fig. 9-3. Carbon isotopic composition of photosynthetically fixed carbon A. Terrestrial plants. B. Known C_3 and C_4 plants. C. Known CAM plants. D. Algae. E. Aquatic plants. F. Marine plants exclusive of plankton. G. Lacustrine plants exclusive of plankton. H. Marine plankton. Data from Bender (1968, 1971), Bender et al. (1973), Brown and Smith (1974), Craig (1953a), Deevey and Stuiver (1964), Degens et al. (1968b), Eadie (1972), Lerman et al. (1969), Lowdon and Dyck (1974), Oana and Deevey (1960), Osmond et al. (1975), Parker (1964), Sackett et al. (1965, 1974a), Smith and Brown (1973), Smith and Epstein (1970, 1971), Stahl (1968a), Troughton (1972), Wickman (1952), and Williams and Gordon (1970).

et al. (1970) and Bender (1971) showed that in a particular growth environment, plants following the C_3 pathway of carbon fixation are enriched in ^{12}C by 12—14‰ over plants following the C_4 pathway (Troughton et al., 1971; Smith and Brown, 1973; Fig. 9-3B).

As one might expect from the discussion of the photosynthetic mechanism available to CAM plants, their isotopic composition covers the whole range of C_3 and C_4 plants. It is interesting to note, however, that there is still a bimodality in the $\delta^{13}C$ distribution of Fig. 9-3C, which might suggest that generally one of the mechanisms predominates or that plants in which both have contributed more or less equally to the carbon fixation have not been analyzed frequently.

The comparatively few data for algae (mainly marine) available (Fig. 9-3D)

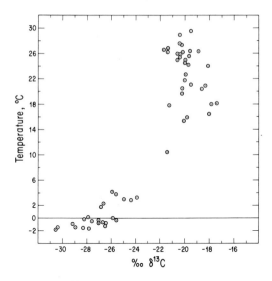

Fig. 9-4. $\delta^{13}C$ of total plankton carbon as a function of surface water temperature at the collection site (after Sackett et al., 1974a).

indicate that their isotopic composition is intermediate between those of C_3- and C_4-type plants, but tends more towards a slight ^{13}C enrichment.

The carbon isotopic composition of aquatic plants covers the same range as that of terrestrial plants (Fig. 9-3E). If lacustrine plant compositions are compared to those of marine plants (excluding plankton) (Fig. 9-3F and 9-3G), there is a suggestion that lacustrine plants show ^{12}C enrichment compared to marine plants. This could be reasonably attributed to a difference in the isotopic composition of the carbon pool from which the two categories of plants derive their carbon. Interestingly marine plankton (Fig. 9-3H) has isotopic compositions which appear to be lighter than other marine plants. The few samples of lacustrine plankton which have been analyzed for ^{13}C (Oana and Deevey, 1960; Deevey and Stuiver, 1964) also tend to be isotopically lighter than other lacustrine plants (−26.4, −38, −32.2, −35.5, and −42‰).

The distribution of carbon isotopes in marine plankton has been studied in more detail (Sackett et al., 1965; Degens et al., 1968a, b; Deuser et al., 1968). Sackett et al. (1965, 1974a) and Eadie (1972) found a correlation between the carbon isotopic composition of plankton and the temperature of the water (growth temperature; latitude) in which the samples were collected (Fig. 9-4). Within the cold water group of samples no difference existed in the isotopic composition of zoo- and phyto-plankton. At about 15°C, which corresponds approximately to the region of Arctic convergence, separating the warm and cold water "spheres" and representing a boundary

TABLE 9-1
Variations of $\delta^{13}C$ within plants (in permil vs. PDB)

Plant	Leaf	Seed	Pollen	Tuber	Root	Stem	Wood	Reference
Zea maize	−11.4	−11.2	−12.1					Troughton (1972)
Potato	−30.0			−27.7				
Tomato	−26.0				−24.4	−25.2		Park and Epstein (1960)
	−24.9				−24.5	−24.6		
Corn	−11.1	−9.7						Lowdon and Dyck (1974)
	−11.0	−9.3						
	−10.9	−9.3						
	−10.9	−9.5						
Panicum maximum	−12.3	−12.6						Smith and Benedict (1974)
Fir, *Abies lasiocarpa* (N.W. Wyoming)	−29.1						−27.9	Craig (1953a)
Fir, *Abies lasiocarpa* (S. Wyoming)	−25.2						−23.8	
Pinus contorta	−27.0						−27.1	
Maple, *Acer saccharium*	−27.7						−27.0	
Willow, *Salix bebbianna*	−26.4						−25.0	
Black locust, *Robinia pseudo-acacia*	−26.4						−25.0	
Osage orange, *Machua*	−26.6						−24.7	
Grass, *Thalassia testudinum*	−10.2				−7.1			Parker (1964)

between different phytoplankton regimes, there is a break. This was attributed to the existence of two types of plankton populations, diatoms in polar regions and dinoflagelates, coccothophores and blue green algae in more tropical waters, the diatomes showing a more pronounced dependence of carbon isotopic composition on temperature.

Compared to the large range in isotopic composition found for different plants, the variations in $\delta^{13}C$ among different parts of the same plant are minor. Table 9-1 shows a few examples of studies in which different morphological parts of the same plant were analyzed, and includes C_3 as well as C_4 plants. The available data suggest that the leaves of plants tend to show lighter isotopic compositions than the other parts. It is interesting to note that the difference between needle and wood isotopic composition is about the same for two firs which grew at different locations and had very different total carbon isotopic compositions. For maple leaves and grass, Lowdon and Dyck (1974) observed a trend of increasing ^{12}C content with progression of the growing season. The reason for this phenomenon, which was noted during several years, is not clear. Differences in the isotopic composition in wood of different portions of tree trunks have also been found.

The ^{13}C content of plant material is correlated with the type of photosynthetic cycle followed by the organism. $\delta^{13}C$ values for C_3 plants fall around $-26^0/_{00}$, for C_4 plants around $-13^0/_{00}$, while those of CAM plants cover the whole range of C_3 and C_4 plants. The isotopic composition of plankton is related to the growth temperature; lowest $\delta^{13}C$ values were measured for samples collected in high latitude waters. There are small ^{13}C content differences between different plant components: leaves appear to be slightly depleted in ^{13}C with respect to the total plant.

The carbon isotopic composition of the chemical components of plant material and respired carbon dioxide

Plant material may be divided into fractions having different solubilities and extraction characteristics when treated with acids, bases or organic solvents. One such extraction scheme has been given in Fig. 9-5. It is apparent (Figs. 9-5 and 9-6) that certain ^{13}C content differences exist between different chemical plant components. While sugar, cellulose and hemi-cellulose show $\delta^{13}C$ values close to the mean plant carbon isotopic composition, pectin appears to be enriched in ^{13}C while lignin and lipids are depleted in this isotope relative to the total plant.

Amino acids as a whole show in most samples studied ^{13}C enrichment with respect to the total plant. Abelson and Hoering (1961) separated different amino acids and demonstrated that carbon isotopic composition differences exist between them. It was also found that the carboxyl group of these acids exhibits a considerable ^{13}C enrichment (average $11.5^0/_{00}$) with respect to the

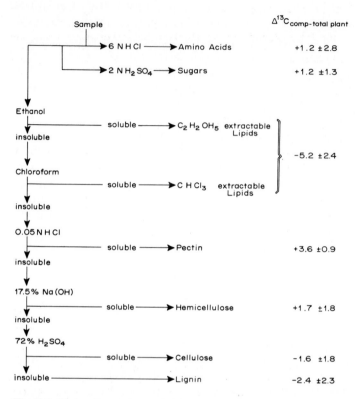

Fig. 9-5. Extraction scheme (after Degens et al., 1968b). The mean $\Delta^{13}C$ values and their standard deviations have been computed on the basis of the data of Fig. 9-6.

remainder of the molecule. There is evidence that this enrichment may differ from one amino acid to another.

Lipids show a consistent ^{12}C enrichment with respect to the total plant; however, the size of the isotopic composition difference varies considerably. Because the lipid fraction includes a wide variety of organic compounds with predominating triglyceride esters of fatty acids and lesser amounts of the hydrocarbons, long-chain alcohols, terpens, pigments, etc., and as each of these classes of compounds may be characterized by its own typical carbon isotopic composition, the range in $\Delta^{13}C$ values for the lipid-plant fractionation is understandable. Degens et al. (1968b) found that the $CHCl_3$ extractable lipids had isotopic compositions very much lighter than C_2H_5OH extractable lipids which would support this point of view.

The isotopic compositions of the fatty acids and total lipid fraction are very close. Different fatty acids from one organism have very similar ^{13}C contents, as one might expect if they were formed through the same synthetic pathway. However, the isotopic composition difference between fatty

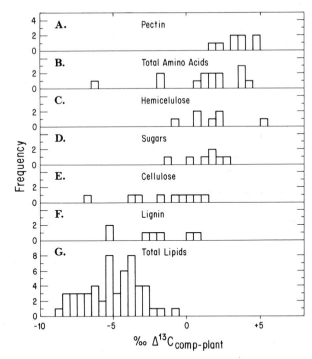

Fig. 9-6. Chemical constituents of plant material. Data from Abelson and Hoering (1960, 1961), Calder and Parker (1973), Degens et al. (1968b), Jacobson et al. (1970b), Park and Epstein (1961), Parker (1962, 1964), Sackett et al. (1974a), Silverman and Epstein (1958), Smith and Benedict (1974), Smith and Epstein (1970), and Whelan et al. (1970).

acids and total plant carbon was found to vary considerably when different organisms were considered (Parker, 1964).

Although plants mainly extract CO_2 from the atmosphere and fix it in the form of organic compounds, a certain amount of CO_2 is also released by them to the surroundings as the result of respiration. Because respiration involves the release of CO_2, it may appear superficially as the reverse of photosynthesis. The two processes are, however, very different in their biochemistry and occur at different cell sites. In the bright light photosynthesis fixes carbon at 10 to 20 times the rate of CO_2 release during dark respiration, however, some CO_2 release usually occurs in the light also. It has been recognized that there are a number of plants which have a special system of respiration known as photo-respiration, characterized by a greatly increased CO_2 evolution in light. Species showing a high rate of photo-respiration are less efficient in terms of CO_2 uptake per unit area and include, e.g. tobacco, bean, soy bean, tomato, and wheat. More efficient photosynthetic plants such as maize, sugar cane and sorghum do not exhibit much photo-respiration.

The carbon isotopic composition of respired CO_2 has not been studied very frequently. Laboratory measurements (Baertschi, 1953; Park and Epstein, 1960, 1961; Smith, 1971) show that for both C_3 and C_4 plants there is little difference in $\delta^{13}C$ between total plant carbon and respired CO_2. If there is any tendency there appears to be a slight depletion in ^{13}C in the CO_2 respired compared to the total plant carbon ($\Delta^{13}C_{plant-CO_2} = 0.5^o/_{oo}$). The fact that the carbon isotopic composition of plant respiration CO_2 should be close to that of the total plant was also deduced by Keeling (1958, 1961a, b).

In the experimental investigations it was noted also that the respired CO_2 showed in some cases $\delta^{13}C$ changes with time (Parker and Epstein, 1960, 1961; Smith, 1971). The observation can be explained if it is suggested that the metabolic substrate of respiration changes during the experiment. Jacobson et al. (1970a, b) found that the carbon isotopic composition of respired CO_2 was within $1.9^o/_{oo}$ of the compounds which served as respiration substrate. Because different plant components which may serve as substrate differ in their isotopic composition, changes in $\delta^{13}C$ of respired CO_2 can be understood. The difference in $\delta^{13}C$ between CO_2 of photo- and dark respiration ($\delta^{13}C_{photo} - \delta^{13}C_{dark} \sim 13^o/_{oo}$) observed by Hsu and Smith (1972) for peanut and sunflower shoots could also be interpreted in this framework. The shift to a different metabolic substrate might result in carbon isotopic composition changes of the released CO_2 either because the new substrate differs in ^{13}C content or that it opened the way for the occurrence of isotope effects, e.g. decarboxylation reactions, which might result in a ^{13}C enrichment in the ultimately released CO_2. As an alternative refixation of some of the CO_2 formed by respiration while it was still in the plant should be considered.

Carbon isotope fractionation in photosynthesis and its dependence on environmental conditions

There has been some effort made to understand why plants show the characteristic isotopic compositions that they have. In experiments Baertschi (1953) found that plants discriminated systematically against ^{13}C and measured a fractionation factor of 1.026 between source CO_2 and photosynthetically formed plant material. Baertschi (1952) also demonstrated that the difference in the rate of collision of the two isotopic species of carbon dioxide ($^{12}CO_2$, and $^{13}CO_2$) against the absorbing surface of the leaf cannot account for this fractionation. Craig (1954a) proposed that several steps of carbon isotope fractionation existed in the reactions of plant metabolism, and that the size of these fractionations might be influenced by environmental conditions.

Park and Epstein (1960, 1961) were the first to outline a model for the carbon isotope fractionation during photosynthesis. The authors proposed that the major fractionation is due to two steps, the first involves the preferential uptake of $^{12}CO_2$ from the atmosphere and the second the preferential conversion of ^{12}C-rich dissolved CO_2 to phosphoglyceric acid. Subsequent metabolic reactions are thought to be connected with isotope effects also, however, these are considered to determine only the isotopic composition of individual plant components and not the overall plant isotopic composition. The fractionation in the first step is maximal if the CO_2 absorption occurs without back diffusion, as in the absorption of CO_2 by barium hydroxide solutions. The absorbed CO_2 is depleted by $14^0/_{00}$ in ^{13}C compared to the gaseous CO_2 (Baertschi, 1952). Minimum fractionation exists if isotopic equilibrium between the atmospheric CO_2 and the CO_2 dissolved in the plant leaves is established ($+6^0/_{00}$). From their own experiments the authors evaluate an average fractionation of about $-7^0/_{00}$ for step I.

If all of the CO_2 taken up in step I would be converted to photosynthetic products, no further fractionation would occur. However, if only a fraction of it is incorporated into the plant, the size of further carbon isotope fractionation would be expected to depend on the efficiency of CO_2 removal. The size of the fractionation between the CO_2 dissolved in the plant leaves and the carbon fixed into sugar was experimentally determined to be $-17^0/_{00}$. This leads to a total fractionation of step I and II of $-24^0/_{00}$. With the aid of this model Park and Epstein were able to explain among other things the difference in isotopic composition between atmospheric CO_2 and plant carbon, as well as the ^{13}C variations observed in naturally occurring plants.

Whelan et al. (1973) further investigated the mechanism of carbon isotope fractionation during photosynthesis. Experiments by these authors indicate that the enzymatic synthesis of PGA from RUDP and CO_2 (Fig. 9-1) at $24°$ and $37°C$ resulted in a carbon isotope fractionation of 33.7 and $18.3^0/_{00}$, respectively, and it is concluded that the fractionation during the enzymatic carboxylation is probably large enough to account for the $\delta^{13}C$ values of C_3 plants. The study also showed that the metabolic intermediates formed by PEP carboxylase in C_4 plants are $2-3^0/_{00}$ lighter than the source bicarbonate and the authors propose that the measured fractionations permit an explanation of the isotopic composition of C_4 plants.

The range in carbon isotopic composition of plankton and its dependence on water temperature can be understood on the basis of a model proposed by Deuser et al. (1968). Degens et al. (1968a) had shown that plankton utilizes CO_2 rather than HCO_3^- and fractionates carbon isotopes with respect to it by a maximum of $18-19^0/_{00}$, the plankton being depleted in the heavy isotope. The metabolic fractionation is independent of ambient temperature but depends on the rate of CO_2 supply. With decreasing availability of CO_2 fractionation decreases from the maximum value to a minimum of $6^0/_{00}$, which is reached at the point where the CO_2 supply is reduced to the limiting value required for continued growth. If the fractionation is measured with respect to HCO_3^-, rather than to CO_2 it is found that the fractionation between source carbon and plankton is by 7—10‰ larger than with respect to CO_2 and changes as a function of temperature. The temperature dependence is identical to that of the HCO_3^--CO_2 fractionation.

In the ocean HCO_3^- constitutes the major carbon reservoir ($\delta^{13}C$ close to +1‰) and CO_{2aq} in isotopic equilibrium with it is depleted in ^{13}C with respect to it. The degree of depletion decreases with increasing temperature; at the highest temperatures generally prevailing in the oceans it is about 7‰, while at the lowest temperatures it reaches about 10‰. Hence, the $\delta^{13}C$ range of plankton becomes understandable. Least ^{13}C-depleted plankton would be expected in warm waters with limited CO_2 supply and could be as heavy as —13‰. The largest ^{13}C depletions would be expected for cold water plankton in areas where there is an ample CO_2 supply; here values as light as —28‰ might be encountered.

The ^{12}C enrichment in organic matter has generally been attributed to the existence of kinetic isotope effects in the extraction of carbon from the inorganic carbon pool and synthesis of organic compounds from it. Galimov (1974) has suggested that the observed isotope distribution in organic compounds may actually be the result of isotopic equilibria, and proposed a method to compute partition function ratios for organic molecules. A number of these were evaluated by Galimov (1973), some of which served as a basis for Fig. 9-7. A hypothetical plant was constructed with a mean isotopic composition of —27‰, which was assumed to have grown from CO_2 of —7‰. Using the difference between total plant carbon and different plant components (Fig. 9-5) the $\delta^{13}C$ values of individual compounds and their fractionation relative to atmospheric CO_2 were computed. These were then plotted against the fractionations between CO_2 and different plant components predicted by the partition function ratios of Galimov (1974) (Fig. 9-7). A highly significant positive correlation between the two parameters is encountered. It should be noted that whereas the intercept of the regression line depends on the $\delta^{13}C$ value assumed for the total plant carbon (—27‰) this is not true for the slope of the relationship. If perfect agreement between Galimov's model and the observed $\delta^{13}C$ distribution existed the slope of the regression line should equal 1; the actual slope found is, however, 0.3 and deviates significantly from the expected value. The upper part of Fig. 9-7 shows the relationship between observed and predicted ^{13}C fractionation between other amino acids and glutamic acid. Again a significant correlation is found, indicating that the isotopic composition differences observed among amino acids are related to their chemical composition, however, in this case also the predicted fractionations tend to be larger than the actually observed ones. Galimov's (1974) model also predicts that the carboxyl group of amino acids should show ^{13}C enrichment with respect to the rest of the amino acid, and that this fractionation should be such that the difference between source CO_2 and carboxyl group would be relatively

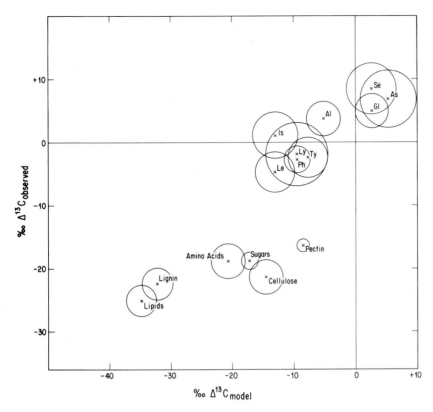

Fig. 9-7. Carbon isotopic composition differences between plant components compared to a fractionation model proposed by Galimov (1974); for the fractionation between amino acids glutamine is used as a reference (data for amino acids, from Abelson and Hoering, 1961; for other plant components, Fig. 9-6). Abbreviations: As = aspartic acid, Se = serine, Gl = glycine, Al = alanine, Is = isoleucine, Le = leucine, Ly = lysine, Ty = tyrosine, Ph = phenylalanine.

small. Both points have actually been observed (Abelson and Hoering, 1961). The model suggests in addition that the fractionation between carboxyl group and the remainder of the amino acid should differ from one acid to another; there is some evidence in the natural data that this is the case. Information collected to date indicates that the model correctly predicts the direction of ^{13}C enrichment observed in different plant components, however, the size of the isotopic composition differences computed on the basis of Galimov's fractionation factors are significantly larger than those actually observed.

The ^{13}C content of carbon fixed photosynthetically by a plant may depend on environmental factors such as temperature, light level, pH of the water in which aquatic plants grow, the availability of carbon dioxide, oxygen and in the case of terrestrial plants, of water. It can be anticipated that the isotopic composition of the carbon acting as source for the photosynthetically fixed carbon will be of major importance. This factor has been made responsible for the observed difference in $\delta^{13}C$ between marine and land plants. It has also been suggested that major changes in the $\delta^{13}C$ of the atmospheric CO_2 have occurred due to industrial pollution and are reflected in photosynthetically fixed carbon.

In an aqueous environment the fractionation between source carbon and plant may depend on (1) in which chemical species (CO_2 or HCO_3^-) most of the carbon is present, and (2) which of these is taken up by the plant. In so far as the relative distribution of CO_2 and HCO_3^- varies with pH, the carbon isotope fractionation in an aqueous environment may under some circumstances be pH dependent. Calder and Parker (1973) studied the effect of pH on the carbon fractionation of blue green algae and found no difference in $\delta^{13}C$ of the plant material accumulated at a pH of 7 and 8.

Because isotope fractionations in chemical reactions change with temperature, one might expect that the carbon isotopic composition of plants should be a function of environmental temperature. In 1952 Wickman had made a comparison of the $\delta^{13}C$ of plant species grown at two different localities and concluded that there was no definite relationship between isotopic ratio and latitude, and hence, temperature. The conclusion was based on a mixture of analyses of C_3 and C_4 plants; however, if one considers only low-$\delta^{13}C$ plants which come from locations differing significantly in their latitude one finds the following:

Plant	Low-latitude $\delta^{13}C$ (‰)	High-latitude $\delta^{13}C$ (‰)
Lycopodium	−24.3	−26.9
Dryopteris	−22.1	−26.5
Stellaria media	−24.6	−25.2

Although the data are very limited they are suggestive and the question whether there are significant $\delta^{13}C$ differences between plants growing at different latitudes may deserve further attention. Troughton (1972) cites experimental evidence which shows that there was no clear correlation between growth temperature and plant isotopic composition. For the blue-green algae grown at 30° and 39°C, respectively, Calder and Parker (1973) did not observe a significant difference in $\delta^{13}C$. On the other hand, Degens et al. (1968a) found that the fractionation between HCO_3^- and plankton was −25 to −27‰ and decreased systematically by 0.35‰ per degree Celsius temperature rise. The effect was attributed to the temperature sensitivity of the HCO_3^--CO_2 isotope fractionation, rather than to that of enzymatic reactions.

One might also expect that the availability of light would influence the carbon isotope fractionation. However, for tomato plants (C_3 plant) Park and Epstein (1960) found no difference in the $\delta^{13}C$ for plants grown at two significantly different light levels. Although the C_3 and C_4 metabolic pathway carbon isotope fractionation may not show a significant light-level dependence (for C_4 plants no studies have been made to date), one can anticipate that the isotopic composition of CAM plants might show measurable changes with photo period. This is so because the photo periodism controls the relative importance of the C_3 and C_4 pathways of carbon fixation, which are characterized by different fractionation factors. Lerman and Queiroz (1974) showed experimentally that the carbon isotopic composition of CAM plants grown with a short-day photo period was about $8.5‰$ heavier when compared with that of the same species grown with a long-day photo period. This suggests a switch from PEPC-dominated activity to RUDPC-dominated activity when the photo period is changed from short to long days. Similar results were obtained by Allaway et al. (1974).

Several authors have found that with increasing CO_2 supply plants show larger fractionations relative to the source carbon (more negative $\delta^{13}C$ values; Park and Epstein, 1960; Deuser et al., 1968). The dependence of algal carbon isotope fractionation on CO_2 supply was investigated by Abelson and Hoering (1960, 1961). Chlorella grown on 5% CO_2 exhibited a fractionation of $-25.8‰$ with vigorous aeration and only $-11.3‰$ with poor aeration. The photosynthetic fractionation of carbon by blue-green algae was studied by Calder and Parker (1973) and found to depend on the CO_2 concentration in the feed gas. The difference in isotopic composition between cells and feed gas CO_2 was about $-8‰$ at very low CO_2 levels, reached a minimum close to $0‰$ when the CO_2 concentration in the air was about 0.2% and increased to about $-18‰$ at a CO_2 concentration of 3% in the feed gas.

The oxygen concentration in the atmosphere also influences the carbon isotope fractionation of some plants. Troughton (1972) states that while *Atriplex rosa*, a C_4 plant, showed no effect of the oxygen content of the ambient atmosphere on the carbon isotope composition of the plant, this was not true for *Atriplex patula* ssp. *hastata*, a C_3 plant. The latter showed a decrease in $\delta^{13}C$ of $0.25‰$ per 1% increase in the oxygen content of the atmosphere (O_2 range studied 4—21%).

The influence of the availability of water on the carbon isotopic composition has been studied for CAM plants. In experiments Allaway et al. (1974) found that the dependence of CAM plants on dark CO_2 fixation was further increased by drought, the $\delta^{13}C$ values of the fixed carbon was substantially less negative than otherwise. Osmond et al. (1975) propose that under field conditions it is probable that water availability is the principal factor limiting CO_2 uptake in the light by CAM plants. The authors observed that the species showing acidification during the dark period (CAM plants) in their natural habitat exhibited a dependence of their $\delta^{13}C$ values on water supply. Plants showed systematically heavier isotopic compositions with decreasing availability of water, which is attributed to a larger contribution of dark CO_2 fixation via PEPC in relation to light CO_2 fixation via RUDPC. Species showing no pronounced dark acidification had more negative $\delta^{13}C$ values and no clear dependence of $\delta^{13}C$ on the availability of water was observed, indicating that dark CO_2 fixation via PEPC does not play a significant role in their carbon metabolism. These observations suggest that while there is a strong dependence of $\delta^{13}C$ of CAM plants on the availability of water, this is apparently not true for C_3 plants.

In summary, environmental conditions may significantly influence the ^{13}C content of photosynthetically fixed carbon. The carbon isotopic composition of the source carbon as well as its availability are important for all plants. Less well established or only observed for certain plant types are the effects of temperature, light level oxygen and water availability. As a group CAM plants show the highest sensitivity of their carbon isotopic composition to environmental conditions. This is readily understood in view of the effects of environmental factors on the intensity of the CAM operation.

THE CARBON ISOTOPIC COMPOSITION OF ORGANIC MATTER IN SEDIMENTS

Humic substances and kerogen, factors determining their isotopic composition

The carbon compounds synthesized by plants are stabilized against the outside during their life by the life processes of the organism. Upon death these no longer exist and the highly complex organic compounds become unstable and disintegrate mechanically and chemically. The biopolymers (e.g. polysacharides, proteins, nucleic acids) are attacked by micro organisms and broken down to soluble compounds. Parts of the decay products are oxidized to form CO_2 and H_2O, some are used to build up new living organic material by organisms living on them; other parts of the degradation products polymerize and form heterogeneous random polymers, i.e. humic substances (molecular weight 3×10^3 to 3×10^5). These are usually defined as base soluble organic material of brown color extractable from soils and sediments and are subdivided according to extractability into: (1) humic acids (insoluble in acids), and (2) fulvic acids (soluble in acids). With increasing molecular weight and degree of cross linkage the polymers change to colloids and gels, and during compaction and dewatering to amorphous solids (geopolymers) which are no longer extractable. This loosely terminates the process of diagenesis of organic matter. The form in which carbon is present in sediments at this stage has been termed kerogen; it is finely dispersed organic matter, insoluble in organic solvents, and composed of coaly particles and unstructured diagenetically formed diffuse organic substances.

The process during which further changes in the composition of organic matter occur as a result of burial has been termed katagenesis. In it functional groups of the organic compounds are lost and CO_2 and H_2O is produced, at the same time hydrocarbons are formed. Their production decreases towards the final stages of katagenesis during which more and more methane is evolved. The most intense alteration of organic matter occurs during metamorphism and as its final product carbon is present in elemental form as graphite.

Marine and continental environment can be envisioned as reflecting the limiting conditions under which organic matter may be incorporated into sediments. Under purely marine conditions, the only source would be autochthonous plankton consisting mainly of proteins, carbohydrates and lipids. These materials are characterized by higher H/C ratios and are more parafinic. Under purely continental conditions the organic matter is the detritus of higher plants and is composed mainly of lignin, cellulose, and sklero proteins, their H/C ratio is lower and the compounds are more aromatic. The carbon isotopic composition of kerogen in recent sediments is governed by four major variables, which themselves are influenced by a number of factors:

(1) The $\delta^{13}C$ value and relative amounts of endogenous carbon.

(2) The $\delta^{13}C$ value and relative amounts of exogenous carbon.

(3) Isotope effects in the transformation of living tissue to humic substances, before the organic matter becomes part of the sediment.

(4) Isotope effects during diagenetic changes after the organic matter has become part of the sediment.

In older sediments, isotope effects during katagenesis and metamorphism have to be considered, also.

The isotopic composition of endogenous carbon (carbon fixed by photosynthesis and contributed to the sediment in the environment in which the sediment is formed) is influenced by that of the local flora and fauna and in aquatic environments by that of the humic substances formed in situ from dissolved organic carbon. As the $\delta^{13}C$ values of photosynthesizing organisms may differ, the make up of a plant community may be of consequence as well as the resistance of individual species to subsequent decay. The carbon isotopic composition of individual species depends in turn on environmental conditions and the isotopic composition of the source carbon. The latter may be assumed to be fairly constant for terrestrial plants, may show, however, by comparison extreme variations for lake plant communities, where it may be affected by the decay of organic matter, exchange with atmospheric CO_2, respiration and the introduction of carbon from the dissolution of limestones. Under marine conditions, the ^{13}C content variability of the source carbon can be expected to be reduced, although the same factors may still influence it.

An important contribution to the $\delta^{13}C$ variations of the organic matter in sediments will come from the additions of exogenous carbon. Their influence will be most noticeable in aquatic sediments and be of less consequence in soils and peat. In lake sediments the relative contribution of terrestrial plant debris will be important, in marine sediments the size of the addition of organic carbon formed on the continents. Exogenous carbon additions to the sediment may also occur through the recycling of kerogen, i.e., the detritus from older carbon-bearing sediments, and discharge of anthropogenic wastes.

The effects of diagenesis on the carbon isotopic composition of sedimentary organic carbon have been considered among others by Sackett and Thompson (1963), Sackett (1964), Degens (1969), Behrens and Frishman (1971) and Johnson and Calder (1973). Between the time of photosynthetic fixation and incorporation into the sediment the organic carbon is exposed to many oxidative processes, and actually only a very small fraction of the produced carbon compounds passes into sediments. Factors that may be of importance in determining the carbon isotopic composition changes include: (1) isotope effects during bacterial degradation of organic matter, (2) preferential elimination of compound groups and preferential preservation of others which differ significantly in $\delta^{13}C$ from the average plant material, and (3) decarboxylation reactions, which would remove ^{13}C-enriched groups from the organic material leading to ^{13}C depletion in the residual. Rosenfeld and Silverman (1959) and Kaplan and Rittenberg (1964) have demonstrated experimentally that bacteria utilize ^{12}C in preference to ^{13}C which leads to ^{13}C enrichment in the residual organic substrate. Sackett and Thompson (1963) found that inorganic oxidation will have a similar effect. Hence, in principle, there are processes that might lead to ^{13}C or ^{12}C enrichment during diagenesis.

Recent continental sediments

There are few analyses available for soil organic matter; the data have been compiled in Fig. 9-8. No detailed study has been carried out to show how closely the $\delta^{13}C$ of the soil humics are related to the local plant cover, however, it has been found (Nissenbaum, 1974) that the two are similar. Soils covered by C_3-type vegetation showed $\delta^{13}C$ values of C_3 plants, while humics from cane field soil from Hawaii gave $\delta^{13}C = -14.8°/_{oo}$ and from a papyrus swamp $-17°/_{oo}$.

Separates of fulvic and humic acid from the same soil showed ^{13}C enrichment in the fulvic component compared to the humic acid fraction. The fulvic acid is closer in $\delta^{13}C$ to plant carbon and is considered by Nissenbaum and Schallinger (1974) to be an intermediate in the humification process in which the plant lignins are transformed to humic acids. The ^{13}C depletion in the humic acid (HA) compared to the fulvic acid (FA) is attributed to the loss of ^{13}C-rich carboxyl groups in this process. A study of $\Delta^{13}C_{FA-HA}$ in a

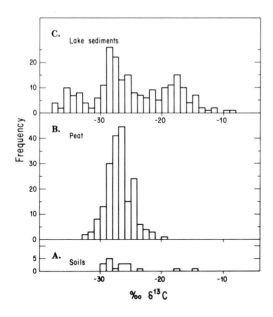

Fig. 9-8. Carbon isotopic composition of organic matter in continental sediments. Data from Craig (1953a), Fischer et al. (1968), Nakai (1972), Nissenbaum (1974), Nissenbaum and Kaplan (1972), Nissenbaum and Schallinger (1974), Oana and Deevey (1960), Stahl (1968a, c), and Stuiver (1964, 1975).

soil profile (Nissenbaum and Schallinger, 1974) indicates that the difference in isotopic composition between the two components may exist only near the surface.

The decay of organic matter adds to the CO_2 present in the soil atmosphere, for which there are three principal sources: decomposition of organic material, plant root respiration, and atmospheric carbon dioxide. In view of the fact that in most cases the partial pressure of CO_2 in the soil gas is very much higher than in the main atmosphere, the latter source may generally be ignored. Exceptions to this have been noted by Pearson and Hanshaw (1970). The isotopic composition of soil CO_2 will then generally be determined by that produced by decay and respiration. As decomposed wood, peat and lignite have approximately the same $\delta^{13}C$ values as fresh wood (Wickman, 1952; Craig, 1953a, 1954a, b), and respiration CO_2 is isotopically similar to the total plant carbon in C_3 and C_4 plants (Baertschi, 1953; Park and Epstein, 1960, 1961; Smith, 1971), one can expect that soil CO_2 from these sources has carbon isotopic compositions close to that of the local plant cover.

Available $\delta^{13}C$ measurements of soil CO_2 have been compiled in Fig. 9-9 which reveals the same bimodality observed for the carbon isotopic composition of plant carbon, and results from the collection of CO_2 samples from

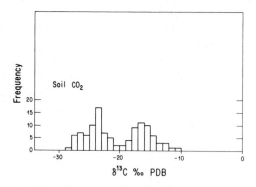

Fig. 9-9. Carbon isotopic composition of soil CO_2. Data from Broecker and Olson (1960), Galimov (1966), Hendy (1971), Kunkler (1969), Lebedev et al. (1969), Lerman (1972), Rightmire and Hanshaw (1973), and Sears (1976).

sites at which C_3 and C_4 or CAM plants predominate. Lerman (1972) studied the isotopic composition of soil CO_2 in localities where large areas were planted with succulent plants on soils which originated from forested regions. While the ^{13}C content of the soil organic matter ($\delta^{13}C = -25‰$) matched that of the previous plant cover (C_3 plants), that of the soil CO_2 ($\delta^{13}C = -15‰$) coincided with that of the present one. In the particular situation studied there seems to be little CO_2 derived from the decomposition of soil organic matter.

Few measurements of the carbon isotopic composition of soil CH_4 have been carried out (Fig. 9-15). The available data indicate a considerable depletion in ^{13}C compared to the associated CO_2 (Lebedev et al., 1969). The $\delta^{13}C$ values range from -30 to $-74‰$ and average $-56‰$.

Most of the peat isotopic analyses shown in Fig. 9-8B are from temperate humid climates and are due to Stahl (1968a, c). The bimodality of the distribution results from the fact that peat formed under different environmental conditions is characterized by different $\delta^{13}C$ values (terrestrial high moors (Hochmoore) $\delta^{13}C \sim -26‰$, semi terrestrial to telmatic forest and reed peats $\delta^{13}C \sim -28.5‰$). If the peat formation commences from an infilling lake, a marsh or swamp is formed with reed peat (telmatic conditions, $\delta^{13}C = -28.8 \pm 0.5‰$). Similar isotopic compositions ($-28.8 \pm 0.2‰$) are found for forest peat which develops subsequently with increasing accumulation of organic matter. Slowly the peat may grow above the groundwater level and a high moor develops. In the lower portions of such a moor $\delta^{13}C = -27.6 \pm 0.8‰$ are measured and in the higher portion $\delta^{13}C = -24.5 \pm 0.4‰$. The systematic difference in the carbon isotopic composition of different peat types may be related to (1) differences in the carbon isotopic composition of the plant source material, and (2) to different conditions of decomposition and reconstitution of the organic matter.

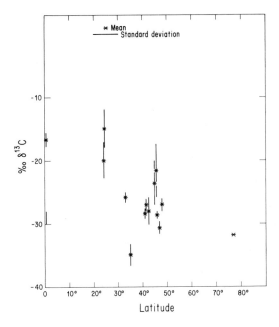

Fig. 9-10. Variations of the carbon isotopic composition of organic matter in lake sediments as a function of latitude (data from references of Fig. 9-8).

Lake sediments show a wide range in carbon isotopic composition (-8 to $-38‰$, Fig. 9-8) resembling that of land and aquatic plants. The factors enumerated above will interact differently at different locations and cause characteristic isotopic compositions in muds of individual lakes; the trimodality of Fig. 9-8C can be attributed to this. Examination of the data plotted in this figure shows also that there is a suggestion of a correlation of $\delta^{13}C$ in the lake mud with latitude, which has been shown in Fig. 9-10. Because $\delta^{13}C$ values of the plant communities producing the organic matter of the lake muds were not measured, it is not clear which of the mentioned factors is determinative, although it would be tempting to attribute the carbon isotopic composition variations to a change in vegetation type. The repeated samplings of one lake indicated in Fig. 9-10 come from drill cores. Measurements of $\delta^{13}C$ along drill cores together with ^{14}C and pollen studies have led Nakai (1972) and Stuiver (1975) to the conclusion that larger changes in $\delta^{13}C$ values of the organic matter in the sediments of some lakes are associated with climatic changes. Lower $\delta^{13}C$ values were observed in periods of colder climate, with increasing paleotemperature they become heavier. Nakai (1972) found that this trend was accompanied by an increase of the organic carbon content of the sediment. The explanation for these changes may lie either in the fact that at higher temperatures the rate of photosynthesis increases and the kinetic fractionation factor decreased with increasing rate, or that a change in the availability of dissolved CO_2 occurred.

Very few river sediments have been analyzed for the ^{13}C content of their organic matter; the determinations available fall in the range of lake muds. Data for sediments from the Pedernales area in the Orinoco River delta by Eckelmann et al. (1962) average $-26.6‰$ (three samples). Sackett and Thompson (1963) investigated the isotopic composition of organic carbon from several rivers in southern Mississippi and Alabama and obtained an average of $-26.6‰$ for eighteen samples. Sediments formed under lagoonal and open marine conditions were studied by these authors also who found a systematic trend towards heavier isotopic composition from river to lagoonal to marine environment. Similarly, Hunt (1970) observed drastic changes in the $\delta^{13}C$ of river sediments close to their mouth. The analyses of bottom sediment organic matter in thirty-two rivers from Maine to Florida had an average $\delta^{13}C$ value of $-26‰$ more than 15 km upstream from the mouth, which changed to $-20‰$ at the mouth of the stream.

The carbon isotopic composition of soil humus and soil CO_2 reflects that of the local plant cover. In peat deposits changes in facies may be accompanied by small variations in $\delta^{13}C$ of the accumulated organic matter. A wide range in ^{13}C content ($30‰$) is observed in lake muds and there is a trend towards lighter isotopic compositions with increasing latitude. In some lakes ^{13}C concentration changes with depth of the sediment have been related to climatic changes; lower $\delta^{13}C$ values are observed during the Pleistocene. Recent river sediments show on the average ^{13}C contents close to $-26‰$. At river mouths either gradual or abrupt changes towards heavier isotopic compositions are encountered.

Recent marine sediments

A sizeable fraction of the isotopic analyses of organic carbon in marine sediments available in the literature has been compiled in Fig. 9-11A. The $\delta^{13}C$ values vary from -10 to $-30‰$; sediments lighter than this appear to be rare. The shape of the distribution in Fig. 9-11A is influenced by: (1) the factors determining the ^{13}C content of the sedimentary organic matter, and (2) by the selection of sample sites. There are relatively few sites or areas from which samples have been collected; thus, although the plot comprises well over 1600 individual sediment samples it cannot be interpreted as representing a random sample of the $\delta^{13}C$ values of organic carbon in the ocean basins. It is interesting that more than 90% of all sediments sampled to date show $\delta^{13}C$ values between -20 and $-27‰$, indeed a very restricted range. All isotopic compositions heavier than $-14‰$ come from recent organic muds from the coasts of Florida, Texas, British Honduras and the Bahamas (Craig, 1953a; Parker, 1964; Scalan and Morgan, 1970). Many of these samples show isotopic compositions similar to those of plants and animals

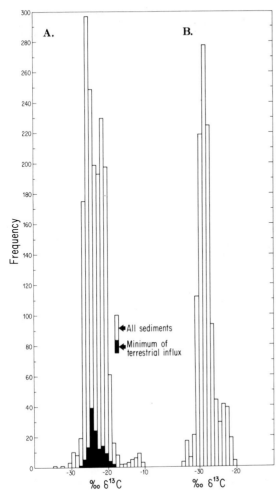

Fig. 9-11. Carbon isotopic composition of recent marine sediments and petroleum. A. Recent marine sediments. Data from Calder et al. (1974), Craig (1953a), Deuser (1972), Eckelmann et al. (1962), Emery (1960), Erlenkeuser et al. (1974), Hedges (1974), Hunt (1970), Johnson and Calder (1973), Landergren and Parwel (1954), Newman et al. (1973), Parker (1964), Rogers and Koons (1969), Rogers et al. (1972), Sackett (1964), Sackett et al. (1965, 1974a), Scalan and Morgan (1970), Silverman and Epstein (1958), and Vinogradov and Galimov (1970). B. Petroleum. Data from Craig (1953a), Eckelmann et al. (1962), Galimov (1973), Krejci-Graf and Wickman (1960), Kvenvolden and Squires (1967), Lebedev (1964), Monster (1972), Nagy (1966), Orr (1974), Park and Dunning (1961), Rogers et al. (1974), Silverman and Epstein (1958), Vinogradov and Galimov (1970), Vredenburgh and Cheney (1971), Wickman (1956), and Williams (1974).

living in the area of collection. All the sediments lighter than $-29‰$ were reported by Calder et al. (1974) for Indian Ocean samples, the only study reporting isotopic compositions from this area.

In order to evaluate whether there is a characteristic $\delta^{13}C$ range for the organic matter produced in the sea, those sediments for which terrestrial contributions are possibly of significance should be eliminated. This has been done in Fig. 9-11A. The peak between -24 and $-25^o/_{oo}$ for sediments showing minimal terrestrial influx arises mainly from the work of Rogers et al. (1972), on a North Atlantic core and data from Sackett et al. (1974a) for cores from the Antarctic. The bimodality of the $\delta^{13}C$ frequency distribution of all sediments is also of interest. Contributions to the peak between -24 and $-26^o/_{oo}$ arise on the one hand from high-latitude samples for which a minimum or absence of terrestrial contribution is likely, on the other hand from samples from the Gulf of Mexico for which a maximum of terrestrial contamination has been claimed. Samples in the -20 to $-22^o/_{oo}$ range come mainly from Gulf of Mexico samples for which minimum amounts of terrestrial contamination have been suggested. Hence, there appears to be a parallelism in the trends of the carbon isotopic composition of endogenous organic matter of marine sediments and of plankton as a function of latitude (Fig. 9-4).

If the $\delta^{13}C$ distribution of organic carbon in marine sediments is compared to that of terrestrial sediments (Fig. 9-8) several differences are apparent. Terrestrial sediments show a wider range in $\delta^{13}C$ values which completely covers that of marine sediments and extends past it both towards heavier and lighter isotopic compositions. Whereas most marine sediment analyses fall within a narrow range and, hence, show a highly peaked frequency distribution, that of terrestrial sediments is more platykurtic. Another distinct difference between the two distributions lies in the very sharp drop in the number of marine sediments which have smaller $\delta^{13}C$ values than $-27^o/_{oo}$; there are continental sediments that are $10^o/_{oo}$ lighter than this. The $\delta^{13}C$ distributions for sediments from the two environments may show characteristic differences, however, it is apparent that a simplistic model suggesting a systematic ^{13}C enrichment in marine compared to non-marine sedimentary organic carbon does not hold.

Variations in the carbon isotopic composition of organic matter in marine sediments have been noted as a function of the distance from the continents (e.g., Sackett and Thompson, 1963; Sackett, 1964; Newman et al., 1973) and also with depth in sediment cores. The interpretation of the observed variations has not always led to the same conclusions. There is general agreement that any one or a combination of the factors that have been discussed above can contribute towards the $\delta^{13}C$ variability observed in the sediments, however, their relative importance is not always clear.

The isotopic composition of endogenous photosynthetic carbon sources has been summarized above. The carbon isotopic composition of animals reflects that of the plant diet (Smith and Epstein, 1970), and no evidence has been found for successively large fractionation along the food chain, plant to herbivore to carnivore (Parker, 1964; Sondheimer et al., 1966).

DeNiro and Epstein (1976) observed that the animal carbon is generally 0—2‰ heavier than that of its food; hence, the given compilation of marine plant $\delta^{13}C$ values will fairly well reflect the $\delta^{13}C$ values of the organic matter contributed to the sediment by plant and animals. Other endogenous carbon additions to marine sediments involving photosynthetically produced organic carbon more indirectly have been suggested by Nissenbaum and Kaplan (1972) and Nissenbaum (1974). Dissolved organic matter in seawater occurs in a variety of compounds only a small fraction of which has been identified; most of it is present in the form of "Gelbstoff" or "Humus", high-molecular-weight polymers. Sources for these polymers that have been considered include: terrestrial humic substances brought into the sea by rivers; decomposition products of planktonic cellular material; and formation from extracellular material (exudates) of marine organisms. Studies by Williams (1968), Calder and Parker (1968) and Jeffrey (1969) of waters from the Pacific Ocean and the Gulf of Mexico showed that such dissolved organic matter has a $\delta^{13}C$ range from —20 to —23‰, most values being close to —23‰. The uniformity of $\delta^{13}C$ suggests that the dissolved organic carbon is formed from a larger reservoir which may be plankton. The slight depletion in ^{13}C compared to warm water plankton (—20 to —21‰) has been attributed to diagenetic processes (Nissenbaum, 1974). Humic substances separated from marine sediments have $\delta^{13}C$ values in the range —20 to —22‰ (Nissenbaum and Kaplan, 1972; Nissenbaum, 1974) and it has been suggested by these authors that humic substances in marine sediments are formed in situ from dissolved organic matter in seawater.

Exogenous carbon contributions (organic debris of terrestrial plants) were thought by Craig (1953a) to be responsible for the higher ^{12}C content of organic matter in marine sediments compared to that of the local flora and fauna. Similarly, the systematic decrease in $\delta^{13}C$ values of organic carbon in clastic sediments from around —20‰ several tens of miles off the eastern Gulf coast to values around —24 to —26‰ close to the coast was interpreted by Sackett and Thompson (1963) as the result of mixing of marine- and land-derived carbon differing in $\delta^{13}C$. Cores from recent sediments from the Gulf of Mexico abyssal plain show a range of variation from —16 to —28‰ (Sackett, 1964; Newman et al., 1973) which has been attributed also to additions from terrestrial organic matter. Newman et al. (1973) found that in about 50% of the cores there appeared to be a systematic shift towards more negative values from —19 at the surface to —25‰ at depth. This change occurs across the Pleistocene-Holocene boundary, and the authors suggest that it is due to an increased influx of terrestrial carbon during periods of glaciation.

The relative contribution of marine and terrestrial organic matter to sediments may be evaluated from $\delta^{13}C$ measurements if the isotopic compositions of the two end members are known and are different. In view of the variability of their ^{13}C content Hedges (1974) proposed a method based on a determination of the lignin component of sediments for this purpose. Lignin compounds occur in vascular plants, they are, however, absent in marine plankton. Hence, if $\delta^{13}C$ changes in the sediment are related to an influx of terrestrial material, $\delta^{13}C$ and lignin content of a series of marine sediments of varying terrestrial contributions should be correlated. Hedges (1974) found in three traverses extending from the shore into the Gulf of Mexico to a depth of about 100 m significant correlations between $\delta^{13}C$ and total lignin content of the surface sediment, indicating that the variations of $\delta^{13}C$ of the sediments were due to terrestrial carbon additions, and that these were noticeable at a considerable distance from the shore line. The lignin-free organic matter (presumably produced in Gulf water) is estimated to have $\delta^{13}C$ values of about $-19‰$ in agreement with values of surface sediments in the Gulf of Mexico from locations at which terrestrial influx is minimal (Sackett, 1964; Hedges, 1974).

Whereas, studies of surface sediments from the Gulf of Mexico have shown that ^{13}C variations due to terrestrial carbon additions can be noticed at some distance from the shore line, different conclusions were reached by Hunt (1970) for sediments from the Atlantic coast. There is a fairly abrupt change in $\delta^{13}C$ in the river sediments from values around $-26‰$ at about 15 km upstream from the mouth of the river to around $-20‰$ very near or at the mouth of the river. From there across the shelf little further systematic change in $\delta^{13}C$ is observed. In bays such as the Delaware and Chesapeake, sediments have $\delta^{13}C$ values in the range -18 to $-22‰$ as far as 80 km upstream. Hence, in the environments studied by Hunt (1970) there appears to be little contribution of carbon derived from the continent to the sediments accumulating on the shelf. The same conclusion was reached by Degens (1969).

There are other exogenous carbon sources which have been considered important for sediments. Sackett et al. (1974a, b) studied $\delta^{13}C$ and petrography of sediment cores from the Ross Sea, Antarctica, and concluded that their data can be best interpreted as resulting from a recycling of kerogen from old sediments, which is isotopically heavier than the plankton living in the environment of deposition. It is suggested that up to 90% of the organic matter in Ross Sea sediments is derived from igneous and metamorphic rocks that are glacially eroded. Erlenkeuser et al. (1974) have attributed a decrease in $\delta^{13}C$ in the top (20 cm) layer of some Baltic Sea sediments to the dumping of sewage sludge.

Prior to 1969 a number of authors had noted changes in $\delta^{13}C$ values in sediment cores; the question whether or not to interpret these as a result of exogenous carbon additions is a matter of debate. Rogers and Koons (1969) suggested that some of the variations may be related to climatic changes. The authors studied the carbon isotopic composition of Quaternary marine sequences in the Gulf of Mexico, observed that the relative variations in $\delta^{13}C$ along the length of the cores correlated with warm and cold water periods as deduced from the coiling direction and relative abundance of certain planktonic foraminifera, and concluded that the principal reason for the change in $\delta^{13}C$ is the variation in water temperature in the photosynthetic zone during interglacial and glacial periods rather than an addition of terrestrial organic matter to the sediments. The observation that the $\delta^{13}C$ values of plankton vary as a function of temperature (Fig. 9-4) provided the background on which this interpretation was made. To test this suggestion further, Rogers et al. (1972) investigated a core in an area where an influx of terrestrial material is absent. Preliminary interpretation of their data suggests that $1-2‰$ variations between positive and more negative $\delta^{13}C$ values are related to the surface water temperature at the time of deposition.

Sackett and Rankin (1970) and Newman et al. (1973), although agreeing with the basic concept that ocean water temperatures may affect the carbon isotopic composition of organic matter accumulated in marine sediments, do not believe that the variations in $\delta^{13}C$ in sediments from the Caribbean/Gulf of Mexico system are related to temperature-controlled isotopic composition of the plankton living in the over-lying waters. The $\delta^{13}C$ of the organic carbon and $\delta^{18}O$ of *Globigerinoides ruba* measured on the same core are correlated; highest $\delta^{18}O$ values in the carbonate (lowest temperatures) are associated with lowest $\delta^{13}C$ values of the organic matter. A change of $\delta^{18}O$ of $2.2‰$ is accompanied by a change of about $5‰$ in $\delta^{13}C$. The authors suggest that if the difference in $\delta^{13}C$ was due to temperature alone it would require a very much larger temperature variation than is indicated by the $2.2‰$ change in $\delta^{18}O$ of the carbonate, and hence, they consider it unlikely that temperature is responsible for the observed $\delta^{13}C$ changes in Gulf sediments. It should be noted, however, that a translation of carbon and oxygen isotopic composition differences into temperature variations is not without difficulties and that the data of Sackett et al. (1974a) (Fig. 9-4) tend to indicate that a change of $5‰$ in $\delta^{13}C$ may correspond to a smaller temperature range than had been estimated by Sackett and Rankin (1970).

There is evidence that the $\delta^{13}C$ values of marine sediments laid down in some areas of the ocean during the Pleistocene appear to be depleted in ^{13}C compared to sediments free of terrestrial contributions that are deposited at low latitudes today. There is also evidence that today at higher latitudes in areas where terrestrial carbon contributions are absent, sediments are deposited which contain organic matter depleted in ^{13}C compared to organic matter photosynthesized in the sea at lower latitudes. Climatic variations, hence, appear to influence the carbon isotopic composition of the organic matter photosynthesized in the sea and incorporated into the sediments. In view of the fact that at lower growth temperatures the $\delta^{13}C$ values of pure marine organic carbon appear to be very similar to terrestrial organic carbon, it is impossible to determine from $\delta^{13}C$ measurements alone how much of the $\delta^{13}C$ variations in a particular sediment is related to a varying influx of terrestrial carbon due to climatic changes and how much to a change in $\delta^{13}C$

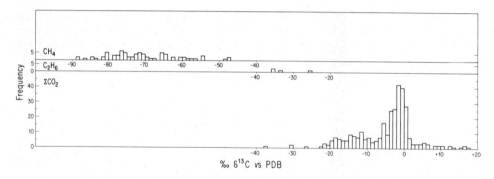

Fig. 9-12. Carbon isotopic composition of carbon-bearing compounds in the interstitial water of deep-sea sediments. Data from Claypool et al. (1973), Lyon (1973), Nissenbaum et al. (1972), Presley and Kaplan (1968; 1971a, b; 1972a, b), Presley et al. (1972a, b; 1973a, b, c, d, e), and Van Donk and Mathieu (1973).

of the photosynthesized material under changed temperature conditions. Study of the abundance and ^{13}C content of different chemical components of the organic matter in conjunction with carbon and oxygen isotope analyses of the sedimentary carbonate may help to clarify the problem.

Systematic trends of the ^{13}C content of organic matter with depth could also be related to diagenetic effects. Although there are numerous marine sediment cores which show depth-related $\delta^{13}C$ changes, there are also many which do not, suggesting that these carbon isotopic composition variations may be more related to the phenomena described rather than to diagenesis, which could be expected to affect the ^{13}C content of sediments more universally. The balance of the evidence collected to date (e.g., Johnson and Calder, 1973; Behrens and Frishman, 1971) suggests that if carbon isotopic composition changes occur during diagenetic processes in sediments they result only in minor differences between the source and the residual carbon. In order that the ^{13}C content of the residual organic matter remain unaffected, the total carbon removed during diagenesis must be more or less equal in isotopic composition to the source carbon. Important products of the process which incorporate carbon are CO_2 and CH_4. The isotopic composition of CO_2 and CH_4 weighted according to their relative proportions should hence equal that of the decaying organic matter. In an actual sediment it may be difficult to establish this balance, however, because the system may be open to CH_4 and CO_2 which may be lost by diffusion or carbonate precipitation. Isotopic composition measurements of CO_2 and CH_4 in interstitial waters in sediments have shown that there is a very large range of ^{13}C concentrations for both CO_2 and CH_4 (Fig. 9-12).

Presley and Kaplan (1968) investigated changes in the concentration of dissolved carbonate and its $\delta^{13}C$ values in interstitial waters of near-shore sediments. It was found that the total dissolved inorganic carbon increased with depth and at the same time $\delta^{13}C$ decreased to about $-20\%_{00}$. An opposite trend was observed by Nissenbaum et al. (1972) in a study of sediments in a reducing fjord, Saanich Inlet, British Columbia. Here also the total dissolved CO_2 increased strongly with depth, but became significantly enriched in ^{13}C (about $+18\%_{00}$). There are apparently various carbon sources and processes which contribute to the large range in $\delta^{13}C$ of CO_2 observed in Fig. 9-12. Inorganic carbon dissolved in seawater which constitutes at least initially part of the interstitial water has $\delta^{13}C$ values close to zero and is responsible for the strong maximum at $0\%_{00}$ (Fig. 9-12). If varying amounts of CO_2 produced by decay of organic matter are added, values between 0 and that characteristic for diagenetic CO_2 can result. The fact that $\delta^{13}C$ values of dissolved CO_2 lighter than $-22\%_{00}$ are few can be interpreted to indicate that the CO_2 produced in the diagenetic processes rarely has $\delta^{13}C$ values significantly lighter than the source carbon. The heavy $\delta^{13}C$ values of the dissolved CO_2 which have been observed by a number of authors may appear peculiar at first. Nissenbaum et al. (1972) suggested that the presence of CH_4 may offer an explanation for them. Of the three possibilities: (1) isotope exchange between CH_4 and CO_2 (Bottinga, 1969), (2) formation of CH_4 and CO_2 by fermentation (Rosenfeld and Silverman, 1959), and (3) reduction of preformed CO_2 by methane-forming bacteria (Nissenbaum et al., 1972; Games and Hayes, 1974, 1976) the authors prefer the third as an explanation. Originally CO_2 formed with an isotopic composition similar to that of the source organic matter. Subsequently, part of it was reduced to ^{12}C-enriched CH_4; this would leave the residual CO_2 pool enriched in ^{13}C. The occasional very light carbon isotopic compositions of dissolved inorganic carbon have been attributed to an oxidation of first-formed CH_4. Methane of the interstitial waters shows extremely light isotopic composition, which is expected on the basis of what is known about bacterial isotope effects (Table 9-2).

If the dissolved carbonate content of the interstitial water rises sufficiently, the solubility product of calcite may be exceeded and $CaCO_3$ is precipitated in the sediment; such diagenetic carbonates should show an isotopic composition range similar to that of the carbon dissolved in the pore waters. A very large $\delta^{13}C$ range has been observed for such diagenetic carbonates ($-64\%_{00}$ to $+21\%_{00}$) in marine as well as in continental sediments (Hodgson, 1966; Murata et al., 1967; Allen et al., 1969; Hathaway and Degens, 1969; Deuser, 1970; Fritz et al., 1971; Roberts and Whelan, 1975).

The carbon isotopic composition of kerogen of 90% of the marine sediments analyzed to date falls between -20 and $-27\%_{00}$, and the data indicate a decrease in the ^{13}C content of endogenous carbon in marine sediments with increasing latitude. The decrease in ^{13}C content towards some shore lines has been explained as result of a dilution of endogenous marine carbon compounds by increased contributions of terrestrial organic debris. The change towards lighter $\delta^{13}C$ values in drill cores across the Holocene-Pleistocene

TABLE 9-2

Carbon isotope fractionation in the decomposition of organic matter at low temperature

Fractionation mechanism	$\delta^{13}C_{CO_2}$	$\delta^{13}C_{CH_4}$	$\delta^{13}C_{CO}$	$\delta^{13}C_{org}$	T (°C)	$\alpha_{org\text{-}CH_4}$	$\alpha_{CO_2\text{-}CH_4}$	$\alpha_{CO_2\text{-}org}$	References
Bacteria	—	−116.8	—	−45.2	30	1.081	—	—	Rosenfeld and Silverman (1959)
	−42.3	−82.6	—	—	40	—	1.044	—	Games and Hayes (1976)
	−42.0	−95.6	—	—	40	—	1.059	—	Games and Hayes (1976)
	−41.4	−64.9	—	—	65	—	1.025	—	Games and Hayes (1976)
Lab reactor, 30–60 days	−5 to −10	−40 to −47	—	—	25	—	1.036	—	Games and Hayes (1976)
			—	—	25	—	1.039	—	Games and Hayes (1976)
Lab reactor, 85–120 days	+10 to +17	−56 to −58	—	—	25	—	1.070	—	Games and Hayes (1976)
			—	—	25	—	1.080	—	Games and Hayes (1976)
Landfill	+16.6	−52.1	—	−25	20	1.029	1.072	1.043	Games and Hayes (1976)
	+16.1	−48.5	—	−25	20	1.025	1.068	1.042	Games and Hayes (1976)
	+20	−48	−13.3	−24	20	1.025	1.071	1.045	Games and Hayes (1976)
Sewage sludge	−5.5	−49.1	—	−25	60	1.025	1.046	1.020	Games and Hayes (1976)
	+4.1	−47.1	—	−22.6		1.026	1.054	1.027	Nissenbaum et al. (1972)
Lake muds	0	−75	—	−30		1.049	1.081	1.031	Oana and Deevy (1960)

boundary has been attributed either to an increased influx of terrestrial plant matter during the Pleistocene or to a change in $\delta^{13}C$ of photosynthesized organic compounds in response to lower environmental temperatures. The carbon isotopic composition changes during diagenesis are small, the resulting carbon-bearing phases contribute to the isotopic variability of CO_2 and CH_4 dissolved in interstitial waters whose isotopic composition ranges are considerable. Diagenetic carbonates that incorporate carbon from these phases show likewise a very large ^{13}C concentration range (-64 to $+21‰$).

Consolidated sediments and metasediments

The older sediments and metasediments (mainly marine) cover approximately the same isotopic composition range as recent marine sediments, but show a slight displacement towards lighter isotopic compositions (Fig. 9-13). Apparently just as diagenesis, lithification does not drastically alter their ^{13}C content.

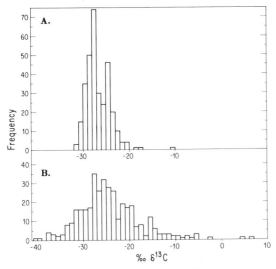

Fig. 9-13. Carbon isotopic composition of kerogen and graphite in sediments, metasediments, and Precambrian rocks. A. Kerogen in sediments and metasediments. Data from Andreae (1974), Baker and Claypool (1970), Barker and Friedman (1969), Craig (1953a), Galimov et al. (1968), Gavelin (1957), Hahn-Weinheimer (1960a, b, 1966), Hahn-Weinheimer et al. (1969), Krejci-Graf and Wickman (1960), Landergren (1953, 1955, 1957), Leythaeuser (1973), McKirdy and Powell (1974), Sackett and Menendez (1972), Silverman and Epstein (1958), and Williams (1974). B. Reduced carbon in Precambrian rocks. Data from Andreae (1974), Barker and Friedman (1969), Craig (1953a, 1955), Eichman and Schidlowski (1975), Gavelin (1957), Hahn-Weinheimer (1966), Hoefs and Schidlowski (1967), Hoering (1962, 1963, 1967a, b), Jackson et al. (1976), Jeffrey et al. (1955), Landergren (1957), McKirdy and Powell (1974), Oehler et al. (1972), Roblot et al. (1964), Schidlowski et al. (1976), Schopf et al. (1971), and Smith et al. (1970).

Several authors have attempted to elucidate changes in $\delta^{13}C$ in the organic matter in sedimentary rocks during metamorphism. Early studies by Landergren (1955) and Gavelin (1957) came to the conclusion that metamorphism would not significantly modify the ^{13}C content of the organic matter in sediments. Later investigations by Hahn-Weinheimer (1960a, b, 1966), Hahn-Weinheimer et al. (1969), Barker and Friedman (1969), McKirdy and Powell (1974), Baker and Claypool (1970) and Andreae (1974) indicate that metamorphism may lead generally to a slight, however, under some circumstances to a considerable, enrichment in ^{13}C in the reduced carbon. In view of this one would expect that the distribution of $\delta^{13}C$ values for consolidated older sediments and metasediments of Fig. 9-13A should be shifted to slightly higher $\delta^{13}C$ values with respect to modern sediments, however, the opposite is observed (Fig. 9-16). An interpretation of this effect would have to consider the possible existence of small changes towards more negative $\delta^{13}C$ values during lithification, which have not been uncovered as yet, as well as possible variations of $\delta^{13}C$ in the surface reservoir of carbon during the geologic past (Silverman, 1962; Weber, 1967; Keith and Weber, 1964; Welte, 1970; Welte et al., 1975).

Organic matter and graphite in Precambrian strata

The carbon isotope distribution of reduced carbon in Precambrian rocks overlaps that of more recent sediments but shows a very much wider range in $\delta^{13}C$ than younger marine and continental sediments combined (Fig. 9-13B), reflecting the large number of factors which ultimately may influence the ^{13}C content of the reduced carbon in these rocks. Many of the samples showing unusual enrichments in ^{13}C are associated with carbonates (Landergren, 1953; McKirdy and Powell, 1974; Andreae, 1974; Hahn-Weinheimer, 1966; Hahn-Weinheimer et al., 1969; Craig, 1953a). Because sedimentary carbonates generally show a considerable ^{13}C enrichment (by 20—25°/oo) over carbon fixed by photosynthesis in the same environment, they have been looked upon as source of the ^{13}C. Compared to younger marine sediments a relatively larger number of Precambrian rocks show a higher ^{13}C depletion and enrichment, so that the shapes of the two distributions bear little resemblance to one another.

THE CARBON ISOTOPIC COMPOSITION OF FOSSIL FUELS

Coal

According to the peat to anthracite theory of coal formation there is a continuous progression in the degree of coalification from peat, which represents the initial stage, through lignite, an intermediate stage of the process, to bituminous coal and anthracite, the advanced stages. The alteration of the

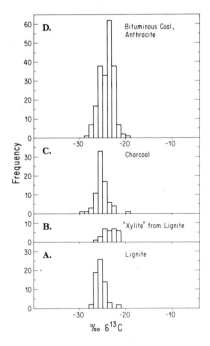

Fig. 9-14. Isotopic composition of solid fossil fuels. Data from Colombo et al. (1968), Compston (1960), Craig (1953a), Fischer et al. (1968, 1970), Garcia-Loygorri et al. (1974), Jeffrey et al. (1955), Stahl (1968a, b), Teichmüller et al. (1970), and Wickman (1953).

photosynthesized material commences upon death of the organism first through biochemical processes, later, with increasing burial geochemical reactions become more and more important. During the formation of coal a relative enrichment of carbon occurs in the deposit compared to the plant material. At the same time a depletion in the elements hydrogen, oxygen and nitrogen takes place, which are given off in the form of water, carbon dioxide, methane and higher hydrocarbons, and nitrogen. This brutto reaction, which is actually a summary of many individual reactions is commonly termed coalification. Because the hydrogen and oxygen split off during this process are bonded to carbon, which is removed from the residual molecules through splitting of C—C bonds, one might expect isotope effects to occur during coalification.

The isotopic composition of peat, under this theory the source of the organic matter of lignite, has been discussed above. Fig. 9-14A gives $\delta^{13}C$ values of some lignites, mainly from eastern Germany. Comparison with Fig. 9-8B shows that in general there is no very drastic shift in $\delta^{13}C$ progressing from peat to lignite. However, the existence of small isotopic composition changes can only be unveiled if more detailed comparisons are made. It is apparent from the work of Stahl (1968a, c) that the $\delta^{13}C$ of the accumulated peat can differ slightly with the facies, hence, in order to detect small changes in $\delta^{13}C$ in the transition from peat to lignite, one cannot make a comparison of peat and lignite per se, but must restrict oneself to a particular set of environmental conditions (Fischer et al., 1968).

Carbon isotope fractionations during coalification up to the lignite stage were experimentally studied by Geissler and Belau (1971). The organic material (moss, $-27.4\%_{oo}$) was heated (200—275°C) in the presence of water. After the termination of the experiment (20—700 minutes) four carbon-bearing phases were recognized; a gaseous phase consisting of CO_2 and CH_4, a water-soluble phase, and a water-insoluble phase. The solid residue had an average carbon isotopic composition of $-28.1\%_{oo}$ and the water-soluble fraction of $-25.3\%_{oo}$, no trend in their ^{13}C content with intensity of coalification was observed. The weighted sum of the $\delta^{13}C$ values of these two components was $-27.5\%_{oo}$ showing no significant difference from the starting material. The carbon isotopic composition of methane liberated first in the reaction is about $-35\%_{oo}$ (i.e., at lower temperatures and shorter times of heating), and that of simultaneously formed carbon dioxide is more or less identical to that of the starting material. With increasing reaction length and temperature the $\delta^{13}C$ of CH_4 changes to $-28.2\%_{oo}$ and that of CO_2 becomes $1.5\%_{oo}$ heavier than the starting material. The difference in isotopic composition between CH_4 and CO_2 was in no case anywhere near the fractionations expected on the basis of the computations of Bottinga (1969). The experiments show in coincidence with the study of natural peat and lignite deposits that if carbon isotope effects occur during the initial stages of coalification these are small and that the carbon isotopic composition of the total gas phase liberated is close to that of the starting material.

In Fig. 9-14D a compilation of $\delta^{13}C$ values of coals of higher ranks is given; there is little if any difference in $\delta^{13}C$ compared to peat or lignite. Hence, there are also no major changes in ^{13}C content in the more advanced stages of coalification; whether smaller differences are produced is a question more difficult to answer. The possible existence of carbon isotopic composition changes during the coalification process has been discussed by a number of authors (Wickman, 1953; Craig, 1953b; Jeffrey et al., 1955; Cook, 1961; Stahl, 1968a, c; Fischer et al., 1968). Evidence for a small ^{13}C enrichment during coalification was presented by Colombo et al. (1968, 1970a, b) and Teichmüller et al. (1970) who measured volatile and ^{13}C content of coals from Germany and found a small but significant shift in $\delta^{13}C$ across the coalification break. At this point, which lies at 29.5% volatile content, also other chemical properties of the coal change; e.g., a stronger removal of hydrogen in the form of methane occurs. Coals of higher volatile content than the coalification break have significantly lower $\delta^{13}C$ values ($\delta^{13}C = -25.4\%_{oo}$, $s_x = 0.6\%_{oo}$, $n = 29$) than coals of lower volatile content ($\delta^{13}C = -24.2\%_{oo}$, $s_x = 0.5\%_{oo}$, $n = 40$).

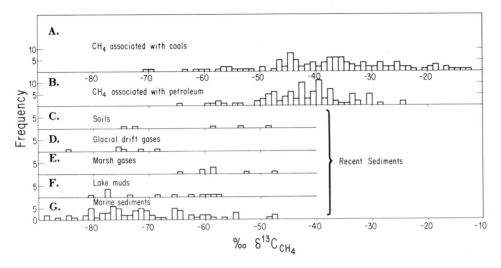

Fig. 9-15. Carbon isotopic composition of CH$_4$ from various sources. A. Coal gases. B. Methane associated with petroleum. C. Soil gases. D. Glacial drift gases. E. Marsh gases. F. Lake sediments. G. Marine sediments. Data from Colombo et al. (1968, 1970a, b), Dubrova and Nesmelova (1968), Galimov (1969), Lebedev (1964), Lebedev et al. (1969), Oana and Deevey (1960), Silverman (1964a, b), Stahl (1968a, b), Teichmüller et al. (1970), Vinogradov and Galimov (1970), Wasserburg et al. (1963), and Fig. 9-12.

Considerable volumes of methane are formed during coal maturation, even at the anthracite stage, as well as during the thermal metamorphism of organic matter in sediments. Only a fraction of such gas can be held absorbed in the source rock, much of it migrates away and may be trapped in reservoirs under impermeable strata and form natural gas deposits. In order to develop a better understanding of the origin of natural gas deposits as well as the processes that occur during coalification, it is of interest to investigate the carbon isotopic composition of gases that are directly associated with coal (Teichmüller et al., 1970; Colombo et al., 1968, 1970a, b). The δ^{13}C values of methane absorbed in coal show a very wide range (−12 to −71‰, Fig. 9-15A) and are independent of the degree of coalification of the coal from which the gas was released. Generally, the methane is isotopically lighter than the coal, an exception being the very heavy gases (−12 to −21‰) which were liberated from some anthracites lying at great depth.

If there is a pressure release on the coal seams (tectonic uplift, erosion) gas will be desorbed from the coal, first methane and later also higher hydrocarbons. It has been observed in the Ruhr section of Germany that at a depth of 600—750 m below the surface of the Carboniferous there are practically no higher hydrocarbons present in the absorbed coal gas (desorption zone); 750 m and below the surface of the Carboniferous (1000 m below the landsurface) higher hydrocarbons constitute a significant fraction of the coal gas (primary zone). In sections where the whole coal measure lies below the desorption zone the higher hydrocarbons disappear one after another with increasing depth, first hexane, then pentane, butane and ethane until only methane is found in very deeply buried anthracites (3000 m below surface). Their decreasing abundance has been attributed to their instability under increasing pressure and temperature conditions to which the coal is exposed. The methane isotopic composition is related to this change in coal gas chemistry. At the top of the Carboniferous, in the desorption zone, $\delta^{13}C$ values between -60 and $-70‰$ are observed which become more positive with depth ($-32‰$). In the primary zone $\delta^{13}C$ values of -30 to $-20‰$ and more positive are generally encountered. Excepting the samples from large depth, the increase in $\delta^{13}C$ in the methane going from the desorption to the primary zone is accompanied by an increase in the abundance of the higher hydrocarbons. These regularities suggest (Teichmüller et al., 1970) that initially CH_4 and higher hydrocarbons were formed throughout the coal measure. Due to pressure release locally, seams that are now in the desorption zone lost all of their higher hydrocarbons and methane. The methane now found in the desorption zone is the result of upward migration from lower coal measures. Because $^{12}CH_4$ diffuses and desorbs more readily than $^{13}CH_4$, the CH_4 now in the desorption zone is isotopically lighter than that in the primary zone. New methane or higher hydrocarbons have not been formed in these measures after the desorption occurred and the CH_4 in them now is in a state of transit from depth to the surface. Hence, a distinction can be made between allochthonous methane in the desorption zone which is isotopically light and autochthonous CH_4 in the primary zone which is isotopically heavy.

Isotope effects in processes that are important in determining the $\delta^{13}C$ of coal gases were studied experimentally by Colombo et al. (1968), who observed that the desorbed CH_4 was about $5‰$ lighter than the CH_4 absorbed in the coal. Similar findings were made by Friedrich and Jüntgen (1972); methane liberated in the progressive degassing of coal changed its isotopic composition. The methane, desorbed at ambient temperatures from coal, was always $4.4-8‰$ depleted in the heavy isotope compared to the CH_4 desorbed upon moderate heating (120°C). The isotopic composition of CH_4 formed during experimental pyrolysis of coal was measured by Friedrich and Jüngten (1973) and Sackett et al. (1970). At 500°C Sackett et al. (1970) observed a fractionation of $-11‰$ between methane and coal. Friedrich and Jüntgen (1973) carried out coal pyrolysis experiments at temperatures from 400 to 800°C. The maximum fractionations measured between coal and CH_4 amount to about $-10‰$. The application of the experimental results to the coalification process is difficult, however, because the rate of heating, which is important for the onset of chemical degradation reactions, varies by about 10 orders of magnitude between laboratory experiment and geologic conditions. Friedrich and Jüntgen

(1973) extrapolate from their data a $\delta^{13}C$ value of $-70^0/_{00}$ for CH_4 liberated during coalification at $100°C$. Assuming the $\delta^{13}C$ of the starting material to be $-26^0/_{00}$ and a removal of 12.6% of carbon in the form of CH_4 ($-60^0/_{00}$) the isotopic composition of the residual coal is evaluated to be $-21^0/_{00}$. The authors suggest that the initial methane formed during coalification had $\delta^{13}C$ values between -55 and $-70^0/_{00}$ and that heavier coal-methane isotopic compositions are due to migration and desorption processes.

If this concept would hold there should be coal methane samples showing $\delta^{13}C$ values lighter than $-70^0/_{00}$, these have not been observed. It should also be noted that only 10% of the coal-methane samples measured to date fall in the range -55 to $-70^0/_{00}$, considered by Friedrich and Jüntgen (1973) as characteristic for primary CH_4 values and less than 6% of the coal $\delta^{13}C$ values are heavier than $-22^0/_{00}$, a ^{13}C enrichment expected to be common if the coalification process was indeed associated with carbon isotope fractionations as large as those extrapolated by Friedrich and Jüntgen (1973).

Progressive coalification leads only to minor changes in the carbon isotopic composition of the coal, and the weighted ^{13}C content of the gaseous phases formed in the process is similar to that of the source material. A ^{13}C enrichment of about $1^0/_{00}$ has been observed across the coalification break. It has been recognized also that conditions during coal formation as well as desorption and migration will affect the isotopic composition of coal gas methane, however, the relative importance of these has not been agreed upon (Sackett, 1968; Friedrich and Jüntgen, 1973). Pyrolysis experiments have demonstrated relatively small fractionations between coal and methane (about $-10^0/_{00}$) and as 50% of the measured $\delta^{13}C$ values for coal gas methane fall between -25 and $-40^0/_{00}$ it can be suggested that a large fraction of the $\delta^{13}C$ values of coal gas methane can be understood by postulating a relatively small carbon isotope fractionation during coalification. This fractionation may be augmented by diffusion, desorption as well as reservoir effects.

Petroleum

The carbon isotopic composition of crude oils has received considerable attention in attempts to clarify the origin of oil deposits. Early studies (West, 1945; Craig, 1953a; Silverman and Epstein, 1954, 1958; Wickman, 1956) established that the carbon isotopic composition of petroleum was considerably lighter than that of marine carbonates and fell approximately in the range of terrestrial plants.

Most oils show $\delta^{13}C$ values in the range from -21 to $-32^0/_{00}$ (Fig. 9-11B). The frequency distribution shows a slow rise from $-20^0/_{00}$, a suggestion of bimodality, to a strong maximum at -28 to $-29^0/_{00}$ and a very sharp drop in frequency towards lighter $\delta^{13}C$ values. If, in the frequency distribution of marine sediments, those deposited under special conditions and showing very

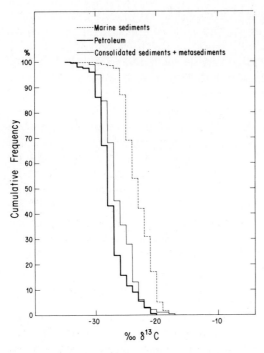

Fig. 9-16. Comparison of the carbon isotopic composition of organic matter in recent marine sediments, consolidated sediments and petroleum. Data from Figs. 9-11 and 9-13.

heavy $\delta^{13}C$ values are for the moment ignored, there is a certain degree of similarity in shape between the $\delta^{13}C$ frequency distribution of the organic carbon from marine sediments and that of petroleum, with a shift of the petroleum $\delta^{13}C$ frequency distribution towards lighter $\delta^{13}C$ values by about $3^0/_{00}$ (Fig. 9-16; Table 9-3). Compared to older sediments petroleum shows again a similar frequency distribution, but it is also systematically displaced

TABLE 9-3

Comparison of the ^{13}C frequency distribution of petroleum and organic matter from recent marine sediments

Compared feature	Sediment (‰)	Petroleum (‰)
5% tile	−19.5	−22.0
First mode	−21.5	−23.5
Second mode	−25.5	−28.5
95% tile	−26.5	−30.0

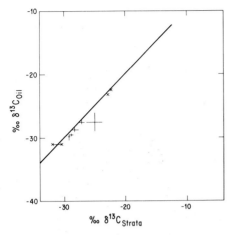

Fig. 9-17. Comparison of carbon isotopic composition of petroleum and suspected source rock. Solid line represents a one-to-one correspondence between source rock and petroleum isotopic composition. Data from Krejci-Graf and Wickman (1960), Orr (1974), Silverman and Epstein (1958), and Williams (1974).

towards lighter isotopic compositions with respect to it. Hence, petroleum is on the average isotopically slightly lighter than the kerogen in sediments. In the few cases where it has been studied (Fig. 9-17), it has been found that this holds true also for oil and supposed source rock. In comparison with coal it appears that in spite of a considerable overlap of the distributions petroleum (mode -27 to $-30°/_{00}$) shows on the average lower ^{13}C contents than coal (mode -23 to $-26°/_{00}$).

Although the carbon isotopic composition of all oils combined covers more than $10°/_{00}$, it has been found that within a particular oil field the carbon isotopic composition variations can be much more restricted (less than $1-2°/_{00}$) (Kvenvolden and Squires, 1967; Vredenburgh and Cheney, 1971; Williams, 1974; Koons et al., 1974). Hence, in combination with other oil characteristics $\delta^{13}C$ measurements may be used to characterize and correlate oils and can represent a useful tool to the explorationist. Likewise, $\delta^{13}C$ determinations may be of help to trace the source of major oil spills.

One factor that has been held responsible for a major part of the variability observed in the carbon isotopic composition of petroleum is the depositional environment. Early investigations by Silverman and Epstein (1958) had suggested that there was a systematic difference in $\delta^{13}C$ between oils derived from continental sediments (lower $\delta^{13}C$ values) compared to those derived from marine sediments (higher $\delta^{13}C$ values). Later work by Krejci-Graf and Wickman (1960) failed to substantiate this supposition, and in view of what has been learned since about the variability of $\delta^{13}C$ in marine and continental sediments, it is doubtful that such a distinction can be made in general.

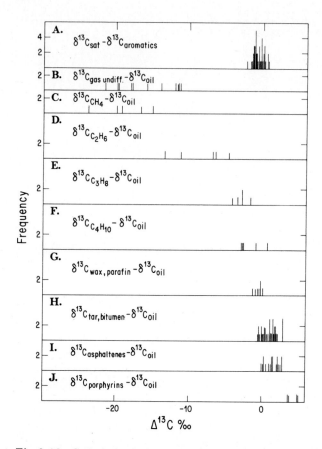

Fig. 9-18. Carbon isotopic composition differences between petroleum fractions. Data from Bailey et al. (1973), Koons et al. (1974), Monster (1972), Park and Dunning (1961), Rogers et al. (1974), Silverman (1964a,b, 1967), Silverman and Epstein (1958), Vinogradov and Galimov (1970), and Welte (1969).

As petroleum is a complex mixture of organic compounds part of the variability in the $\delta^{13}C$ values of petroleum might be related to carbon isotopic composition differences between the different chemical components of crude oil. There have been relatively few studies in which separated chemical compounds or compound groups have been analyzed. In some of these only very small $\delta^{13}C$ differences between petroleum fractions were observed, while in others systematic trends in $\delta^{13}C$ variations were discovered. Some of the results have been compiled in Figs. 9-18 and 9-19. If the gaseous hydrocarbons and porphyrins are excluded it is apparent that the variation of carbon isotopic composition between different compound groups within a given oil is on the order of a few permil. Hence, the $\delta^{13}C$ range observed for petroleum cannot be attributed to chemical differences coupled with iso-

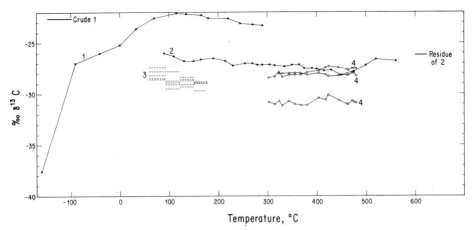

Fig. 9-19. Carbon isotopic composition of petroleum distillates and separated compounds. *Curves 1* (Silverman, 1967) and *3* (Vinogradov and Galimov, 1970): distillation experiments, the isotopic composition of the separated fraction is plotted at the distillation temperature. *Curve 2* (Silverman, 1967): fractions separated by vacuum distillation achieving an atmospheric pressure equivalent endpoint of 635°C without exceeding the cracking temperature of petroleum. The isotopic composition of the vacuum distillates is plotted at the equivalent atmospheric boiling temperature. *Curve 4* (Welte, 1969): *n*-paraffins separated by gas chromatography vs. their atmospheric boiling points.

topic composition differences of the components. There is a systematic pattern to be observed in Fig. 9-18. Asphaltenes, which contain the highest molecular weight molecules, and tars show a slight ^{13}C enrichment with respect to the total oil; wax and paraffin, containing somewhat lower molecular weight molecules show a slight ^{13}C depletion with respect to the total oil. For the gaseous components butane, propane, ethane and methane (see also Fig. 9-15B) the ^{13}C depletion increases with decreasing molecular weight. Another feature of Fig. 9-18 worth noting is that in most of the oils studied the saturates are systematically depleted in ^{13}C with respect to the aromatic compounds.

Similar trends can be noted in Fig. 9-19:

(1) For distillates obtained below about 100°C the ^{13}C content rises with the increasing distillation temperature, a reflection of the ^{13}C depletion of the low-molecular-weight components.

(2) A maximum ^{13}C content appears in the distillates at about 100°C or somewhat below this temperature.

(3) Above 100°C there is a slow decline in ^{13}C content.

(4) The variation in δ^{13}C of different distillates from the same oil above 100°C are generally smaller than 2‰.

(5) Silverman (1967) found that a pronounced minimum in δ^{13}C occurs in distillates obtained around 450°C. The average molecular weight of this fraction is about 420.

(6) Separated n-paraffins from three oils (curves 4) were found by Welte (1969) to show minima in ^{13}C content in the range from 27 to 31 carbon atoms; n-paraffins of this chain length have atmospheric boiling points in the range 423—463°C. Hence, the minima mimic that observed by Silverman (1967).

The distribution of carbon isotopes in petroleums and their fractions is a result of a number of factors. The more important ones include: (1) the ^{13}C content of the organic matter which serves as a source, this is in turn influenced by a series of other parameters which have been discussed above; (2) diagenetic carbon isotopic composition changes; (3) conditions of accumulation; (4) maturation, thermal alteration due to increased temperature; (5) deasphalting, i.e., precipitation of the heavy asphaltene molecules caused by the addition of natural gas generated within or outside the reservoir; and (6) degradation by processes of water washing, bio-degradation, and oxidation. In view of the multitude of possibilities it will be difficult to isolate one factor as the most relevant.

Compared to any total organic carbon fixed today by photosynthesis and accumulated in recent sediments petroleum shows on the average a depletion in ^{13}C, as pointed out above, older sediments show a similar pattern. This lower ^{13}C content could be attributed to any one or combination of the following: (1) ^{13}C content changes in the surface CO_2 reservoir; (2) changes in the fractionation of photosynthesizing organisms with time; and (3) ^{13}C changes occurring during post-depositional alteration of the organic matter. If the latter should be the case, the change must have occurred soon after the inclusion of the organic matter in the sediment, because both old sediments and oils show similar carbon isotopic compositions. It might be brought about either through the removal of ^{13}C-enriched decomposition products or the preferential retention of particular fractions of the original organic matter. If the preferentially retained compounds would be isotopically lighter than the total organism, then the difference between organic source and petroleum isotopic composition might be explained. Silverman (1964a) suggested that lipids might represent such a fraction. He pointed out that the stability of lipids may enable them to survive adverse geologic conditions better than other organic compounds, and that in terms of elemental composition and structural molecular configuration lipid components more closely resemble petroleum constituents than any other biological compounds. It has been recognized that low-molecular-weight hydrocarbons (aliphatics and aromatics) present in petroleum can be reasonably explained as decomposition fragments of lipid-related compounds. As lipids show the largest ^{13}C depletion of all separated plant components (Fig. 9-6), their preferential retention to form kerogen of sediments and petroleum would be compatible with chemical and isotopic evidence.

There have been several investigations in which rising ^{13}C contents in petroleum of a particular source have been attributed to increasing maturity of the oil in response to a subjection to elevated temperatures (Vredenburgh and Cheney, 1971; Koons et al., 1974; Rogers et al., 1974; Orr, 1974). Sackett et al. (1970) performed experiments evaluating carbon isotope effects during thermal cracking of (n-paraffins): propane, n-butane, n-heptane and n-octane. It was found that the fractionation between liberated methane and source hydrocarbon increased with increasing chain length. In the cracking C—C bonds are broken, and relatively less energy is required to rupture ^{12}C—^{12}C bonds compared to ^{13}C—^{12}C bonds in the same structural position of a molecule. The effect of the difference in energy between these two types of bonds will be more important if the C—C bond dissociation energy is low; with decreasing energy of the C—C bond broken the carbon isotope effect should increase. Study has shown that the terminal bond dissociation energy decreases in a series such as ethane-propane-butane, and the experimentally observed fractionations are hence plausible. It may also be anticipated that as larger and larger molecules are cracked more possibilities exist for the formation of low-energy terminal carbon bonds, and the fractionation effects might consequently increase. Methane formed by thermal cracking of oil at 500°C showed a fractionation of —25.8‰ with respect to the crude oil; a fractionation of —24.4‰ was found for methane produced from a shale under the same conditions. Because the methane liberated in the cracking process is depleted in ^{13}C relative to the starting material the residual will be enriched in ^{13}C and the δ^{13}C trends with increasing thermal alteration can be understood.

Further insight into the oil maturation process has come from studies of Silverman (1964a, b; 1967) (Fig. 9-19). The observed changes in δ^{13}C with boiling point are taken by Silverman as evidence that the lower-molecular-weight hydrocarbons were formed by the decomposition of higher-molecular-weight compounds. The mechanism of light hydrocarbon formation is a dehydrogenation process in which, by introduction of undersaturation and of ring-closure in the parent molecules, low-molecular-weight aromatics might be formed. A polymerization process might also be indicated by which the highly polycyclic compounds such as asphaltenes are produced. The concept is illustrated in Fig. 9-20. On the basis of these ideas one could expect that the aromatics might show a certain ^{13}C enrichment with respect to the source molecules, so should high-molecular-weight secondary products such as tars or certain asphaltenes. Inspection of Fig. 9-18 shows that this trend exists in petroleum separates. If the model is correct one would expect furthermore (Silverman, 1967) that there should be a fraction in the petroleum of high-molecular-weight which shows a minimum ^{13}C content and which could be interpreted as representing the primary organic material. Vacuum distillation experiments by Silverman (1967) revealed a fraction of minimum ^{13}C content at 450°C. The average molecular weight of this fraction is estimated to be 420, and is within the range of that of some lipid components. It is hence suggested that this fraction represents the original source organic

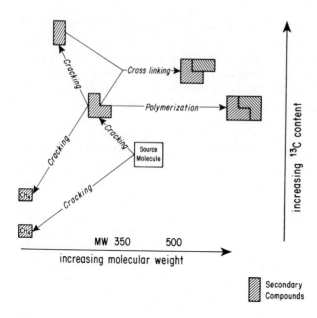

Fig. 9-20. Petroleum maturation (after Silverman, 1964, 1967).

matter in the form of lipid components or lipid-related compounds. Similarly, Welte (1970) observed minimum ^{13}C contents in compounds with 27 to 31 carbon atoms and interpreted these minima tentatively as representing higher concentrations of biologically produced carbon skeletons in this particular molecular weight range.

While carbon isotopic composition changes during thermal alteration appear to be small (1—2‰), leading to ^{13}C enrichment in the residual petroleum, those occurring during deasphalting, water washing and biodegradation appear to be even smaller (Rogers et al., 1974; Bailey et al., 1973).

Most petroleum samples analyzed to date fall in the range —21 to —32‰. In an overall comparison they show generally ^{13}C depletion with respect to most of the total plant carbon photosynthesized, organic matter incorporated into recent sediments, kerogen of older sediments, kerogen in supposed source rocks of the oils, and coals. One possible explanation for the lower ^{13}C contents is the preferential retention of lipid components to form oil; chemical, isotopic and experimental evidence is in accord with this suggestion. Although the total range in $\delta^{13}C$ of oils is sizeable, within a given deposit it is much smaller (1—2‰). Small but systematic $\delta^{13}C$ differences exist between the liquid fractions of oils; in the associated gases the ^{13}C content decreases significantly with decreasing atomic weight. Isotope effects during the maturation process appear to be minor, once accumulated petroleum has a stable isotopic composition.

Natural gas

The term natural gas is usually used to describe the mixture of gaseous compounds, including hydrocarbons and non-hydrocarbon gases which are found in subsurface rock reservoirs. The deposits of natural gas need not necessarily be associated with petroleum occurrences, although they often are. If the latter is the case, such deposits are called associated gases, and their carbon isotopic composition is shown in Fig. 9-15B. In this section we consider the carbon isotopic composition of non-associated natural gases, i.e., those which are not directly connected to a petroleum reservoir.

The hydrocarbons present in such gases include as a major constituent methane (CH_4); smaller amounts of ethane (C_2H_6), propane (C_3H_8), *n*-butane and isobutane (C_4H_{10}) are often, but by no means always, present. Also hydrocarbon compounds that are liquid at normal pressure and temperature are sometimes observed, including iso-pentane, *n*-pentane (C_5H_{12}), *n*-hexane (C_6H_{14}) and *n*-heptane (C_7H_{16}). If appreciable amounts of these compounds are present, the natural gases are termed "wet", if they are absent, or present in small concentrations, the gases are termed dry. In addition to the hydrocarbons other non-combustible gases are present, which generally reduce the quality of the natural gas as fuel. Such gases include CO_2, N_2, H_2S, argon, and helium. In a few occurrences concentrations of CO_2 or helium are high enough to warrant their commercial extraction.

At first sight the occurrence of hydrocarbons may appear most simply related to the decomposition of organic material under thermal stress. However, closer examination of their composition and $^{13}C/^{12}C$ ratios shows that such a derivation may be quite complex. The origin of the extraneous gases is not yet resolved. The presence of helium and argon is attributed to the radioactive decay of uranium, thorium and potassium, however, the manner in which the gases become associated with the natural gas is not always clear. The argon content of natural gases is generally low; some of the higher helium contents have been attributed to the occurrence of radioactive petroleum residual deposits. The H_2S present has been thought to be the product of the activity of sulfate-reducing bacteria which may extend to considerable depth below the present surface; as an alternative the reducing action of liquid bitumen on gypsum has been evoked.

Carbon dioxide can amount to a considerable fraction of natural gas and a multiplicity of sources have been suggested for its origin. Considered have been: emanations from igneous activity, contact and regional metamorphism of carbonates, as well as the oxidation of organic matter by bacteria.

Nitrogen may also be a principal constituent of some natural gas occurrences, and three main sources have been proposed for it: a release from deep-seated magmatic processes, trapped atmospheric nitrogen, release from organic compounds incorporated in the sediments due to bacterial activity or bombardment with alpha particles. Their relative importance is generally not known.

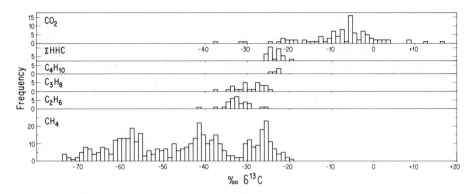

Fig. 9-21. Carbon isotopic composition of carbon-bearing gases of natural gas. Data from Boigk et al. (1976), Brooks et al. (1974), Colombo et al. (1965, 1966, 1969), Dubrova and Nesmelova (1968), Galimov (1969), Müller and Wienholz (1967), Nakai (1960), Sackett (1968), Silverman and Epstein (1958), Stahl (1968a, b), Stahl and Tang (1971), Stahl and Carey (1975), Wasserburg et al. (1963), and Zartman et al. (1961).

Methane is the most abundant species in most natural gases and carbon isotopic composition studies have been undertaken to elucidate its origin. The gas is known to be produced in considerable quantities through bacterial decomposition of cellulose in recent sediments under reducing conditions, and this appears to be one viable source. Others include its production during katagenesis, and metamorphism of coal and dispersed organic matter in sediments as well as the formation during the petroleum maturation process.

In Fig. 9-21 the carbon isotopic composition measured for a number of natural gas occurrences have been compiled. A broad range of $\delta^{13}C$ values is observed for CH_4 (−21 to −76‰) very much wider than that for the other two fossil fuels, petroleum and coal. The tri-modality is in part related to the intense sampling of a few natural gas occurrences. It would be interesting to determine whether it would persist if a larger number of additional gas fields were sampled in detail.

The similarity in $\delta^{13}C$ between CH_4 found in recent sediments and some of the natural gases (−60 to −70‰, Fig. 9-15) in conjunction with the low $\Sigma C_n/C_1$ ratio in both of them has led to the suggestion (Colombo et al., 1966) that such natural gas methanes may be the product of bacterial processes. The isotopically heavy methanes of natural gases (−25‰) fall into the $\delta^{13}C$ range of some of the CH_4 samples obtained from coal samples. This in conjunction with the low $\Sigma C_n/C_1$ ratios of the gases from which this methane was liberated has led to the suggestion that they were formed during the coalification process (Stahl, 1968a, b). Methane of intermediate isotopic composition might be related to the petroleum maturation process. Considering the chemical and isotopic composition variability of gases

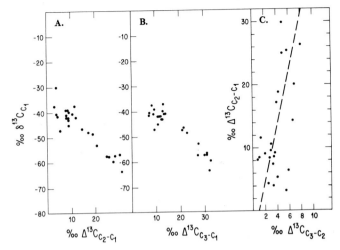

Fig. 9-22. Carbon isotopic composition relationships between methane and higher hydrocarbons in natural gases (data from references to Fig. 9-21).

known to be derived from these sources (Fig. 9-15) it is apparent, however, that such interpretations are not without ambiguities.

Relationships between the methane isotopic composition and the abundance and isotopic composition of higher hydrocarbons, carbon dioxide and nitrogen, as well as the depth of production have been observed and attempts have been made to use these to elucidate the history of natural gas occurrences.

For higher hydrocarbons in natural gas there is a very much smaller variation of $\delta^{13}C$ and a suggestion of a systematic increase in ^{13}C content with increasing hydrocarbon chain length (Fig. 9-21). This is similar to the trend observed for the gases associated with petroleum (Fig. 9-18). The relative constancy of the carbon isotopic composition of the higher hydrocarbons is remarkable compared to the high variability in $\delta^{13}C_{CH_4}$. This is not the result of fortuitous sampling in which $\delta^{13}C_{C_n}$ analyses were carried out only on samples which showed a small range in $\delta^{13}C_{CH_4}$. In Fig. 9-22A, B $\delta^{13}C_{CH_4}$ is compared with $\Delta^{13}C_{C_3-C_1}$ and $\Delta^{13}C_{C_2-C_1}$, significant correlations are observed which result largely from the fact that while $\delta^{13}C_{CH_4}$ varies by about $30^0/_{00}$, $\delta^{13}C$ of C_2H_6 and C_3H_8 vary only by about $8^0/_{00}$. The relationships were understandable if CH_4, C_2H_6 and C_3H_8 were formed in a process in which there was only a small isotope fractionation between C_2H_6 and C_3H_8 and the source organic matter which was relatively insensitive to changes in temperature, whereas the fractionation between CH_4 and the source carbon was large and showed a stronger dependence on temperature. In this case one would expect a correlation between $\Delta^{13}C_{C_2-C_1}$ and $\Delta^{13}C_{C_3-C_2}$. Fig. 9-22 shows a weak, but significant correlation between these variables; one might

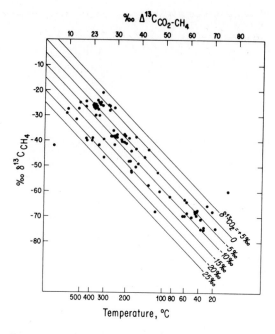

Fig. 9-23. Relationship between methane isotopic composition and the carbon isotope fractionation between CH_4 and CO_2 in non-associated natural gases (data from references to Fig. 9-21). The relationship between carbon isotope fractionation and temperature was derived from the work of Bottinga (1969).

hence suggest that the variation in $\delta^{13}C$ of CH_4 may be in part related to its temperature of formation.

Carbon dioxide shows the largest ^{13}C enrichment of any of the carbon-bearing compounds in natural gases. This trend follows the expectations formed on theoretical grounds (Craig, 1953a; Bottinga, 1969; and Galimov, 1974). Whereas, the lighter isotopic compositions of CO_2 have been readily linked to a derivation from organic matter, values around $-5‰$ have been considered to be indicative of juvenile CO_2, while values around $0‰$ have been attributed to metamorphic decarbonation reactions of sedimentary carbonates. It is very doubtful that such a simplified interpretation can be generally valid. The carbon isotopic composition variability of CO_2 is much less than that of the associated CH_4, so that a significant correlation results if $\delta^{13}C_{CH_4}$ is plotted against $\Delta^{13}C_{CO_2-CH_4}$ (Fig. 9-23). If the two gases undergo isotope exchange the carbon isotope fractionation between them changes as a function of temperature (Craig, 1953a; Bottinga, 1969). For some natural gas occurrences there is an approximate correspondence between the temperatures at which the gases occur or can be expected to have been formed and the temperature deduced from the carbon isotopic composition

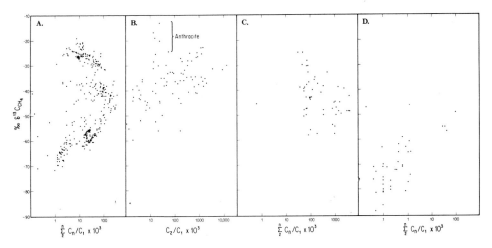

Fig. 9-24. Relationship between methane isotopic composition and the ratio of higher hydrocarbons to methane (data from references to Fig. 9-21). A. Non-associated natural gases. B. Coal gases. C. Petroleum associated gases. D. Gases from recent sediments.

difference between the associated methane and carbon dioxide. In other cases, the deduced temperatures are unreasonably high.

Methane isotopic composition and the abundance of higher hydrocarbons in natural gases may be correlated. Colombo et al. (1966, 1969) observed in Italian natural gases that the methane isotopic composition became systematically heavier with increasing concentrations of higher hydrocarbons (lower part of Fig. 9-24A). The reverse trend was observed by Stahl (1968a, b) for natural gases in Germany (upper part of Fig. 9-24A). A significant difference between the two occurrences is that the CH_4 from the Italian gases generally had $\delta^{13}C$ values below $-50‰$, whereas the methanes from Germany showed ^{13}C enrichments falling above $-30‰$. The two cited studies form the flanks of a broad maximum of higher hydrocarbon contents which occurs in gases with methane isotopic compositions around -35 to $-45‰$. For comparison the corresponding plots for coal gases, petroleum associated gases and deep-sea sediment gases are given in Fig. 9-24B, C, D. Coal gases show a systematic increase in $\delta^{13}C_{CH_4}$ with increasing C_2/C_1 ratio, however, anthracites which are characterized by the highest $\delta^{13}C_{CH_4}$ values contain lowest contents of higher hydrocarbons. Petroleum-associated gases show compositional tendencies similar to non-associated gases, while the gases from deep-sea sediments show generally only very light $\delta^{13}C$ values and hence only one flank of the trend is observed for them. In Fig. 9-25 contours outline the maximum $\Sigma C_n/C_1$ ratio observed for each $\delta^{13}C$ value for the different types of gases. The similarity in the general shape of the curves is notable, and it would be tempting to relate the shift towards heavier $\delta^{13}C$ values of the

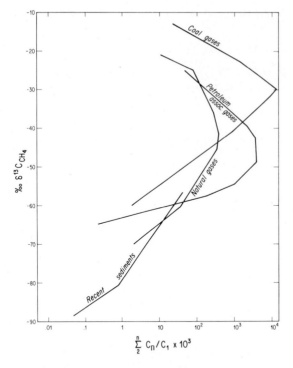

Fig. 9-25. Maximum $\Sigma C_n/C_1$ ratios and methane carbon isotopic compositions for non-associated gases, coal gases, petroleum-associated gases and gases from recent sediments.

curve for coal gases compared to that of the petroleum-associated gases to the difference in source material isotopic composition (coal vs. petroleum, compare Figs. 9-14 and 9-11). However, whether there are true systematic differences remains to be established through more detailed studies. The comparison shows also that in none of the non-associated natural gases analyzed to date have $\Sigma C_n/C_1$ ratios been encountered which are as high as those found in petroleum-associated gases or coal gases. There is a difference of more than one order of magnitude in the maximum $\Sigma C_n/C_1$ ratio between associated and non-associated gases.

Relationships between the methane isotopic composition and N_2/CH_4 ratio of the natural gas have also been found in several gas fields; generally with increasing N_2/C_1 ratio the methane becomes richer in ^{13}C (Fig. 9-26). A broad correlation between $\delta^{13}C_{CH_4}$ and N_2/CH_4 ratio is observed for all analyses, however, this becomes very much tighter if individual gas fields are considered separately, e.g. the Italian gases (Colombo et al., 1966, 1969) or the gases from Germany (Stahl, 1968a, b; Boigk et al., 1976). It is interesting to note that whereas N_2/CH_4 ratios show a positive correlation with $\delta^{13}C_{CH_4}$

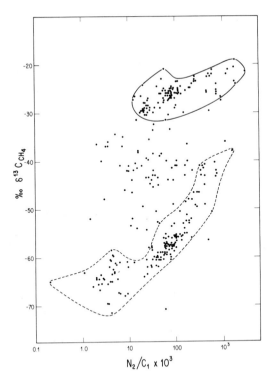

Fig. 9-26. Relationship between methane isotopic composition and nitrogen/methane ratio in natural gases (data from references to Fig. 9-21). Dashed line encloses natural gases from Italy, solid line, natural gases from northwestern Germany.

in both natural gas occurrences, this is not true for the relationship between $\Sigma C_n/C_1$ and $\delta^{13}C_{CH_4}$.

The fact that in some natural gas fields $\delta^{13}C$ of the methane increases with depth has been remarked upon by several authors. Galimov (1969) suggested that this was a general phenomenon and that depth provided a major control on the carbon isotopic composition of natural gas methanes. In Fig. 9-27 $\delta^{13}C_{CH_4}$ has been plotted as a function of production depth. There is a very weak correlation of $\delta^{13}C_{CH_4}$ with depth, however, much less well defined than originally proposed by Galimov (1969).

The isotopic composition of natural gas methane will depend on the ^{13}C content of the source material, the manner in which the methane is produced (bacterial decomposition vs. thermal cracking), the pressure and temperature at which the gas is formed, i.e., the thermal stress to which the source material is subjected as well as the partial pressure of oxygen. Modifications of the isotopic composition of the gases formed may occur during the process of gas migration and accumulation. Although the significance of

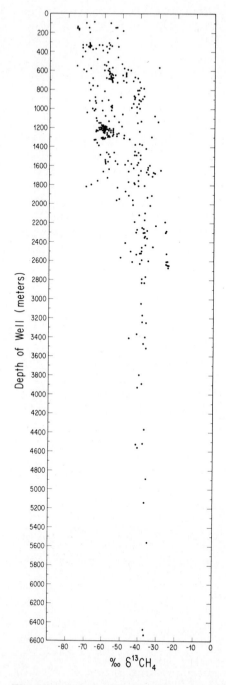

Fig. 9-27. Relationship between methane isotopic composition and depth of production of natural gas (data from references to Fig. 9-21).

all of these factors is generally recognized, their relative importance in the determination of the final chemical and isotopic make-up of the natural gas is not agreed upon.

Colombo et al. (1966) considered two hypotheses to be compatible with the positive correlations between ^{13}C content of the methane and the N_2 and higher hydrocarbon contents observed in natural gases from Italy (lower portions of Figs. 9-24A and 9-26). In the first hypothesis the gases were considered to be a mixture of isotopically light CH_4 with a very low content of higher hydrocarbons of bacterial origin and chemically formed CH_4, isotopically heavier, and associated with larger amounts of higher hydrocarbons. The second proposed hypothesis interprets the observed variations as a result of a separation process occurring during gas migration in which $^{12}CH_4$ migrated faster compared to $^{13}CH_4$. The authors cite experiments in which they found isotope effects in diffusion. Isotope fractionation was largest when diffusion took place through a clay column; smaller but measurable effects were found with sand, limestone or dolomite columns. The lighter isotopic molecule always diffused faster than the heavy one.

For a second Italian gas field Colombo et al. (1969) considered it not likely that the range in $\delta^{13}C_{CH_4}$ observed in conjunction with the gas chemistry changes was the result of mixing of two types of methane. Gaseous diffusion would lead to the observed relationships; however, the authors believe that the range of C_2/C_1 ratios observed in the gases is too wide compared to the range in methane isotopic composition to be explained by gaseous diffusion processes alone. As an alternative it is proposed that diffusion occurred in an aqueous medium in which it is expected that the isotope separation effects are reduced. Hence, the authors suggest that from a qualitative point of view diffusion in an aqueous medium appears to explain satisfactorily the chemical and isotopic relationships observed in the gas field.

Whereas, Colombo et al. (1966, 1969) propose that $\delta^{13}C$ variations are related to the faster rate of diffusion of $^{12}CH_4$, Galimov (1967), May et al. (1968), and Gunter and Gleason (1971) have given evidence that under some circumstances $^{13}CH_4$ may migrate faster than $^{12}CH_4$.

Galimov (1967) studied the carbon isotopic composition of methane which had migrated from an artificial methane storage reservoir. He observed that with increasing distance from the reservoir the ^{13}C content of the methane increased leading him to the suggestion that $^{13}CH_4$ moved more rapidly than $^{12}CH_4$. The phenomenon is thought to be related to an interaction of methane with water. The author suggests that the CH_4 dissolved in water is isotopically lighter than the total methane and he interprets the observed isotopic composition gradient as a result of a preferential dissolution of the $^{12}CH_4$ molecule along the pathway of migration.

May et al. (1968) performed experiments to clarify the carbon isotopic fractionation during the migration of CH_4. N_2/CH_4 mixtures were diffused through a bed of molecular sieve, following the methods of frontal chromatography. Argon was used as eluting gas. The chemical and isotopic composition of the gas leaving the bed were measured as a function of time. It was observed that nitrogen appeared before CH_4 and that the first CH_4 was enriched in $^{13}CH_4$ compared to the initial methane. The proposed explanation for this phenomenon is a preferential adsorption of $^{12}CH_4$ on the molecular sieve. The manner in which the isotopic molecules of methane are separated during gas migration will depend on whether or not adsorption is important in the process; if it is, a $^{13}CH_4$ enrichment is expected in the direction of migration; if the gas can effuse freely, and adsorption is unimportant, $^{12}CH_4$ enrichment will occur in the direction of diffusion. The question whether the isotope effects during adsorption are similar for all materials remains open. Teichmüller et al. (1970) observed in their study of coal gases that the $^{12}CH_4$ is preferentially desorbed and that $^{13}CH_4$ remained preferentially adsorbed on coal. The same observation was made by Friedrich and Jüntgen (1972) who found that CH_4 desorbed from coal is enriched in ^{12}C and that with increasing degree of desorption the desorbed CH_4 becomes isotopically heavier, suggesting a tighter bonding of $^{13}CH_4$ on the coal.

The changes of chemical and isotopic composition of natural gases that have been measured and attributed to migration are manifold, some of the directions of fractionation observed have been summarized in Table 9-4. Correlations between these parameters and noble gas contents have been noted also (Wollanke et al., 1974). The processes affecting the chemical and isotopic composition of gases during migration are complex and not completely understood to date.

Whereas, the authors of the studies mentioned so far have suggested gas migration as a prime factor in determining the carbon isotopic composition

TABLE 9-4

Isotope and chemical effects during gas migration

Parameter	Direction of migration	Reference
$\delta^{13}C_{CH_4}$	heavy $\longrightarrow$ light	Colombo et al. (1969)
	light $\longrightarrow$ heavy	May et al. (1968), Galimov (1967)
$\delta^{15}N_{N_2}$	light $\to$ heavy $\to$ light	Hoering and Moore (1958), Craig (1968)
	light $\longrightarrow$ heavy	Boigk et al. (1976)
	heavy $\longrightarrow$ light	May et al. (1968)
$\sum_{2}^{n} C_n/C_1$	high $\longrightarrow$ low	Colombo et al. (1969), Teichmüller et al. (1970)
N_2/C_1	low $\longrightarrow$ high	May et al. (1968)

of natural gas methane, views of other authors differ (Sackett, 1968; Frank and Sackett, 1969; Frank et al., 1974; Stahl, 1968a,b, 1974a,b, 1975; Boigk and Stahl, 1970; Boigk et al., 1976). Sackett and his co-workers have suggested that migration is not an important factor in the determination of the ^{13}C content of natural gases, and that kinetic isotope effects during the thermal cracking of higher hydrocarbons are of prime importance. It has been shown that in this process $^{12}C-^{12}C$ bonds are broken in preference to $^{12}C-^{13}C$ bonds so that the methane formed during the thermal cracking is enriched in ^{12}C compared to the source material. Although at high temperatures (500—600°C) the fractionation does not vary appreciably with temperature it is suggested that at lower temperatures thought to be characteristic for the maturation of oils the fractionation would be considerably increased and a dependence of the fractionation factor on temperature would have to be considered. Data on the thermal cracking of octadecane led Frank et al. (1974) to propose that there is not a single equivalent source of the CH_4 produced, but that it is probable that the first CH_4 formed comes from the terminal methyl groups of the source molecule, then the long-chain-carbon residue is broken up into smaller fragments. During the decomposition of the long-chained hydrocarbons branching can be expected to occur, which would result in a multitude of terminal methyl groups of varying bond strength and hence isotope fractionation properties. The authors suggest that the thermal cracking of octadecane may serve as a model for the decomposition of the source matter of natural gas methane and explain the isotopic variability of its CH_4.

In a number of studies (Stahl, 1968a, 1974a,b, 1975; Boigk and Stahl, 1970; Stahl and Koch, 1974; Stahl and Carey, 1975; Boigk et al., 1976; Stahl and Tang, 1971), the carbon isotopic composition of natural gas methane has been related to the degree of thermal stress to which the supposed source material has been subjected. Gases studied by these authors generally have $\delta^{13}C$ values heavier than about $-45^0/_{00}$. For the cases for which it has been evaluated the vitrinite reflectance of the supposed source rocks, which increases with increasing thermal stress to which a rock has been exposed, is positively correlated with the ^{13}C content of methane which is thought to have been derived from this organic matter; at the same time the $\Sigma C_n/C_1$ ratio decreases (upper section of Fig. 9-24A). Increasing nitrogen concentration and $\delta^{15}N$ values of the natural gas have also been ascribed to increasing degrees of coalification of the source substance (Stahl and Koch, 1974; Boigk et al., 1976). It should be noted that many of the analyses in the upper portions of Figs. 9-24A and 9-26 (i.e. heavier isotopic compositions) come from the natural gases of northwestern Germany which are thought to be derived in the coalification process from the underlying Carboniferous strata.

Stahl (1974a, b) and Boigk et al. (1976) have developed the following concept for the interpretation of the chemical and isotopic composition of natural gases.

(1) The very young natural gases can be recognized by high CH_4 contents and highest ^{12}C enrichments in the methane.

(2) With increasing thermal stress the released gases show increased higher hydrocarbon and ^{13}C contents.

(3) The thermocatalytic stress on the parent material is determined by subsidence as well as length of maximum burial, and hence there should be some dependence of the chemical and isotopic composition of natural gas on these parameters.

(4) An endpoint to this development is expected when $\delta^{13}C_{CH_4}$ is about $-44‰$.

(5) For natural gases derived from coal the ^{13}C content of methane and the ^{15}N content of nitrogen increase with increasing degree of coalification due to the preferential breaking of bonds involving the light isotopes. At the same time the relative abundance of the higher hydrocarbons decreases and the nitrogen content increases. The unusually strong ^{13}C enrichment of CH_4 in North German gases is explained by the suggestion that the gases which were formed during the early coalification stages of the Carboniferous strata and were isotopically light and rich in higher hydrocarbons were released to the atmosphere prior to the deposition of Permian sediments.

(6) Both thermal maturation as well as migration will influence the final chemical and isotopic composition of natural gases. The nitrogen isotopic composition is thought to reflect mainly the migration history of the gases while the carbon isotopic composition of methane is mainly an indicator of the maturity of the source material.

Natural gases show a very wide compositional range and chemical and isotopic variations may be correlated. The existence of different gas sources, thermal stress and migration are recognized as important factors in establishing the chemical and isotopic make up of natural gases; their relative importance is a matter of active investigation. Chemical and isotopic effects during natural gas formation and migration are manifold and the direction and size of the fractionations may vary depending on factors not yet well understood. However, within restricted fields, compositional variations are more narrowly defined and detailed studies combining chemical and isotopic analyses with proper geologic control may be of considerable value in elucidating the origin and migration of their gases.

THE CARBON ISOTOPIC COMPOSITION OF ATMOSPHERIC COMPOUNDS

Atmospheric carbon-containing gases include CO_2 (320 ppm), CH_4 (1—2 ppm), CO (0.1—1 ppm), CCl_4 (150 ppt) as well as compounds of the Freon group — polyhalogenated derivatives of methane and ethane — (Freon-11, CCl_3F, 100 ppt) (1 ppm = 1 part in 10^6, 1 ppt = 1 part in 10^{12}). The given concentrations represent only approximate ranges; it is known that they may vary with location and season. The sources and sinks of atmospheric CCl_4 are poorly understood. Freon compounds have no known natural sources; they are released to the atmosphere through their use as aerosol propellants and refrigerants. CO_2, CH_4, CO all have a number of natural as well as anthropogenic sources. Although the concentrations of carbon compounds in the atmosphere are low, their effect on the environmental conditions at the surface of the earth may be profound. Hence, a considerable effort has been devoted to develop an understanding of the natural and anthropogenic sources of these compounds and the manner in which they are removed from the atmosphere. Isotopic composition studies have aided in the elucidation of their geochemistry.

The long-term carbon content of the atmosphere is governed by the general geochemical cycle of carbon. Most of the carbon supplied to the earth's surface is by erosion and metamorphism of sediments. Additions of carbon from the mantle are thought to be small; a mechanism for such additions might be volcanic and geothermal activity. The chemical composition of high-temperature gas emanations varies considerably. The gases contain in addition to H_2O, CO_2, CO, H_2, HCl, HF, H_2S, SO_2, N_2, CH_4, He, and Ar other minor constituents. On the basis of hydrogen and oxygen isotopic composition studies one has established that the largest fraction of the water is of meteoric origin (e.g. White, 1974); carbon isotopic composition measurements have been carried out to establish the source of CO_2 and CH_4 of high-temperature gases. The results of such measurements have been

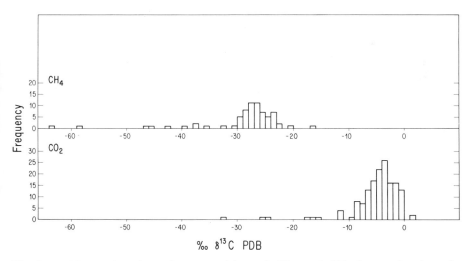

Fig. 9-28. The carbon isotopic composition of CO_2 and CH_4 from volcanic and geothermal areas. Data from Cheminee et al. (1968), Craig (1953b, 1963), Ferrara et al. (1963), Gunter and Musgrave (1971), Hulston and McCabe (1962a, b), Naughton and Terrada (1954), Stahl (1971), Taylor et al. (1967), and Wasserburg et al. (1963).

compiled in Fig. 9-28. In spite of a considerable range in ^{13}C content (−33 to +2‰), most CO_2 samples show carbon isotopic compositions between 0 and −10‰. δ^{13}C values below −12 are rare. If these gases are excluded, the average δ^{13}C value for CO_2 given off in high-temperature discharges is −4.2‰, s = 2.6‰), indistinguishable from that of carbon thought to be derived from deep-seated sources, i.e. carbonate from carbonatites (δ^{13}C = −5.1‰, s = 1.4‰) kimberlites (δ^{13}C = −4.7‰, s = 1.2‰) as well as of diamonds (δ^{13}C = −5.8‰, s = 1.8‰) (Deines and Gold, 1973). Whereas, such a coincidence in carbon isotopic composition is interesting to note, it cannot be taken as evidence of a commonality of origin of these different forms of carbon (Deines and Gold, 1969, 1973). Especially the heavier CO_2 isotopic compositions have been attributed to a decomposition of marine limestones (δ^{13}C = +0.53‰, Keith and Weber, 1964) (Craig, 1953b, 1963, 1963; Hulston and McCabe, 1962a, b; Ferrara et al., 1963); decomposition of continental limestones (δ^{13}C = −4.93‰, Keith and Weber, 1964) and addition of organic carbon have been suggested as explanations for the observed lighter isotopic compositions.

The methane of high-temperature emanations shows a systematic depletion in ^{13}C compared to the CO_2 with which it is associated. This is expected on theoretical grounds (Craig, 1953a; Bottinga, 1969) and has led to attempts to evaluate the ^{13}C fractionation between the two gases in terms of temperatures of equilibration (Craig, 1953a; Hulston and McCabe, 1962a, b; Bottinga, 1969; Gunter and Musgrave, 1971). The actual δ^{13}C values of CH_4 are restricted mainly to a range covering that of reduced carbon in sediments, so that CH_4 could conceivably be derived from such a source. Hence, based on the carbon isotopic composition data alone one may not make a convincing argument whether the CO_2 and the CH_4 of high-temperature emanations result from: (1) juvenile CO_2 with which CH_4 is in isotopic equilibrium; (2) decomposition of carbonates, and formation of CH_4 which is in isotopic equilibrium with the liberated CO_2; (3) decomposition of limestone to form CO_2 and production of CH_4 from the organic matter contained in it. In the latter case, the isotopic fractionation between CO_2 and CH_4 would be a remnant of the fractionation established in the sedimentary environment. Although the mode of origin of CO_2 and CH_4 of volcanic and geothermal gases is debated it appears that as far as additions to the atmosphere are concerned most measurements collected to date indicate that on the average the carbon dioxide from this source has δ^{13}C values between −4 and −5‰ and the methane has on the average carbon isotopic compositions of about −27‰.

The carbon dioxide of the atmosphere (6.2×10^{17} g C) constitutes only a fraction of the total surface carbon reservoir of the earth; most of it is present in the ocean (3.9×10^{19} g C) and the biosphere (1.6×10^{18} g C). The rate at which anthropogenic CO_2 additions enter this reservoir is a matter of active investigation (e.g. Bacastow and Keeling, 1973). The atmospheric CO_2

content estimated by Ekdahl and Keeling (1973) to be 322—323 ppm in 1971 is not constant but shows daily, seasonal as well as secular variations (Keeling, 1958, 1961a; Ekdahl and Keeling, 1973). While the daily and seasonal cycles have been attributed to cycles in the photosynthetic activity, the secular increase of the atmospheric CO_2 content of 0.8 ppm/year (1958 to 1968) is believed to be due to retention of industrial CO_2 in the atmosphere; this has been reduced also from a systematic decrease in the initial ^{14}C content of wood grown since the beginning of the industrial revolution (Suess effect, Suess, 1955; Fergusson, 1958; Houtermans et al., 1967; Lerman et al., 1969). It has been estimated that the present CO_2 level of the atmosphere has been elevated over the pre-industrial concentration (290 ppm) by some 30 ppm, and that about 49% of the contemporary production of CO_2 from the combustion of fossil fuels and kilning of limestone is retained in the atmosphere, whereas, 51% is incorporated into the ocean and land biota (Ekdahl and Keeling, 1973). The total industrial addition of carbon to the surface reservoir is in the order of 1.1×10^{17} g C (Keeling, 1973). The rate at which this addition occurred was 3.6×10^{15} g/year in 1960 and it has been deduced (Bacastow and Keeling, 1973) that the CO_2 addition is growing at 4% per year. In so far as the increased addition of industrial CO_2 may lead to climatic changes secular variations of CO_2 content have attracted considerable attention. However, to predict future climatic changes from predicted future CO_2 concentration increases in the atmosphere is found to be a very uncertain proposition (Machta, 1973).

Early measurements of the carbon isotopic composition of atmospheric CO_2 were carried out by Craig (1953a) on Chicago air, and ranged from -9.9 to $-7.4^0/_{00}$. The lighter values are probably due to industrial contamination; Craig (1953a) assumed that a reasonable value for atmospheric CO_2 might be $-7^0/_{00}$. More detailed studies were undertaken by Keeling (1958, 1961a, b; see also Craig and Keeling, 1963) for air in rural areas over forest and grasslands as well as over marine waters. He observed that the amount of CO_2 in the atmosphere is negatively correlated with the ^{13}C content of the CO_2 (Fig. 9-29). Furthermore, diurnal CO_2 content and isotopic composition variations occur at locations where plant cover is important; minimum $\delta^{13}C$ values and maximum CO_2 contents are attained at night. The correlation may be explained by an admixing of respiration CO_2 with $\delta^{13}C$ of about -26 to $-21^0/_{00}$ to air of 310 ppm CO_2 of $\delta^{13}C = -7^0/_{00}$. Kroopnick (1974) pointed out that average atmospheric CO_2 is in isotopic equilibrium with the dissolved carbon in marine surface waters. Alteration of the normal carbon isotopic composition of atmospheric CO_2 due to anthropogenic sources was observed by Friedman and Irsa (1967). Significant differences in the carbon isotopic composition and CO_2 content were found in samples taken at centrally located points at street level in the lower Manhattan business district. An increase in atmospheric CO_2 was accompanied by an increase in ^{12}C concentration, as a result of additions from automobile exhaust.

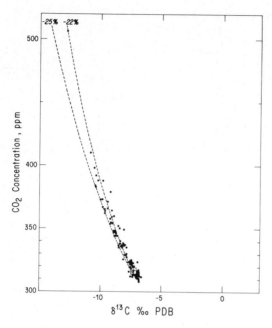

Fig. 9-29. Concentration and carbon isotopic composition of atmospheric CO_2 (Data of Keeling, 1958, 1961a, b). Square represents 77 samples.

In so far as present atmospheric CO_2 possesses an isotopic composition of $-7‰$ and the CO_2 added from the combustion of fossil fuels can be expected to have an isotopic composition between -25 and $-30‰$ one could suggest that the rise in atmospheric CO_2 content due to anthropogenic additions should be accompanied by a parallel increase in ^{12}C content. This rise might be followed if carbon fixed at known times in the past was analyzed, e.g. by the study of the ^{13}C contents of tree rings of known age. A number of authors have consequently investigated $\delta^{13}C$ variations in cross sections through trees grown during the period of industrial expansion. Some of the results are given in Fig. 9-30. The work of Libby and Pandolfi (1974) is not shown as there appears to be a question about the size of the observed effects (Freyer and Wiesberg, 1975). Although the basic principle is simple enough, the detailed evaluation of $\delta^{13}C$ variations of tree rings is not without difficulties.

The definition of the form of carbon analyzed is very important, because it is known that resin channels permeate the wood tissue and contain organic matter which cannot be correlated with the age of the wood structure and may be emplaced at considerably later dates (Freyer and Wiesberg, 1974; Cain and Suess, 1976). As $\delta^{13}C$ differences exist between lignin and cellulose, their relative abundances will influence the measured isotopic composition,

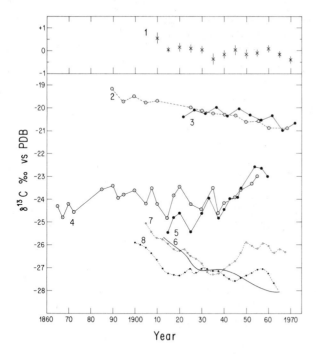

Fig. 9-30. Carbon isotopic composition variations in cross sections of tree trunk. *1* = Freyer and Wiesberg (1975); crosses represent mean of six trees, bars indicate standard deviation, data relative to the mean of six wood samples; *2* = Rebello and Wagener (1976), 0.65 m above ground; *3* = Robello and Wagener (1976), 12.6 m above ground; *4*, *5* = Galimov (1976); *6* = Dequasie and Grey (1970); *7*, *8* = Farmer and Baxter (1974a, b).

if total wood samples are analyzed. It has been found also that even within a single tree ring variations in the order of 1‰ may occur depending on the radial direction in which the sample is taken from the tree cross section. Changes of ^{13}C content with height at constant radial direction are apparently negligible (Freyer and Wiesberg, 1975). The possibility of changes in the isotopic fractionation in the formation of cellulose with age of the tree has also to be evaluated.

Based on the rate of industrial CO_2 additions one might expect a decline in $\delta^{13}C$ of atmospheric CO_2 by $-2.6‰$ since the middle of the last century (Freyer and Wiesberg, 1975). Whereas, the analyses of total wood samples by Dequasie and Grey (1970) and Farmer and Baxter (1974a) might indicate such a drop, the data for tree ring cellulose of Freyer and Wiesberg (1973, 1975) and Rebello and Wagener (1976) appear to indicate a much smaller change. Trends in the opposite direction, i.e. the formation of isotopically heavier (total) wood in the recent past is suggested by the work of Galimov (1976) and Jansen (1962). Similarly Craig (1954a) and Jansen (1962) had observed for trees grown over very much longer time spans that the center portions of the trees appeared to be isotopically lighter than their outermost parts. From the results Jansen (1962) concluded that the ^{13}C variations observed in the tree cross sections are not so much related to ^{13}C variations in the atmospheric CO_2 but to differences in the carbon isotope fractionation during the growth of the trees. Hence, although one can expect that the ^{12}C content of atmospheric CO_2 might increase as a result of anthropogenic additions, the size of the effect and its significance in terms of atmospheric CO_2 concentration changes are still a matter of debate (Freyer and Wiesberg, 1974; Farmer and Baxter, 1974b).

The abundance of atmospheric methane, as well as its sources and sinks have attracted attention because its oxydation in the stratosphere contributes about 50% of the water vapor in this region (Ehhalt, 1973) and hence, influences the temperature in stratosphere and atmosphere. In addition CH_4 oxidation may lead to the formation of CO as an intermediate product, hence, the CH_4, CO and H_2 cycles appear to be closely related. As the ^{14}C activity of atmospheric CH_4 prior to the explosion of nuclear devices was about 80% of modern carbon (Ehhalt, 1973) it has been suggested that the major source of the methane is biogenic and that only 20% results from volcanic activity and anthropogenic sources, which contribute CH_4 free of ^{14}C. The biogenic production occurs mainly in paddy fields, swamps and humid tropical areas and has been estimated (Ehhalt, 1973) to be in the order of $5.2-15.1 \times 10^{14}$ g/year. Available measurements suggest that the gas is in a steady state in the atmosphere with a turnover time of a few years, hence it must be removed at the same rate. The removal is thought to occur mainly through oxidation in the stratosphere and the troposphere.

It is conceivable that $\delta^{13}C$ measurements might aid in the further elucidation of the CH_4 cycle; their number has been, however, very limited (Bainbridge et al., 1961; Ehhalt, 1973). The available three analyses indicate a $\delta^{13}C$ value around $-40^0/_{00}$. If, as suggested, the atmospheric CH_4 results mainly from bacterial decomposition of modern organic matter ($\delta^{13}C = -25^0/_{00}$) a fractionation of only $15^0/_{00}$ would be implied for the bacterial CH_4 production. This value is, however, very much smaller than the measured fractionations for bacterial CH_4 formation (see Table 9-2). The discrepancy may be understood if isotopic effects favoring the light isotope occurred in the CH_4 oxidation.

After the discovery of CO in the atmosphere it was initially thought to be exclusively of anthropogenic origin and a major problem of accumulative global air pollution. In the interim several natural sources and sinks of CO have been identified. The anthropogenic CO production (estimates range from 2×10^{14} to 6.4×10^{14} g/year; Seiler and Zankl, 1976) results mainly from the incomplete combustion of carbon fuels. Natural sources of CO, which add about 2×10^{14} g/year include the microbiological activity in the surface layers of the ocean, the photochemical oxidation of higher hydro-

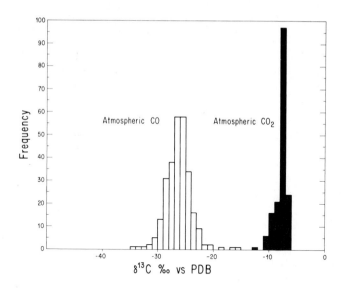

Fig. 9-31. The carbon isotopic composition of atmospheric CO and CO_2 (data from Keeling (1958, 1961a, and Stevens et al., 1972).

carbons in the atmosphere as well as the CO production of higher plants (see Seiler and Zankl, 1976). Very much higher rates of CO production are attributed by some authors (Weinstock and Nicki, 1972; Wofsy et al., 1972; Levy, 1972) to the photochemical oxidation of methane by OH radicals in the troposphere ($20-40 \times 10^{14}$ g/year). There are also several mechanisms by which CO is removed from the lower atmosphere, which include CO consumption by microorganisms in soils (4.5×10^{14} g/year), losses to the stratosphere (1.1×10^{14} g/year) and oxidation in the troposphere by (OH) radicals ($20-50 \times 10^{14}$ g/year, Weinstock and Nicki, 1972; Wofsy et al., 1972; Levy, 1972). It would then appear that the CO cycle is mainly controlled by photochemical processes occurring in the atmosphere. Seiler and Zankl (1976) point out, however, that the estimates of photochemical CO production and destruction are based exclusively on model calculations and may be in actuality much smaller than suggested.

Stevens et al. (1972) attempted to use concentration as well as carbon and oxygen isotopic composition data of atmospheric CO in order to isolate different sources of atmospheric CO. The world average carbon isotopic composition for engine CO, the most important source of anthropogenic CO, is found to be $-27.4 \pm 0.3‰$. Although a significant fraction of the atmospheric CO samples show $\delta^{13}C$ values similar to it (see Fig. 9-31), it is observed by Stevens et al. (1972) that if carbon and oxygen isotopic composition of the gas are considered in combination, it has in most cases compositions which differ significantly from that of CO from automobile exhaust. It

TABLE 9-5
Varieties of atmospheric CO (after Stevens et al., 1972)

Variety	$\delta^{18}O$ (⁰/₀₀ vs. SMOW)	$\delta^{13}C$ (⁰/₀₀ vs. PDB)	Principal seasonal or meridional occurrence	Probable source
I	+5	−30	major species occurring during summer in northern midlatitudes	atmospheric CH_4(?)
II	+5	−24	occurs in varying amounts with I in northern midlatitudes; most abundant in winter and spring; also occurs in marine air of low northern latitudes	unknown
III	+16 to +18	~−28	lesser abundant heavy-oxygen variety occurring during summer in northern midlatitudes	unknown
IV	+26 to +33	−26 to −22	major variety occurring during autumn in northern midlatitudes	degradation of chlorophyll
V	+20 to +25	~−27	major variety occurring during winter and spring in northern midlatitudes	consists partially of anthropogenic emissions; other species unknown

appears that the isotopic composition of atmospheric CO in all seasons is more variable than can be explained by assuming various mixtures of only two isotopic species. On the basis of their measurements the authors identify five isotopically distinct varieties of CO (see Table 9-5). In view of the multitude of factors which may influence them, isotopic composition variations of atmospheric CO are only beginning to be understood. The data of Stevens et al. (1972) show, however, that there are several natural sources of CO, one or two of which combined (I and II in Table 9-5) are much greater than the anthropogenic emissions.

Although the concentration of carbon-bearing compounds in the atmosphere is low they may have profound influence on climatic conditions. As their abundance can be augmented through anthropogenic additions it is of interest to develop an understanding of their sources and sinks. In so far as the general geochemical cycle establishes a given concentration and isotopic composition of them in the atmosphere, and anthropogenic additions, which result mainly through the use of fossil fuels, may be characterized by different and known $\delta^{13}C$, $\delta^{18}O$, or δD ranges, isotopic composition studies may aid in the elucidation of atmospheric pollution problems. The actual application of this principle may, however, not be a simple matter.

REFERENCES

Abelson, P.H. and Hoering, T.C., 1960. The biogeochemistry of the stable isotopes of carbon. *Carnegie Inst. Washington Yearb.*, 59: 158—165.

Abelson, P.H. and Hoering, T.C., 1961. Carbon isotope fractionation in formation of amino acids by photosynthetic organisms. *Proc. Natl. Acad. Sci. U.S.A.*, 47: 623—632.

Allaway, W.G., Osmond, C.B. and Troughton, J.H., 1974. Environmental regulation of growth, photosynthetic pathway and carbon isotope discrimination in plants capable of crassulacean acid metabolism. In: R.L. Bielski, A.R. Ferguson and M.M. Cresswell (Editors), *Mechanisms of Regulation of Plant Growth. R. Soc. N.Z., Bull.*, 12: 195—202.

Allen, R.C., Gavish, E., Friedman, G.M. and Sanders, J.E., 1969. Aragonite-cemented sandstone from outer continental shelf off Delaware Bay: submarine lithification mechanism yields products resembling beachrock. *J. Sediment. Petrol.*, 39: 136—149.

Andreae, M.O., 1974. Chemical and stable isotope composition of the high grade metamorphic rocks from the Arendal area, southern Norway. *Contrib. Mineral. Petrol.*, 47: 299—316.

Bacastow, R. and Keeling, C.D., 1973. Atmospheric carbon dioxide and radiocarbon in the natural carbon cycle, II. Changes from A.D. 1700 to 2070 as deduced from a geochemical mode. In: G.M. Woodwell and E.V. Pecan (Editors), *Carbon and the Biosphere. USAEC Conf.*, 720510: 86—135.

Baertschi, P., 1952. Die Fraktionierung der Kohlenstoffisotopen bei der Absorption von Kohlendioxyd. *Helv. Chim. Acta*, 35: 1030—1036.

Baertschi, P., 1953. Die Fraktionierung der natürlichen Kohlenstoffisotopen im Kohlendioxydstoffwechsel grüner Pflanzen. *Helv. Chim. Acta*, 36: 773—781.

Bailey, N.J.L., Jobson, A.M. and Rogers, M.A., 1973. Bacterial degradation of crude oil — comparison of field and experimental data. *Chem. Geol.*, 11: 203—221.

Bainbridge, A.E., Suess, E. and Friedman, I., 1961. Isotopic composition of atmospheric hydrogen and methane. *Nature*, 192: 648—649.
Baker, D.R. and Claypool, G.E., 1970. Effects of incipient metamorphism on organic matter in mudrock. *Bull. Am. Assoc. Pet. Geol.*, 54: 456—468.
Barker, F. and Friedman, I., 1969. Carbon isotopes in pelites of the Precambrian Uncompahgre formation, Needle Mountains, Colorado. *Geol. Soc. Am. Bull.*, 80: 1403—1408.
Bassham, J.A., 1971. Photosynthetic carbon metabolism. *Proc. Natl. Acad. Sci. U.S.A.*, 68: 2877—2882.
Behrens, E.W. and Frishman, S.A., 1971. Stable carbon isotopes in blue-green algal mats. *J. Geol.*, 79: 94—100.
Bender, M.M., 1968. Mass spectrometric studies of carbon-13 variations in corn and other grasses. *Radiocarbon*, 10: 468—472.
Bender, M.M., 1971. Variations in the $^{13}C/^{12}C$ ratios of plants in relation to the pathway of photosynthetic carbon dioxide fixation. *Phytochemistry*, 10: 1239—1244.
Bender, M.M., Rouhani, I., Vines, H.M. and Black, C.C., 1973. $^{13}C/^{12}C$ ratio changes in crassulacean acid metabolism. *Plant Physiol.*, 52: 427—430.
Boigk, H. and Stahl, W., 1970. Zum Problem der Entstehung nordwestdeutscher Erdgaslagerstätten. *Erdöl, Kohle, Erdgas, Petrochem.*, 23: 325—333.
Boigk, H., Stahl, W., Teichmüller, M. and Techmüller, R., 1971. Inkohlung und Erdgas. *Fortschr. Geol. Rheinl. Westfalen*, 19: 101—108.
Boigk, H., Hagemann, H.W., Stahl, W. and Wollanke, G., 1976. Zur Herkunft und Migration des Stickstoffs Nordwestdeutscher Erdgase aus Oberkarbon und Rotliegend. *Erdöl, Kohle, Erdgas, Petrochem., Ergänzungsband*, pp. 112—139.
Bottinga, Y., 1969. Calculated fractionation factors for carbon and hydrogen isotope exchange in the system calcite—carbon dioxide—graphite—methane—hydrogen—water vapor. *Geochim. Cosmochim. Acta*, 33: 49—64.
Broecker, W.S. and Olson, E.A., 1960. Radiocarbon from nuclear tests, 2. *Science*, 132: 712—721.
Brooks, J.M., Gormly, J.R. and Sackett, W.M., 1974. Molecular and isotopic composition of two seep gases from the Gulf of Mexico. *Geophys. Res. Lett.*, 1: 213—216.
Brown, W.V. and Smith, B.N., 1974. The Kranz syndrome in Uniola *(Gramineae). Bull. Torrey Botan. Club*, 101: 117—120.
Cain, W.F. and Suess, H.E., 1976. Carbon-14 in tree rings. *J. Geophys. Res.*, 81: 3688—3694.
Calder, J.A. and Parker, P.L., 1968. Stable carbon isotope ratios as indices of petrochemical pollution of aquatic systems. *Environ. Sci. Technol.*, 2: 535—539.
Calder, J.A. and Parker, P.L., 1973. Geochemical implications of induced changes in ^{13}C fractionation by blue-green algae. *Geochim. Cosmochim. Acta*, 37: 133—140.
Calder, J.A., Horvath, G.J., Shultz, D.J. and Newman, J.W., 1974. Geochemistry of the stable carbon isotopes in some Indian Ocean sediments. In: *Initial Reports of the Deep Sea Drilling Project, 26.* pp. 613—617.
Cheminee, J., Létolle, R. and Olive, Ph., 1968. Premières données isotopiques sur des fumarolles de volcans Italiens. *Bull. Volcanol.*, 32: 469—475.
Claypool, G.E., Presley, B.J. and Kaplan, I.R., 1973. Gas analyses in sediment samples from legs 10, 11, 13, 14, 15, 18, and 19. In: *Initial Reports of the Deep Sea Drilling Project, 29.* pp. 879—884.
Colombo, U., Gazzarrini, F., Sironi, G., Gonfiantini, R. and Tongiorgi, E., 1965. Carbon isotope composition of individual hydrocarbons from Italian natural gases. *Nature*, 205: 1303—1304.
Colombo, U., Gazzarrini, F., Gonfiantini, R., Sironi, G. and Tongiorgi, E., 1966. Measurements of C^{13}/C^{12} isotope ratios on Italian natural gases and their geochemical inter-

pretation. In: G.D. Hobson and M.C. Louis (Editors), *Advances in Organic Geochemistry 1964*. pp. 279—292.

Colombo, U., Gazzarrini, F., Gonfiantini, R., Kneuper, G., Teichmüller, M. and Teichmüller, R., 1968. Das Verhältnis der stabilen Kohlenstoff-Isotope von Steinkohlen und kohlenbürtigem Methan in Nordwestdeutschland. *Z. Angew. Geol.*, 14: 257—265.

Colombo, U., Gazzarrini, F., Gonfiantini, R., Tongiorgi, E. and Caflisch, L., 1969. In: P.A. Schenck and I. Havenaar (Editors), *Advances in Organic Geochemistry 1968*. pp. 499—516.

Colombo, U., Gazzarrini, F., Gonfiantini, R., Kneuper, G., Teichmüller, M. and Teichmüller, R., 1970a. Das C^{12}/C^{13}-Verhältnis von Kohlen und kohlenbürtigem Methan. *C.R. 6e Congr. Int. Stratigr. Geol. Carboniferous, Sheffield, 1967*, 2: 557—574.

Colombo, U., Gazzarrini, F., Gonfiantini, R., Kneuper, G., Teichmüller, M. and Teichmüller, R., 1970b. Carbon isotope study of methane from German coal deposits. In: G.B. Hobson and G.C. Speers (Editors), *Advances in Organic Geochemistry 1966*. pp. 1—26.

Compston, W., 1960. The carbon isotopic compositions of certain marine invertebrates and coals from the Australian Permian. *Geochim. Cosmochim. Acta*, 18: 1—22.

Cook, A.C., 1961. The carbon isotopic compositions of certain marine invertebrates and coals from the Australian Permian. *Geochim. Cosmochim. Acta*, 22: 289—290.

Craig, H., 1953a. The geochemistry of the stable carbon isotopes. *Geochim. Cosmochim. Acta*, 3: 53—92.

Craig, H., 1953b. Isotopic geochemistry of hot springs. *Geol. Soc. Am. Bull.*, 64: 1410.

Craig, H., 1954a. Carbon-13 in plants and the relationships between carbon-13 and carbon-14 variations in nature. *J. Geol.*, 62: 115—149.

Craig, H., 1954b. Carbon-13 variations in sequoia rings and the atmosphere. *Science*, 119: 141—143.

Craig, H., 1955. Geochemical implications of the isotopic composition of carbon in ancient rocks. *Geochim. Cosmochim. Acta*, 6: 186—196.

Craig, H., 1963. The isotopic geochemistry of water and carbon in geothermal areas. In: E. Tongiorgi (Editor), *Nuclear Geology on Geothermal Areas, Spoleto 1963*. CNR, Pisa, pp. 17—53.

Craig, H., 1968. Isotope separation by carrier diffusion. *Science*, 159: 93—96.

Craig, H. and Keeling, C.D., 1963. The effects of atmospheric N_2O on the measured isotopic composition of atmospheric CO_2. *Geochim. Cosmochim. Acta*, 27: 549—551.

Deevey, E.S. and Stuiver, M., 1964. Distribution of natural isotopes of carbon in Linsley Pond and other New England lakes. *Limnol. Oceanogr.*, 9: 1—11.

Degens, E.T., 1969. Biogeochemistry of stable carbon isotopes. In: G. Eglinton and M.T.J. Murphy (Editors), *Organic Geochemistry*. Springer-Verlag, Berlin, pp. 304—329.

Degens, E.T., Guillard, R.R.L., Sackett, W.M. and Hellebust, J.A., 1968a. Metabolic fractionation of carbon isotopes in marine plankton, I. Temperature and respiration experiments. *Deep-Sea Res.*, 15: 1—9.

Degens, E.T., Behrendt, M., Gotthardt, B. and Reppmann, E., 1968b. Metabolic fractionation of carbon isotopes in marine plankton, II. Data on samples collected off the coasts of Peru and Ecuador. *Deep-Sea Res.*, 15: 11—20.

Deines, P. and Gold, D.P., 1969. The change in carbon and oxygen isotopic composition during contact metamorphism of Trenton limestone by the Mount Royal pluton. *Geochim. Cosmochim. Acta*, 33: 421—424.

Deines, P. and Gold, D.P., 1973. The isotopic composition of carbonatite and kimberlite carbonates and their bearing on the isotopic composition of deep-seated carbon. *Geochim. Cosmochim. Acta*, 37: 1709—1733.

DeNiro, M.J. and Epstein, S., 1976. You are what you eat (plus a few $^0/_{00}$): the carbon isotope cycle in food chains. *Geol. Soc. Am., Abstr. Progr.*, 8: 834—835.

Dequasie, H.L. and Grey, D.C., 1970. Stable isotopes applied to pollution studies. *Mer. Lab.*, Dec. 1970, pp. 19—27.
Deuser, W.G., 1970. Extreme $^{13}C/^{12}C$ variations in Quaternary dolomites from the continental shelf. *Earth Planet. Sci. Lett.*, 8: 118—124.
Deuser, W.G., 1972. Late-Pleistocene and Holocene history of the Black Sea as indicated by stable-isotope studies. *J. Geophys. Res.*, 77: 1071—1077.
Deuser, W.G., Degens, E.T. and Guillard, R.R.L., 1968. Carbon isotope relationships between plankton and sea water. *Geochim. Cosmochim. Acta*, 32: 657—660.
Dubrova, N.V. and Nesmelova, Z.N., 1968. Carbon isotope composition of natural methane. *Geochem. Int.*, 5: 872—876.
Eadie, B.J., 1972. *Distribution and Fractionation of Stable Carbon Isotopes in the Antarctic Ecosystem*. Ph.D. Dissertation, Texas A&M University, College Station, Texas, 119 pp.
Eckelmann, W.R., Broecker, W.S., Whitlock, D.W. and Allsup, J.R., 1962. Implications of carbon isotopic composition of total organic carbon of some recent sediments and ancient oils. *Bull. Am. Assoc. Pet. Geol.*, 46: 699—704.
Ehhalt, D.H., 1973. Methane in the atmosphere. In: G.M. Woodwell and E.V. Pecan (Editors), *Carbon and the Biosphere* USAEC Conf., 720510: 144—158.
Eichmann, R. and Schidlowski, M., 1975. Isotopic fractionation between coexisting organic carbon- carbonate pairs in Precambrian sediments. *Geochim. Cosmochim. Acta*, 39: 585—595.
Ekdahl, C.A. and Keeling, C.D., 1973. Atmospheric carbon dioxide and radiocarbon in the natural carbon cycle, 1. Quantitative deductions from records at Mauna Loa observatory and at the South Pole. In: G.M. Woodwell and E.V. Pecan (Editors), *Carbon and the Biosphere*. USAEC Conf., 720510: 51—85.
Emery, K.O., 1960. *The Sea off Southern California*. Wiley, New York, N.Y., 366 pp.
Erlenkeuser, H., Suess, E. and Willkomm, H., 1974. Industrialization affects heavy metal and carbon isotope concentrations in recent Baltic Sea sediments. *Geochim. Cosmochim. Acta*, 38: 823—842.
Farmer, J.G. and Baxter, M.S., 1974a. Atmospheric carbon dioxide levels as indicated by the stable isotope record in wood. *Nature*, 247: 273—275.
Farmer, J.G. and Baxter, M.S., 1974b. Dendrochronology and ^{13}C content in atmospheric carbon dioxide — reply to comments. *Nature*, 252, 757.
Fergusson, G.J., 1958. Reduction of atmospheric radiocarbon concentration by fossil fuel carbon dioxide and the mean life of carbon dioxide in the atmosphere. *Proc. R. Soc. London, Ser. A*, 234: 561—574.
Ferrara, G.C., Ferrara, G. and Gonfiantini, R., 1963. Carbon isotopic composition of carbon dioxide and methane from steam jets of Tuscany. In: E. Tongiorgi (Editor), *Nuclear Geology on Geothermal Areas, Spoleto 1963*. CNR, Pisa, pp. 277—284.
Fischer, W., Maass, I., Sontag, E. and Süss, M., 1968. $^{12}C/^{13}C$-Untersuchungen an petrologisch, chemisch und technologisch charakterisierten-Braunkohleproben. *Z. Angew. Geol.*, 14: 182—187.
Fischer, W., Maass, I. and Sontag, E., 1970. $^{12}C:^{13}C$-Untersuchungen an Braunkohleinhaltsstoffen. *Z. Angew. Geol.*, 16: 126—130.
Frank, D.J. and Sackett, W.M., 1969. Kinetic isotope effects in the thermal cracking of neopentane. *Geochim. Cosmochim. Acta*, 33: 811—820.
Frank, D.J., Gromly, J.R. and Sackett, W.M., 1974. Revaluation of carbon-isotope compositions of natural methanes. *Bull. Am. Assoc. Pet. Geol.*, 58: 2319—2325.
Freyer, H.D. and Wiesberg, L., 1973. ^{13}C-decrease in modern wood due to the large-scale combustion of fossil fuels. *Naturwissenschaften*, 60: 517—518.
Freyer, H.D. and Wiesberg, L., 1974. Dendrochronology and ^{13}C content in atmospheric carbon dioxide — comments. *Nature*, 252: 757.

Freyer, H.D. and Wiesberg, L., 1975. Anthropogenic carbon-13 decrease in atmospheric carbon dioxide as recorded in modern wood. In: *Isotope Ratios as Pollutant Source and Behaviour Indicators.* IAEA, Vienna, pp. 49—62.

Friedman, L. and Irsa, A.P., 1967. Variations in the isotopic composition of carbon in urban atmospheric carbon dioxide. *Science*, 158: 263—264.

Friedrich, H.U. and Jüntgen, H., 1972. Some measurements of the $^{12}C/^{13}C$-ratio in methane or ethane desorbed from hard coal or released by pyrolysis. In: H.R. von Gaertner and H. Wehner (Editors), *Advances in Organic Geochemistry 1971.* pp. 639—646.

Friedrich, H.U. and Jüntgen, H., 1973. Aussagen zum $^{13}C/^{12}C$-Verhälnis des bei der Inkohlung gebildeten Methans aufgrund von Pyrolyse-Versuchen. *Erdöl Kohle, Ergas, Petrochem.*, 26: 636—639.

Fritz, P., Binda, P.L., Folinsbee, R.E., and Krouse, H.R., 1971. Isotopic composition of diagenetic siderites from Cretaceous sediments in western Canada. *J. Sediment Petrol.*, 41: 282—288.

Galimov, E.M., 1966. Carbon isotopes in soil CO_2. *Geochem. Int.*, 3: 889—897.

Galimov, E.M., 1967. ^{13}C enrichment of methane during passage through rocks. *Geochem. Int.*, 4: 1180—1181.

Galimov, E.M., 1969. Die Isotopenzusammensetzung des Kohlenstoffs in den Gasen der Erdkruste. *Z. Angew. Geol.*, 15: 63—70.

Galimov, E.M., 1973. *Carbon Isotopes in Oil and Gas Geology.* Izdatel'stvo Nedra, Moscva, 384 pp.

Galimov, E.M., 1974. Organic geochemistry of carbon isotopes. In: B. Tissot and F. Bienner (Editors), *Advances in Organic Geochemistry 1973.* pp. 439—452.

Galimov, E.M., 1976. Variations of the carbon cycle at present and in the geological past. In J.O. Nriagu (Editor), *Environmental Biogeochemistry, 1. Carbon, Nitrogen, Phosphorus, Sulfur and Selenium Cycles.* pp. 3—11.

Galimov, E.M., Mamchur, G.P. and Kuznetsova, N.G., 1968. C^{13}/C^{12} ratio in bitumens disseminated in the sedimentary rocks of a Ciscarpathian native sulfur deposit. *Geochem. Int.*, 5: 619—623.

Games, L.M. and Hayes, J.M., 1974. Carbon in ground water at the Columbus, Indiana landfill. In: D.P. Waldrip and R.V. Ruhe (Editors), *Solid Waste Disposal by Land Burial in Southern Indiana.* Indiana University Water Resources Research Center, Bloomington, Ind., pp. 81—110.

Games, L.M. and Hayes, J.M., 1976. On the mechanisms of CO_2 and CH_4 production in natural anaerobic environments. In: J.O. Nriagu (Editor), *Environmental Biogeochemistry, 1. Carbon, Nitrogen, Phosphorus, Sulfur and Selenium Cycles.* pp. 51—73.

Garcia-Loygorri, A., Bosch, B. and Marcé, A., 1974. Étude isotopique du carbone de différentes couches du bassin houiller central des Asturies (Espagne). In: B. Tissot and F. Bienner (Editors), *Advances in Organic Geochemistry 1973.* pp. 859—873.

Gavelin, S., 1957. Variations in isotopic composition of carbon from metamorphic rocks in northern Sweden and their geologic significance. *Geochim. Cosmochim. Acta*, 12: 297—314.

Geissler, C. and Belau, L., 1971. Zum Verhalten der stabilen Kohlenstoffisotope bei der Inkohlung. *Z. Angew. Geol.*, 17: 13—17.

Gunter, B.D. and Gleason, J.D., 1971. Isotope fractionation during gas chromatographic separations. *J. Chromatogr. Sci.*, 9: 191—192.

Gunter, B.D. and Musgrave, B.C., 1971. New evidence on the origin of methane in hydrothermal gases. *Geochim. Cosmochim. Acta*, 35: 113—118.

Hahn-Weinheimer, P., 1960a. Alterseinstufung von eklogitischen Gesteinen mit Hilfe des C^{12}/C^{13}-Isotopenverhältnisses von Graphit- und Karbonat-Kohlenstoff. *Geol. Rundsch.*, 49: 308—314.

Hahn-Weinheimer, P., 1960b. Bor und Kohlenstoffgehalte basischer bis intermediärer Metamorphite der Münchberger Gneissmasse und ihre C^{12}/C^{13} Isotopenverhältnisse. *21st Int. Geol. Congr., Norden*, pp. 431—442.

Hahn-Weinheimer, P., 1966. Die isotopische Verteilung von Kohlenstoff und Schwefel in Marmor und anderen Metamorphiten. *Geol. Rundsch.*, 55: 197—209.

Hahn-Weinheimer, P., Markl, G. and Raschka, H., 1969. Stable carbon isotope compositions of graphite and marble in the deposits of Kropfmühl/NE Bavaria. In: P.A. Schenck and I. Havenaar (Editors), *Advances in Organic Geochemistry 1968*. pp. 517—533.

Hall, R.L., 1967. Those late corn dates: isotopic fractionation as a source of error. *Mich. Archeol.*, 13 (3).

Hathaway, J.C. and Degens, E.T., 1969. Methane-derived marine carbonates of Pleistocene age. *Science*, 165: 690—692.

Hedges, J., 1974. Lignin compounds as indicators of terrestrial organic matter in marine sediments. *Papers, Geophys. Lab. Carnegie Inst. Washington*, 1655: 581—590.

Hendy, C.H., 1971. The isotopic geochemistry of speleothems, I. The calculation of the effects of different modes of formation on the isotopic composition of speleothems and their applicability as paleoclimatic indicators. *Geochim. Cosmochim. Acta*, 35: 801—824.

Hodgson, W.A., 1966. Carbon and oxygen isotope ratios in diagenetic carbonates from marine sediments. *Geochim. Cosmochim. Acta*, 30: 1223—1233.

Hoefs, J. and Schidlowski, M., 1967. Carbon isotope composition of carbonaceous matter from the Precambrian of the Witwatersrand System. *Science*, 155: 1096—1097.

Hoering, T.C., 1962. The stable isotopes of carbon in the carbonate and reduced carbon of Precambrian sediments. *Carnegie Inst. Washington Yearb.*, 61: 190—191.

Hoering, T.C., 1963. The stable isotopes of carbon and the organic compounds in Precambrian sedimentary rocks. *Natl. Acad. Sci., Natl. Res. Counc., Publ.*, 1075: 196—208.

Hoering, T.C., 1967a. Criteria for suitable rocks in Precambrian organic geochemistry. *Papers, Geophys. Lab. Carnegie Inst. Washington*, 1480: 365—372.

Hoering, T.C., 1967b. The organic geochemistry of Precambrian rocks. In: P.H. Abelson (Editor), *Researches in Geochemistry*, 2. Wiley, New York, N.Y., pp. 87—111.

Hoering, T.C., 1974. The isotopic composition of carbon and hydrogen in organic matter of recent sediments. *Papers, Geophys. Lab. Carnegie Inst. Washington*, 1655: 590—595.

Hoering, T.C. and Moore, H.E., 1958. The isotopic composition of the nitrogen in natural gas and associated crude oils. *Geochim. Cosmochim. Acta*, 13: 225—232.

Houtermans, J., Suess, H.E. and Munk, W., 1967. Effect of industrial fuel combustion on the carbon-14 level of atmospheric CO_2. *IAEA Publ.*, SM-87153: 57—67.

Hsu, J.G. and Smith, B.N., 1972. $^{13}C/^{12}C$ ratios of carbon dioxide from peanut and sunflower seedlings and tobacco leaves in light and in darkness. *Plant Cell Physiol.*, 13: 689—694.

Hulston, J.R. and McCabe, W.J., 1962a. Mass spectrometer measurements in the thermal areas of New Zealand, 1. Carbon dioxide and residual gas analyses. *Geochim. Cosmochim. Acta*, 26: 383—397.

Hulston, J.R. and McCabe, W.J., 1962b. Mass spectrometer measurements in the thermal areas of New Zealand, 2. Carbon isotopic ratios. *Geochim. Cosmochim. Acta*, 26: 399—410.

Hunt, J.M., 1970. The significance of carbon isotope variations in marine sediments. In: G.B. Hobson and G.C. Speers (Editors), *Advances in Organic Geochemistry 1966*. pp. 27—35.

Jackson, T.A., Fritz, P. and Drimmie, R., 1976. Stable carbon isotope ratios and chemical properties of kerogen and extractable organic matter in various Pre-Phanerozoic and

Phanerozoic sediments — their interrelations and paleobiological significance (manuscript).
Jacobson, B.S., Smith, B.N., Epstein, S. and Laties, G.G., 1970a. The prevalence of carbon-13 in respiratory carbon dioxide as an indicator of the type of endogeneous substrate. *J. Gen. Physiol.*, 55: 1—17.
Jacobson, B.S., Laties, G.G., Smith, B.N., Epstein, S. and Laties, B., 1970b. Cyanide introduced transition from endogeneous carbohydrate to lipid oxidation as indicated by the carbon-13 content of respiratory CO_2. *Biochim. Biophys. Acta*, 216: 295—304.
Jansen, H.S., 1962. Depletion of carbon-13 in young kauri tree. *Nature*, 196: 84—85.
Jeffrey, L.M., 1969. *Development of a Method for Isolation of Gram Quantities of Dissolved Organic Matter from Sea Water and Some Chemical and Isotopic Characteristics of the Isolated Material*. Ph.D. Thesis, Texas A&M University, College Station, Texas.
Jeffrey, P.M., Compston, W., Greenhalgh, D. and De Laeter, J., 1955. The carbon-13 abundance of limestone and coals. *Geochim. Cosmochim. Acta*, 7: 255—286.
Johnson, R.W. and Calder, J.A., 1973. Early diagenesis of fatty acids and hydrocarbons in a salt marsh environment. *Geochim. Cosmochim. Acta*, 37: 1943—1955.
Junge, C., Seiler, W. and Warneck, P., 1971. The atmospheric ^{12}CO and ^{14}CO budget. *J. Geophys. Res.*, 76: 2866—2879.
Kaplan, I.R. and Rittenberg, S.C., 1964. Carbon isotope fractionation during metabolism of lactate by desulfovibrio desulfuricans. *J. Gen. Microbiol.*, 34: 213—217.
Keeling, C.D., 1958. The concentration and isotopic abundance of atmospheric carbon dioxide in rural areas. *Geochim. Cosmochim. Acta*, 13: 322—334.
Keeling, C.D., 1961a. The concentration and isotopic abundances of carbon dioxide in rural and marine air. *Geochim. Cosmochim. Acta*, 24: 277—298.
Keeling, C.D., 1961b. A mechanism for cyclic enrichment of carbon-12 by terrestrial plants. *Geochim. Cosmochim. Acta*, 24: 299—313.
Keeling, C.D., 1973. Industrial production of carbon dioxide from fossil fuels and limestone. *Tellus*, 25: 174—198.
Keith, M.L. and Weber, J.N., 1964. Carbon and oxygen isotopic composition of selected limestones and fossils. *Geochim. Cosmochim. Acta*, 28: 1787—1816.
Koons, C.B., Bond, J.G. and Peirce, F.C., 1974. Effects of depositional environment and post-depositional history on chemical composition of lower Tuscaloosa oils. *Bull. Am. Assoc. Pet. Geol.*, 58: 1272—1280.
Krejci-Graf, K. and Wickman, F.E., 1960. Ein geochemisches Profil durch den Lias alpha (zur Frage der Entstehung des Erdöls). *Geochim. Cosmochim. Acta*, 18: 259—272.
Kroopnick, P., 1974. Correlation between ^{13}C and ΣCO_2 in surface waters and atmospheric CO_2. *Earth Planet. Sci. Lett.*, 22: 397—403.
Kunkler, J.L., 1969. The sources of carbon dioxide in the zone of aeration of the Bandelier Tuff, near Los Alamos, New Mexico. *U.S. Geol. Surv., Prof. Paper*, 650-B: 185—188.
Kvenvolden, K. and Squires, R., 1967. Carbon isotopic composition of crude oils from Ellenburger Group (Lower Ordovician), Permian Basin, West Texas and eastern New Mexico. *Bull. Am. Assoc. Pet. Geol.*, 51: 1293—1303.
Landergren, S., 1953. Über die Gleichgewichtserscheinungen im Austausch der stabilen Kohlenstoffisotope in marinen Sedimenten. *Z. Naturforsch.*, 8B: 537—541.
Landergren, S., 1955. A note on the isotope ratio $^{12}C/^{13}C$ in metamorphosed alum shale. *Geochim. Cosmochim. Acta*, 7: 240—241.
Landergren, S., 1957. Preliminary note on the isotopic composition of carbon in some Swedish rocks. *Geol. Fören. Stockholm Förhand.*, 79: 274—275.
Landergren, S. and Parwel, A., 1954. On the relative abundance of stable carbon isotopes in marine sediments. *Deep-Sea Res.*, 1: 98—120.

Lebedev, V.S., 1964. Isotopic composition of carbon in petroleum and natural gas. *Geochem. Int.*, 1964: 1075—1082.

Lebedev, V.S., Ovsyannikov, V.M. and Mogilevskiy, G.A., 1969. Separation of carbon isotopes by microbiological processes in the biochemical zone. *Geochem. Int.*, 6: 971—976.

Lebedew, W.C., Owsjannikow, W.M., Mogilewskij, G.A. and Bogdanow, W.M., 1969. Fraktionieuring der Kohlenstoffisotope durch mikrobiologische Prozesse in der biochemischen Zone. *Z. Angew. Geol.*, 15: 621—624.

Lerman, J.C., 1972. Soil-CO_2 and groundwater: carbon isotopic compositions. In: *Proc. 8th Int. Conf. Radiocarbon Dating, Wellington, N.Z., 1972*, pp. D93—D105.

Lerman, J.C. and Queiroz, O., 1974. Carbon fixation and isotope discrimination by a crassulacean plant: Dependence on the photoperiod. *Science*, 183: 1207—1209.

Lerman, J.C., Mook, W.G. and Vogel, J.C., 1969. C-14 in tree rings from different localities. In: I.U. Olsson (Editor), *Radiocarbon Variations and Absolute Chronology, Nobel Symposium 12*. Wiley, New York, N.Y., pp. 275—301.

Levy, H., 1972. Photochemistry of the lower troposphere. *Planet. Space Sci.*, 20: 919—939.

Leythaeuser, D., 1973. Effects of weathering on organic matter in shales. *Geochim. Cosmochim. Acta*, 37: 113—120.

Libby, L.M. and Pandolfi, L.J., 1974. Temperature dependence of isotope ratios in tree rings. *Proc. Natl. Acad. Sci. U.S.A.*, 71: 2482—2486.

Lowdon, J.A. and Dyck, W., 1974. Seasonal variations in the isotope ratios of carbon in maple leaves and other plants. *Can. J. Earth Sci.*, 11: 79—88.

Lyon, G.L., 1973. Interstitial water studies, Leg 15 — chemical and isotopic composition of gases from Cariaco Trench sediments. In: *Initial Reports of the Deep Sea Drilling Project, 20*. pp. 773—774.

Machta, L., 1973. Prediction of CO_2 in the atmosphere. In: G.M. Woodwell and E.V. Pecan (Editors), *Carbon and the Biosphere. USAEC Conf.*, 720510: 21—31.

May, F., Freund, W., Müller, E.P. and Dostal, K.P., 1968. Modellversuche über Isotopenfraktionierung von Erdgaskomponenten während der Migration. *Z. Angew. Geol.*, 14: 376—380.

McKirdy, D.M. and Powell, T.G., 1974. Metamorphic alteration of carbon isotopic composition in ancient sedimentary organic matter: new evidence from Australia and South Africa. *Geology*, 2: 591—595.

Monster, J., 1972. Homogeneity of sulfur and carbon isotope ratios S^{34}/S^{32} and C^{13}/C^{12} in petroleum. *Bull. Am. Assoc. Pet. Geol.*, 56: 941—949.

Müller, P. and Wienholz, R., 1967. Bestimmung der natürlichen Variationen der Kohlenstoffisotope in Erdöl und Erdgas Komponenten und ihre Beziehung zur Genese. *Z. Angew. Geol.*, 13: 456—461.

Müller, P. and Wienholz, R., 1968. Isotopengeochemie und Erdölgenese. *Z. Angew. Geol.*, 14: 176—182.

Murata, K.J., Friedman, I. and Madsen, B.M., 1967. Carbon-13-rich diagenetic carbonates in Miocene formations of California and Oregon. *Science*, 156: 1484—1486.

Murphey, B.F. and Nier, A.O., 1941. Variations in the relative abundance of the carbon isotopes. *Proc. Am. Phys. Soc.*, 59: 771—772.

Nagy, B., 1966. A study of the optical rotation of lipids extracted from soils, sediments, and the Orgueil carbonaceous meteorite. *Proc. Acad. Sci. U.S.A.*, 56: 389—398.

Nakai, N., 1960. Carbon isotope fractionation of a natural gas in Japan. *J. Earth Sci.*, 8: 174—180.

Nakai, N., 1972. Carbon isotopic variation and the paleoclimate of sediments from Lake Biwa. *Proc. Jpn. Acad.*, 48: 516—521.

Naughton, J.J. and Terada, K., 1954. Effect of eruption of Hawaiian volcanoes on the composition and carbon isotope content of associated volcanic and fumarolic gases. *Science*, 120: 580—581.

Newman, J.W., Parker, P.L. and Behrens, E.W., 1973. Organic carbon isotope ratios in Quaternary cores from the Gulf of Mexico. *Geochim. Cosmochim. Acta*, 37: 225—238.

Nier, A.O. and Gulbransen, E.A., 1939. Variations in the relative abundance of the carbon isotopes. *J. Am. Chem. Soc.*, 61: 697.

Nissenbaum, A., 1974. The organic geochemistry of marine and terrestrial humic substances: Implications of carbon and hydrogen isotope studies. In: B. Tissot and F. Bienner (Editors), *Advances in Organic Geochemistry 1973*. pp. 39—52.

Nissenbaum, A. and Kaplan, I.R., 1972. Chemical and isotopic evidence for the in situ origin of marine humic substances. *Limnol. Oceanogr.*, 17: 570—582.

Nissenbaum, A. and Schallinger, K.M., 1974. The distribution of stable carbon isotopes ($^{13}C/^{12}C$) in fractions of soil organic matter. *Geoderma*, 11: 137—145.

Nissenbaum, A., Presley, B.J. and Kaplan, I.R., 1972. Early diagenesis in a reducing fjord, Saanich Inlet, British Columbia, I. Chemical and isotopic changes in major components of interstitial water. *Geochim. Cosmochim. Acta*, 36: 1007—1027.

Oana, S. and Deevey, E.S., 1960. Carbon-13 in lake waters, and its possible bearing on paleolimnology. *Am. J. Sci.*, 258-A: 253—272.

Oehler, D.Z., Schopf, J.W. and Kvenvolden, K.A., 1972. Carbon isotopic studies of organic matter in Precambrian rocks. *Science*, 175: 1246—1248.

Orr, W.L., 1974. Changes in sulfur content and isotopic ratios of sulfur during petroleum maturation — study of Big Horn Basin Paleozoic oils. *Bull. Am. Assoc. Pet. Geol.*, 58: 2295—2318.

Osmond, C.B., 1975. Environmental control of photosynthetic options in crassulacean plants. In: R. Marcelle (Editor), *Environmental and Biological Control of Photosynthesis*. Junk, The Hague, pp. 299—309.

Osmond, C.B., Allaway, W.G., Sutton, B.G., Troughton, J.H., Queiroz, O., Lüttge, U. and Winter, K., 1973. Carbon isotope discrimination in photosynthesis of CAM plants. *Nature*, 246: 41—42.

Osmond, C.B., Ziegler, H., Stichler, W. and Trimborn, P., 1975. Carbon isotope discrimination in Alpine succulent plants supposed to be capable of crassulacean acid metabolism (CAM). *Oecologia*, 18: 209—217.

Park, R. and Dunning, H.N., 1961. Stable carbon isotope studies of crude oils and their porphyrin aggregates. *Geochim. Cosmochim. Acta*, 22: 99—105.

Park, R. and Epstein, S., 1960. Carbon isotope fractionation during photosynthesis. *Geochim. Cosmochim. Acta*, 21: 110—126.

Park, R. and Epstein, S., 1961. Metabolic fractionation of C^{13} and C^{12} in plants. *Plant Physiol.*, 36: 133—138.

Parker, P.L., 1962. The isotopic composition of the carbon of fatty acids. *Carnegie Inst. Washington Yearb.*, 61: 187—190.

Parker, P.L., 1964. The biogeochemistry of the stable isotopes of carbon in a marine bay. *Geochim. Cosmochim. Acta*, 28: 1155—1164.

Paulitsch, P. and Hahn-Weinheimer, P., 1961. ^{12}C- und ^{13}C-Isotope in Metamorphiten. *Naturwissenschaften*, 48: 597—598.

Pearson, F.J. and Hanshaw, B.B., 1970. Sources of dissolved carbonate species in groundwater and their effects on carbon-14 dating. In: *Isotope Hydrology, 1970*. IAEA, Vienna, pp. 271—286.

Presley, B.J. and Kaplan, I.R., 1968. Changes in dissolved sulfate, calcium and carbonate from interstitial water of near-shore sediments. *Geochim. Cosmochim. Acta*, 32: 1037—1048.

Presley, B.J. and Kaplan, I.R., 1971a. Interstitial water chemistry: Deep Sea Drilling Project, Leg 7. In: *Initial Reports of the Deep Sea Drilling Project, 17*. pp. 883—887.

Presley, B.J. and Kaplan, I.R., 1971b. Interstitial water chemistry: Deep Sea Drilling Project, Leg 8. In: *Initial Reports of the Deep Sea Drilling Project, 8*. pp. 853—856.

Presley, B.J. and Kaplan, I.R., 1972a. Interstitial water chemistry: Deep Sea Drilling Project, Leg 9. In: *Initital Reports of the Deep Sea Drilling Project, 9*. pp. 841—844.

Presley, B.J. and Kaplan, I.R., 1972b. Interstitial water chemistry: Deep Sea Drilling Project, Leg 11. In: *Initial Reports of the Deep Sea Drilling Project, 11*. pp. 1009—1012.

Presley, B.J., Petrowski, C. and Kaplan, I.R., 1972a. Interstitial water chemistry: Deep Sea Drilling Project, Leg 10. In: *Initial Reports of the Deep Sea Drilling Project, 10*. pp. 613—614.

Presley, B.J., Petrowski, C. and Kaplan, I.R., 1972b. Interstitial water chemistry: Deep Sea Drilling Project, Leg 14. In: *Initial Reports of the Deep Sea Drilling Project, 14*. pp. 763—765.

Presley, B.J., Culp, J., Petrowski, C. and Kaplan, I.R., 1973a. Interstitial water studies, Leg 15 — major ions, Br, Mn, NH_3, Li, B, Si, and $\delta^{13}C$. In: *Initial Reports of the Deep Sea Drilling Project, 20*. pp. 805—809.

Presley, B.J., Culp, J.H., Petrowski, C. and Kaplan, I.R., 1973b. Interstitial water chemistry, Leg 17. In: *Initial Reports of the Deep Sea Drilling Project, 17*. pp. 515—516.

Presley, B.J., Petrowski, C. and Kaplan, I.R., 1973c. Interstitial water chemistry: Deep Sea Drilling Project, Leg 13. In: *Initial Reports of the Deep Sea Drilling Project, 13*. pp. 809—811.

Presley, B.J., Petrowski, C. and Kaplan, I.R., 1973d. Interstitial water chemistry, Leg 16. In: *Initial Reports of the Deep Sea Drilling Project, 16*. pp. 573—574.

Presley, B.J., Petrowski, C. and Kaplan, I.R., 1973e. Interstitial water chemistry, Leg 12. In: *Initial Reports of the Deep Sea Drilling Project, 16*. pp. 891—892.

Rebello, A. and Wagener, K., 1976. Evaluation of ^{12}C and ^{13}C data on atmospheric CO_2 on the basis of a diffusion model for ocean mixing. In: J.O. Nriagu (Editor), *Environmental Biogeochemistry, 1. Carbon, Nitrogen, Phosphorus, Sulfur and Selenium Cycles*. pp. 13—23.

Rightmire, C.T. and Hanshaw, B.B., 1973. Relation between the carbon isotope composition of soil CO_2 and dissolved carbonate species in groundwater. *Water Resour. Res.*, 9: 958—967.

Roberts, H.H. and Whelan, T., 1975. Methane-derived carbonate cements in barrier and beach sands of a subtropical delta complex. *Geochim. Cosmochim. Acta*, 39: 1085—1089.

Roblot, M., Chaigneau, M. and Majzoube, M., 1964. Détermination du rapport des isotopes stables du carbone dans des phtanites précambriens. *C.R. Acad. Sci. Paris, Ser. D*, 258: 253—255.

Rogers, M.A. and Koons, C.B., 1969. Organic carbon $\delta^{13}C$ values from Quaternary marine sequences in the Gulf of Mexico: a reflection of paleotemperature changes. *Trans. Gulf Coast Assoc. Geol. Soc.*, 19: 529—534.

Rogers, M.A., Van Hinte, J.E. and Sugden, J.G., 1972. Organic carbon $\delta^{13}C$ values from Cretaceous, Tertiary and Quaternary marine sequences in the North Atlantic. In: *Initial Reports of the Deep Sea Drilling Project, 12*. pp. 1115—1126.

Rogers, M.A., McAlary, J.D. and Bailey, N.J.L., 1974. Significance of reservoir bitumens to thermal-maturation studies, Western Canada Basin. *Bull. Am. Assoc. Pet. Geol.*, 58: 1806—1824.

Rosenfeld, W.D. and Silverman, S.R., 1959. Carbon isotope fractionation in bacterial production of methane. *Science*, 130: 1658—1659.

Sackett, W.M., 1964. The depositional history and isotopic organic carbon composition of marine sediments. *Mar. Geol.*, 2: 173—185.

Sackett, W.M., 1968. Carbon isotope composition of natural methane occurrences. *Bull. Am. Assoc. Pet. Geol.*, 52: 853—857.

Sackett, W.M. and Menendez, R., 1972. Carbon isotope study of the hydro carbons and kerogen in the Aquitaine basin, southwest France. In: H.R. von Gaertner and H. Wehner (Editors), *Advances in Organic Geochemistry 1971*. pp. 523—533.
Sackett, W.M. and Rankin, J.G., 1970. Paleotemperatures for the Gulf of Mexico. *J. Geophys. Res.*, 75: 4557—4560.
Sackett, W.M. and Thompson, R.R., 1963. Isotopic organic carbon composition of recent continental derived clastic sediments of Eastern Gulf Coast, Gulf of Mexico. *Bull. Am. Assoc. Pet. Geol.*, 47: 525—528.
Sackett, W.M., Eckelmann, W.R., Bender, M.L. and Bé, A.W.H., 1965. Temperature dependence of carbon isotope composition in marine plankton and sediments. *Science*, 148: 235—237.
Sackett, W.M., Nakaparksin, S. and Dalrymple, D., 1970. Carbon isotope effects in methane production by thermal cracking. In: G.B. Hobson and G.C. Speers (Editors), *Advances in Organic Geochemistry 1966*. pp. 37—53.
Sackett, W.M., Eadie, B.J. and Exner, M.E., 1974a. Stable isotope composition of organic carbon in recent Antarctic sediments. In: B. Tissot and F. Bienner (Editors), *Advances in Organic Geochemistry 1973*. pp. 661—671.
Sackett, W.M., Poag, C.W. and Eadie, B.J., 1974b. Kerogen recycling in the Ross Sea, Antarctica. *Science*, 185: 1045—1047.
Scalan, R.S. and Morgan, T.D., 1970. Isotope ratio mass spectrometer instrumentation and application to organic matter contained in recent sediments. *Int. J. Mass Spectrom. Ion Phys.*, 4: 267—281.
Schidlowski, M., Eichmann, R. and Fiebiger, W., 1976. Isotopic fractionation between organic carbon and carbonate carbon in Precambrian banded ironstone series from Brazil. *Neues Jahrb. Mineral., Monatsh.*, 8: 344—353.
Schopf, J.W., Oehler, D.Z., Horodyski, R.J. and Kvenvolden, K., 1971. Biogenicity and significance of the oldest known stromatolites. *J. Paleontol.*, 45: 477—485.
Sears, S.O., 1976. *Inorganic and Isotopic Geochemistry of the Unsaturated Zone in a Carbonate Terrane*. Ph.D. Thesis, Pennsylvania State University, University Park, Pa.
Seiler, W. and Zankl, H., 1976. Man's impact on the atmospheric carbon monoxide cycle. In: J.O. Nriagu (Editor), *Environmental Biogeochemistry, 1. Carbon, Nitrogen, Phosphorus, Sulfur, and Selenium Cycles*. pp. 25—37.
Silverman, S.R., 1961. Evidence for an age effect in the carbon isotopic composition of natural organic materials. *Geol. Soc. Am., Spec. Publ.*, 68: 272.
Silverman, S.R., 1962. Effect of evolution of land plants on C^{13}/C^{12} ratios of natural organic materials. *J. Geophys. Res.*, 67: 1657.
Silverman, S.R., 1963. Carbon isotope geochemistry of petroleum and other natural organic materials. *Wissensch. Tagung Erdölbergbau, Budapest, 1962*, 2: 328—341.
Silverman, S.R., 1964a. Investigations of petroleum origin and evolution mechanisms by carbon isotope studies. In: H. Craig, S.L. Miller and G.J. Wasserburg (Editors), *Isotopic and Cosmic Chemistry*. North-Holland, Amsterdam, pp. 92—102.
Silverman, S.R., 1964b. Carbon isotope geochemistry of petroleum. *Bull. Am. Assoc. Pet. Geol.*, 48: 547.
Silverman, S.R., 1967. Carbon isotopic evidence for the role of lipids in petroleum formation. *J. Am. Oil Chem. Soc.*, 44: 691—695.
Silverman, S.R. and Epstein, S., 1954. Isotopic composition of carbon in petroleum and other organic constituents of sediments. *Geol. Soc. Am. Bull.*, 65: 1305.
Silverman, S.R. and Epstein, S., 1958. Carbon isotopic compositions of petroleums and other sedimentary organic materials. *Bull. Am. Assoc. Pet. Geol.*, 42: 998—1012.
Smith, B.N., 1971. Carbon isotope ratios of respired CO_2 from caster bean, corn, peanut, pea, radish, squash, sunflower and wheat seedlings. *Plant Cell Physiol.*, 12: 451—455.
Smith, B.N., 1972. Natural abundance of the stable isotopes of carbon in biological systems. *BioScience*, 22: 226—231.

Smith, B.N. and Benedict, C.R., 1974. Carbon isotopic ratios of chemical constituents of *Panicum maximum* L. *Plant Cell Physiol.*, 15: 949—951.

Smith, B.N. and Brown, W.V., 1973. The Kranz syndrome in the gramineae as indicated by carbon isotopic ratios. *Am. J. Bot.*, 60: 505—513.

Smith, B.N. and Epstein, S., 1970. Biogeochemistry of the stable isotopes of hydrogen and carbon in salt marsh biota. *Plant Physiol.*, 46: 738—742.

Smith, B.N. and Epstein, S., 1971. Two categories of $^{13}C/^{12}C$ ratios for higher plants. *Plant Physiol.*, 47: 380—384.

Smith, J.W., Schopf, J.W. and Kaplan, I.R., 1970. Extractable organic matter in Precambrian cherts. *Geochim. Cosmochim. Acta*, 34: 659—675.

Sondheimer, E.W.A., Dence, L.R., Mattick, L.R. and Silverman, S.R., 1966. Composition of combustible concretions of the alewife, *Alosa pseudoharengus*. *Science*, 152: 221—223.

Stahl, W., 1967. Zur Methodik der $^{12}C/^{13}C$ Isotopen- Untersuchungen an Erdgasen. *Erdöl Kohle, Erdgas, Petrochem.*, 20: 556—559.

Stahl, W., 1968a. *Kohlenstoff-Isotopen Analysen zur Klärung der Herkunft nordwestdeutscher Erdgase*. Ph.D. Thesis, Technische Hochschule Clausthal, Clausthal, 98 pp.

Stahl, W., 1968b. Zur Herkunft nordwestdeutscher Erdgase. *Erdöl Kohle, Erdgas, Petrochem.*, 21: 514—518.

Stahl, W., 1968c. Die Verteilung der C^{13}/C^{12}-Isotopenverhältnisse von Torf, Holz, und Holzkohle. *Brennst.-Chem.*, 49: 69—71.

Stahl, W., 1971. Isotopen-Analysen an Carbonaten und Kohlendioxid-Proben aus dem Einflussbereich und der weiteren Umgebung des Bramscher Intrusivs und an hydrothermalen Carbonaten aus dem Siegerland. *Fortschr. Geol. Rheinl. Westfalen*, 18: 429—438.

Stahl, W., 1974a. Carbon isotope ratios of German natural gases in comparison with isotope data of gaseous hydrocarbons from other parts of the world. In: B. Tissot and F. Bienner (Editors), *Advances in Organic Geochemistry 1973*. pp. 453—462.

Stahl, W., 1974b. Carbon isotope fractionations in natural gases. *Nature*, 251: 134—135.

Stahl, W., 1975. Kohlenstoff-Isotopenverhältnisse von Erdgasen Reifezeichen ihrer Muttersubstanzen. *Erdöl Kohle, Erdgas, Petrochem.*, 28: 188—191.

Stahl, W. and Carey, B.D., 1975. Source-rock identification by isotope analyses of natural gases from fields in the Val Verde and Delaware Basins, West Texas. *Chem. Geol.*, 16: 257—267.

Stahl, W. and Koch, J., 1974. $^{13}C/^{12}C$-Verhältnis norddeutscher Erdgase. *Erdöl Kohle, Erdgas, Petrochem.*, 27: 623.

Stahl, W. and Tang, C.H., 1971. Carbon isotope measurements of methane, higher hydrocarbons, and carbon dioxide of natural gases from northwestern Taiwan. *Pet. Geol. Taiwan*, 8: 77—91.

Stevens, C.M. and Krout, L., 1972. Method for the determination of the concentration and of the carbon and oxygen isotopic composition of atmospheric carbon monoxide. *Int. J. Mass Spectrom. Ion Phys.*, 8: 265—275.

Stevens, C.M., Krout, L., Walling, D., Venters, A., Engelkemeir, A. and Ross, L.E., 1972. The isotopic composition of atmospheric carbon monoxide. *Earth Planet. Sci. Lett.*, 16: 147—165.

Stuiver, M., 1964. Carbon isotopic distribution and correlated chronology of Searles Lake sediments. *Am. J. Sci.*, 262: 377—392.

Stuiver, M., 1975. Climate versus changes in ^{13}C content of the organic component of lake sediments during the Late Quaternary. *Quat. Res.*, 5: 251—262.

Stuiver, M. and Deevey, E.S., 1962. Yale natural radiocarbon measurements, VII. *Radiocarbon*, 4: 250—262.

Suess, H.E., 1955. Radiocarbon concentration in modern wood. *Science*, 122: 414—417.

Tauber, H., 1967. Copenhagen radiocarbon measurements, VIII. Geographic variations in atmospheric C^{14} activity. *Radiocarbon*, 9: 246—257.
Taylor, H.P., Frechen, J. and Degens, E.T., 1967. Oxygen and carbon isotope studies of carbonatites from the Laacher See district, West Germany and the Alnö district, Sweden. *Geochim. Cosmochim. Acta*, 31: 407—430.
Teichmüller, R., Teichmüller, M., Colombo, U., Gazzarrini, F., Gonfiantini, R. and Kneuper, G., 1970. Das Kohlenstoff-Isotopen-Verhältnis im Methan von Grubengas und Flözgas und seine Abhängigkeit von den geologischen Verhältnissen. *Geol. Mitt.*, 9: 181—206.
Tregunna, E.B., Smith, B.N., Berry, J.A. and Downton, W.J.S., 1970. Some methods for studying the photosynthetic taxonomy of the angiosperms. *Can. J. Bot.*, 48: 1209—1214.
Troughton, J.H., 1972. Carbon isotope fractionation by plants. In: *Proc. 8th Int. Conf. Radiocarbon Dating. Wellington, N.Z., 1972*, pp. E20—E57.
Troughton, J.H., Hendy, C.H. and Card, K.A., 1971. Carbon isotope fractionation in *Atriplex* spp. *Z. Pflanzenphysiol.*, 65: 461—464.
Van Donk, J. and Mathieu, M., 1973. Interstitial water studies, Leg 15 — isotopic measurements on CO_2 gas from pockets in deep sea cores, Site 147. In: *Initial Reports of the Deep Sea Drilling Project, 20*. pp. 775—776.
Vinogradov, A.P. and Galimov, E.M., 1970. Carbon isotopes and the origin of petroleum. *Geochem. Int.*, 7: 217—235.
Vredenburgh, L.D. and Cheney, E.S., 1971. Sulfur and carbon isotopic investigation of petroleum, Wind River Basin, Wyoming. *Bull. Am. Assoc. Pet. Geol.*, 55: 1954—1975.
Wasserburg, G.J., Mazor, E. and Zartman, R.E., 1963. Isotopic and chemical composition of some terrestrial natural gases. In: J. Geiss and E.D. Goldberg (Editors), *Earth Science and Meteoritics*, North-Holland, Amsterdam, pp. 219—240.
Weber, J.N., 1967. Possible changes in the isotopic composition of the oceanic and atmospheric carbon reservoir over geologic time. *Geochim. Cosmochim. Acta*, 31: 2343—2351.
Weinstock, B. and Nicki, H., 1972. Carbon monoxide balance in nature. *Science*, 176: 290—292.
Welte, D.H., 1969. Determination of $^{13}C/^{12}C$ isotope ratios of individual higher *n*-paraffins from different petroleums. In: P.A. Schenk and I. Havenaar (Editors), *Advances in Organic Geochemistry 1968*, pp. 269—277.
Welte, D.H., 1970. Organischer Kohlenstoff und die Entwicklung der Photosynthese auf der Erde. *Naturwissenschaften*, 57: 17—23.
Welte, D.H., Kalkreuth, W. and Hoefs, J., 1975. Age-trend in carbon isotopic composition in Paleozoic sediments. *Naturwissenschaften*, 62: 482—483.
West, S.S., 1945. The relative abundance of the carbon isotopes in petroleum. *Geophysics*, 10: 406—420.
Whelan, T., Scakett, W.M. and Benedict, C.R., 1970. Carbon isotope discrimination in a plant possessing the C_4 dicarboxylic acid pathway. *Biochem. Biophys. Res. Commun.*, 41: 1205—1210.
Whelan, T., Sackett, W.M. and Benedict, C.R., 1973. Enzymatic fractionation of carbon isotopes by phosphoenolpyruvate carboxylase from C_4 plants. *Plant Physiol.*, 51: 1051—1054.
White, D.E., 1974. Diverse origins of hydrothermal ore fluids. *Econ. Geol.*, 69: 954—973.
Wickman, F.E., 1952. Variations in the relative abundance of the carbon isotopes in plants. *Geochim. Cosmochim. Acta*, 2: 243—254.
Wickman, F.E., 1953. Wird das Häufigkeitsverhältnis der Kohlenstoffisotopen bei der Inkohlung verändert? *Geochim. Cosmochim. Acta*, 3: 244—252.
Wickman, F.E., 1956. The cycle of carbon and stable carbon isotopes. *Geochim. Cosmochim. Acta*, 9: 136—153.

Williams, J.A., 1974. Characterization of oil types in Williston Basin. *Bull. Am. Assoc. Pet. Geol.*, 58: 1243—1252.

Williams, P.M., 1968. Stable carbon isotopes in the dissolved organic matter of the sea. *Nature*, 219: 152—153.

Williams, P.M. and Gordon, L.I., 1970. Carbon-13: carbon-12 ratios in dissolved and particulate organic matter in the sea. *Deep-Sea Res.*, 17: 19—27.

Wofsy, S., McConnell, J.C. and McElroy, M.B., 1972. Atmospheric CH_4, CO and CO_2. *J. Geophys. Res.*, 77: 4477—4495.

Wollanke, G., Behrens, W. and Horgan, T., 1974. Stickstoffisotopenverhältnis von Erdgasen des Emslandes. *Erdöl Kohle, Erdgas, Petrochem.*, 27: 523.

Zartman, R.E., Wasserburg, G.J. and Reynolds, J.H., 1961. Helium, argon, and carbon in some natural gases. *J. Geophys. Res.*, 66: 277—306.

Chapter 10

NITROGEN-15 IN THE NATURAL ENVIRONMENT

RENÉ LÉTOLLE

INTRODUCTION

Nitrogen possesses two stable isotopes: ^{14}N and ^{15}N. The average natural abundance of ^{15}N in air is 0.3663% (Sweeney et al., 1976) and is constant within analytical precision. Therefore it is used as a standard for nitrogen isotope analysis. Several radioactive isotopes are also known to exist, especially: ^{12}N (half-life 11 ms); ^{13}N (half-life 10 min); ^{16}N (7.11 s); ^{17}N (4.16 s); ^{18}N (0.63 s). However, because of their too short half-life, their usefulness in environmental studies is limited and, instead, ^{15}N-enriched compounds are widely used for tracer studies in the biological nitrogen cycle. This application resulted in the publication of a great number of papers primarily focussing on biological and agronomical problems.

In this presentation an attempt is made to summarize the present knowledge on the natural abundance of ^{15}N in organic matter, in soils and the aqueous environment. The need for systematic studies of the natural concentration of ^{15}N and the fractionation factors associated with the most common biochemical reactions was emphasized by Bremner et al. (1963). However, the prime reason for rather extensive studies during the late 60's and early 70's was the question whether ^{15}N concentrations in aqueous nitrogen compounds, primarily nitrate, were indicative of their origin; in other words, whether it was possible to distinguish fertilizer nitrogen from compounds of natural origin (Kohl et al., 1971).

Beyond these studies relatively little has been done despite the fact that nitrogen compounds are widely distributed in terrestrial substances (Fig. 10-1, Table 10-1). However, their concentrations are rather low and both abundance and isotope compositions are, at least for the upper lithosphere, essentially controlled by biological rather than inorganic thermodynamic processes. This may explain the relative lack of interest by geochemists in nitrogen isotope studies. Short summary accounts of nitrogen isotopes were presented by Wlotzka (1972), Hoefs (1973), Létolle (1974) and Sweeney et al. (1976).

The natural concentrations of ^{15}N are usually given as permil ($^o/_{oo}$) differences from atmospheric molecular N_2 and written in the classical delta nota-

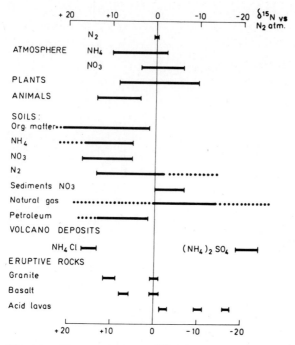

Fig. 10-1. The abundance of ^{15}N in terrestrial materials.

tion where

$$\delta^{15}N = \left[\frac{(^{15}N/^{14}N)_{sample}}{(^{15}N/^{14}N)_{air}} - 1 \right] \times 1000$$

Isotope analyses should be carried out on classical double inlet/double collection mass spectrometers. Interferences for impurities (especially from CO molecules) are to be avoided. Some published data were obtained through single inlet/single collector mass spectrometers and it is possible that they were quoted with somewhat optimistic accuracies.

Preparation of samples should follow the methods given by Bremner and his colleagues in a series of papers (see References), plus the paper by Ross and Martin (1970). The best way for conversion of organic nitrogen is to follow Bremner's suggestions for the use of the Kjeldahl procedure. Nitrate conversion to ammonium also must follow Bremner's procedure, and Devarda's alloy seems to be preferred to Arndt's as a reducing agent. Ammonium sulfate is the best intermediate on which to carry out conversion to molecular nitrogen. Oxidation of ammonium is best carried out by the use of lithium hypobromide, freshly prepared in a helium atmosphere. Traces of water are to be rigorously avoided in the last steps of the purification of molecular N_2,

TABLE 10-1

Relative abundance of nitrogen in various media *

	Relative abundance (ppm)
Cosmos	2—6
Sun	3000
Meteorites	1
Ultrabasic rocks	1 **
Basic rocks	20
Granitic rocks	20
Lithosphere	20
Schists	20
Carbonates	60—720
Soils	1000
Seawater	0.5
with NO_3	0.003—0.7
organic	0.03—0.2
NO_2	$< 10^{-4}$
NH_4	0.05—0.005

Abundance on earth (absolute, in tons)

Hydrosphere	$\simeq 10^{10}$
Lithosphere	$\simeq 2 \times 10^{15}$
Fresh sediments and soils	$\simeq 4.5 \times 10^{12}$
Atmosphere	$\simeq 3.86 \times 10^{15}$

Abundance in organic compounds (%)

Phytoplankton	3
Zooplankton	10
Land plants	2.8
Humus	3.5
Lignite	0.8
Wood	1
Coals	0.8
Fats	0.6
Proteins	18

Atmospheric compounds (volume %)

N_2	75.53
N_2O	$2.5 - 6 \times 10^{-5}$
NO_2	3×10^{-7}
NH_3	1.9×10^{-6}

* Essentially from Rösler and Länge (1972).
**From Becker and Clayton (1977, q.v.).

which must evidently be carried out sheltered from the atmosphere. However, some corrections may be applied to data to correct for minor atmospheric contamination detectable through argon peaks. Such procedures should not be commonly employed. Overall analytical precision can be better than ± 0.2‰.

^{15}N IN NATURE

Neglecting the deep-seated rocks for which almost nothing is known relative to nitrogen geochemistry, atmospheric nitrogen can be considered the principal reservoir for all nitrogen cycles, and as such is an isotopic buffer, essentially in the form of molecular N_2. Small quantities of NH_3 and nitrogen oxides exist also in the atmosphere. In the aqueous environment nitrogen occurs usually as NO_3^-, NO_2^-, NH_4^+ and aminoacids. The same compounds occur in soils where organic matter of heavy molecular weight and its derivates play a most important role.

Sedimentary rocks generally contain small quantities of nitrogen, the chemical form and the isotope composition of which are virtually unknown. An exception are the salpeter deposits found in some deserts, especially in northern Chile, the isotope composition of which is shown in Fig. 10-1. Ammonium chloride is a minor component of volcanic fumaroles and the few isotope analyses available indicate $\delta^{15}N$ values close to +10‰. Other mineral forms of nitrogen are very scarce and have not been studied.

Nitrogen (N_2) associated with natural gas has very variable ^{15}N contents with $\delta^{15}N$ values between —20 and +45‰. Despite these large variations it has been shown that individual gas fields have characteristic ^{15}N contents. Smith and Hudson (1951), Pilot (1963), Bokhoven and Theuwen (1966), Eichmann et al. (1971), Behrens et al. (1971), Wollanke et al. (1974), Stahl et al. (1975) coupled in their investigations of natural gas fields in northwestern Germany ^{15}N and ^{13}C determinations. They observed three principal groups with $\delta^{15}N$ values of about —10, +5 and +17‰ (Fig. 10-3). Their interpretation is that if a natural gas is still associated with its parent organic matter then the N_2 originating during thermal cracking increases its ^{15}N content as the maturity of the organic matter increases. Such a nitrogen has isotope compositions around —10‰. The migration of gas within porous media results in decreasing ^{15}N contents with increasing migration distance, since diffusion has a fractionating effect (see below). Two different migration pathways lead to the clustering of values around +5‰ and +17‰.

Almost nothing is known about nitrogen isotopes in magmatic and metamorphic rocks, although a few data have been published by Mayne (1957) and Volynets et al. (1967) (Fig. 10-1). A large scatter is observed and, if real, may be attributed to migration of a gas phase as a consequence of magma decompression. This would result in a preferential loss of the lighter isotope and would explain why nitrogen in deep-seated rocks (basalts, garnet lherzo-

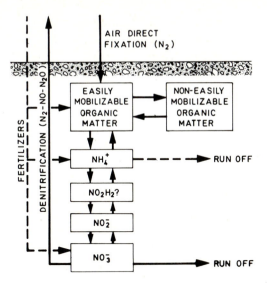

Fig. 10-2. Diagrammatic representation of the geochemical cycle of nitrogen in the upper lithosphere.

lites) are enriched by about 17‰ (Becker and Clayton, 1977).

Many more data are at present available for the biochemically controlled part of the natural nitrogen cycle (Fig. 10-2). The cycle functions primarily in soils and the hydrosphere where most of the reactions between the primary source of nitrogen (the atmosphere) and its ultimate sink (the atmosphere, too) occur. The various steps within the cycle are biochemically controlled whereby the external appearance of the cycle indicates an overall steady state. However, each reaction within the cycle may bring about isotope fractionations with the result that the different compounds produced during different steps may also have distinct isotopic compositions. This does not signify that the isotopic composition of each compound is characteristic to the point where it can be used for direct source and process identification, although, under favorable circumstances, it may be possible to trace the history of a compound through its isotopic composition.

In the following the discussion will essentially be limited to isotope effects and ^{15}N abundances in reactions and compounds of the "biologic" nitrogen cycle. Too little is known about the controls of ^{15}N concentrations in various rocks; and fossil fuel studies warrant a more comprehensive analysis than can be offered here, including carbon and hydrogen isotope analyses.

TABLE 10-2
Equilibrium fractionation factors for various nitrogen species

	Enriched species						
	NO_3^- (aq)	NO_2 (gas)	NO_2^- (aq)	NO (gas)	N_2O (gas)	N_2 (gas)	NH_3 (gas)
NO_2 (gas)	1.073 (25) [a] 1.053 (25) [a] 1.057 (25) [b]						
NO_2^- (aq)	1.0744 (0) [k] 1.0722 (10) [k] 1.0685 (15) [k] 1.0664 (20) [k] 1.050 (25) [d] 1.090 (25) [a] 1.071 (25) [e]	1.015 (25) [a] 1.034 (25) [a]					
NO (gas)	1.096 (25) [a] 1.086 (0) [f] 1.060 (25) [c] 1.099 (25) [b]	1.021 (25) [a] 1.040 (25) [a] 1.0447 (0) [h] 1.0390 (25) [h]	1.066 (25) [a] 1.027 (25) [g,**] 1.024 (25) [g,***]				
N_2O (gas)		1.00098 (0) [h] 1.00063 (25) [h]					
N_2 (gas)	1.0772 (0) [k] 1.0678 (20) [k]	1.0295 (0) [h] 1.0250 (25) [h]		1.016 (0) [i] 1.007 (327) [i]	1.0285 (0) [h] 1.0243 (25) [h]		
NH_3 (gas)		1.0414 (0) [h] 1.0355 (25) [h]			1.0318 (0) [h] 1.0349 (25) [h]	1.0115 (0) [h] 1.0103 (25) [h]	
NH_4^+ (gas)	1.0485 (0) [k] 1.0453 (10) [k] 1.0438 (15) [k] 1.0422 (20) [k] 1.041 (25) [k]			1.042 (0) [i] 1.036 (25) [i] 1.027 (127) [i]		1.025 (0) [i] 1.022 (25) [i] 1.015 (127) [i]	1.039 (0) [i] 1.031 (25) [j] 1.024 (127) [i]

* Numbers between parentheses refer to the temperature in °C.
** Calculated.
*** Measured.

[a] Spindel (1954), two values; [b] Mahenc (1965); [c] London (1961); [d] Brown and Drury (1968); [e] Begun and Fletcher (1960); [f] Stern et al. (1960); [g] Brown and Drury (1969); [h] Richet (1976); [i] Urey (1947); [j] Thode and Urey (1939); [k] in Myake and Wada (1971).

ISOTOPE FRACTIONATIONS

Isotope fractionations under equilibrium conditions

For a limited number of reactions of geochemical importance the thermodynamically controlled isotope fractionation factors have been determined either theoretically or experimentally. They are listed in Table 10-2. Considerable discrepancies exist between different authors, primarily because only reactions used in the preparation of ^{15}N-enriched compounds have been thoroughly investigated. Much work has thus to be done for natural systems (Fig. 10-2) and at present the theoretical values given by Richet (1976) and Richet et al. (1977), which are based on recent spectrographic data and improved mathematical treatment are to be preferred over the limited number of experimentally determined isotope effects.

Kinetic isotope effects

In preceding chapters it has been shown that isotope fractionations caused by kinetic processes as they occur in biological systems can be quite different from theoretically predicted ones. The isotope effects discussed below are those between reactants and instantaneous products and are given as the ratio of reaction rates K_1 and K_2 between each isotope species. The fractionation factors are defined as $\alpha = K_1/K_2$, independant of reaction order, and the enrichment factor is defined as $\epsilon = (\alpha-1) \times 1000$.

Inorganic physicochemical effects. Diffusion through porous and aqueous media, dissolution of molecular nitrogen in water, evaporation of volatile nitrogen compounds and ion exchange are the principal isotope fractionating processes to be considered here.

It has been mentioned above that the migration of natural gas results in decreasing ^{15}N contents with the migration distance, and similar effects have been observed for ^{13}C and ^2H variations in hydrocarbons of natural gas (Colombo et al., 1965). These isotope effects are strictly mass-dependent and the kinetic fractionation factors are defined as $\alpha = (M_1/M_2)^{1/2}$, where M_1 and M_2 are the masses of light and heavy molecules respectively.

The diffusion of nitrogen in porous materials has been studied experimentally by Craig (1968) who demonstrated that substantial fractionations were produced when nitrogen was laterally injected in a flowing stream of another gas, e.g. CO_2. This carrier diffusion effect has been invoked by Stahl et al. (1975) in their study of natural gas quoted above (Fig. 10-3).

It was suggested above that a large isotopic difference may exist between the nitrogen in mantle rocks and the atmosphere, which may have to be explained by diffusive degassing since early Precambrian times, and one may search for a similar explanation to describe the reason for the apparently un-

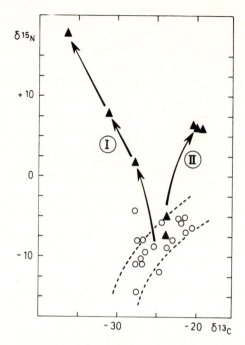

Fig. 10-3. Evolution of ^{15}N and ^{13}C of natural gas with migration in reservoirs from the Rothliegende (triangles) and Upper Carboniferous (circles) from northern Germany. Path I represents an essentially horizontal migration from the original source (Upper Carboniferous) to the present reservoir, with important isotope fractionation. Path II represents a much shorter migration way, in a different geological setting, with an increase of both nitrogen and methane content of the gas.

usual isotopic composition of nitrogen in the atmosphere of Mars (McElroy et al., 1976). Due to the low gravity at the surface of this planet, which favors its degassing, most of the lighter gases of the primeval atmosphere have been lost, leading to the present $^{15}N/^{14}N$ ratio of 0.6 ± 0.1% (earth atmosphere: 0.366%).

The dissolution of molecular nitrogen in water has been studied by several investigators and small fractionations effects have been observed. Benson and Parker (1961) and Klots and Benson (1963) observed that dissolved N_2 is enriched by 0.85 ± 0.10‰ with respect to the gaseous phase. The same value was observed by W.G. Mook (personal communication). Indirect determinations by Myiake and Wada (1967) gave an enrichment of 0.99‰ at ordinary temperatures. Although small, this effect will be of significance for studies relative to denitrification nitrogen in waters, although such investigations have yet to be undertaken.

Freyer and Aly (1976) document the importance of NH_3 volatilization during the application of ammonium fertilizers. If NH_3 "evaporates" from a NH_4^+ solution, it will be depleted in ^{15}N with respect to the original solution.

This escape is unidirectional and cannot be compared with the evaporation of water since no exchange between liquid and gas phases occurs. The isotope fractionations show very large variations which depend on the NH_4 content of the solution and its physicochemical conditions. Preliminary data are available and, for example, A. Mariotti (personal communication) measured an ϵ of $-20^0/_{00}$ at $20°$ C, which is close to a value deductible from observations by Kreitler (1974). However, these data are the results of a combination of two reactions: the oxidation of urea under the influence of *Micrococcus ureae* then ammonia volatilization.

Ion exchange in soils is also possibly responsible for some fractionation effects affecting NH_4^+. Contrarily to nitrite or nitrate ions, NH_4^+ readily exchanges with clays or the so-called "humic-clay complex" of soils. Isotope fractionations between exchange resins and a liquid phase are known to exist, with an enrichment between 5 and $25^0/_{00}$ for the solid phase (London, 1961). The only data available for natural systems are those of Delwiche and Steyn (1970). Their study on fractionation effects between aqueous solutions and kaolinite clays are, however, not entirely conclusive about the importance of isotopic discrimination in such inorganic natural systems.

Biochemical processes. Data available to date indicate that the $^{15}N/^{14}N$ ratio in biological compounds differs considerably from those one would expect if isotope equilibria were existing. Furthermore those reactions of the biological nitrogen cycle which have been investigated often yield contradictory data and the discussion is difficult since the kinetics of most biochemical reactions involving nitrogen is yet poorly understood. However, some of the better known isotope effects and processes which cause nitrogen isotope fractionations in biologic systems are presented below.

Kinetic fractionation effects depend on the relative speed of the chemical reaction for the two isotopes. These are a function of environmental conditions, such as temperature, humidity, oxygen and water content of the soil, its content in organic matter and mineralogical composition. Furthermore many of the isotope effects observed in experiments and in nature are "overall" fractionations and the resultant of several successive reactions.

The fixation of molecular nitrogen by living organisms was first investigated by Hoering and Ford (1960). They could not detect within their analytical precision any fractionation between atmospheric nitrogen and synthetized organic matter. However, later experiments by Delwiche and Steyn (1970) show depletion in ^{15}N by about $5^0/_{00}$ for organic matter; Laishley et al. (1975) found spurious enrichments in ^{15}N from +5 to $-8^0/_{00}$ and Ohmori (quoted by Wada et al., 1975, p. 143) notes enrichments from -1 to $-3^0/_{00}$. When *Leguminosae* symbiotically fix atmospheric nitrogen, fresh organic matter is impoverished by $1^0/_{00}$ (Amarger et al., 1977). Although available data are scarce, it appears that direct assimilation of N_2 slightly favors the lighter isotope. Indirect evidence is provided by the negative ^{15}N

values found in fresh litter on forest soils where the use of N_2 as a nitrogen source supersedes that of nitrate or ammonium (Riga et al., 1971).

When nitrate is available as nitrogen source, the freshly synthetized organic matter also shows an impoverishment of several permil (Meints et al., 1975). This result has to be confirmed.

As for catabolic pathways, the fractionation between organic matter and NH_4^+ was investigated through soil incubation experiments (Feigin et al., 1974a; Freyer and Aly, 1975), which showed that NH_4^+ produced by "organic matter mineralization" is depleted in ^{15}N by about $5-7^0/_{00}$. Focht (1973) arrived to a theoretical depletion of $4.6^0/_{00}$, assuming that $\alpha = (M_1/M_2)^{1/2}$, where M_1 and M_2 are, respectively, the masses of the light and heavy molecules of tryptophane, the heaviest natural amino acid. Modelization of an experiment by Mariotti et al. (1977a) predicts a somewhat greater depletion with an ϵ-value of about $-10^0/_{00}$.

During the oxidation from NH_4^+ to NO_3^- a further depletion in ^{15}N by several permil has been observed (Delwiche and Steyn, 1970; Freyer and Aly, 1975). Similarly, Myiake and Wada (1971) noticed depletions between -5 and $-21^0/_{00}$ in NO_3^- produced during the oxidation of ammonia by marine bacteria.

This oxidation involves intermediate steps, during which nitrite is formed. It is difficult to investigate this step since it is almost impossible to extract nitrite without modifying it. Theoretical considerations by Shearer et al. (1974a), however, predict that the step $NO_2^- \rightarrow NO_3^-$ should be non-fractionating. On the other hand, Delwiche and Steyn (1970) found that during the oxidation of ammonia by *Nitrosomonas europea* NO_2^{2-} was depleted by $26^0/_{00}$ with respect to the ammonium source. Therefore, one could suggest that during the conversion of NH_4^+ to NO_3^- only the first step $NH_4^+ \rightarrow NO_2^-$ is responsible for the fractionation of nitrogen isotopes.

Denitrification, which involves the transformation of NO_3^- into molecular N_2, is carried out by several bacteria, especially *Thiobacillus denitrificans*. Several intermediate steps are known to exist, during which different nitrogen oxides are formed. Isotope effects during denitrification were studied by Cook et al. (1970) and Wellman et al. (1968), who found complicated kinetic effects. The mean overall fractionation effect was about $-20^0/_{00}$, but it appears that this reaction, as many others, is step-controlled. Various still unknown isotope effects must occur at different reaction levels. Unpublished data by the author obtained on dissolved nitrogen and nitrate in an aneorobic pond environment also led to a value of about $-20^0/_{00}$ at 20°C, whereas Myiake and Wada (1971) and Wada et al. (1975) found values ranging from -1 to $-20.7^0/_{00}$. The latter data indicate that environmental conditions have a significant influence on the process and the associated fractionation effects, which is supported by data from Cline and Kaplan (1975) who found for the kinetic fractionation factor in east tropical North Pacific waters a much larger ϵ-value of about $-40^0/_{00}$.

Blackmer and Bremner (1977) studied the evolution of nitrate in soils treated with glucose under an helium atmosphere and found an isotope effect of $-14 \pm 3^0/_{00}$ during the transition of NO_3^- to NO_2^-. Most of the evolved gas occurred in the form of nitrogen oxides and isotope effects as they may occur during the reduction of these oxides to molecular nitrogen have not yet been investigated.

^{15}N IN ORGANIC MATTER AND SOILS

Animal matter is usually enriched in ^{15}N if compared to vegetal matter (Fig. 10-1; Hoering, 1955; Gaebler et al., 1963). This is due to the catabolic pathways which favor the elimination of small nitrogen molecules depleted in ^{15}N (Fig. 10-4). On the other hand, as seen above, vegetal organic matter is normally slightly depleted in ^{15}N with respect to the nitrogen source. This depletion is small since it appears that direct assimilation of N_2 by plants through symbiotic bacteria or free micro-organisms would give "fresh" organic matter. However, if the nitrogen source is nitrate or ammonia, with variable ^{15}N content from different assimilation pathways, a full spectrum of $\delta^{15}N$ values for plants can be expected.

Decaying plant material is the main nitrogen source in natural soils, whereas nitrate contributes only about 1% of the "total soil nitrogen"; ammonia and nitrite occur in normal well aerated soils only as transient components in very small concentrations.

All isotope data available for organic matter in soils refer to "total organic matter" (Table 10-3) and little or no information about isolated fractions has yet been published. However, during decay, when complex organic matter is destroyed, and smaller organic molecules are extracted from it, this process may lead to the enrichment in ^{15}N of the remaining complex organic matter, for isotopic balance reasons.

Soils of different origins have been studied by Cheng et al. (1964) and Bremner and Tabatabai (1973), who found a large scatter of values, from -4.5 to $+17^0/_{00}$. These data were complemented by Shearer et al. (1974b), Rennie and Paul (1975), Riga et al. (1971), Freyer and Kasten (1976), Delwiche and Steyn (1970), Domenach and Chalamet (1977), Bardin et al.

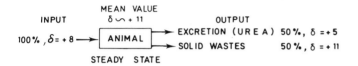

Fig. 10-4. A possible model to explain how animal organic matter is enriched in ^{15}N due to the elimination of low-^{15}N urea.

TABLE 10-3

Total soil ^{15}N analyses

Number of analyses	Mean	Extreme values	
28	+ 6.3 ± 5.2	−1.0, +17.0	Cheng et al. (1964)
16	− 0.2 ± 2.1	−4.4, +3	Bremner and Tabatabai (1973)
29	+ 7.8 ± 1.9	−4.3, +11.9	Shearer et al. (1974b)
11	+10.9 ± 2.1	+4.6, +12.2	Rennie and Paul (1975)
11	+ 4.3 ± 1.0	+1.8, +5.2	Freyer and Kasten (1976)
17		+6.5, +3	Riga et al. (1971)
14		+4.6, +8.5	Wada et al. (1975)
50		+2.5, +8.3	Mariotti et al. (1977)

(1977), Black and Waring (1977), and Mariotti et al. (1977b). On the whole, these authors find a more narrow spectrum of values, whether one considers the types of soils or their differentiation with depth.

Fig. 10-5A shows typical soil profiles, with low δ^{15}N values at the surface and increasing ^{15}N contents with depth. Similar trends were found in most studies, and regardless of the δ^{15}N value in surface horizons, a common value between +5 and +7°/$_{00}$ is approached with depth.

Mariotti et al. (1977b) undertook a detailed study of soils developed on the same substratum (flysch), but at different altitudes. Good correlations between δ^{15}N of the total soil organic matter with total carbon, total nitrogen and C/N ratios were observed; however, these relations differ for different types of humification processes (Fig. 10-5B,C). The observed profiles are interpreted as being the result of two main factors:

(1) A different state of "diagenesis" of organic matter in soils: decay produces isotopically light nitrogen compounds and the remaining organic matter is therefore enriched in ^{15}N. The more evolved the remaining organic matter is, the higher appears their ^{15}N content, which is supported by preliminary data on isolated fractions of soil organic matter (Mariotti, personal communication).

(2) The migration and accumulation of decay products, such as humin, fulvin, humic and fulvic acids among others, in different horizons.

Farming practices appear to enhance bacterial activity and thus influence the sequence and intensity of events observed in virgin soils. Therefore cultivated areas are in general somewhat richer in ^{15}N than uncultivated ones.

An important problem is the impact of artificial fertilizers may have on

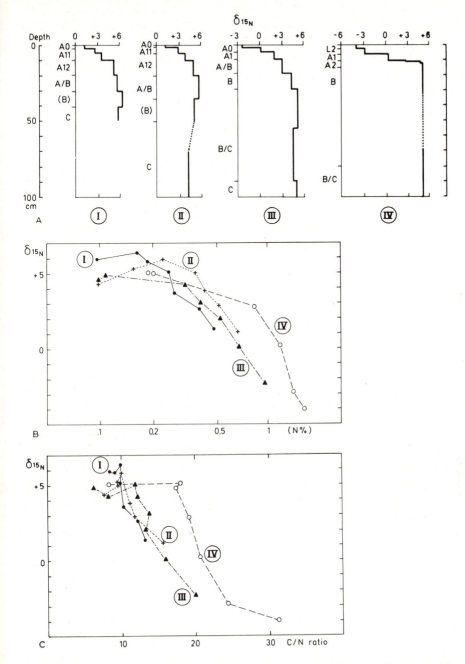

Fig. 10-5. Evolution of $\delta^{15}N$ of organic matter in soils grown on the same substratum (flysch) at different altitudes (Avoriaz, France). A. Evolution with depth for each profile, with horizons A to C. B. Relationship between $\delta^{15}N$ and nitrogen contents of soil organic matter. C. Relationship between soil organic matter and the C/N ratio. Explanation of Roman numerals: I = brown eutrophic (altitude 800 m); II = brown mesotrophic (1100 m); III = brown ochreous (1400 m); IV = ochreous podzolic (1800 m).

the isotopic composition of soil organic matter. Fertilizers* have lower $\delta^{15}N$ than most natural nitrogen sources available to plants (+5 to +7°/oo) and one might expect that in the long run the $\delta^{15}N$ of soil organic matter would show a shift towards lower values.

Meints et al. (1975) studied parcels cultivated during several years and found that fresh organic matter (corn or soybean) was systematically depleted in ^{15}N by about 3°/oo with respect to the average ^{15}N value of the soil organic matter. However, the length of application of fertilizers did not seem to matter as far as the soils were concerned, since the ^{15}N of the soils remained very constant. Based on a mass balance analysis, they concluded that other effects such as nitrate seepage to the aquifer interfere and therefore the time of artificial fertilizing (up to 10 years) was not long enough to permit the recognition of a possible isotope shift towards lower ^{15}N concentrations. In fact, detailed studies in progress on areas cultivated for about one century under controlled various agricultural practices (I.N.A., Grignon, France) show also in this case no difference in $\delta^{15}N$ between total organic matter from soils cultivated with or without fertilizers of any type (Mariotti, personal communication).

^{15}N IN NITRATES

Nitrates in soils

Detailed studies in progress on areas cultivated under various agricultural practices show no systematic differences in ^{15}N between total organic matter and the evolved nitrate. Such observations have been made by Feigin et al. (1974b), Rennie and Paul (1975), and Freyer and Aly (1975). However, contradictory data have been presented by Bremner and Tabatabai (1973), Cheng et al. (1964), and others (Fig. 10-6).

Incubation experiments of soils show that after a short transient state, the first nitrate produced has a low $\delta^{15}N$, which subsequently increases towards a steady-state value (Fig. 10-7; Feigin et al., 1974a; Freyer and Aly, 1975; Mariotti et al., 1977). In incubation experiments, this steady-state composition of nitrate is not related to the composition of the soil organic matter, but some correlation seems to exist under natural conditions. An explanation for this difference may be that the low-$\delta^{15}N$ nitrate produced in experiments represents "fresh" nitrate produced directly by the degradation of the organic matter, whereas nitrate with a higher ^{15}N content found in incuba-

---

*Artificial fertilizers are systematically impoverished in ^{15}N in comparison with soil organic matter: $-1 \pm 3°/_{oo}$ (NH_4^+); $+2 \pm 2°/_{oo}$ (NO_3^-), due to their synthesis from air (see Shearer et al., 1974b; Freyer and Aly, 1974; Mariotti and Létolle, 1977). Urea is also near zero (insufficient data). Note also that fresh nitrate formed by beginning decay of "green fertilizers" is about $+2°/_{oo}$ vs. atmospheric nitrogen.

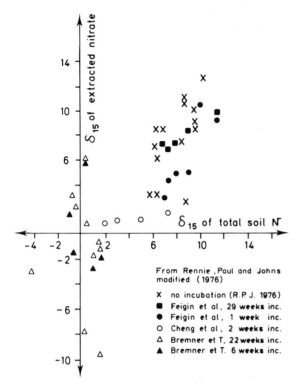

Fig. 10-6. Relationship between $\delta^{15}N$ of organic matter and nitrate in soils (from Feigin et al., 1974a; Cheng et al., 1964; and Bremner and Tabatabai, 1973).

tions of long duration, or extracted from natural soils represents a balance between the input of the "fresh", low-^{15}N nitrate and the loss of nitrate depleted in ^{15}N through utilization in the synthesis of new organic matter (Fig. 10-7) and/or denitrification: in all cases, kinetic reactions favor the lighter isotope and if the starting nitrate has a lower ^{15}N content than the organic matter from which it originates, then denitrification and/or re-use in metabolic processes will increase the $\delta^{15}N$ of the remaining nitrate.

Models based on such considerations (Fig. 10-8) were developed by Shearer et al. (1974a), as well as by Létolle and Mariotti (1974). ^{15}N enrichment in nitrate during denitrification was investigated by Rees (1973), and Focht (1973) developed a complete model including all processes involving mineral nitrogen species. As might be expected, model considerations show that the ^{15}N content of the steady-state nitrate depends also on the relative importance of the biochemical reactions involved (Figs. 10-7 and 10-8).

Fig. 10-7 shows data for an experiment on a renzine chalk soil, together with a modelization. This soil possesses little organic matter. Through the monitoring of the $\delta^{15}N$ of nitrate, it is possible to detect whether denitrification or reassimilation took place, especially if such measurements are combined with determinations of ^{15}N contents in molecular nitrogen and N_2O. In this specific case denitrification was not recognized. Yet denitrification certainly will have to be considered at least in hydromorphic soils, although some agronomers discard it as a significant process in soils.

Two more additional factors have to be considered for the interpretation of ^{15}N data relative to soil nitrates:

(1) The limiting effect of one in a sequence of successive reactions where

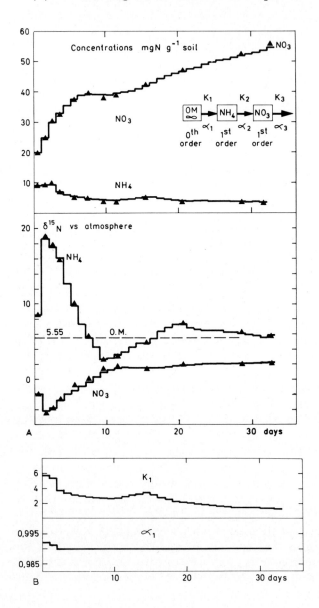

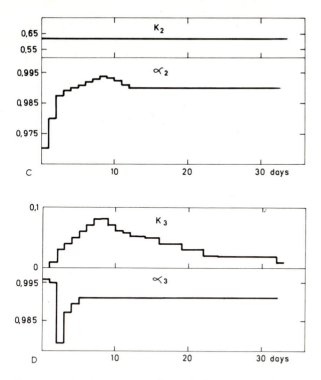

Fig. 10-7. Evolution of NH_4^+ and NO_3^- in a chalk lysimeter and model interpretation. The K values are the speed constants for reactions and α the fractionation factors. A. Evolution of NH_4^+ and NO_3^- concentrations and $\delta^{15}N$ values. Triangles are experimental data. Simulation is done along the model given in the upper right of the figure. B, C, and D, Best fit of parameters as shown on the model. The simulation is not possible with constant values for K_i and α_i.

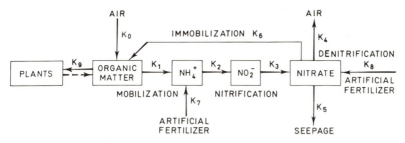

Fig. 10-8. Model for the ^{15}N evolution in soils and water; the K values are the speed constants. K_0 is 0th order, with isotope fractionation $\alpha_0 \simeq 0$ (see text); K_1 is 0th or 1st order (along with the disposable or mobilizable organic matter (see Wong Chong and Loehr, 1975), $\alpha_1 \simeq 0.980$–0.990; K_2 is 1st order, with $\alpha_2 \simeq 0.980$–0.990; K_3 is 1st order, with α_3 unknown (perhaps 1); K_4 is 1st order, with $\alpha_4 \simeq 0.960$–0.980; K_5 may be 0th order, and not fractionating; K_6 is 1st order, fractionation unknown ($\alpha < 1$).

each possesses its own fractionation effect for nitrogen. Kemp and Thode (1968) have demonstrated that during the reduction of sulfate by bacteria the SO_3^{2-} produced showed a variable impoverishment in ^{34}S. This is due to the fact that this reduction is the combination of two kinetically controlled reactions: the entry of sulfate into the cells and the breaking of the S—O bond in the sulfate ion. Each reaction possesses its own fractionation factor and if the first reaction is faster than the second the isotope effect of the second is predominant, and vice versa. Rees (1973) presented a detailed interpretation of the observations made by Kemp and Thode (1968). The system organic matter—ammonia nitrate can be considered in a similar way which can explain why the overall fractionation effect in the production of nitrate from organic matter with a given ^{15}N content may change from an incubation experiment to another.

(2) The "seepage effect", which is presented here as a variant of the well-known "reservoir effect". In the above discussion, it was assumed that no loss of nitrate through drainage occurred (K_5 in Fig. 10-8). Although this phenomenon, the loss of nitrate by seepage through the soil to subsurface or surface drainage systems probably does not fractionate isotopes, it does decrease the quantity of nitrate in the system and therefore modify the balance of nitrogen isotopes.

In summary, nitrate production, denitrification, seepage and reassimilation by organisms monitor the quantity of available nitrate and its isotopic composition. Therefore it is imperative that the hydrogeological conditions be well known before an interpretation of natural ^{15}N concentrations measured in a given location can be attempted. An area with poor drainage will not yield the same $\delta^{15}N$ values as a well-drained one; all other conditions being the same. As always in the studies of dynamic systems, the ^{15}N of organic matter and nitrate in soils is the balance between inputs and outputs which may or may not fractionate isotopes.

^{15}N IN THE HYDROSPHERE

Hoering (1955) reported ^{15}N values between 0 and $+9^0/_{00}$ for ammonia in rain and measured values of $0 \pm 4^0/_{00}$ for associated nitrates. He concluded that these nitrogen compounds were synthetized in the atmosphere and that the NH_4^+/NO_3^- isotope equilibrium was approached (see Table 10-2). However, these data were not approved by Wada et al. (1975), and Freyer and Aly (1976) observed that NH_4^+ in rains over western Germany had $\delta^{15}N$ between -5 and $-10^0/_{00}$ with the associated NO_3^- ranging from $+5$ to $-5^0/_{00}$. Since similar values were found by A. Mariotti (personal communication) for precipitations over Paris (France), one might suggest that a seasonal effect exists, with summer rains having lower $\delta^{15}N$ values. Freyer and Aly (1976) suggest that the ^{15}N-depleted ammonia mostly originates from NH_3 "evaporating" from fields, whereas Wada et al. (1975) interpret low ^{15}N concentrations in atmospheric nitrogen compounds as a consequence of fossil fuel burning.

As nitrogen from precipitations is now considered as an important source for soil nitrogen, it appears that further studies are needed in this field as well.

Surface and groundwaters. The organic matter dissolved in oceans (Wada and Hattori, 1975) and in lakes (Pang and Nriagu, 1976, 1977) has been analyzed and did not show significant differences from ^{15}N concentrations in soils. However, the most abundant nitrogen species in water is nitrate, and ammonia is noticeable only in reducing environments such as peat bogs, sewers, etc. The increase of nitrate concentrations in cultivated areas as well as in heavily populated regions has especially been noticed since 1950 along with health accidents attributed to excess nitrate in drinking water. The increasing nitrate contents of surface waters is also linked to the eutrophication of lakes and rivers. One may note that present purification plants for sewage waters do not suppress nitrogen in their discharges, but only change its chemical form to nitrate.

The first investigation of ^{15}N in nitrate dissolved in ground and surface waters was carried out by Kohl et al. (1971). On the basis of their analyses, they attributed the origin of nitrogen in drainage waters of the Sangamon basin (Illinois, U.S.A.) to the mixing of nitrate produced by the normal biochemical activity of soil and unmodified fertilizer nitrate which dissolved directly in soil water and was subsequently discharged in the subsurface drainage system. This conclusion was criticized by Hauck et al. (1972) who argued that fractionation effects would modify the isotope balance of nitrogen fertilizer through metabolism, and therefore too simple mass balances as presented by Kohl et al. were not possible. Their view was supported also by Edwards (1973) and Bremner and Tabatabai (1973) who note specifically that the isotope content of nitrate in aquifers shows a great variability with values ranging from zero to $+25^0/_{00}$. A similar range was also observed by Kohl et al. (1971), Freyer and Aly (1975), Ben Halima (1977), and Mariotti et al. (1976).

The following case is presented to emphasize the complexity of ^{15}N abundances in groundwaters and surface water systems. Mariotti et al. (1975) and Mariotti and Létolle (1977) investigated a small agricultural basin of a few square kilometers some 50 km east of Paris (Melarchez experimental basin), which is drained by a small creek, Ru du Fossé Rognon. In this basin, the phreatic shallow aquifer is tile-drained at about 0.8 m depth, a deeper aquifer exists in the underlying silts. Domestic and agricultural wastes, as well as the drains, discharge into the creek. The entire area is cultivated and artificial fertilizers are applied at the beginning of spring. Chemical and isotopic data for 1972—1973 are shown in Fig. 10-9. More recent data may be found in Ben Halima (1977).

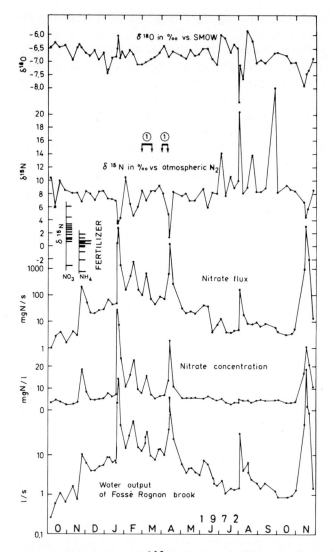

Fig. 10-9. Evolution of $\delta^{15}N$ of nitrate in Melarchez basin (40 km east of Paris) (Mariotti et al., 1975), showing relations between surface discharge, nitrate content and $\delta^{15}N$ of nitrate. Arrows indicate the time when fertilizers are used. The spectrum of $\delta^{15}N$ values of fertilizers is indicated on the left. The nitrate flux is the product of concentration and the river water output. The mean $\delta^{15}N$ value around +7 to +8‰ is interpreted as the steady-state nitrate produced by the normal biological activity of the soil. "Negative peaks" in $\delta^{15}N$, which coincide with surges of nitrate output, are interpreted as elimination of fertilizers: since there is no time lag between the flood and the nitrate peak, the acceleration of nitrate production by soil, with a low $\delta^{15}N$ (transient production) cannot be invoked. The positive $\delta^{15}N$ peaks during summer are due to the intervention of nitrate from fermented organic matter (manure, domestic wastes) (see p. 427). Such nitrates cannot be easily detected in presence of great quantities of nitrate from other sources and other isotopic composition.

Except during major precipitation events, the discharge of nitrate and its $\delta^{15}N$ were quite uniform. The latter is close to $+7.5^0/_{00}$ with perhaps a small seasonal effect. This "steady-state" mean value is the same as in the aquifer underlying the drains and also in the soil of a nearby experimental plot where no fertilizers are used (Ben Halima, 1977). During summer, peaks of high ^{15}N occur during low run-off. These high values are interpreted as being the discharge of nitrate produced during the oxidation of ammonia after a fractional evaporation. This observation is based on observations by Kreitler (1974) and Kreitler and Jones (1973). Such high values are observed only during the summer when the output of nitrate of other origins is low and not masked by nitrate with lower ^{15}N values. Other observations substantiate this interpretation since it has been observed that nitrate rich in ^{15}N is produced during degradation of wastes, manure, etc., during which loss of ammonia through volatilization is noticeable.

The other important features are the negative peaks synchronous with floods. After heavy rains, nitrate discharge increases by several orders of

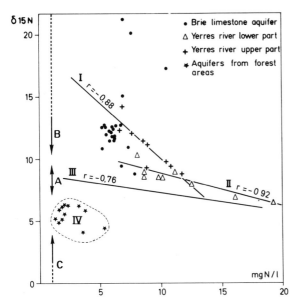

Fig. 10-10. Clustering of values in the ^{15}N vs. nitrate concentrations for the upper Yerres river (triangles and correlation line II), showing predominance of nitrates of agricultural origin; of the lower Yerres river (crosses and correlation line I), where the influence of organogenic nitrogen appears. Line III corresponds to the data shown in Fig. 10-9 for the Melarchez basin. Circles: Brie limestone aquifer, partly fed by the lower Yerres river. Stars (Group IV) represent nitrates from forest soils in the same region. A = domain of isotope values for cultivated soils of the area; B = same for nitrate of organic origin through fermentation; C = same for the fertilizers used in the area.

magnitude. Production of fresh nitrate in the soil does not immediately follow major rain events but shows up with a delay of a few days, yet the elimination of nitrate from upper soil horizons is immediate. Furthermore, the instantaneous production of nitrate from upper soil horizons is several orders of magnitude lower than the observed discharge which implies that the large nitrate discharge immediately following major precipitation events is not freshly produced nitrate but has another origin. The question to be answered is thus whether this other source could be nitrate produced in the soil during a rather long time between major precipitations events and which is discharged only during such events. The answer may be found in analysis of ^{15}N in steady-state soils and the underlying aquifer which is supplied with nitrate by the overlying soil. The isotopic composition of the nitrate produced by the nearby experimental plot (not fertilized), and also in the aquifer just under the tile drains does not change at all during or after rainfalls (Mariotti et al., 1977; Ben Halima, 1977).

The input of artificial fertilizers is well known for this region and has a mean $\delta^{15}N$ of $+2 \pm 2^0/_{00}$ (Fig. 10-9). One may thus conclude that in this precise case the nitrate discharge during precipitation events is a combination of steady-state nitrate with a $\delta^{15}N$ of $+7.5^0/_{00}$ and fertilizer nitrate with a $\delta^{15}N$ close to $+2^0/_{00}$. A mixing line describing this is shown in Fig. 10-10 as line III.

On the same figure data for shallow aquifers from forest soils of the same region are shown (group IV), together with data for a small river, the Yerres, southeast of Paris. In its upper part (line II), it flows on cultivated soils and receives a mixture of nitrate from various origins, especially from biochemical activity of cultivated soils together with unused fertilizers. In its lower part (Fig. 10-10, line I) abundant nitrate discharges from sewers and treatment plants modify the isotope composition of the river nitrate, although dilution occurs. As the river leaks to the underlying Brie limestone aquifer, the latter is polluted, and its isotope composition shows the origin of the pollution to be nitrate evolved from refuses.

Throughout this river system the nitrate originates from three sources: (1) nitrates resulting from the oxidation of organic wastes with ammonia volatilization as a mechanism to explain high δ-values; (2) fertilizer nitrate; and (3) biochemically produced nitrates from soils. The fertilizer contribution to aquifers appears to be directly controlled by the type of drainage system and if, as is the case for Sangamon and Melarchez basins, surface water is quickly eliminated through surface drainage or via a tile-draining system, more of the applied fertilizer will be lost.

The solution of the very important problem of the origin of excess nitrates in waters necessitates a good knowledge of all the ecological parameters of the questioned system. It is not impossible that information gained by the knowledge of ^{18}O content in nitrate may clarify further the question.

ACKNOWLEDGMENTS

Thanks are due to colleagues who presented me with yet unpublished data, especially Dr. A. Mariotti. Drs. J.C. Fontes, P. Fritz and E.A. Paul helped much in discussing several versions of the manuscript.

REFERENCES

Abbott, D., Dodson, M.J. and Powell, H., 1953. Natural abundance of ^{15}N in hemin and plasma protein from normal blood. *Proc. Soc. Exp. Biol. Med.*, 84: 402—404.

Amarger, N., Mariotti, A. and Mariotti, F., 1977. Essai d'estimation du taux d'azote fixé symbiotiquement chez le Lupin par le traçage isotopique naturel (^{15}N). *C.R. Acad. Sci. Paris, Ser. D*, 284: 2179—2182.

Bardin, R., Domenach, A.M. and Chalamet, A., 1977. Rapports isotopiques naturels de l'azote, II. Variations d'abondance du ^{15}N dans les échantillons de sols et plantes; applications. *Rev. Ecol. Biol. Sols*, 3: 395—402.

Becker, R.H. and Clayton, R.N., 1977. Nitrogen isotopes in igneous rocks. *Annu. Meet., Am. Geol. Soc., Abstr.*, 5(137): 536.

Begun, G.M. and Flechter, W.H., 1960. Partition functions ratios for molecules containing nitrogen isotopes. *J. Chem. Phys.*, 33: 1083—1084.

Behrens, W., Eichmann, R., Plate, A. and Kroepelin, H., 1971. Zur Methodik der Bestimmung des Stickstoffisotopenverhältnisses in Erdgasen. *Erdöl Kohle*, 24: 76—78.

Ben Halima, A., 1977. *Apport de la Géochimie Isotopique de l'Azote à la Connaissance des Sources de Pollution par les Nitrates sur l'Exemple de la Brie.* Thesis, Université Pierre et Marie Curie, Paris.

Benson, B.B. and Parker, P.D.M., 1961. Nitrogen/argon and nitrogen isotope ratios in aerobic sea water. *Deep-Sea Res.*, 7: 237—253.

Black, A.S. and Waring, S.A., 1977. The natural abundance of ^{15}N in the soil-water system of a small catchment area. *Aust. J. Soil Res.*, 15: 51—57.

Blackmer, A.M. and Bremner, J.M., 1977. Nitrogen isotope discrimination in denitrification of nitrates in soils. *Soil Biol. Biochem.*, 9: 73—77.

Boigk, H., Hagemann, H.W., Stahl, W. and Wollanke, G., 1976. Zur Herfunkt und Migration des Stickstoffs nordwestdeutscher Erdgase aus Oberkarbon und Rotliegende. *Erdöl Kohle*, 29: 103—112.

Bokhoven, C. and Theuwen, H.J., 1966. Determination of the abundance of carbon and nitrogen isotopes in Dutch coals and natural gas. *Nature*, 211: 927.

Bremner, J.M., 1965. Isotope ratio analysis of nitrogen in nitrogen-15 tracer investigations. In: C.A. Black (Editor), *Methods of Soil Analysis*, American Society of Agronomy, Madison, Wisc., pp. 1256—1286.

Bremner, J.M. and Keeney, D.R., 1966. Determination and isotope ratio analysis of different forms of nitrogen in soils, III. *Soil Sci. Soc. Am. Proc.*, 30: 577—582.

Bremner, J.M. and Tabatabai, M.A., 1973. Nitrogen-15 enrichment of soils and soil-derived nitrate. *J. Environ. Qual.*, 2: 363—365.

Bremner, J.M., Cheng, H.H. and Edwards, A.P., 1963. Assumptions and errors in nitrogen-15 tracer investigations. In: *Nitrogen-15 Tracer Research, Report of the FAO/IAEA Technical Meeting, Brunswick, September 1963.* Pergamon, New York, N.Y., pp. 429—442.

Brown, L.L. and Drury, J.S., 1967. Nitrogen isotope effects in the reduction of nitrate, nitrite and hydroxylamine to ammonia. *J. Chem. Phys*, 46: 2883—2887.

Brown, L.L. and Drury, J.S., 1968. Exchange and fractionation of nitrogen isotopes between NO_2^- and NO_3^-. *J. Chem. Phys.*, 48: 1399—1400.
Brown, L.L. and Drury, J.S., 1969. Nitrogen isotope effects, II. The MgO and $CuSO_4$ systems. *J. Chem. Phys.*, 51: 3771—3775.
Cheng, H.H., Bremner, J.M. and Edwards, A.P., 1964. Variation of nitrogen-15 abundances in soils. *Science*, 146: 1574—1575.
Cline, J.D., 1973. *Denitrification and Isotope Fractionation in Two Contrasting Marine Environments: the Eastern Tropical and North Pacific Ocean and the Cariaco Trench.* Ph.D. Thesis, University of California, Los Angeles, Calif.
Cline, J.D. and Kaplan, I.R., 1975. Isotopic fractionation of dissolved nitrate during denitrification in the eastern tropical North Pacific Ocean. *Mar. Chem.*, 3: 271—289.
Colombo, U., Gazzarrini, F., Sironi, G., Gonfiantini, R. and Tongiorgi, E., 1965. Carbon isotope composition of individual hydrocarbons from Italian natural gases. *Nature*, 205: 1303—1304.
Cook, F.D., Wellman, R.P. and Krouse, H.R., 1970. Nitrogen isotope fractionation in the nitrogen cycle. *Int. Symp. on Hydrogeochemistry and Biochemistry, Tokyo, September 6—12, 1970.* Paper B 1/6, 23 pp.
Craig, H., 1968. Isotope separation by carrier diffusion. *Science*, 159: 93—96.
Delwiche, C.C. and Steyn, P.L., 1970. Nitrogen fractionation in soils and microbial reactions. *Environ. Sci. Technol.*, 4: 929—935.
Domenach, A.M. and Chalamet, A., 1977. Rapports isotopiques naturels de l'azote, I. Premiers résultats: sols des Dombes. *Rev. Ecol. Biol. Sols*, 4: 279—287.
Edwards, A.P., 1973. Isotope tracers techniques for identification of sources of nitrate. *J. Environ. Qual.*, 2: 382—386.
Edwards, A.P., 1975. Isotope effects in relation to the interpretation of $^{15}N/^{14}N$ ratios in tracer studies. In: *Isotope Ratios as Pollutant Source and Behaviour Indicators.* IAEA, Vienna, pp. 455—468.
Eichmann, R., Plate, A., Behrens, W. and Kroepelin, H., 1971. Das Isotopenverhältnis des Stickstoffs in einigen Erdgasen; Erdölgasen und Erdölen Nordwestdeutschands. *Erdöl Kohle*, 24: 2—7.
Feigin, A.D., Kohl, D.H., Shearer, G.B. and Commoner, B., 1974a. Variations of the natural nitrogen-15 abundance in nitrate mineralized during incubation of several Illinois soils. *Soil Sci. Soc. Am., Proc.*, 38: 90—95.
Feigin, A.D., Shearer, G.B., Kohl, D.H. and Commoner, B., 1974b. The amount and nitrogen-15 content of nitrate in soils profiles from two central Illinois fields in a corn-soybean rotation. *Soil Sci. Soc. Am., Proc.*, 38: 465—471.
Focht, D.D., 1973. Isotope fractionation of ^{15}N and ^{14}N in microbiological nitrogen transformations: a theoretical model. *J. Environ. Qual.*, 2: 247—252.
Freyer, H.D. and Aly, A.I.M., 1974. Nitrogen-15 variations in fertilizer nitrogen. *J. Environ. Qual.*, 3: 405—406.
Freyer, H.D. and Aly, A.I.M., 1975. Nitrogen-15 studies on identifying fertilizer excess in environmental systems. In: *Isotope Ratios as Pollutant Source and Behaviour Indicators.* IAEA, Vienna, pp. 21—33.
Freyer, H.D. and Aly, A.I.M., 1976. Seasonal trends of NH_4^+ and NO_3^- nitrogen isotope composition in rain collected in rural air. *ECOG IV, Amsterdam, April 1976* (abstract).
Freyer, H.D. and Kasten, M., 1976. Isotope studies of ammonium, nitrate and total nitrogen in modified loess soils (unpublished manuscript).
Gaebler, O.H., Choitz, H.C., Vitti, T.G. and Vumirovich, R., 1963. Significance of ^{15}N excess in nitrogenous compounds of biological origin. *J. Biochem. Phys.*, 41: 1089—1097.
Hauck, R.D., 1973. Nitrogen tracers in nitrogen cycle studies. *J. Environ. Qual.*, 2: 317—327.

Hauck, R.D., Bartholomew, W.V., Bremner, J.M., Broadbent, F.E., Cheng, H.H., Edwards, A.P., Keeney, D.R., Legg, J.O., Olsen, S.R. and Porter, L.K., 1972. Use of variations in natural nitrogen isotope abundance for environmental studies: a questionable approach. *Science*, 177: 453—454.
Hoefs, J., 1973. *Stable Isotope Geochemistry*. Springer Verlag, New York, N.Y., 189 pp.
Hoering, T.C., 1955. Variations of nitrogen-15 abundance in naturally occurring substances. *Science*, 122: 1233—1234.
Hoering, T.C., 1957. Isotopic composition of the ammonia and the nitrate ions in rain. *Geochim. Cosmochim. Acta*, 12: 97—102.
Hoering, T.C. and Ford, H.T., 1960. Isotope effect in the fixation of nitrogen by Azotobacter. *J. Am. Chem. Soc.*, 82: 376—378.
Hoering, T.C. and Moore, H.E., 1957. The isotopic composition of the nitrogen in natural gases and associated crude oils. *Geochim. Cosmochim. Acta*, 13: 225—232.
Junk, G. and Svec, H.J., 1958. The absolute abundance of the nitrogen isotopes in the atmosphere and compressed gas from various sources. *Geochim. Cosmochim. Acta*, 14: 234—243.
Keeney, D.R. and Bremner, J.M., 1966. Determination and isotope analysis of different forms of nitrogen in soils, IV. *Soil Sci. Soc. Am. Proc.*, 30: 583—587.
Kemp, A.L. and Thode, H.G., 1968. The mechanism of the bacterial reduction of sulphate and sulphite from isotope fractionation studies. *Geochim. Cosmochim. Acta*, 32: 71—91.
Kirshenbaum, I., Smith, J.S., Crowell, T., Graff, J. and Mokee, R., 1947. Separation of the nitrogen isotopes by the exchange reaction between ammonia and solutions of ammonium nitrate. *J. Chem. Phys.*, 15: 440—446.
Klots, C.E. and Benson, B.B., 1963. Isotope effect in the solution of oxygen and nitrogen in distilled water. *J. Chem. Phys.*, 38: 890—893.
Kohl, D.A., Shearer, G.B. and Commoner, B., 1971. Fertilizer nitrogen: contribution to nitrate in surface water in a Corn Belt watershed. *Science*, 174: 1331—1334.
Koike, I., Wada, E., Tsuti, T. and Hattori, A., 1972. Studies on denitrification in a brackish lake. *Arch. Hydrobiol.*, 69: 508—520.
Kreitler, C.W., 1974. *Determining the Source of Nitrate in Groundwater by Nitrogen Isotope Studies*. Ph.D. Thesis, University of Texas, Austin, Texas.
Kreitler, C.W. and Jones, D.C., 1973. Natural soil nitrate: the cause of the nitrate contamination of groundwater in Runnels County, Texas. *Ground Water*, 15: 53—62.
Laishley, E.J., McCready, R.G.L., Bryant, R. and Krouse, H.R., 1975. Stable isotope fractionation by Clostridium pasteurianum. 2nd Int. Symp. Biogechemistry, Hamilton, Ont., 1975. Preprint, 37 pp.
Létolle, R., 1974. Etat actuel des connaissances relatives à la géochimie isotopique de l'azote. *Rev. Géogr. Phys. Géol. Dyn.*, 16: 131—138.
Létolle, R. and Mariotti, A., 1974. Utilisation des variations naturelles d'abondance de l'azote 15 comme traceur en hydrogéologie: premiers résultats. In: *Isotope Techniques in Groundwater Hydrology*. IAEA Vienna, pp. 209—220.
London, M., 1961. *Separation of Isotopes*. Newnes, London, 489 pp.
Mahenc, J., 1965. Influence de la composition de la phase gazeuse sur la séparation isotopique de l'azote 15 par échange chimique entre solution nitrique et vapeurs nitreuses. *J. Chim. Phys.*, 62: 1399—1403.
Mariotti, A. and Létolle, R., 1977. Application de l'étude isotopique de l'azote en hydrologie et hydrogéologie. Analyse des résultats obtenus sur un exemple précis: le bassin de Melarchez (Seine et Marne, France). *J. Hydrol.*, 33: 157—172.
Mariotti, A., Létolle, R., Blavoux, B. and Chassaing, B., 1975. Determination par les teneurs naturelles en ^{15}N de l'origine des nitrates: résultats préliminaires sur le bassin de Melarchez (Seine et Marne). *C.R. Acad. Sci., Paris*, 280: 423—426.

Mariotti, A., Berger, G. and Ben Halima, A., 1976. Apport de l'étude isotopique de l'azote à la connaissance de la pollution des aquifères souterrains par les nitrates, en milieu agricole (Brie, Beauce, France). *Rev. Géogr. Phys. Géol. Dyn.*, 18: 375—384.

Mariotti, A., Muller, J. and Létolle, R., 1977a (in preparation).

Mariotti, A., Pierre, D., Vedy, J.C. and Bruckert, S., 1977b. Nitrogen-15 natural abundance in organic matter of a soil sequence along an altitudinal gradient, Alpes de Savoie, France (in preparation).

Mayne, K.I., 1957. Natural variations, 1. The isotope abundance of nitrogen in igneous rocks. *Geochim. Cosmochim. Acta*, 12: 185—189.

McElroy, M.B., Yung, Y.L. and Nier, A.O., 1976. Isotopic composition of nitrogen: implication for the past history of Mars atmosphere. *Science*, 194: 70—72.

Meints, V.W., Boone, L.V. and Kurtz, L.T., 1975. Natural ^{15}N abundance in soils, leaves and grain as influenced by long term addition of fertilizer N at several rates. *J. Environ. Qual.*, 4: 486—490.

Meints, V.W., Shearer, G., Kohl, D.H. and Kurtz, L.T., 1975. A comparison of unenriched versus ^{15}N enriched fertilizer as a tracer for N fertilizer uptake. *Soil. Sci.*, 119: 421—425.

Miyake, Y. and Wada, E., 1967. The abundance ratio of ^{15}N/^{14}N in marine environments. *Rec. Oceanogr. Works Jpn.*, 9: 37—53.

Miyake, Y. and Wada, E., 1971. The isotope effect on the nitrogen in biochemical oxidation-reductions. *Rec. Oceanogr. Works Jpn.*, 11: 1—6.

Pang, P.C. and Nriagu, J., 1976. Distribution and isotope composition of nitrogen in Bay of Quinte sediments. *Chem. Geol.*, 18: 93—105.

Pang, P.C. and Nriagu, J., 1977. Isotope variation of nitrogen in Lake Superior. *Geochim. Cosmochim. Acta*, 41: 811—814.

Parwell, A., Ryhage, A.R. and Wickman, F.E., 1957. Natural variations in the relative abundance of the nitrogen isotopes. *Geochim. Cosmochim. Acta*, 11: 165—170.

Pilot, J., 1963. Über die massenspektrometrische Isotopenanalyse an Stickstoff aus Erdgasen und Gesteine. *Kernenergie*, 12: 714—717.

Rees, C.E., 1973. A steady state model for sulphur isotope fractionation in bacterial reduction processes. *Geochim. Cosmochim. Acta*, 37: 1141—1162.

Rennie, D.A. and Paul, E.A., 1975. Nitrogen isotope ratios in surface and subsurface soils horizons. In: *Isotope Ratios as Pollutant Source and Behaviour Indicators*. IAEA Vienna, pp. 441—451.

Rennie, D.A., Paul, E.A. and Johns, L.E., 1976. Natural nitrogen-15 abundance of soils and plant samples. *Can. J. Soil Sci.*, 56: 43—50.

Richet, P., 1976. *Calcul des Fractionnements Isotopiques des Molécules Simples d'Intêret Géochimique. Application à Quelques Systèmes Naturels*. Thesis, Université de Paris VII, Paris.

Richet, P., Bottinga, Y. and Javoy, M., 1977. A review of hydrogen, carbon, nitrogen, oxygen sulfur and chlorine stable isotope fractionation among gaseous molecules. *Annu. Rev. Earth Planet. Sci.*, 5: 65—110.

Riga, A., Van Praag, H.T. and Brigode, N., 1971. Rapport isotopique naturel de l'azote dans quelques sols forestiers et agricoles de Belgique soumis à divers traitements culturaux. *Geoderma*, 6: 213—222.

Rösler, H.J. and Lange, H., 1972. *Geochemical Tables*. Elsevier, Amsterdam, 468 pp.

Ross, P.J. and Martin, E.A., 1970. Rapid procedure for preparing gas samples for nitrogen-15 determination. *Analyst*, 95: 817—822.

Shearer, G., Duffy, J., Kohl, D.H. and Commoner, B., 1974a. A steady state model of isotopic fractionation accompanying nitrogen transformations in soils. *Soil Sci. Soc. Am., Proc.*, 38: 315—322.

Shearer, G.B., Kohl, D.H. and Commoner, B., 1974b. The precision of determination of the natural abundance of nitrogen-15 in soils, fertilizers and shelf chemicals. *Soil Sci.*, 118: 308—316.

Silver, A. and Bremner, J.M., 1966. Determination and isotope ratio analysis of different forms of nitrogen in soils, V. *Soil Sci. Soc. Am., Proc.*, 30: 587—594.

Sims, A.P. and Cockin, E.C., 1958. Assay of isotopic nitrogen by mass spectrometer. *Nature*, 181: 474.

Smith, P.V. and Hudson, B.E., 1951. Abundance of ^{15}N in the nitrogen present in crude oil and coal. *Science*, 113: 577.

Spindel, W., 1954. The calculation of equilibrium constants for several exchange reactions of nitrogen 15 between oxycompounds of nitrogen. *J. Chem. Phys.*, 23: 1271—1272.

Stahl, W., Wollanke, G. and Boigk, H., 1975. Carbon and nitrogen isotope data of upper Carboniferous and Rotliegende natural gases from North Germany and their relationship to the maturity of the organic source material. *7th Int. Meet. Org. Chem., Madrid, September 1975* (preprint).

Stern, M.J., Kauder, L.N. and Spindel, W., 1960. Temperature dependence of the fractionation of nitrogen in the NO_2^-—NO_3^- system. *J. Chem. Phys.*, 34: 333—334.

Stroud, L., Meyer, T.O. and Emerson, D.E., 1967. Isotopic abundance of neon, argon and nitrogen in natural gases. *U.S. Dep. Inter., Bur. Mines Rep. Invest.*, 6936: 1—27.

Sweeney, R.E., Liu K.K. and Kaplan, I.R., 1976. Oceanic nitrogen isotopes and their uses in determining the source of sedimentary nitrogen. *Int. Symp. Stable Isotopes Geochem., New Zealand, August 4—6, 1976* (preprint).

Thode, H.G. and Urey, H.C., 1939. Further concentration of ^{15}N. *J. Chem. Phys.*, 7: 34—39.

Urey, H.C., 1947. The thermodynamic properties of isotopic substances. *J. Am. Chem. Soc.*, pp. 562—581.

Volyinets, V.P., Zadorozhny, I.K. and Florenski, K.P., 1967. Isotope composition of nitrogen in the Earth's crust. *Geokhimiya*, 7: 587—593.

Wada, E. and Hattori, A., 1975. Natural abundance of ^{15}N in particulate organic matter in the North Pacific Ocean. *Geochim. Cosmochim. Acta*, 40: 249—251.

Wada, E., Kadonaga, T. and Matsuo, S., 1975. ^{15}N abundance in nitrogen of naturally occurring substances and global assessment of denitrification from isotopic view point. *Geochem. J.*, 9: 139—148.

Wellman, R.P., Cook, F.D. and Krouse, H.R., 1968. Nitrogen-15 microbial alteration of abundance. *Science*, 161: 269—270.

Wlotzka, F., 1972. Nitrogen. In: K.K. Wedepohl, (Editor), *Handbook of Geochemistry*, 11. Springer Verlag, New York, N.Y., Chapter 7.

Wollanke, G., Behrens, W. and Hörgant, T., 1974. Stickstoffisotopenverhältnis von Erdgasen des Emslandes. *Erdöl Kohle*, 27: 523.

Wong Chong, G.M. and Loehr, R.C., 1975. The kinetics of microbial nitrification. *Water Res.*, 9: 1099—1106.

Chapter 11

SULPHUR ISOTOPES IN OUR ENVIRONMENT

H.R. KROUSE

INTRODUCTION

The element sulphur has four stable isotopes, ^{32}S, ^{33}S, ^{34}S and ^{36}S occurring naturally with approximate abundances 95.02, 0.75, 4.21, and 0.02% respectively. Since sulphur exists in many forms and valence states (—2 to +6) in nature, it is most complex and interesting to study from the viewpoint of isotope fractionation. Most reported studies have examined ^{32}S and ^{34}S because of their more favourable abundances and the extensive use of SO_2 gas for mass spectrometric determinations.

Although SO_2 is readily prepared, it has disadvantages for sulphur isotope analyses. It is a "sticky" gas in vacuum systems, particularly in the presence of traces of water. Since $^{32}S^{16}O^{18}O^+$ and $^{34}S^{16}O_2^+$ are not resolved in typical isotope abundance mass spectrometers, corrections are necessary. These oxygen isotope corrections coupled with SO_3 production require consistency in the combustion procedure.

In recent years SF_6 has been gaining favour for sulphur isotope analyses. Although the fluorination and purification is time consuming, the gas, once formed, is extremely stable. Since fluorine has only one stable isotope, the major ion species SF_5^+ yields the abundances desired without corrections. Whereas precisions of one part in 10^4 for ratios of isotope abundances are typical using SO_2, two parts in 10^5 seem routinely possible using SF_6 (C.E. Rees, personal communication, 1974).

Techniques for sulphur isotope analyses have been summarized by Holt (1975).

TERRESTRIAL SULPHUR ISOTOPE ABUNDANCES AND CYCLING OF MOBILE SULPHUR COMPOUNDS

Since sulphur exists in many forms, numerous terrestrial processes alter its isotope abundances. Early work by Thode et al. (1949) established general trends. Subsequently this group and other researchers utilized sulphur isotope abundances to elucidate many problems in the lithosphere, bio-

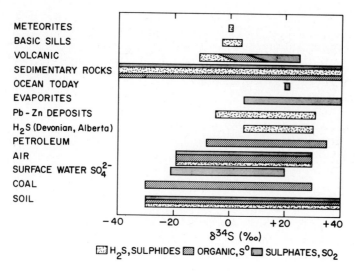

Fig. 11-1. Sulphur isotope abundances in nature.

sphere, hydrosphere and, more recently, the atmosphere. Fig. 11-1 summarizes the range of $\delta^{34}S$ variations encountered terrestrially where $\delta^{34}S$ is expressed on a parts permil scale (‰) by:

$$\delta^{34}S = \left[\frac{(^{34}S/^{32}S)_{sample}}{(^{34}S/^{32}S)_{meteorite}} - 1\right] \times 1000$$

The choice for meteoritic troilite (specifically from the Cañon Diablo meteorite) as the standard arises from its consistency in isotopic composition and its proximity to the mean of the terrestrial range of δ-values (Fig. 11-1). Some terrestrial samples fall outside the range of Fig. 11-1 such as barite concretions at +87‰ (Sakai, 1971), pyrites from Upper Silesia at +67‰ (Gehlen and Nielsen, 1969), pyrite concretions ranging from +70 to —47‰ (Bogdanov et al., 1971) and hydrotroilite at —45‰ (Veselovsky et al., 1969). Deep-seated primary sulphides tend to have narrow $\delta^{34}S$ values near the troilite reference as provided by data from basic sills (Shima et al., 1963), magmatic deposits (Thode et al., 1962; Ryznar et al., 1967; Schwarcz, 1973) and carbonatites (Grinenko et al., 1970; Mitchell and Krouse, 1975). Volcanic gases and rocks tend to have larger spreads in $\delta^{34}S$ values because of isotope exchange processes and the fact that there may be sources other than primary sulphur (Rafter and Wilson, 1963).

Whereas present-day ocean sulphate is remarkably uniform with δ-values slightly greater than +20‰ (Thode et al., 1961) (except near freshwater tributaries), data from evaporite deposits attest to significant temporal isotopic variations in the ancient oceans. These isotopic shifts were worldwide (Nielsen and Ricke, 1964; Thode and Monster, 1965; Holser and Kaplan, 1966; Davies and Krouse, 1975) and have been caused by input of sulphate and sulphur isotope fractionation during removal processes such as biological reduction and evaporite formation. Among others, Rees (1970) has modelled these changes.

Sulphur compounds which have participated in the biological sulphur cycle range widely in their isotopic compositions. Although the more oxidized species tend to be enriched in ^{34}S consistent with the predictions of statistical mechanics (Tudge and Thode, 1950) the situation may be mostly fortuitous since kinetic isotope effects during reduction processes usually favour the lighter isotope. The fact that sulphates are frequently found depleted in ^{34}S as compared to adjacent sulphides as a result of kinetic isotope effects during oxidation suggests that the world's sulphur isotope distribution is far from thermodynamic equilibrium.

Biological isotope effects superimposed on the variations of the ancient oceans effect the large spread found for sedimentary rocks. Deposits related to the ancient oceans such as lead-zinc ores (Campbell et al., 1968; Sasaki and Krouse, 1969; Sangster, 1968) petroleum (Vredenburgh and Cheney, 1971; Orr, 1974) as well as H_2S in sour-gas wells (Hitchon et al., 1975; Krouse, 1977a) are generally enriched in ^{34}S upwards to the values identified with the associated evaporites. Thode and Monster (1965) found petroleum and associated H_2S to be on the average some $15^0/_{00}$ depleted in ^{34}S as compared to related evaporites (Fig. 11-2). (A problem arises in that the oil in a given reservoir rock may have formed during a much earlier geological event.) Recent studies on oil and gas accumulations as reviewed by Krouse (1977a), show that while shallower H_2S may be depleted with respect to contemporaneous seawater sulphate, deeper H_2S and oil differ little isotopically from associated evaporite.

Air and water show large variations in isotopic composition. This is not surprising since the atmosphere and hydrosphere can receive sulphur compounds from many sources. For example, the range for surface water sulphate shown in Fig. 11-1 is encountered in the Mackenzie River system (Hitchon and Krouse, 1972). Reduced sulphur compounds in surface waters may have negative $\delta^{34}S$ values extending beyond the limits of Fig. 11-1 (see section "Sulphur isotope fractionation during transformations of atmospheric and aqueous sulphur compounds").

In summary, Figs. 11-1 and 11-2 show that a wide range of sulphur isotope compositions arise in our environment regardless of man's activities. The variations in the air and water may be related to inputs from many geological sources as in the case of the Mackenzie River system (Hitchon and Krouse, 1972) or to isotopic fractionation during biological activity which depends upon the concentrations of sulphur nutrients, metabolic rates, and the organisms involved.

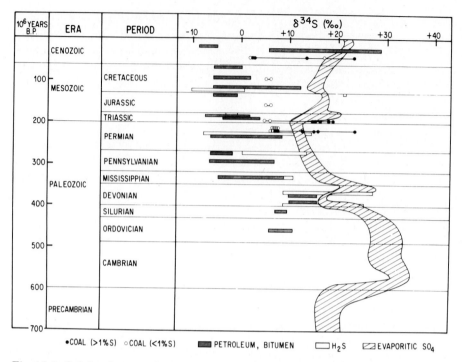

Fig. 11-2. Sulphur isotope abundances in fossil fuels and evaporites of different geological ages. Evaporite data after Kaplan (1975); coal data after Smith and Batts (1974); petroleum data from many sources summarized by Krouse (1977b).

Sulphur isotope fractionation arises in many processes in the overall sulphur cycle (Fig. 11-3). Many sulphur cycles can be found in the literature dependent upon the purpose for which they were designed. The overall sulphur cycle contains many processes not represented in Fig. 11-3 and is made up of many subcycles. The subcycle of oxidation and reduction by microorganisms is designated by the encircled "1" in Fig. 11-3. The cycle designated by the encircled "2" is of agricultural interest where crops devoured by animals are partially cycled as wastes back into the soil to become crop nutrients.

Man has perturbed the fluxes of sulphur compounds in the cycle by bringing raw materials from deep in the earth's crust to the surface, processing them and discharging sulphur-containing compounds to the atmosphere, hydrosphere, and consequently the biosphere. It is not the purpose of this chapter to review the magnitudes of fluxes and reservoirs in the sulphur cycle. Recent analyses have been performed by Erikson (1963), Junge (1963), Robinson and Robbins (1970), Kellogg et al. (1972), Friend (1973), and Granat et al. (1976). Rather, it is the intent to examine sulphur

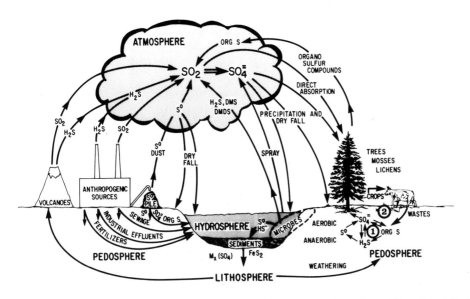

Fig. 11-3. Global cycling of sulphur. The encircled "1" is the microbial cycle while the encircled "2" represents a food cycle with higher animals. Abbreviations: DMS = dimethyl sulphide, DMDS = dimethyl disulphide.

isotopes as a means of delineating natural and anthropogenic fluxes of sulphur compounds.

One aspect of this assessment is the usefulness of sulphur isotopes in source identification. For example, atmospheric particulates arising from the manufacture of wall board from gypsum are expected to have $\delta^{34}S$ values consistent with the evaporite data of Fig. 11-2. The smelting of magmatic sulphides introduces SO_2 into the atmosphere with $\delta^{34}S$ values near $0^0/_{00}$, whereas lead-zinc recovery from sedimentary ores tends to release gaseous sulphur compounds with greater enrichments in $\delta^{34}S$ (Fig. 11-1). Emission from power plants have a wide range in isotopic composition, dependent upon the source of fuel. Another aspect of assessing the capability of sulphur isotopes in environmental studies is determining the extent to which the isotope abundances are altered in natural processes. Source identification becomes difficult if large isotopic selectivities arise in ensuing processes. On the other hand, the isotopic fractionation is a useful tool for elucidating natural processes.

Previous reviews on the use of sulphur isotopes in environmental studies have been given by Nielsen (1974), Holt (1975) and Smith (1975).

ELUCIDATION OF SOURCES, MIXING, AND DISPERSION OF SULPHUR COMPOUNDS

Interpretation of hypothetical plots of $\delta^{34}S$ versus concentration

Plots of $\delta^{34}S$ values versus concentrations can prove valuable in identifying sources and monitoring the fate of sulphur compounds. In this section, hypothetical situations will be considered in the first instance. Then data from atmospheric SO_2 will be examined in the light of the hypothetical models. Finally, the possibility of extending these analyses to sulphate in precipitation and surface waters will be considered briefly.

Fig. 11-4 depicts δ versus concentration plots for a number of hypothetical

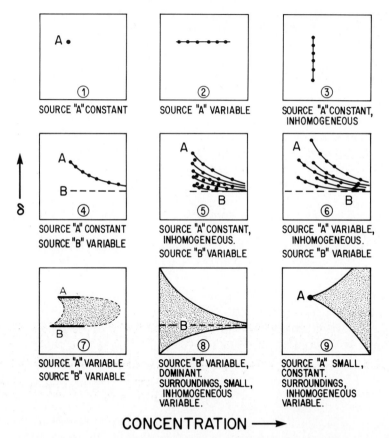

Fig. 11-4. Variation of δ-value with concentration for a number of hypothetical situations. These models assume simple mixing without any isotopic fractionation. The terms "constant" and "variable" refer to the rate of emission while the term "inhomogeneous" refers to variations in the isotopic composition.

source combinations. This is a general analysis, applicable to any element and, in principle, pertains to gases, ions, and fine particulates. Thorough mixing is assumed in the volume of air or water under consideration. In this figure the terms "constant" and "variable" refer to the rate of emission while the isotopic composition of the sources have fixed $\delta^{34}S$ values unless they are designated as "inhomogeneous."

Case 1 is the simplest with a source A which is fixed in both its emission rate and isotopic composition. If a steady-state condition is achieved at the observation site, a plot of δ versus concentration yields a point. Such an ideal situation is seldom observed except for a short interval of time near a dominant source such as an industrial stack.

In case 2, source A retains its fixed δ-value but varies in its emission rate. Data plot as a horizontal line with the extremities corresponding to the minimum and maximum concentrations.

In case 3, source A is constant but varies in its isotopic composition. The idealized plot is a vertical line with the extremities corresponding to the minimum and maximum δ-values. Such a plot could be approached where a dominant industrial source functions steadily but alters its feed stocks.

In case 4, the emission rate for source A is constant while that of B varies. When the emissions from B approach zero, the isotopic composition observed approaches that of source A and corresponds to the minimum concentration. As the emission from B increases, the observed concentration increases and the isotopic composition approaches that of source B. This case is interesting since simple mathematical manipulation permits evaluation of the δ-value of source B. The isotopic balance is given by the approximate expression:

$$C_A \delta_A + C_B \delta_B = C \delta \tag{1}$$

where C_A and C_B are the concentrations due to source A and B, respectively; C is the total concentration; δ_A and δ_B are the δ-values of source A and B; δ is the measured δ-value at the observation site.

Since $C_B = C - C_A$, equation [1] may be rewritten to give:

$$C_A \delta_A + (C - C_A) \delta_B = C \delta \tag{2}$$

Rearrangement gives:

$$\frac{C_A (\delta_A - \delta_B)}{C} = \delta - \delta_B \tag{3}$$

Equation (3) is general. However, case 4 adds the additional constraints that

C_A, δ_A and δ_B are fixed. Then equation (3) takes the form:

$$\delta = \frac{\text{constant}}{C} + \delta_B \qquad [4]$$

Therefore if case 4 applies, a plot of δ versus concentration^{-1} yields a straight line which intercepts the y-axis at the δ-value for source B.

Case 5 resembles case 4 except that source A has a range in its isotopic composition. However, random observations would give points in the shaded area of case 5.

Case 6 adds the further complication that source A varies in its emission rate.

Case 7 has sources A and B varying in their emission rates but fixed in their isotopic compositions. Data points will fall somewhere in the shaded area. The horizontal lines correspond to the δ-values of the sources while the lengths give the range of concentrations attributable to each source. Points on either horizontal line signify that the ambient concentration is due solely to that source.

For cases 4 to 7, the sources A and B were comparable in their emission rates. In case 8, source B is dominant while a large number of smaller sources of various isotopic compositions abound in the surroundings. Since these sources are small, their presence will only be noted when the concentration is low. At higher concentrations, the δ-values converge to that of the dominant source B.

Finally, in case 9, A is a small source which is fixed in its emission rate and δ-value in the midst of a number of variable sources which differ in their δ-values. At the lowest concentration observed, the δ-value is that of A since it is the sole source. With increasing concentration, the various sources contribute so that the range of δ-values departs increasingly from that of source A.

Case 9 has an interesting application within the laboratory where source A corresponds to traces of the element under examination in the reagents, i.e., the "blank". The variable sources correspond to the natural samples being processed. The concentration and the isotopic composition of the blank can be determined by plotting δ versus concentration^{-1} for a number of samples. If a batch of samples have the same isotopic composition but differing concentrations, then their relationship with respect to the laboratory blank is given by case 4.

Other situations or combinations of these cases can be envisaged. The plot realized will depend upon the data chosen. In the surroundings of a known industrial emitter, data may be taken from several stations at one time, one station at several times, or several stations on a number of occasions.

Identification of sources of atmospheric sulphur compounds using $\delta^{34}S$ versus concentration plots

The extent to which the hypothetical cases of Fig. 11-4 are encountered naturally can be illustrated with data from the Province of Alberta, Canada, in the vicinity of sour-gas-processing plants. In these operations, the well-known Claus process is used whereby part of the H_2S is oxidized to SO_2 which is in turn reacted with more H_2S to yield elemental sulphur. Consequently some 1200 tons of sulphur are discharged into the atmosphere daily, mostly as SO_2. Environmental research programmes include the use of lead peroxide exposure cylinders as well as high-volume sampling techniques similar to those described by Forrest and Newman (1973). Data obtained since 1971 show that atmospheric sulphur oxides in the vicinity of these industries have $\delta^{34}S$ values ranging from +5 to +30‰. The problem is to assess the extent to which this variation represents contributions from a number of difference sources.

Fig. 11-5 shows data from a site near a sour-gas plant operation in the Ram River area of Alberta where lead peroxide cylinders have been exposed for one-month periods. Higher δ-values generally correspond to higher concentrations and, with a few exceptions, are consistent with case 4 of Fig. 11-4. One can interpret the data in terms of the background fixed in concentration and isotopic composition (source A) and the industrial source (B) which varies in its emission rate. A plot of δ versus concentration^{-1} yields a straight line for four of the five data and identifies the industrial source (B) as having a δ-value near +22‰. Analyses of stack gases of the nearby industry verified this prediction. For station 5, one high $\delta^{34}S$ value was found at a low concentration and it did not fall on the linear plot. This either represents a month in which the background was very low or the isotopic composition of the background was altered towards the value of the industrial source. In practice, it is exceptional with monthly exposure cylinders to have four consecutive months giving data with the consistency shown in Fig. 11-5.

Data taken over a long time and at different stations around an industrial emitter usually fit case 8 of Fig. 11-4 as shown in Fig. 11-6. The dominant source is the Balzac sour-gas-processing plant which varies in its emitted contributions to any given location. The surroundings consist of many sources whose isotopic compositions and emissions vary spatially and temporally. The histogram of Fig. 11-6 gives the frequency of encountering $\delta^{34}S$ values independent of concentration. The peaking of the histogram near the same δ-value (+17‰) as the data converge to at high concentrations is further evidence of a dominant source with nearly constant $\delta^{34}S$ value. The inhomogeneity of the background is highly suggestive of biological sources of atmospheric SO_2.

Fig. 11-7 summarizes data using ground-level high-volume sampling techniques at three stations around the sour-gas-processing plant near Crossfield,

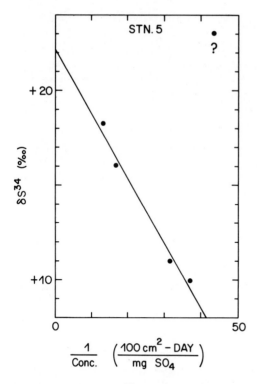

Fig. 11-5. Plot of $\delta^{34}S$ versus concentration^{-1} for monthly exposed lead peroxide cylinders at a site near Ram River, Alberta, 1971 (H.R. Krouse, unpublished data, 1971). With the exception of one sample, the data fit a straight-line plot which extrapolates to a $\delta^{34}S$ value of $+22‰$. This is the δ-value of the variable source B in case 4 of Fig. 11-4 and is identified with a sour gas processing operation in the vicinity.

Alberta. When the plant operation was closed down on June 16, δ-values for SO_2 were in the range +4 to $+10‰$. As the plant resumed operation, the δ-values increased to between +22 and $+28‰$. Plots of δ versus concentration^{-1} yield three lines which extrapolate to $+29‰$ which corresponds to analyses of the stack gases and samples of the plume taken by helicopter. The three lines correspond to the three different stations and signify that the concentration and isotopic compositions of the background were different at each station.

When high-volume samplers were operated randomly at Crossfield over a number of months, the data ceased to conform to the straight lines of Fig. 11-7. The data (Fig. 11-8) are best described by case 8 of Fig. 11-4. It is interesting that with the exception of one point, the δ-values of the background are less positive than the industrial emitter. Hence the histogram is not as symmetric as with the Balzac data of Fig. 11-6 where the industrial

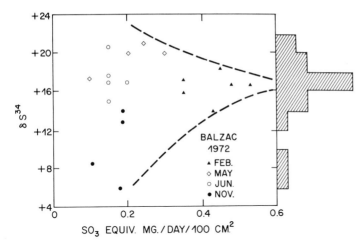

Fig. 11-6. $\delta^{34}S$ versus concentration for monthly exposed lead peroxide cylinders at various stations, Balzac, Alberta, 1972 (H.R. Krouse, unpublished data, 1972). Changing meteorological conditions and biological activity alter the concentrations and isotopic compositions of sources surrounding the industrial emitter. Hence the data fit case 8 of Fig. 11-4. The $\delta^{34}S$ histogram is independent of concentration. The peak at a $\delta^{34}S$ value near $+17^0/_{00}$ is identified with the industry.

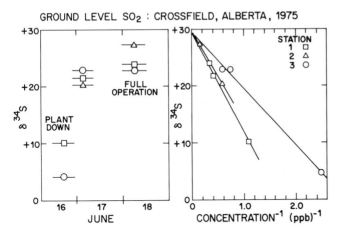

Fig. 11-7. Isotopic monitoring of a sour gas plant start-up with ground-level high-volume sampling of atmospheric SO_2, Crossfield, Alberta, 1975 (H.R. Krouse and H.M. Brown, unpublished UNISUL data, 1975). As the operation commenced, the $\delta^{34}S$ value and the concentration of SO_2 in the air increased. Plots of $\delta^{34}S$ versus concentration^{-1} extrapolate to a $\delta^{34}S$ value near $+29^0/_{00}$ consistent with case 4 of Fig. 11-4. This extrapolated value was identical with the $\delta^{34}S$ value of the plume sampled using high-volume sampling from a helicopter. Three linear plots arise since the contributions of SO_2 from other sources differed at the three stations.

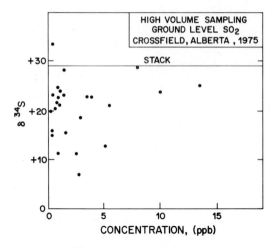

Fig. 11-8. $\delta^{34}S$ values for random ground-level high-volume sampling of atmospheric SO_2 over several months, Crossfield, Alberta, 1975 (H.R. Krouse and H.M. Brown, unpublished UNISUL data, 1975). Changing meteorological conditions and biological activity yield a distribution similar to Fig. 11-6 or case 8 of Fig. 11-4. However in contrast to Fig. 11-6, the $\delta^{34}S$ value of the industrial emissions is much higher (+29°/oo as compared to +17°/oo). Consequently at Crossfield, only one sample gave a $\delta^{34}S$ value higher than that of the plume. This suggests an upper limit in $\delta^{34}S$ value of +30°/oo for various sources in Alberta.

source has a much lower δ-value. This is expected, assuming that there is a maximum $\delta^{34}S$ (in this case +30°/oo) associated with sources in the Province of Alberta.

The examples cited above show that the two-source model (case 4, Fig. 11-4) accounts for the observed δ versus concentration^{-1} relationship in some locations for periods of months but in other locations for periods of only a few hours. Otherwise, case 8 of Fig. 11-4 usually applies. The explanation rests on a number of factors, the dominant one being wind direcion. If different sources exist about the observation point, case 4 of Fig. 11-4 might apply for a specific wind direction. Since case 8 can be considered as a summation of many case 4 situations with different δ_A and C_A values, it follows that changing wind directions can give rise to case 8. Departures from the case 4 model also arise when δ_A and C_A change with season as in the case of biological sources such as swamps or sewage lagoons. Lack of mixing of atmospheric gases is found where trees exert a canopy action and the data are generally inconsistent with Fig. 11-4.

Finally, the cases in Fig. 11-4 do not consider conversions or isotope exchange processes among sulphur compounds. In the atmosphere, H_2S and SO_2 are oxidized eventually to sulphate whereas in the hydrosphere, biological metabolism increases and decreases the valence states of sulphur by forming different chemical compounds. These problems will be considered in the section "Sulphur isotope fractionation during transformations of atmospheric and aqueous sulphur compounds".

It is noteworthy that in all the studies described above, the SO_2 concentrations were one to two orders of magnitude lower than those permitted by environmental regulations. This means that isotopic determinations coupled with concentration measurements can work well for typical environmental assessment problems and identify pollutant sources before hazardous levels are reached.

Polar plots of $\delta^{34}S$ values of atmospheric sulphur compounds versus wind direction

Since atmospheric gases from various sources reaching an observation point depend upon wind direction, polar coordinates can be used to locate emission sources. $\delta^{34}S$ is plotted as the radial coordinate while wind direction corresponds to the angular coordinates: hypothetical examples are considered in Fig. 11-9. In case 1, a source with a $\delta^{34}S$ value close to zero is located to the west of the monitoring site. When the wind is westerly, data points are obtained in the region A of the diagram if dispersion is small.

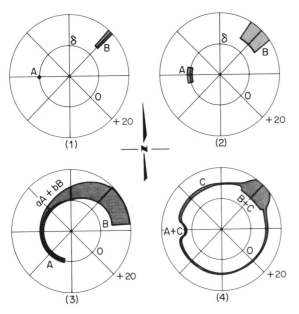

Fig. 11-9. Polar plots of δ versus wind direction for hypothetical cases as discussed in the text. δ is the radial coordinate.

Emission from another source to the northeast varies in $\delta^{34}S$ values from +8 to +20‰. Northeasterly winds provide data points in region B of the plot where the radial elongation provides a measure of the isotopic variations. Again, the small angular spread indicates little dispersion.

In case 2, the sources are the same as in case 1, but the dispersion is larger, signifying that the sources are farther away from the monitoring site and/or the winds are lower in velocity.

Case 3 considers further dispersion to the extent that there is overlap of the two sources on the polar diagram. The winds between westerly and northeasterly bear emissions from both sources and the $\delta^{34}S$ values fall between 0 and +20‰, dependent upon the relative contributions from the two sources. The shape of the shaded area is a function of the contribution from each source as well as the dispersions.

Case 4 adds an additional background component C which has a $\delta^{34}S$ value near +10‰ independent of wind direction. The two sources have dispersions consistent with case 2. An examination of the polar diagram shows that the emissions from source B are much higher than the background since the distribution in the area B and C is very similar to B alone in case 2. In contrast, contributions from source A are comparable to the background since the plot in the region $A + C$ does not inflect inwards to the δ-value of A (0‰) but rather to a value intermediate between A and C.

Fig. 11-10A illustrates how isotopic composition responded to wind direction in a study during September, 1975, near Whitecourt, Alberta. The high-volume sampler was operated on a 16-m-high scaffold in order to study the effects of SO_2 emissions on mid-crown and upper-crown foliage of a mature lodgepole/jack pine stand. On September 18, the wind was northwest and the SO_2 had a high $\delta^{34}S$ value corresponding to emissions from a sour-gas plant operation. The wind shifted easterly on September 19 and the $\delta^{34}S$ value immediately decreased since the SO_2 came from other source(s). From September 20 onwards, the wind swung southerly, then southwesterly to westerly with increases in $\delta^{34}S$ resulting from increased contributions from the industrial source.

Fig. 11-10B depicts the data on a polar diagram. Unfortunately the data are inadequate to plot a more desirably detailed diagram, one factor being uncertainties in the wind direction. However, there can be no doubt that westerly winds transport the industrial emissions to the study site. This plot emphasizes that it is inefficient to collect samples and meteorological data simultaneously and then interpret later since the wind may change direction many times during the acquisition of a sample. One complication is that the wind direction vector can change with altitude so that interpretation of the influence of industrial emissions at a give site is more complicated than suggested by Fig. 11-9. One step in overcoming possible errors in interpretation is to directionally control the data acquisition. A scheme under development at the University of Calgary is an array of high-volume samplers, each of

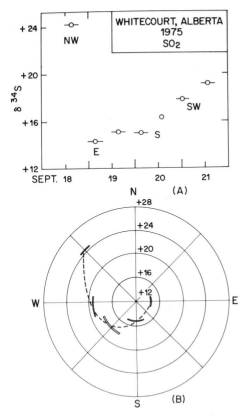

Fig. 11-10. A. Changes in $\delta^{34}S$ values for atmospheric SO_2 with wind direction at a sampling site near Whitecourt, Alberta, September 18—21, 1975. B. Polar plot of δ versus wind direction for A. The industrial operation is to the northwest of the sampling site. (Data of A. Legge and H.R. Krouse, published in reports to the Whitecourt Environmental Study Group, 1976).

which responds to a selected wind direction by magnetic reed switches activated by a magnet on a weather-vane axis.

Extension to identifying sources of sulphate in precipitation and surface waters

Since stable isotopes in the hydrosphere are considered in Chapters 1—4, this section will treat briefly the possibilities of extending the concepts of pp. 440—442 to precipitation and surface waters.

A number of sulphur isotope studies have been carried out on sulphate ions in rainfall (Jensen and Nakai, 1961; Eriksson, 1963; Nakai and Jensen, 1967; Mizutani and Rafter, 1969; Cortecci and Longinelli, 1970; Holt et al.,

1972). These studies have concluded that rainwater sulphate is depleted in ^{34}S with respect to seawater sulphate, the effect being more pronounced in industrial areas. These observations arise since rain and snow contain sulphate from a number of sources. Sea spray sulphate should have a $\delta^{34}S$ value near $+20^0/_{00}$, assuming that only mechanical transport takes place. Sulphate arising from the oxidation of biogenic H_2S or volcanic gases is comparatively depleted in ^{34}S. The industries involved in most studies reported to date have emissions with lower $\delta^{34}S$ values and hence the observation of $\delta^{34}S$ values for rainwater sulphate $\leq +20^0/_{00}$. Consistent with Figs. 11-5 to 11-8, $\delta^{34}S$ values greater than $+20^0/_{00}$ can be found in precipitation where industries such as sour-gas plant operations utilize isotopically heavier sulphur compounds.

Chloride concentrations have been utilized to delineate the seawater spray sulphate component in precipitation. Mizutani and Rafter (1969) utilized the parameter:

$$A = \frac{[SO_4^{2-}]_{seawater}}{[SO_4^{2-}]_{observed}} \times 100 = \frac{[Cl^-]_{observed}}{7.1[SO_4^{2-}]_{observed}} \times 100$$

This relationship assumes that all chloride came from the sea and the chloride/sulphate ratio in the marine component of precipitation is the same as in seawater, i.e., $[Cl^-/SO_4^{2-}] = 7.1$ by weight. By plotting $\delta^{34}S$ versus A for data from Gracefield, New Zealand, Mizutani and Rafter (1969) achieved a linear plot with much scatter which yielded $\delta^{34}S \approx -2$ and $+20^0/_{00}$ for $A = 0$ and 100% respectively. Cortecci and Longinelli (1970) carried out the same analysis for rainwater in Pisa, Italy. The majority of their data gave A values under 10% and one value near 20%. Hence the extrapolation to $\delta = +20^0/_{00}$ at A equals 100% is less certain than in the study by Mizutani and Rafter (1969) where the raw data extended over the total range of A.

The use of chloride content to establish seawater spray sulphate concentrations is an example of a general technique whereby an additional parameter is used to evaluate the contribution from a known source.

With reference to Fig. 11-4, the data of Mizutani and Rafter (1969) can be reassessed. Absolute concentrations of ions in rain and snow are difficult to interpret because of evaporation and other processes. However, case 4 of Fig. 11-4 can be tested by dealing with ratios of concentrations attributed to sources A and B. For the analysis it is necessary to fix the concentration identified with source A, the seawater sulphate, in all samples by a normalization procedure. Since Mizutani and Rafter (1969) considered one sample with 180 ppm Cl^- and 25.4 ppm SO_4^{2-} to be totally derived from seawater the simplest way to normalize their data is to adjust all samples to contain 180 units of Cl^- or 25.4 units of marine SO_4^{2-}. A plot of $\delta^{34}S$ versus the normalized concentration units is given in Fig. 11-11. When the inverse plot of $\delta^{34}S$ versus (normalized concentration units)$^{-1}$ is examined, all but one point

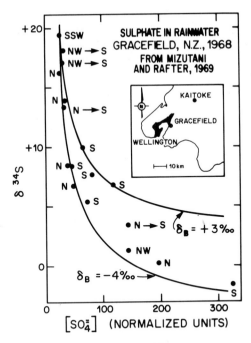

Fig. 11-11. Variation of $\delta^{34}S$ with sulphate concentration for rainfall, Gracefield, New Zealand, 1968 (raw data from Mizutani and Rafter, 1969). The sulphate concentrations have been normalized so that each sample contains 180 units of Cl^- and therefore 25.4 units of SO_4^{2-} which is derived from seawater. A plot of $\delta^{34}S$ versus (normalized concentration units)$^{-1}$ shows that all points but one are enclosed by two straight lines with δ_B intercepts of -4 and $+3^0/_{00}$. The curves above correspond to these δ_B values and assume that δ_A corresponds to seawater sulphate. The wind data suggest that $\delta_B = +3^0/_{00}$ is identified with a southerly direction while the value of $-4^0/_{00}$ arises from sources to the north or northwest of the sampling location.

can be enclosed by two straight lines with δ_B intercepts of -4 and $+3^0/_{00}$. The two curves corresponding to these values of δ_B have been drawn in Fig. 11-11. The simplest interpretation is that sulphate in the rainwater arises from a minimum of three sources with $\delta^{34}S$ values near +20 (seawater), +3, and $-4^0/_{00}$. An examination of the wind data generally shows that the $+3^0/_{00}$ is identified with a southerly direction and low Beaufort numbers whereas the source with $\delta^{34}S$ near $-4^0/_{00}$ exists to the north or northwest. Change in wind direction during a sampling period would produce points between the curves. This interpretation suggests the desirability of also using the wind directional sampling concept discussed in the previous section for precipitation.

Although a number of sulphur isotope studies have been conducted on river systems (Veselovsky et al., 1969; Longinelli and Cortecci, 1970; Holt et al., 1972; Hitchon and Krouse, 1972), detailed examinations of cross sec-

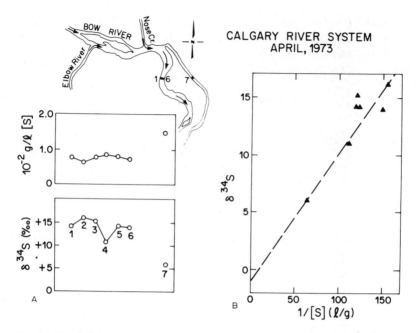

Fig. 11-12. Sulphur isotope variations in aqueous sulphate in a cross section of the Bow River, City of Calgary, April, 1973. Sample 7 in the side stream has a higher sulphate concentration and a much lower δ-value than samples in the mainstream and identifies a source just north of the sampling site. A plot of $\delta^{34}S$ versus $[SO_4^{2-}]^{-1}$ extrapolates to a $\delta^{34}S$ value of $-1‰$ for this source. Scatter about the line plot is related to non-mixing and perhaps biological conversions.

tions of streams are rare in the literature. This is unfortunate since, for example, Krouse and Mackay (1971) showed with oxygen isotopes that the Mackenzie and Liard Rivers required a distance of nearly 400 km for thorough lateral mixing. An appreciation of the slow mixing of flowing water is also illustrated by the data for a cross section of the City of Calgary river system depicted in Fig. 11-12. Both concentration and $\delta^{34}S$ values fluctuate in the main stream. The east branch stream is particularly interesting because of the higher SO_4^{2-} concentration and much lower $\delta^{34}S$ value in comparison to the main stream. These factors pinpoint a source to the east or north of the east branch since contributions from further upstream should have altered sample 6 in the main stream.

As a test of the hypothetical models of Fig. 11-4, a plot of δ versus concentration^{-1} is presented in Fig. 11-12. The isotopic composition of the additional effluent to the east branch can be estimated as near $0‰$ by simply comparing concentrations and δ-values of samples 6 and 7. Fig. 11-12B uses case 4 from Fig. 11-4 to verify this value and further suggests that sample 4 in the mainstream arose from a source which also had a $\delta^{34}S$ value near zero.

However, one must exercise caution in applying the theoretical considerations of Fig. 11-4 to streams since the variations in concentrations and $\delta^{34}S$ in the cross section are consistent with non-mixing while Fig. 11-4 is based on complete mixing of sulphur compounds from various sources. This may account for the scatter of data in Fig. 11-12B. Further, there is the problem of chemical and biological conversions of the dissolved sulphur species with attending isotopic fractionation.

SULPHUR ISOTOPE FRACTIONATION DURING TRANSFORMATIONS OF ATMOSPHERIC AND AQUEOUS SULPHUR COMPOUNDS

Introduction

The transport of sulphur compounds in the overall sulphur cycle involves physical, chemical, and biological processes, many of which are isotopically selective. It is important to understand the isotopic fractionation attending these processes in order to use sulphur isotopes effectively in environmental assessment. Thus laboratory experiments in which sulphur isotope abundances are altered assist in the interpretation of natural phenomena. It will be seen in the following sections that microbiological conversions generally effect larger isotopic fractionations than those associated with physical processes. This section will examine theoretical alterations of the $^{34}S/^{32}S$ abundance ratio which can assist in the interpretation of natural isotopic variations.

Isotopic fractionation phenomena can be considered in terms of isotope exchange reactions and kinetic isotope effects. Isotopic partition function ratios relevant to predicting the equilibrium constant for sulphur isotope exchange reactions have been calculated by Tudge and Thode (1950) and Sakai (1957). Of these, the equilibrium constant for the exchange reaction:

$$H_2{}^{34}S + {}^{32}SO_4^{2-} \overset{K}{\rightleftharpoons} H_2{}^{32}S + {}^{34}SO_4^{2-}$$

is interesting for a number of reasons. It involves the highest and lowest valence states (+6 and —2) of sulphur in nature and the predicted K has a reasonably large value (1.074 at 25°C). Difficulties have been experienced in effecting exchange between aqueous sulphide and sulphate species in the laboratory but studies have shown a favourable exchange rate at low pH values (Igumnov, 1976; Robinson, 1978).

For the understanding of kinetic isotope effects in the environment, it is useful to review the behaviour of a simple one-step, first-order conversion. In reference to Fig. 11-13, the term "kinetic isotope effect" is associated with the reactions:

$$R_{32} + \text{other reactants} \overset{k_{32}}{\longrightarrow} P_{32} + \text{other products}$$

R_{34} + other reactants $\xrightarrow{k_{34}}$ P_{34} + other products

R and P signify the reactant and product respectively which contain a sulphur atom. The subscripts 32 and 34 refer to the sulphur isotopes while k_{32}/k_{34} is the ratio of the isotopic rate constants.

Since k_{32}/k_{34} is usually larger than unity, the initial product is depleted in ^{34}S as compared to the reactant. As the reaction proceeds, the remaining reactant becomes progressively enriched in ^{34}S (Fig. 11-13). The product formed at any instant is depleted in ^{34}S with respect to the remaining reactant by an amount consistent with k_{32}/k_{34}. On the other hand, all of the product up to a given point in the reaction (accumulated P, Fig. 11-13) becomes slowly enriched in ^{34}S and acquires the same isotopic composition as the initial reactant when the conversion is completed. The isotopic difference between the accumulated product and the remaining reactant becomes very large as the percentage conversion increases.

It is readily appreciated that when "isotopic fractionation factors" are measured in nature, it is important to realize whether one is dealing with the accumulated or instantaneous product. An example of the latter case is

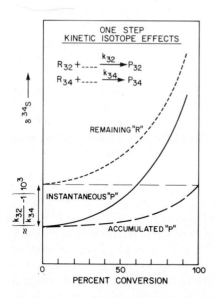

Fig. 11-13. Isotopic behaviour of components during a simple one-step first-order process where the ratio of the rate constants k_{32}/k_{34} is larger than unity. The "instantaneous" product curve refers to the product formed at a given instant in the conversion. It parallels the remaining reactant curve and the difference between the two curves is related to the k_{32}/k_{34} ratio. The "accumulated" product curve refers to all product formed up to a given point in the conversion. At 100% conversion, its isotopic composition must become that of the initial reactant.

biogenic H_2S which is steadily escaping. Unless the isotopic composition of the initial reactant is known, interpretation is very difficult. Mathematical expressions for the behaviour of components in a simple one-step conversion have been derived by Nakai and Jensen (1964). Fig. 11-13 can be presented in another manner in which $\delta^{34}S$ is plotted against the logarithm of the percent of remaining reactant. In this case, the remaining reactant and instantaneous product plot as straight lines (see Introductory Chapter).

Isotope fractionation in physical and chemical transformations

Physical processes such as the solution of atmospheric gases or particulates in raindrops and surface water are expected to be accompanied by small isotopic selectivities. Gases and ions are also adsorbed on solid sources. Nriagu (1974) carried out experiments on the adsorption of SO_4^{2-} by low-carbonate clayey mud from the Bay of Quinte, Lake Ontario. $^{32}SO_4^{2-}$ was preferentially adsorbed to the sediments. The isotopic behaviour is similar to that shown in Fig. 11-13 with the associated solutions becoming enriched in ^{34}S by upwards to $+6‰$ dependent upon the percentage of sulphate adsorbed. The $\delta^{34}S$ values between SO_4^{2-} remaining in solution and the accumulated adsorbed sulphate ranged as high as $22‰$. In waters with low SO_4^{2-} concentration, such adsorption (and desorption) processes might well alter the isotopic composition of the dissolved SO_4^{2-} by a few permil.

A chemical process of significance is the oxidation of H_2S, SO_2, and SO_3 in emissions to sulphate. The extent of SO_2 oxidation in power plant plumes has been pursued using sulphur isotopes by the Bookhaven National Laboratory (Newman et al., 1975a, b, c). In these studies, high-volume sampling techniques were used on board a single-engine aircraft and isotopic ratio measurements were used in conjunction with simultaneous concentration measurements of SO_2 and sulphate. There tended generally to be a small decrease in ^{34}S in the SO_2 with distance downwind of the stack. The data were interpreted in terms of a pseudo-second-order mechanism depending on sulphur dioxide and particulate concentrations. In the case of coal-fired plumes, the oxidation seldom exceeded 5% for distances up to 50 km, whereas in the case of oil-fired power plants, conversions of up to 13% were found over much shorter distances. It was proposed that vanadium compounds originating from the oil served to catalyze oxidation in the latter case. In such studies, it is difficult to determine the extent of the reaction. With distance from the stack, more contributions of sulphur compounds may arise from other sources. One approach is to inject a conservative tracer such as SF_6 (Newman et al., 1975c) or gold dust (UNISUL, unpublished work, 1976) into the stack. Assuming that the SO_2 or other emissions and the conservative tracer are dispersed with the same characteristics, then the ratio of their concentrations should provide a good measure of the extent of SO_2 oxidation with distance from the stack.

Harrison and Thode (1975) found that the isotope fractionation during inorganic chemical sulphate reduction at temperatures below 100°C could be approximated by a one-step first-order process (assumed to be the initial S—O bond rupture) with $k_{32}/k_{34} = 1.022$. Recent observations (C. Downes and T. Donnelly, unpublished data, 1976) with other reducing agents have realized smaller fractionations suggesting that processes other than S—O bond rupture can exert some control on the reaction rate. From the environmental viewpoint, the reduction of sulphate at low temperatures appears to be exclusively effected by microorganisms. It will be seen in the following sections that the isotopic fractionation in these conversions is much more complex than in the chemical reduction.

Isotopic selectivity by microorganisms in laboratory experiments

Fig. 11-14 summarizes a number of conversions for which data on sulphur isotope selectivity have been obtained in the laboratory. The figure represents a microbiological sulphur cycle with reduction processes (downward arrows) on the left side and oxidation processes (upward arrows) on the right side. The inorganic compounds have been ranked in their order of average valence state. This is somewhat unsatisfactory in that sulphur in the polythionates exists in more than one valence state and a number of disproportionation reactions occur. Sulphate and sulphite are metabolized by two distinct pathways; assimilatory and dissimilatory reductions. During assimila-

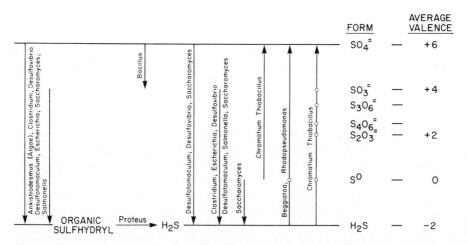

Fig. 11-14. Examples of microbiological conversions of sulphur compounds for which sulphur isotope fractionation data have been obtained in the laboratory. The left-hand side corresponds to assimilatory reduction while downward arrows in the middle represent dissimilatory reduction. Oxidation processes are shown by the upward arrows on the right.

tory reduction (left-hand side of Fig. 11-14) the resultant sulphide is incorporated into required metabolites such as cysteine, methionine, and glutathione. Although large fractionations favouring preferential reaction of $^{32}SO_4^{2-}$ and $^{32}SO_3^{2-}$ were found during assimilatory reduction by *Saccharomyces cerevisiae* (McCready et al., 1974), sulphate assimilation by micro-organisms (as well as plants and animals) usually results in small isotopic selectivity (Kaplan and Rittenberg, 1964; Mekhtieva, 1971). Since assimilatory or biosynthetic sulphate reduction is directed towards nutritional needs, this mechanism contributes little sulphide to the environment until death and decay of the organisms.

In dissimilatory or respiratory reduction, the sulphate acts in the absence of oxygen as a terminal electron acceptor. Dissimilatory sulphate and sulphite reductions are accompanied by a large range of isotope fractionations (Thode et al., 1951; Jones and Starkey, 1957; Harrison and Thode, 1958; Yeremenko and Mekhtieva, 1961; Kaplan and Rittenberg, 1964; Krouse et al., 1967; Krouse and Sasaki, 1968; Kemp and Thode, 1968; Šmejkal et al., 1971a; McCready et al., 1974, 1975; Chambers et al., 1975).

In dissimilatory sulphate reduction, fast rates are usually accompanied by negligible isotope fractionation. Harrison and Thode (1958) interpreted this as evidence that a non-isotopically selective step prior to the initial S—O bond rupture assumed rate control. In contrast, depletions of ^{34}S in the product H_2S of as much as $-46^0/_{00}$ have been reported during sulphate reduction by *Desulphovibrio desulphuricans* in the presence of ethanol (Kaplan and Rittenberg, 1964). Depletions of the same order have been routinely found during sulphite reduction by *Salmonella* sp. (Krouse et al., 1967) and *Saccharomyces cerevisiae* (McCready et al., 1974). It follows that in these reductions, so many bond ruptures and formations are involved that the one-step approximation is inadequate. In other words, the isotopic behaviour of the product is not consistent with that of the disappearing reactant because of the formation of intermediate sulphur compounds in the conversion. Rees (1973) has presented a steady-state model for bacterial reduction of sulphate, which is capable of explaining the above observations.

A very interesting phenomenon was noted when sulphite was reduced by *Clostridium* sp. (Šmejkal et al., 1971a; McCready et al., 1975; Laishley et al., 1976). During initial stages of the conversion, H_2S evolved which was depleted in ^{34}S with respect to the initial SO_3^{2-}, i.e., a normal kinetic isotope effect. After reaching a minimum $\delta^{34}S$ value (as low as $-20^0/_{00}$), the instananeously produced H_2S became isotopically heavier until it reached a point where it was more enriched in ^{34}S than unreacted SO_3^{2-} and intermediates in the reaction vessel, i.e., an inverse isotope effect existed. The enrichment in ^{34}S reached a maximum value (as high as $+45^0/_{00}$) after which it decreased. One simple explanation of these data involves branching during the reduction into two parallel pathways (Laishley et al., 1976).

Krouse et al. (1970) and Šmejkal et al. (1971a) reported that stepwise reductions of sulphate existed in thermal springs whereby one organism, an unidentified *bacillus* reduced sulphate to sulphite, while a *Clostridium* reduced the sulphite to H_2S. Isotopic fractionation realized with such pairs of organisms was discussed by Šmejkal et al. (1971a).

Kaplan and Rittenberg (1964) reported H_2S evolved by hydrolysis of the sulphydryl bond (—C—SH) from cysteine by *Proteus vulgaris* to be depleted in ^{34}S by about $5‰$.

Reduction of elemental sulphur by *Saccharomyces cerevisiae* to H_2S was found to be accompanied by negligible isotopic fractionation (Kaplan and Rittenberg, 1964).

Isotopic fractionation realized during biological oxidation is generally much smaller than that achieved during reduction processes. One problem with oxidation experiments is the delineation of the chemical and the microbiological effects. In experiments of sulphide oxidation with *Beggiatoa* there were slight depletions in ^{34}S in the product S^0 and SO_4^{2-}. However, the same fractionations were also realized in the control vessel with no bacteria. The *Beggiatoa* definitely thrived on the oxidation since the growth was orders of magnitude larger than in the control vessel with no sulphide. Further, the presence of the bacteria enhanced the oxidation rate by a factor of five. However, the organisms apparently exerted little influence on the isotopic selectivity (H.R. Krouse, R. Lewin and A. Sasaki, unpublished data, 1967).

By chemosynthetic or respiratory oxidation by members of the genus *Thiobacillus*, sulphide can be oxidized to S^0, polythionates, and SO_4^{2-} in the presence of oxygen. Using growing cultures of *Thiobacillus concretivorus*, Kaplan and Rittenberg (1964) found the isotopic composition of S^0 to be similar to the reactant sulphide, the polythionates enriched by up to $19‰$ in ^{34}S and SO_4^{2-} depleted in ^{34}S by as much as $-18‰$.

During photosynthetic oxidation of sulphide by *Chromatium* sp., Kaplan and Rittenberg (1964) found the polythionates again enriched in ^{34}S (up to $+11‰$), the S^0 depleted in ^{34}S (down to $-10‰$) and the SO_4^{2-} almost identical isotopically to the starting sulphide. In contrast, with the photosynthesizing purple bacteria *Rhodopseudomonas* sp., Mekhtieva and Kondrat'eva (1966) found the sulphide to become progressively lighter (reaching $-2.7‰$), the elemental sulphur slightly lighter (reaching $-1.0‰$) and the SO_4^{2-} heavier ($+3.1‰$) as the oxidation proceeded.

The microbiological oxidation of sulphur compounds is understood even less than reduction processes. Goldhaber and Kaplan (1974) discuss possible mechanisms and list the characteristics of many organisms which participate in the oxidative processes.

Microbiological fractionation in the hydrosphere

Microbiological isotope fractionation in springs has been reported by a number of authors (Kaplan et al., 1960; Krouse et al., 1970; Schoen and

Rye, 1970; Šmejkal et al., 1971b). Data from worldwide locations have been summarized by Krouse (1976) and for waters in the U.S.S.R. by Rabinovich (1969). Either of two processes may dominate. Sulphides may be initially dissolved and oxidized to elemental sulphur and sulphate with or without biological participation and almost always with negligible isotopic fractionation. This situation was found at Yellowstone (Schoen and Rye, 1970). More frequently, sulphate is dissolved from evaporite strata and upon transport to the surface undergoes biological reduction (see Chapter 6). The H_2S product is depleted by 0 to $-50^0/_{00}$ in ^{34}S with respect to the sulphate which is the same range found for laboratory experiments. Krouse and McCready (1979) plotted $\Delta\delta^{34}S$ for SO_4^{2-} and H_2S (corrected to zero conversion as in Fig. 11-13) for 52 thermal springs and found the most frequent $\Delta\delta$ value to be $30^0/_{00}$.

In the case of some Flysch waters in Czechoslovakia, the isotopic fractionation is consistent with the sequence:

$$\text{evaporitic } SO_4^{2-} \xrightarrow{\text{reduction}} \text{sulphide} \xrightarrow{\text{oxidation}} SO_4^{2-} \xrightarrow{\text{reduction}} \text{sulphide}$$

with $\delta^{34}S$ values as low as $-62^0/_{00}$ for the currently produced sulphide (Šmejkal et al., 1971b).

The actual site of SO_4^{2-} reduction tends to be close to the water sediment interface where conditions seem to be optimal in terms of organic matter and sulphate supply.

Sulphur isotope fractionations of the same magnitude as those encountered in springs have been observed during sulphate reduction in an estuarine basin (Orr and Gaines, 1973), the Dead Sea (Nissenbaum and Kaplan, 1976), the Black Sea (Vinogradov et al., 1962), stratified Lake Vanda, Antarctica (Nakai et al., 1975) and numerous marine sediment cores — off Southern California (Kaplan et al., 1963), Sites 26 and 27, Leg 4, JOIDES Deep Sea Drilling Project (Presley and Kaplan, 1970), Saanich Inlet, British Columbia (Nissenbaum et al., 1972).

With respect to Fig. 11-13, difficulties are encountered when assessing sulphur isotope fractionation during sulphate reduction in nature. If H_2S escapes or precipitates with iron, then the instantaneous product curve applies whereas if all the H_2S is retained, the accumulated product curve is appropriate. In practice, the situation usually lies somewhere between the two curves. In addition, there are input and removal processes for the sulphate. For example, oilfield formation waters have often been recharged by fresh surface waters and thus the sulphate concentration cannot be used to evaluate the fraction of sulphate reduced. In marine sediments, there is usually excess sulphur as discussed by Berner (1964), Strakhov (1972) and Goldhaber and Kaplan (1974). The amounts of reduced sulphur are as much as an order of magnitude greater than the sulphate originally enclosed by the sediment in the pore water. Movement of SO_4^{2-} into the sediment must have occurred during or after burial.

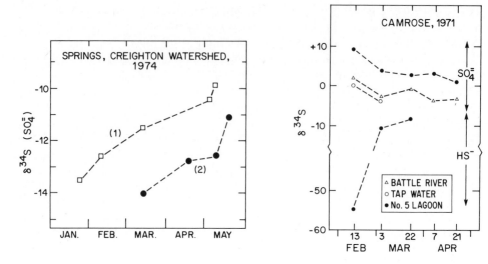

Fig. 11-15. Isotopic data from springs in the Creighton watershed, Saskatchewan, demonstrate higher ^{34}S values for the unconverted SO_4^{2-} as the season becomes warmer because of the increased activity of sulphate-reducing bacteria (H.R. Krouse and H. Steppuhn, unpublished data, 1974).

Fig. 11-16. Isotopic behaviour of sulphur compounds in the Camrose, Alberta sewer system (D.K. Robertson and H.R. Krouse, unpublished data, 1971). When the lagoons are ice-covered, anaerobic sulphate-reducing bacteria are more active and produce H$_2$S greatly depleted in ^{34}S. When the ice breaks up, most of this sulphide is reoxidized and aerobic bacteria become more prevalent.

The extent of SO_4^{2-} reduction and the isotopic fractionation in the environment can be seasonally dependent as illustrated for two small springs in Fig. 11-15. As the weather becomes warmer, there is increased conversion of SO_4^{2-} to H$_2$S and possibly methylsulphides with the result that the unreacted SO_4^{2-} decreases in concentration and increases in its ^{34}S content.

Fig. 11-16 shows the opposite trend with the coming of warmer weather in the City of Camrose water-sewer system. In the winter, waters were ice-covered so that oxygen for sewage oxidation tended to be derived from inorganic ions such as SO_4^{2-}. Consequently copious quantities of aqueous sulphide ions with large depletions of ^{34}S (—55‰) formed and the unreacted SO_4^{2-} became isotopically heavier. When breaks occurred in the ice cover, aerobic bacteria which utilized atmospheric oxygen became more active in processing the sewage. Under these conditions some of the sulphide formed by the sulphate reducers was probably re-oxidized. Hence the net sulphide production decreased and the sulphate and sulphide isotopic composition tended to become isotopically lighter and heavier respectively as spring approached. The changing temperature in all likelihood altered the isotopic selectivity of the sulphate reducers. This factor, coupled with sulphide oxidation, means that the situation cannot be described simply by the curves of Fig. 11-13.

H_2S generated by biological sulphate reduction in the hydrosphere is oxidized with biological assistance to elemental sulphur and sulphate. In springs of western Canada, sulphate minerals with δ-values near zero can be found attached to algae in waters where the δ-value for SO_4^{2-} is $+25°/_{00}$. The explanation is that *Beggiatoa* in association with the algae were able to oxidize ^{34}S-depleted HS^- and concentrate the product SO_4^{2-} to such an extent that precipitation occurs on a very localized scale.

Tufa on the ceilings of caves or gypsum deposits near thermal springs may be depleted in ^{34}S to the same extent as sulphide ions in the associated water. This arises by the oxidation of the H_2S released to the atmosphere. In one set of springs in northern Canada (Van Everdingen and Krouse, 1977) sulphate in the waters had δ-values near $+20°/_{00}$ while dissolved sulphide species were near $-30°/_{00}$. Sulphate minerals near the spring had δ-values lower than $-20°/_{00}$ and it was postulated that they arose from oxidation of the biogenic H_2S. This was verified by high-volume sampling of the atmosphere and the collection of copious amounts of H_2S and SO_2 which also had δ-values near $-30°/_{00}$. Further, sulphate rinsed from the surface of surrounding carbonate rocks was found to be similarly depleted in ^{34}S.

Baas-Becking (1925) first used the term "sulphuretum" to recognize the fact that a variety of micro-organisms oxidize and reduce sulphur compounds simultaneously in natural settings. The combination of sulphate reducers and *Beggiatoa* in association with algae described above represents a miniature sulphuretum. Larger sulphureta which have been studied using sulphur isotopes include the Cyrenaican Lakes in the Libyan desert (Macnamara and Thode, 1951), the Black Sea (Vinogradov et al., 1962) and the Kona sulphur deposit on the coast near Masulipatam, Madras, India (Kaplan et al., 1960). The enormous salt dome sulphur deposits of Texas and Louisiana can be described as large dried-up sulphureta. Some of the earliest sulphur isotope work verified that this sulphur was of biogenic origin (Thode et al., 1954).

SULPHUR ISOTOPE FRACTIONATION IN THE PEDOSPHERE

Sulphur compounds may enter the upper soil from deep sources or may penetrate from the surface. Organosulphur compounds can arise from degradation of dead plant and animal remains or by subsurface bacterial conversions. The behaviour of the water table influences the net upward or downward movement of sulphur compounds although biological factors are also important. Whereas upward transport usually involves sulphate ions in solution, Fig. 11-17 suggests that this is not the only mechanism. In this case

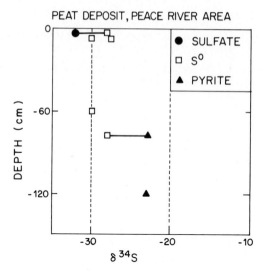

Fig. 11-17. The isotope composition of sulphur compounds in peat of the Peace River area, Canada (H.R. Krouse, unpublished data, 1969). The most plausible explanation is that reduced forms of sulphur migrated upwards and the near surface sulphate and elemental sulphur are isotopically lighter than the deeper sulphides because of kinetic isotope effects during oxidation.

sulphide at depth appears to be the source and either H_2S or HS^- has been oxidized as it approached the surface. The order of $\delta^{34}S$ enrichments of sulphate < elemental sulphur < sulphide is consistent with a kinetic isotope effect in which the lighter species were preferentially oxidized.

The extent to which atmospheric sulphur compounds penetrate the soil is dependent upon the vegetation covering the surface, the water table, the concentration of atmospheric compounds and the time duration. This is summarised in Fig. 11-18 which is based on several hundred analyses by the author in the Province of Alberta. Soil profiles for low (solid line) and high (dashed plot) penetration of atmospheric compounds are presented assuming the virgin soil had a $\delta^{34}S$ value near $0^0/_{00}$ and the atmospheric gases are around $+20^0/_{00}$. Where the soil is damp, well covered with mossy layers and organic duff, and the ambient SO_2 is low, penetration is negligible. Where the water table is low and the soil dry and lacking plant cover, sulphur compounds of industrial origin have been found as deep as 40 cm. The $\delta^{34}S$ profiles in soil in Fig. 11-18 are based on total sulphur. Since Lowe et al. (1971) found variations in the sulphur isotope composition of different chemical separates from the same soil specimen, the relative concentrations of individual sulphur compounds with depth contribute to the complexity of the isotopic profile.

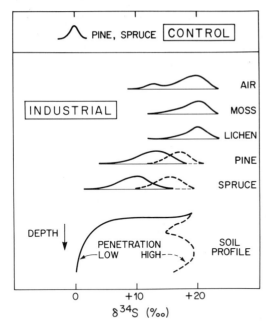

Fig. 11-18. Isotopic variations in total sulphur of trees, mosses, lichens, and soil profiles where the virgin soil has $\delta^{34}S$ values near zero and an industrial source emits SO_2 with $\delta^{34}S$ values near $+20‰$. The solid and dashed lines refer to low and high soil penetration respectively of the compounds of industrial origin.

SULPHUR ISOTOPES ELUCIDATE UPTAKE OF INDUSTRIAL SULPHUR COMPOUNDS BY VEGETATION

Fig. 11-18 also portrays the distributions in $\delta^{34}S$ values which arise for vegetation, assuming that the $\delta^{34}S$ value for virgin soil was near zero and industrial SO_2 had a δ-value near $+20‰$. For other locations, different δ-values will arise for the soil and atmosphere but the general conclusions should be the same. (For example, near Sudbury and Wawa, Ontario, the emissions have δ-values closer to zero.)

As discussed previously, the air at various locations around the industrial operation has long-term isotopic variations according to case 8 of Fig. 11-4. When mosses, lichens, and atmospheric SO_2 have been randomly sampled at various intervals for periods of two or more years, the histograms for $\delta^{34}S$ distributions are similar to those of air (Krouse, 1977b). This signifies that lichens and mosses take up sulphur from the atmosphere by rather direct means. One process is direct interaction with the gaseous sulphur compounds. Passage of rained-out sulphate formed by SO_2 oxidation and solutions of particulate sulphate on bark represent other possible mechanisms by which lichens and mosses may take up sulphur oxides from industrial sources. Fig. 11-19 shows the uniform isotopic composition of moss. The rhizoids are slightly depleted in ^{34}S compared to the upper parts but this may reflect attached organic debris of lower δ-value (which is difficult to remove).

If conifer needles are also randomly sampled for long periods of time, they have a sulphur isotope distribution similar to that of atmospheric sulphur oxides but shifted towards soil values (Krouse, 1977b). This shift is dependent upon whether the industrial sulphur compounds have penetrated the soil as discussed in the previous section. (Compare dashed and solid plots in Fig. 11-18.) Penetration results in the needles having δ-values closer to the atmospheric sulphur oxide distribution. If conifer needles are examined in a control area, i.e., essentially devoid of industrial emissions, their δ-values are closer to those of the soil. In contrast to lichens and mosses whose rhizines and rhizoids respectively serve for attachment purposes, trees have dynamic root systems which transport SO_4^{2-} upwards. Therefore it is not surprising that needles and leaves possess $\delta^{34}S$ values intermediate

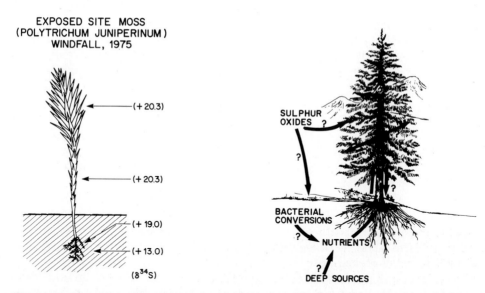

Fig. 11-19. Isotopic variations of total sulphur in moss exposed to emissions from a sour gas plant operation. The surrounding organic debris has a lower $\delta^{34}S$ value because of contributions from leaves. The slightly lower $\delta^{34}S$ value for the rhizoids as compared to the upper portions may relate to incomplete removal of the organic matter from these hair-like structures.

Fig. 11-20. Possible movements of sulphur compounds in a tree and its surroundings. Hopefully sulphur isotope abundances can evaluate which movements dominate under various environmental conditions.

between those of sulphur compounds in the atmosphere and in the soil. Fig. 11-20 shows the various uptake and transport possibilities for sulphur compounds in a tree and its environment. Many questions arise. How does the rate of atmospheric uptake compare to transportation from the root system? Do they vary with position on the tree? Is the net movement of sulphur upwards or downwards? Can sulphur in one needle find its way to another needle elsewhere on the tree? Is the sulphur isotope composition "fixed" in older needles? Stable isotope investigations elucidate many of these questions. For example, at a given location, pine needles are usually isotopically closer to the atmospheric sulphur oxides than spruce needles, indicating that the former derives proportionately more atmospheric sulphur through its needle surfaces. Generally upper needles are isotopically nearer to the ambient SO_2 than the lower needles, suggesting that the latter are shielded by the canopy of upper growth. The question also arises as to whether some of the processes in Fig. 11-20 are isotopically selective. One approach to these problems is to set out seedlings with soil sulphate of different concentrations and varying isotopic compositions (much heavier and much lighter than the atmospheric sulphur oxides). After some months, the isotopic composition of the needles, soil and air should delineate simple mixing and isotopically selective processes.

Sulphur isotope histograms for atmospheric sulphur oxides often have a secondary maximum as indicated around $+12°/_{00}$ in Fig. 11-18. In interpreting this peak, it must be recalled that the data are from lead peroxide exposure cylinders at a height of only one or two meters above the ground. Therefore a possible explanation is the emission of sulphur compounds from adjacent vegetation.

EVALUATION OF ANTHROPOGENIC AND NATURAL SOURCES OF SULPHUR COMPOUNDS

The sulphur cycle (Fig. 11-3) reveals that sulphur compounds arising from anthropogenic and natural sources are mixed in the atmosphere and hydrosphere. Since natural sulphur isotope abundances vary so greatly, assessment of man's contributions of sulphur compounds to the atmosphere by $^{34}S/^{32}S$ measurements (or other techniques) is difficult. If an industrial source dominates, the assessment is easier as seen by the plots of Fig. 11-4. Further, if the industrial operation shuts down and starts up (Fig. 11-7), a better appreciation of the fate of the industrial emissions can be obtained. Grey and Jensen (1972) reported sulphur isotope changes for a similar situation near Salt Lake City, Utah, when copper workers were on strike during July, 1971. With resumption of smelting, the air and precipitation $\delta^{34}S$ values decreased from 6 to $2°/_{00}$ as a result of inputs from the smelter plume (δ-values from -3.8 to -3.4).

The opposite extreme to assessment in close proximity of a known industrial emitter is the evaluation of biological contributions far removed from any processing plant. Hitchcock (1976) reported data obtained by the Brookhaven National Laboratory on gaseous and particulate sulphur collected from the tradewinds in Hawaii. The object was to establish biogenic contributions from the ocean to the atmosphere since the site was some 4000 km away from upwind anthropogenic sources. The $\delta^{34}S$ value for the atmospheric SO_2 was $+6^0/_{00}$ and Hitchcock (1976) concluded that it arose from oxidation of H_2S generated during bacterial reduction of oceanic sulphate. The extent to which the majority of marine biogenic H_2S can be considered as near $+6^0/_{00}$ is in question. Sulphides of far greater depletions in ^{34}S are precipitated in oceanic sediments. Much larger isotopic selectivities have been found with fresh waters.

Sulphur compounds present in an aerosol layer in the lower stratosphere have been correlated to volcanic eruptions on the basis of sulphur isotope measurements and known eruptions. The background which exists at times of volcanic inactivity was judged as biogenic and had an average $\delta^{34}S$ value near $2^0/_{00}$ (Castleman et al., 1973, 1974).

A few major projects in the literature have attempted to account isotopically for inputs and outputs of sulphur compounds to large systems. One such study concerned Lake Erie (Nriagu, 1975). Isotopic determinations and concentrations of sulphur compounds in cores revealed distinct differences among three major basins of the lake. These were generally correlated to differences in bacterial populations. Among the general conclusions were: (1) the SO_4^{2-} due to pollution sources had increased at the rate of 2 mg/l per decade since 1850 A.D. with a levelling off in recent years; (2) anthropogenic sources now accounted for 60—70% of the current SO_4^{2-} in Lake Erie; (3) less than 2% of the sulphur inputs were retained in the sediments.

SUMMARY

The examples cited in this chapter attest to the ability of sulphur isotopes to identify sources of sulphur compounds and elucidate their transformations in our environment. As with any technique, the success of the approach depends upon the effort expended in a given study. The success of a given investigation also rests on a number of "do's" and "don'ts" and it is appropriate to consider a few of these in summary.

Do: Obtain data for as many parameters in the system as possible. In the case of atmospheric studies, meteorological data and accurate concentration measurements are important. Make use of other techniques where possible. Oxygen isotope measurements on sulphates can assist in distinguishing seawater and anhydrite sulphate from that formed by oxidation of biogenic and industrial emissions (Mizutani and Rafter, 1969; Longinelli and Cortecci, 1970). Concentration measurements of other elements help to assess contributions to the environment from various sources, e.g. SO_4^{2-}/Cl^- ratios as discussed on p. 450. In hydrology, $^{18}O/^{16}O$ and/or D/H measurements aid in assessing lateral and vertical mixing and represent in combination with the sulphur isotopes, one of the few potential means of following the dispersion and transformations of aqueous sulphur compounds.

Don't: Assume that data can be transferred among locations. The isotopic "backgrounds" of our globe vary with location and season. For any subject of study, e.g. a lake, the δ-values for atmospheric and biological inputs must be determined for that system. Incorporation of data from other sites is risky.

ACKNOWLEDGEMENTS

The author's research on sulphur isotope abundances in the environment over the years has been funded by the National Research Council of Canada, Atmospheric Environment Service of Canada, Canadian Forestry Service, Aquitaine Co. Ltd., the Alberta Environment Research Trust and the University of Calgary Interdisciplinary Sulphur Research Group (acronym UNISUL, established as a negotiated development grant from the National Research Council of Canada).

REFERENCES

Baas-Becking, L.G.M., 1925. Studies on the sulphur bacteria. *Ann. Bot.*, 39: 613—650.
Berner, R.A., 1964. Distribution and diagenesis of sulphur in some sediments from the Gulf of California. *Mar. Geol.*, 1: 117—140.
Bogdanov, Yu.V., Golubchina, M.N., Prilutsky, R.E., Joksubayev, A.I. and Geoktistov, V.P., 1971. Some characteristics of sulphur isotopic composition of iron sulphides in Paleozoic sedimentary rocks at Dzhezkazgan. *Geokhimiya*, 11: 1376—1378 (in Russian). English translation, *Geochem. Int.*, 1971: 856—858.
Campbell, F.A., Evans, T.L. and Krouse, H.R., 1968. A reconnaissance study of some western Canadian lead-zinc deposits. *Econ. Geol.*, 63: 349—359.
Castleman, A.W., Munkelwitz, H.R. and Manowitz, B., 1973. Contribution of volcanic sulphur compounds to the stratospheric aerosol layer. *Nature*, 244: 345—346.
Castleman, A.W., Munkelwitz, H.R. and Manowitz, B., 1974. Isotopic studies of the sulphur component of the stratospheric aerosol layer. *Tellus*, 26: 223—234.
Chambers, L.A., Trudinger, P.A., Smith, J.W. and Burns, M.S., 1975. Fractionation of sulfur isotopes by continuous cultures of *Desulfovibrio desulfuricans*. *Can. J. Microbiol.*, 21: 1602—1607.
Cortecci, G. and Longinelli, A., 1970. Isotopic composition of sulphate in rainwater, Pisa, Italy. *Earth Planet. Sci. Lett.*, 8: 36—40.
Davies, G.R. and Krouse, H.R., 1975. Sulphur isotope distribution in Paleozoic evaporites, Canadian Arctic Archipelago. *Geol. Surv. Can. Paper*, 75-1 (B): 221—225.

Eriksson, E., 1963. The yearly circulation of sulphur in nature. *J. Geophys. Res.*, 68: 4001—4008.

Forrest, J. and Newman, L., 1973. Sampling and analysis of atmospheric sulphur compounds for isotope ratio studies. *Atmos. Environ.*, 7: 561—573.

Friend, J.P., 1973. The global surfur cycle. In: S.I. Rasool (Editor), *Chemistry of the Lower Atmosphere*. Plenum, New York, N.Y., pp. 177—201.

Gehlen, K. and Nielsen, H., 1969. Schwefel-Isotope aus Blei-Zink-Erzen von Oberschlesien. *Mineral. Deposita*, 4: 308—310.

Goldhaber, M.B. and Kaplan, I.R., 1974. The sulfur cycle. In: E.D. Goldberg (Editor), *The Sea, 5. Marine Chemistry*. Wiley-Interscience, New York, N.Y., pp. 569—655.

Granat, L., Hallberg, R.O. and Rodhe, H., 1976. The global sulphur cycle. In: B.H. Svensson and R. Söderlund (Editors), *Ecological Bulletins, 22. Nitrogen, Phosphorous and Sulphur — Global Cycles. SCOPE Rep.*, 7: 89—134.

Grey, D.C. and Jensen, M.L., 1972. Bacteriogenic sulfur in air pollution. *Science*, 177: 1099—1100.

Grinenko, L.N., Kononova, V.A. and Grinenko, V.A., 1970. Isotopic composition of sulphide sulphur in carbonatites. *Geochem. Int.*, 6: 45—53.

Harrison, A.G. and Thode, H.G., 1957. The kinetic isotope effect in the chemical reduction of sulphate. *Trans. Faraday Soc.*, 53: 1648—1651.

Harrison, A.G. and Thode, H.G., 1958. Mechanism of bacterial reduction of sulphate from isotope fractionation studies. *Trans. Faraday Soc.*, 54: 84—92.

Hitchcock, D.R., 1976. Microbiological contributions to the atmospheric level of particulate sulphate. In: J.O. Nriagu (Editor), *Environmental Biogeochemistry*. Ann Arbor Sci. Publ., Ann Arbor, Mich., pp. 351—367.

Hitchon, B. and Krouse, H.R., 1972. Hydrogeochemistry of the surface waters of the Mackenzie River drainage basin, Canada, III. Stable isotopes of oxygen, carbon and sulphur. *Geochim. Cosmochim. Acta*, 36: 1337—1357.

Hitchon, B., Brown, H.M. and Krouse, H.R., 1975. Stable isotope geochemistry of natural gases from Devonian strata, Alberta, Canada. *O.G.D. Symp. Organic Geochemistry of Natural Gas, Geol. Soc. Am. Meet., Salt Lake City, Utah, October 1975* (oral presentation).

Holser, W.T. and Kaplan, I.R., 1966. Isotope geochemistry of sedimentary sulfates. *Chem. Geol.*, 1: 93—135.

Holt, B.D., 1975. Determination of stable sulfur isotope ratios in the environment. *Prog. Nucl. Energy, Anal. Chem.*, 12: 11—26.

Holt, B.D., Engelkemeir, A.G. and Venters, A., 1972. Variations of sulfur isotope ratios in samples of water and air near Chicago. *Environ. Sci. Technol.*, 6: 338—341.

Igumnov, S.A., 1976. Experimental study of isotope exchange between sulphide and sulphate sulphur in hydrothermal solution. *Geokhimiya*, 4: 497—503.

Jensen, M.L. and Nakai, N., 1961. Sources and isotopic composition of atmospheric sulfur. *Science*, 134: 2102—2104.

Jones, G.E. and Starkey, R.L., 1957. Some necessary conditions for fractionation of stable isotopes of sulfur by *Desulfovibrio desulfuricans*. *Appl. Microbiol.* 5: 111—118.

Junge, C.E., 1963. Sulfur in the atmosphere. *J. Geophys. Res.*, 68: 3975—3976.

Kaplan, I.R., 1975. Stable isotopes as a guide to biogeochemical processes. *Proc. R. Soc. London, Ser. B*, 189: 183—211.

Kaplan, I.R. and Rittenberg, S.C., 1964. Microbiological fractionation of sulfur isotopes. *J. Gen. Microbiol.*, 34: 195—212.

Kaplan, I.R., Rafter, T.A. and Hulston, J.R., 1960. Sulphur isotopic variations in nature, 8. Application to some biogeochemical problems. *N.Z. J. Sci.*, 3: 338—361.

Kaplan, I.R., Emery, K.O. and Rittenberg, S.C., 1963. The distribution and isotopic abundance of sulfur in recent marine sediments off Southern California. *Geochim. Cosmochim. Acta*, 27: 297—331.

Kellogg, W.W., Cadle, R.D., Allen, E.R., Lagrus, A.L. and Martell, E.A., 1972. The sulfur cycle. *Science*, 175: 587—596.

Kemp, A.L.W. and Thode, H.G., 1968. The mechanism of the bacterial reduction of sulphate and of sulphite from isotope fractionation studies. *Geochim. Cosmochim. Acta*, 32: 71—91.

Krouse, H.R., 1976. Sulfur isotope variations in thermal and mineral waters. *Proc. Int. Symp. on Water-Rock Interaction (IAGC), Prague, 1974*. Geological Survey, Prague, pp. 1340—1347.

Krouse, H.R., 1977a. Sulphur isotope studies and their role in petroleum exploration. *J. Geochem. Explor.*, 7: 189—211.

Krouse, H.R., 1977b. Sulphur isotope abundances elucidate uptake of atmospheric sulphur emissions by vegetation. *Nature*, 265: 45—46.

Krouse, H.R. and Mackay, J.R., 1971. Application of H_2O^{18}/H_2O^{16} abundances to the problem of lateral mixing in the Liard-Mackenzie River system. *Can. J. Earth Sci.*, 8: 1107—1109.

Krouse, H.R., and McCready, R.G.L., 1979. Reductive processes in sulfur cycling. In: P.A. Trudinger and D. Swaine (Editors), *Biological Factors in Mineral Cyling*. Elsevier, Amsterdam, pp. 315—368.

Krouse, H.R. and Sasaki, A., 1968. Sulphur and carbon isotope fractionation by *Salmonella heidelberg* during anaerobic sulphite reaction in trypticase soy broth medium. *Can. J. Microbiol.*, 14: 417—422.

Krouse, H.R., McCready, R.G.L., Husain, S.A. and Campbell, J.N., 1967. Sulphur isotope fractionation by *Salmonella* sp. *Can. J. Microbiol.*, 13: 21—25.

Krouse, H.R., Cook, F.D., Sasaki, A. and Šmejkal, V., 1970. Microbiological isotope fractionation in springs of western Canada. In: *Recent Developments in Mass Spectroscopy*. University of Tokyo, Tokyo, pp. 629—639.

Laishley, E.J., McCready, R.G.L., Bryant, R. and Krouse, H.R., 1976. Stable isotope fractionation by *Clostridium pasteurianum*. In: J.O. Nriagu (Editor), *Environmental Biogeochemistry*. Ann Arbor Sci. Publ., Ann Arbor, Mich., pp. 327—349.

Longinelli, A. and Cortecci, G., 1970. Isotopic abundances of oxygen and sulphur in sulphate ions from river water. *Earth Planet. Sci. Lett.*, 7: 376—380.

Lowe, L.E., Sasaki, A. and Krouse, H.R., 1971. Variations of sulphur-34 : sulphur-32 ratios in soil fractions in western Canada. *Can. J. Soil Sci.*, 51: 129—131.

Macnamara, J. and Thode, H.G., 1951. The distribution of sulphur-34 in nature and the origin of native sulphur deposits. *Research (London)*, 4: 582—583.

McCready, R.G.L., Kaplan, I.R. and Din, G.A., 1974. Fractionation of sulfur isotopes by the yeast *Saccharomyces cerevisiae*. *Geochim. Cosmochim. Acta*, 38: 1239—1253.

McCready, R.G.L., Laishley, E.J. and Krouse, H.R., 1975. Stable isotope fractionation by *Clostridium pasteuranium*, 1. $^{34}S/^{32}S$: inverse isotope effects during SO_4^{2-} and SO_3^{2-} reduction. *Can. J. Microbiol.*, 21: 235—244.

Mekhtieva, V.L., 1971. Isotope composition of sulfur of plants and animals from reservoirs of different salinity. *Geokhimiya*, 6: 725—730 (in Russian).

Mekhtieva, V.L. and Kondrat'eva, E.N., 1966. Fractionation of stable isotopes of sulfur by photosynthesizing purple bacteria *Rhodopseudomonas* sp. *Dokl. Akad. Nauk SSSR*, 166: 465—468.

Mitchell, R.H. and Krouse, H.R., 1975. Sulfur isotope geochemistry of carbonatites. *Geochim. Cosmochim. Acta*, 39: 1505—1513.

Mizutani, Y. and Rafter, T.S., 1969. Isotopic composition of sulphate in rain water. Gracefield, New Zealand. *N.Z. J. Sci.*, 12: 69—80.

Nakai, N. and Jensen, M.L., 1964. The kinetic isotope effect in the bacterial reduction and oxidation of sulfur. *Geochim. Cosmochim. Acta*, 28: 1893—1912.

Nakai, N. and Jensen, M.L., 1967. Sources of atmospheric sulfur compounds. *Geochem. J.*, 1: 199—210.

Nakai, N., Wada, H., Kiyosu, Y. and Takimoto, M., 1975. Stable isotope studies on the origin and geological history of water and salts in the Lake Vanda area, Antarctica. *Geochem. J.*, 9: 7—24.

Newman, L., Forrest, J. and Manowitz, B., 1975a. Determining the extent of oxidation of sulfur dioxide in power plant plumes from isotopic ratio measurements. In: *Isotope Ratios as Pollutant Source and Behaviour Indicators*. IAEA, Vienna, pp. 63—76.

Newman, L., Forrest, J. and Manowitz, B., 1975b. The application of an isotopic ratio technique to a study of the atmospheric oxidation of sulfur dioxide in the plume from an oil-fired power plant. *Atmos. Environ.*, 9: 959—968.

Newman, L., Forrest, J. and Manowitz, B., 1975c. The application of an isotopic ratio technique to a study of the atmospheric oxidation of sulfur dioxide in the plume from a coal fired power plant. *Atmos. Environ.*, 9: 969—977.

Nielsen, H., 1974. Isotope composition of the major contributors to atmospheric sulfur. *Tellus*, 26: 213—221.

Nielsen, H. and Ricke, W., 1964. Schwefel-Isotopenverhältnisse von Evaporiten aus Deutschland; ein Beitrag zur Kenntnis von $\delta^{34}S$ in Meerwasser-Sulfat. *Geochim. Cosmochim. Acta*, 28: 577—591.

Nissenbaum, A. and Kaplan, I.R., 1976. Sulfur and carbon isotopic evidence for biogeochemical processes in the Dead Sea ecosystem. In: J.O. Nriagu (Editor), *Environmental Biogeochemistry*. Ann Arbor Sci. Publ., Ann Arbor, Mich., pp. 309—325.

Nissenbaum, A., Presley, B.J. and Kaplan, I.R., 1972. Early diagenesis in a reducing fjord, Saanich Inlet, British Columbia, I. Chemical and isotopic changes in major components of interstitial water. *Geochim. Cosmochim. Acta*, 36: 1007—1027.

Nriagu, J.O., 1974. Fractionation of sulfur isotopes by sediment adsorption of sulfate. *Earth Planet. Sci. Lett.*, 22: 366—370.

Nriagu, J.O., 1975. Sulfur isotopic variations in relation to sulfur pollution of Lake Erie. In: *Isotope Ratios as Pollutant Source and Behaviour Indicators*. IAEA, Vienna, pp. 77—93.

Orr, W.L., 1974. Changes in sulfur content and isotopic ratios of sulfur during petroleum maturation — study of Big Horn basin Paleozoic oils. *Bull. Am. Assoc. Pet. Geol.*, 50: 2295—2318.

Orr, W.L. and Gaines, A.G., 1973. Observations on rate of sulfate reduction and organic matter oxidation in the bottom waters of an estuarine basin: the upper basin of the Pettaquamseutt River (Rhode Island). In: *Advances in Organic Geochemistry*. Éditions Technip, Paris, pp. 791—812.

Presley, B.J. and Kaplan, I.R., 1970. Interstitial water chemistry: Deep Sea Drilling Project Leg 4. In: *Initial Reports of the Deep Sea Drilling Project IV*. U.S. Government Printing Office, Washington, D.C., pp. 415—430.

Rabinovitch, A.L., 1969. Stable isotopes of sulfur in surface waters (review). *Gidrokhim. Mater.*, 51: 98—105.

Rafter, T.A. and Wilson, S.H., 1963. The examination of sulphur isotopic ratios in the geothermal and volcanic environment. In: E. Tongiorgi (Editor), *Nuclear Geology in Geothermal Areas, Spoleto, 1963*. Consiglio Nazionale Delle Ricerche Laboratorio di Geologia Nucleare, Pisa, pp. 139—172.

Rees, C.E., 1970. The sulfur-isotope balance of the ocean, an improved model. *Earth Planet. Sci. Lett.*, 7: 366—370.

Rees, C.E., 1973. A steady-state model for sulfur isotope fractionation in bacterial reduction processes. *Geochim. Cosmochim. Acta*, 37: 1141—1162.

Robinson, B.W., 1978. Sulphur isotope equilibria between sulphur solute species at high temperature. In: B.W. Robinson (Editor), *Stable Isotopes in the Earth Sciences. N.Z. Dep. Sci. Ind. Res., Bull.*, 220: 203—206.

Robinson, E. and Robbins, R.C., 1970. Gaseous sulfur pollutants from urban and natural sources. *J. Air Pollut. Control Assoc.*, 20: 233—235.

Ryznar, G., Campbell, F.A. and Krouse, H.R., 1967. Sulfur isotopes and the origin of the Quemont ore body. *Econ. Geol.*, 62: 664—678.
Sakai, H., 1957. Fractionation of sulfur isotopes in nature. *Geochim. Cosmochim. Acta*, 12: 150—169.
Sakai, H., 1971. Sulfur and oxygen isotopic study of barite concretions from banks in the Japan Sea off the northeast Honshu, Japan. *Geochem., J.*, 5: 79—93.
Sangster, D.F., 1968. Relative sulphur isotope abundances of ancient seas and stratabound sulphide deposits. *Geol. Assoc. Can. Proc.*, 19: 79—91.
Sasaki, A. and Krouse, H.R., 1969. Sulfur isotopes and the Pine Point lead-zinc mineralization. *Econ. Geol.*, 64: 718—730.
Schoen, R. and Rye, R.O., 1970. Sulfur isotopes distribution in Solfataras, Yellowstone National Park. *Science*, 170: 1082—1084.
Schwarcz, H.P., 1973. Sulphur isotope analysis of some Sudbury, Ontario, ores. *Can. J. Earth Sci.*, 10: 1444—1459.
Shima, M., Gross, W.H. and Thode, H.G., 1963. Sulfur isotope abundances in basic sills, differentiated granites, and meteorites. *J. Geophys. Res.*, 68: 2835—2847.
Šmejkal, V., Cook, F.D. and Krouse, H.R., 1971a. Studies of sulfur and carbon isotope fractionation with microorganisms isolated from springs of western Canada. *Geochim. Cosmochim. Acta*, 35: 787—800.
Šmejkal, V., Michalícek, M. and Krouse, H.R., 1971b. Sulfur isotope fractionation in some springs of the Carpathian mountain system in Czechoslovakia. *Cas. Mineral Geol.*, 16: 275—283.
Smith, J.W., 1975. Stable isotope studies and biological element cycling. In: G. Eglinton Senior Reporter), *Environmental Chemistry, 1*. Chemical Society, London, pp. 1—21.
Smith, J.W. and Batts, B.D., 1974. The distribution and isotopic composition of sulfur in coal. *Geochim. Cosmochim. Acta*, 38: 121—133.
Strakhov, N.M., 1972. The balance of reducing processes in the sediments of the Pacific Ocean. *Litol. Polezn. Iskop.*, 4: 65—92.
Thode, H.G. and Monster, J., 1965. Sulfur-isotope geochemistry of petroleum, evaporites, and ancient seas. In: A. Young and J.E. Galley (Editors), *Fluids in Subsurface Environments. Am. Assoc. Pet. Geol., Mem.*, 4: 367—377.
Thode, H.G., Macnamara, J. and Collins, C.B., 1949. Natural variations in the isotopic content of sulphur and their significance. *Can. J. Res.*, 27: 361—373.
Thode, H.G., Kleerekoper, H. and McElcheran, D., 1951. Isotope fractionation in the bacterial reduction of sulphate. *Research (London)*, 4: 581—582.
Thode, H.G., Wanless, R.K. and Wallouch, R., 1954. The origin of native sulfur deposits from isotope fractionation studies. *Geochim. Cosmochim. Acta*, 5: 286—298.
Thode, H.G., Monster, J. and Dunford, H.B., 1961. Sulfur isotope geochemistry. *Geochim. Cosmochim. Acta*, 25: 159—174.
Thode, H.G. Dunford, H.B. and Shima, M., 1962. Sulfur isotope abundances in rocks of the Sudbury District and their geological significance. *Econ. Geol.*, 57: 565—578.
Tudge, A.P. and Thode, H.G., 1950. Thermodynamic properties of isotopic compounds of sulphur. *Can. J. Res., Sect. B*, 28: 567—578.
Van Everdingen, R.O. and Krouse, H.R., 1977. Sulphur isotope geochemistry of springs in the Frankling Mountains, N.W.T., Canada. *Proc. 2nd Int. Symp. on Water-Rock Interaction, Strasbourg, August 1977* (in press).
Veselovsky, N.V., Rabinovitch, A.L. and Putintseva, V.S., 1969. Isotopic compositions of sulphur in sulphate ion of the Kuma River and some of its tributaries. *Gidrokhim. Mater.*, 51: 112—119.
Vinogradov, A.P., Grinenko, V.A. and Ustinov, V.I., 1962. Isotopic composition of sulfur compounds in the Black Sea. *Geokhimiya*, 10: 851—873.
Vredenburgh, L.D. and Cheney, E.S., 1971. Sulfur and carbon isotopic investigation of petroleu, Windy River Basin, Wyoming. *Bull. Am. Assoc. Pet. Geol.*, 55: 1954—1975.
Yeremenko, N.A. and Mekhtieva, V.L., 1961. The role of microorganisms in fractionation of stable sulfur isotopes. *Geokhimiya*, 2: 174—180.

Chapter 12

ENVIRONMENTAL ISOTOPES AS ENVIRONMENTAL AND CLIMATOLOGICAL INDICATORS

B. BUCHARDT and P. FRITZ

INTRODUCTION

The ^{18}O contents in glacier ice and the carbonate tests and skeletons of marine micro-organisms and mollusks are widely recognized as indicators for climatic changes. In precipitations which form glaciers and the ice sheets of Antarctica and Greenland the ^{18}O contents decrease with decreasing temperature, but the ^{18}O contents of marine biogenic carbonates usually decrease with increasing temperature. This differential response is due to the fact that the isotopic composition of glacier ice does not only depend on isotope fractionation effects between vapour-liquid or vapour-solid but more on the "reservoir effects" experienced by the vapour reservoir during progressive moisture removal (Chapter 1). Since the solid or liquid phases are enriched in heavy isotopes as compared to the vapour the ^{18}O or deuterium contents of the latter decrease with progressing condensation at increasingly lower temperatures. Such effects are not important during carbonate precipitation in the ocean and the carbonate is enriched in ^{18}O with respect to the ocean water. This isotope effect increases with decreasing temperature and so does the ^{18}O in the carbonate (Chapter 6). However, the ^{18}O in marine carbonate skeletons depends not only on fractionation effects between carbonate and water but as well on the melting or formation of arctic ice masses which in turn influence the isotope composition of the oceans. Melting of ice masses caused by climatic improvements will bring about a decrease in ^{18}O in the oceans due to the inflow in ^{18}O-depleted melt water. The variation in ^{18}O contents of ocean water between maximum and minimum glaciation will be close to 1.5‰ (Craig, 1965). This can be more important than temperature for the isotopic composition of marine carbonates, such as foraminifera in tropical oceans, since the temperature of their environment may change very little during glaciation or deglaciation.

These ^{18}O temperature effects cause about 0.7‰ increase in ^{18}O in precipitations per degree Celsius rise in temperature. However, isotopic differences between shells formed during maximum or minimum glaciation in equatorial Atlantic surface waters and in the Caribbean amount to only about 1.6‰. This shows that freshwater systems respond much more vigorously to climatic changes than does a marine system and it is not unreasonable to expect (in continental regions) variations in $\delta^{18}O$ of more than 10‰ between glacial and interglacial periods. It has therefore been suggested that it might be possible to document climatic changes which are not recognizable on marine samples through analyses of freshwater carbonates, especially mollusks which presumably form their shells in isotopic equilibrium with their environment.

This chapter presents a summary of paleoclimatic and paleoenvironmental studies based on ^{18}O and ^{13}C variations in the carbonates of freshwater shells. Since both the inorganic dissolved carbon and the carbon of allochthonous and autochthonous organic matter in a lake or river system potentially contribute to the carbon in shells, a discussion of carbon isotopes in freshwater systems is also warranted. The isotopic composition of these components and their relative and absolute abundances are dependent on climatic and environmental parameters and, as a result, the organic matter in lake sediments may through its ^{13}C contents reflect significant environmental changes which affected a lake system.

The last section of this chapter presents a brief discussion on the use of deuterium in organic matter as an indicator for paleoclimatic conditions. This section does thus not directly deal with freshwater sediments, yet the deuterium contents in plant materials are dependent on the deuterium contents of the available freshwater and recent studies were judged important enough to be included in this chapter on paleoclimatology.

Not included in this chapter is a discussion of travertine and other spring deposits. Although the occurrence of such carbonates is very widespread and both their abundance and isotopic compositions reflect paleoenvironmental and paleohydrologic conditions, it has been shown that their deposition usually does not occur under isotopic equilibrium conditions (Gonfiantini et al., 1968). The complexity of the formation of freshwater carbonates as well as the lack of detailed investigations, which include considerations about their present-day formation, make it premature to discuss 'travertines as a tool in paleoclimatic studies''.

A similar argument can be brought forward for speleothems. Hendy (1971) showed how complex their formation is and how difficult the interpretation of isotope data from such cave carbonates can be. However, recent work by Schwarcz et al. (1976) on fluid inclusions may lead to the identification of controlling parameters and then permit a more reliable interpretation of carbonate isotope data (see Volume 2 of this series).

THE CARBON AND OXYGEN ISOTOPE COMPOSITION OF FRESHWATER SHELLS

A great number of studies dealing with the use of stable isotope contents in marine biogenic carbonates as paleoclimatic tool have been published since the pioneering work by Urey et al. (1951) and Epstein et al. (1951). Much less attention has been paid to freshwater environments since the interpretation of such data is considerably more complex.

In the preceding chapters it has been shown how the carbon and oxygen isotope composition of water and dissolved inorganic carbon in a freshwater system is controlled by several factors. In summary, the ^{18}O content of a given freshwater system is a function of (1) the average ^{18}O content of the local precipitation which in turn is dependent on climatic parameters (Dansgaard, 1964), (2) the hydrological regimes in a drainage basin, which can involve the mixing of water masses originating from areas with different meteorological parameters (Mook, 1970), and (3) the local degree of evaporation and residence time of a water mass within a basin (Craig, 1961). The ^{13}C compositions of the aqueous carbonate species are controlled by a delicate interaction between (1) CO_2 exchange with the atmosphere, (2) processes involving photosynthesis and respiration, (3) decay of ^{13}C-depleted organic matter, and (4) dissolution of solid calcium carbonate (Oana and Deevey, 1960; Mook, 1970; Fritz and Poplawski, 1974).

Carbonate minerals precipitated from a given freshwater system under isotopic equilibrium conditions (Chapter 9) will have an isotopic composition that reflects the ^{13}C and ^{18}O contents of the system. Thus the stable isotope compositions of fossil limestones and shells from freshwater deposits may carry information about ancient environments. Two criteria must be met, however, before a given freshwater shell can be used for such studies: (1) the organism in question must have formed its shell in isotopic equilibrium with the surrounding water, and (2) the original ^{13}C and ^{18}O ratios in the shell must not have been changed by any post-depositional processes. The following paragraphs discuss the possibilities for utilizing freshwater shell carbonate as a tool in paleoenvironmental and paleoclimatological research in the light of these conditions.

The oxygen isotope composition

The oxygen isotope composition of calcium carbonate precipitated under equilibrium conditions is a function of temperature and isotopic composition of the water (McCrea, 1950). The equation relating these variables is:

$$t = 16.9 - 4.2\,(\delta^{18}O_c - \delta^{18}O_w) + 0.13\,(\delta^{18}O_c - \delta^{18}O_w)^2$$

(Epstein et al., 1953; Craig, 1965), where $\delta^{18}O_c$ is the isotopic composition of CO_2 produced by phosphoric acid reaction of the carbonate and $\delta^{18}O_w$ is the isotopic composition of CO_2 in equilibrium at $25°C$ with the water from which the carbonate was precipitated; t is the environmental temperatue in $°C$.

This equation was developed for paleotemperature studies and can be used only when the isotopic composition of the water is known or can be assessed. Seawater has a relatively constant $\delta^{18}O$ value close to zero on the SMOW scale and paleotemperature calculations based on marine shell carbonate are normally carried out assuming that the isotope contents of oceans do not significantly change with time. The oxygen isotope composition of freshwater systems, on the other hand, is known to vary up to ten times as much as that of seawater, a fact which is due to the extensive fractionation of oxygen isotopes in the meteoric water cycle as described in Chapter 1. Direct paleotemperature determinations based on the oxygen isotope composition of carbonates are therefore limited to marine environments with relatively constant water isotopic compositions.

Inorganic carbonate precipitation under non-equilibrium conditions will in most cases lead to reaction products relatively depleted in ^{18}O due to kinetic isotope fractionation effects (Gonfiantini et al., 1968). However, as mentioned above, little work has been done on such carbonates (e.g. travertines) and more attention has been paid to biogenic shell-forming processes. There, kinetic effects can also affect the isotopic compositions and several groups of marine calcium carbonate-secreting organisms exhibit non-equilibrium or "vital effects" (as defined by Urey et al., 1951). Among these echinoderms (Weber and Raup, 1968) and corals (Weber, 1973, among others) are most important. Good evidence exists, however, that mollusks both from marine and freshwater environments form their shells at or close to oxygen isotope equilibrium (Epstein et al., 1951; Fritz and Poplawski, 1974). Besides mollusks, ostracods are the only calcium carbonate-shelled group of any importance in freshwater systems, and studies by Fritz et al. (1975) and Durazzi (1977) seem to indicate shell deposition under isotopic equilibrium conditions or close to it in this group, too.

The oxygen isotope composition of biogenic calcium phosphates has been investigated by Longinelli (1966), Longinelli and Sordi (1966) and Longinelli and Nuti (1965, 1973). Of the marine groups examined brachiopods, mollusks and, to a certain degree, vertebrates contain phosphate with a $^{18}O/^{16}O$ ratio related to temperature and isotopic composition of the water. Crustaceans on the other hand probably deposit their shells under non-equilibrium conditions. Only a single ^{18}O measurement of calcium phosphate from freshwater shell material was reported (the pelecypod *Unio pictorum* from the Tiber River in Italy; Longinelli and Nuti, 1973), and this result supported the conclusion reached for marine mollusks.

Because of the existence of such ^{18}O isotopic equilibria and since the growth temperature for most mollusks does not vary by more than about $10°C$ (which corresponds to $2-2.5‰$ variation in ^{18}O content) it is possible to deduce from the ^{18}O in shells a substantial amount of information on the isotopic composition of the water in their habitat and thus the paleoenvironment in which they existed. It should, however, be noted that aragonite is about $0.6‰$ richer in ^{18}O (at $25°C$) than calcite formed under the same conditions (Tarutani et al., 1969). In very detailed studies this should be taken into account.

The carbon isotopic composition

The study of carbonate isotope fractionations during shell deposition is complicated by the fact that different sources of carbon are available to the shell-forming organism. Besides one or more of the dissolved species of carbonate carbon the organism has the possibility of utilizing carbon from the food, or — as may be the case for gastropods — the ingestion of limestone particles.

The carbon isotope content of inorganically precipitated calcium carbonate is a function of temperature-dependent fractionation effects and the isotope content of the aqueous carbonate reservoir. Deines et al. (1974) have summarized the fractionation factors for the formation of calcite and presented the following equations:

$$1000 \ln \alpha_1 = -0.01 + 0.0063 \times 10^6 T^{-2}$$
$$1000 \ln \alpha_2 = -4.54 + 1.099 \times 10^6 T^{-2}$$
$$1000 \ln \alpha_3 = -3.4 + 0.87 \times 10^6 T^{-2}$$
$$1000 \ln \alpha_4 = -3.63 + 1.194 \times 10^6 T^{-2}$$

where α_1, α_2, α_3 and α_4 are the isotope fractionation factors between CO_2 (aq), HCO_3^-, CO_3^{2-} and calcite respectively, and CO_2 gas. T is the absolute temperature.

In most freshwater systems bicarbonate will be the dominant aqueous carbonate species. The isotope fractionation between bicarbonate and crystalline carbonate in the temperature range considered here leads to about 2.0‰ enrichment in ^{13}C in the carbonate, i.e., the precipitated calcite will have somewhat more ^{13}C than the dissolved inorganic carbon in the water. Similar to the isotope effects governing the ^{18}O contents in carbonates the $^{13}C/^{12}C$ ratios in precipitated calcite will increase as the temperature decreases within an isotopically constant environment (Emrich et al., 1970). A measurable fractionation between calcite and aragonite has also been reported (Rubinson and Clayton, 1969). These effects, however, are normally concealed by the far greater variation in the $^{13}C/^{12}C$ ratio of bicarbonate in natural freshwater systems. Thus the ^{13}C content of freshwater calcium carbonates precipitated in isotopic equilibrium with the dissolved inorganic carbon cannot be applied for paleotemperature calculations, but valuable information can be obtained about type and origin of the dissolved bicarbonate in the system. An exception would be a situation where one can demonstrate that an isotopic equilibrium with the atmosphere exists as was proposed, for example, for inorganic carbonate deposits in Lake Abbé (Djibouti) (Fontes and Pouchan, 1975).

The ^{13}C compositions of freshwater mollusks were first studied by Keith and Weber (1964) in a comparison of marine and freshwater shells. They found a variation in $\delta^{13}C$ exceeding 15‰ in the material examined, an observation they thought partly to be due to variations in the carbon isotope composition of the dissolved inorganic carbon and in part due to uptake of carbon from the food web. Unfortunately it is impossible to evaluate the extent of the postulated food effect as no information about the $^{13}C/^{12}C$ ratio of the environmental bicarbonate was given. A study of the ^{13}C and ^{18}O distribution in mollusk shells in Dutch estuaries by Mook and Vogel (1968) and Mook (1971) overcame this problem. When bicarbonate and shell data were compared, no significant deviations from the expected isotopic equilibrium values were found, a fact that, at least for the samples studied so far, contradicted the existence of vital effects in mollusks.

Other attempts to understand the nature of carbon vital effects have been made through ^{14}C analyses of molluskal shell carbonate. Several authors have reported ^{14}C concentrations in modern shells yielding radiometric ages in excess of 2000 years (e.g. Keith and Anderson, 1963; Tamers, 1970). The effect has been explained by (1) a possible incorporation of ^{14}C-deficient carbon from humus probably through the food web, (2) uptake of fossil carbonate carbon during rasping activities, or (3) utilization of ^{14}C-deficient bicarbonate from the water. The problem is of great importance as fossil shell carbonate could be well suited for ^{14}C age determinations. However, as for ^{13}C investigations, lack of information about isotopic compositions of the dissolved inorganic carbon in the snail habitats made it impossible to reach any final conclusions.

The most obvious method of clarifying the mechanisms of ^{13}C and ^{14}C uptake during shell-building processes is through tank experiments where the ^{13}C and ^{14}C sources can be monitored. Such experiments have been performed to a very limited extent. Rubin et al. (1963) observed ^{14}C deficiencies in regenerated shell carbonate from damaged mollusk shells, claiming ^{14}C-free food or limestone particles as the carbon source. Contrary to this Fritz and Poplawsky (1974) reported shell ^{13}C of three gastropod and one bivalve species studied in tanks to be dominated by the $^{13}C/^{12}C$ ratio of the dissolved inorganic carbon which was mostly bicarbonate. The latter experiment was undertaken with undamaged shells bred in the laboratory and may reflect natural conditions more closely. Minor fluctuations in the ^{13}C content within the populations in the different tanks can probably be explained by small contributions of food carbon or by variations in growth rate and thus in carbon fixation between different specimens.

In summary, there is good reason to assume that the carbon isotopic composition of mollusk shells is controlled by the dissolved inorganic carbon (DIC) in their habitat and not by vital effects or fractionations in the mollusks themselves. This is in accordance with the conclusions reached by Mook (1971) concerning carbon fixation in marine mollusks. As a consequence

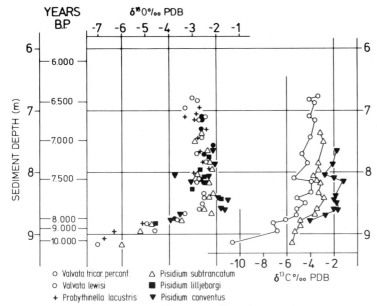

Fig. 12-1. Composition of $\delta^{18}O$ and $\delta^{13}C$ values obtained on different mollusk species from the same Lake Erie core. Note the systematic difference in ^{13}C contents which is not recognizable in the ^{18}O data. For further discussion of this core, see Fritz and Poplawski (1974).

^{13}C analyses of fossil freshwater shells can be applied to determine the $^{13}C/^{12}C$ composition of the dissolved inorganic carbon. Noteworthy here is that a rather consistent difference in $^{13}C/^{12}C$ ratios has been found to exist between different mollusk species from the same population (Fritz and Poplawski, 1974; Fritz et al., 1975). An example is shown in Fig. 12-1. Following the above reasoning about equilibrium deposition these differences indicate that in the lakes studied each species lives in a rather well defined micro-environment.

Diagenesis and preservation of the isotope compositions of carbonates

Before stable isotope analyses of fossil shell material can be applied to paleoenvironmental studies it is necessary to evaluate the degree of post-depositional alterations which may have affected the shell chemistry. The preservation of stable isotope ratios in marine shell material has been studied by several authors (e.g. Lowenstam and Epstein, 1953; Weber and La Roque, 1963; Stahl and Jordan, 1969; Tan and Hudson, 1974; Veizer, 1974; and Buchardt, 1977).

In principle post-depositional, diagenetic processes can change the original isotopic composition of a fossil carbonate shell in two ways: (1) through recrystallization, and (2) through solid state processes.

Since most studies with freshwater shells focus on the very recent history of geologic environments the latter process is thought to be insignificant. Weber and La Roque (1963) report that marine shells exposed for 4500 years to running water apparently remained isotopically unaltered.

However, recrystallization within sedimentary environments is known to occur and the recrystallized carbonates will have an isotopic composition reflecting that of the surrounding interstitial waters. Marine shell carbonates which normally have original $\delta^{18}O$ values reflecting isotopic equilibrium with seawater therefore exhibit characteristic isotopic depletions when exposed to recrystallization in meteoric water environments depleted in ^{18}O (Gross, 1964). Freshwater shells on the other hand have original isotopic compositions reflecting that of local meteoric water, and the ^{13}C and ^{18}O contents of recrystallized freshwater shell carbonate will be close to those of the original carbonate. This makes it impossible to assess on isotopic criteria alone any diagenetic impact on freshwater shells. Similar arguments can be applied to the trace element distribution between original and recrystallized freshwater carbonates. Measurements of the strontium and magnesium concentrations — which are important when diagenetic modifications of marine shell carbonate have to be estimated — are thus of no value.

Fortunately most freshwater mollusk shells are composed of aragonite, a mineral which is metastable in most surface and near-surface environments. Consequently, any diagenetic change will lead to either total dissolution or recrystallization in the form of calcite — the stable $CaCO_3$-polymorph under these circumstances. This implies that any fossil freshwater shells still composed of aragonite have survived diagenetic modifications, and such material probably yields unchanged $\delta^{13}C$ and $\delta^{18}O$ values. Thus X-ray analyses of the aragonite/calcite ratios combined with investigations of possible destructions of the shell ultrastructure are the only means of testing the preservation of the isotopic composition of fossil freshwater shell carbonate (Buchardt, 1977).

Identification of ancient freshwater environments

In well-preserved shell material and sedimentary carbonates one might expect that differences in isotope contents between marine and freshwater material would be large enough to use them as tools for the identification of ancient freshwater environments. Although many freshwater carbonates of biogenic or inorganic origin have lower ^{13}C and ^{18}O contents than their marine equivalents, this is not always the case. Freshwater shells from tropical environments, where extensive evaporation leads to ^{18}O enrichment of waters in rivers and lakes, may often show oxygen isotope compositions identical to or even enriched in ^{18}O when compared to marine shells. For example, *Melania* shells from Lake Chad have $\delta^{18}O = +3.6‰$ which corresponds to a deposition under equilibrium conditions or close to it from a water with $\delta^{18}O = +6.3‰$ at the average lake temperature of 27°C (Gasse et al., 1974). Similar problems exist for the interpretation of $^{13}C/^{12}C$ ratios. As demonstrated by Mook (1970) and by Fritz and Poplawski (1974) the carbon isotopic composition of bicarbonate in stagnant lakes approaches or exceeds the ^{13}C contents of dissolved inorganic carbon in marine environments. A distinction between marine and non-marine shells based on isotopic analyses is then no longer possible but can often be done on the basis of trace elements.

In summary, only strongly ^{13}C- and ^{18}O-depleted shell carbonates indicate non-marine environments — as long as vital effects and recrystallization can be ignored. Fossil shells with ^{13}C and ^{18}O compositions characteristic for marine conditions may, unfortunately, originate from marine as well as freshwater environments.

Clayton and Degens (1959), Keith and Weber (1964), Allen and Keith (1965), Tan and Hudson (1971, 1974), Allen et al. (1973), Rothe et al. (1974) and Dodd and Stanton (1975) applied ^{13}C and ^{18}O data from fossil molluskal shell carbonates and inorganic precipitates in an effort to distinguish between freshwater and marine environments. Increased age of material seems to be a critical limitation to the method as recrystallization and other isotopic exchange mechanisms eliminate the original isotopic differences. Nevertheless, as demonstrated by Tan and Hudson (1974), it is possible to identify characteristic stable isotope differences between well defined marine and non-marine mollusk shell carbonate as old as of Middle Jurassic age. This indicates that precise knowledge of the ^{13}C and ^{18}O composition of ancient freshwater systems is important when refined paleoenvironmental studies are undertaken, and the isotopic composition of shell carbonate may often be the only way to obtain such information.

In brackish environments which are transitional between an open marine and freshwater milieu the isotope contents of water, dissolved inorganic carbon and shell material are less well defined (Mook and Vogel, 1968; Mook, 1970). As stated by Mook (1970) the isotopic composition of water in estuaries with small residence time is solely determined by the mixing ratio of fresh- and seawater. The relation of $\delta^{18}O$ with chlorinity is linear; that of $\delta^{13}C$ depends on the relative carbon content of the unmixed components. Long residence time leads to additional ^{13}C enrichment due to growth of plankton and isotopic exchange with the atmosphere.

Based on this relationship it is possible to estimate the salinity conditions for mollusk shells grown in an estuarine, brackish environment by measuring the stable isotope composition of the shell carbonate — provided that temperature as well as ^{13}C and ^{18}O parameters of local fresh- and seawater are known or can be inferred. Studies following these lines have been published by Dodd and Stanton (1975) and Mook and Eisma (1976) dealing with Pliocene deposits in California and interglacial Eem deposits in Holland, respectively.

Paleoclimatological and paleoenvironmental studies

Shell carbonate from freshwater environments is in most cases inadequate for isotopic paleotemperature studies. Nevertheless, such material may carry isotopic information of paleoclimatological importance. As already mentioned, the $^{18}O/^{16}O$ ratio in a freshwater shell grown under isotopic equilibrium conditions reflects the isotopic composition of the local freshwater reservoir. This implies that long-term fluctuations in the oxygen isotope composition of such systems may be preserved in stratigraphic successions of shell material grown in the water.

A climatic "situation" at a locality S can be described by temperature and precipitation parameters. The oxygen isotope composition of the corresponding freshwater reservoir is related to these variables through ^{18}O contents of precipitations and degree of evaporation. In temperate and arctic areas where evaporation is small the ^{18}O composition of local precipitation is a function of temperature difference $\Delta T = T_{oc} - T_S$ between site of formation (the subtropical oceans with temperature T_{oc}) and site of precipitation (locality S with temperature T_S; Dansgaard, 1964). In North Atlantic coastal areas a drop in annual mean temperature of 1°C is followed by an approximate decrease in the $^{18}O/^{16}O$ ratio in local precipitations of $0.7^0/_{00}$ (latitude effect), while the decrease may be somewhat larger or smaller in non-coastal areas (continental effect) and elevated areas (altitude effect). As indicated by ^{18}O analyses of deep-sea cores from tropical and subtropical oceans T_{oc} as well as $\delta^{18}O_{oc}$ have been almost constant during postglacial time. This implies that climatically induced fluctuations in T_S (and therefore in $\Delta T = T_{oc} - T_S$) in temperate and arctic regions will be reflected in ^{18}O content of local precipitations. If these fluctuations are preserved in a suitable medium, the climatic history of the area can be inferred. The important consequences for paleoclimatological investigations were pointed out already by Dansgaard (1954) and Epstein (1956) and culminated in the very detailed studies of ice core material from Antarctica and the Greenland ice cap (Gonfiantini et al., 1963; Dansgaard et al., 1973, 1975). The same principles can be applied to freshwater shell material. Unfortunately only very few systematic studies on freshwater mollusks separated from lake sediments have been undertaken. Stuiver (1968, 1970) presents data on shells collected in lakes in the northern United States and covers the timespan between about 10,500 years B.P. and present, Covich and Stuiver (1974) discuss changes in tropical lake levels and molluskan populations in the Yukatan Peninsula during the past 9000 years and Fritz et al. (1975) studied samples from a core from central Lake Erie and cover the time from about 14,000 years B.P. to present. One of the difficulties one faces in quantitative evaluations of freshwater mollusk data are the seasonal variations of the isotopic composition of a lake. Not only might it be enriched in heavy isotopes because of significant evaporation effects, but one might also observe seasonal variations dependent on the balance between the addition of isotopically very variable precipitations, the inflow of groundwater, evaporation and finally the residence time of the water in a lake. Most shallow lakes show pronounced seasonal variations in their ^{18}O contents with the highest values observed during the summer months whereas little or no seasonal variations are observed in large deep lakes. However, both shallow and deep lakes will isotopically respond to major climatic or other events which affect the residence times of their water.

Similar observations are possible for the differences in ^{13}C contents in mollusk shells and the inorganic aqueous carbonate (mostly bicarbonate) although micro-environments and possibly minor "vital effects" cause some species-dependent differences in ^{13}C contents of a mollusk population within a lake. Parameters controlling the ^{13}C of the dissolved inorganic carbon in

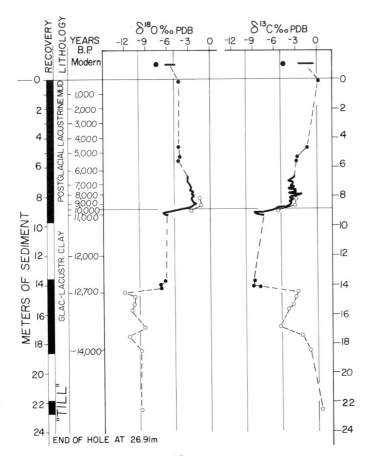

Fig. 12-2. Average ^{18}O and ^{13}C contents from mollusks separated from Lake Erie sediments (after Fritz et al., 1975).

freshwater systems are discussed in detail in the following section.

The principal conclusion of the studies mentioned was that the oxygen and carbon isotopic composition of freshwater mollusk shells do not vary at random but systematic and reproducible trends are observed. This is illustrated in Fig. 12-1 in which six mollusk species from Lake Erie sediments are compared.

Stuiver (1968, 1970) reported that both ^{18}O and ^{13}C contents in molluskal carbonate are lower before about 19,000 years B.P. than after and that it is possible to recognize from such data the hypsithermal interval between about 9000 and 5000 years B.P. which has been extensively documented through pollen studies. Furthermore, he shows that atmospheric circulation patterns over the Great Lakes region were rather stable for the past 9000 years. Furthermore, on the basis of the ^{18}O data obtained by Fritz et al. (1975) and shown in Fig. 12-2, it is possible to recognize at least two major events which affected Lake Erie and its predecessor: between 13,000 and 12,000 years B.P. and between 10,500 and 9000 years B.P. the ^{18}O content in the lake water rose sharply probably in response to major climatic improvements. Changes which occurred at about 6000 years B.P. are possibly related to changes in waterflow through Lake Erie since at that time the Niagara River opened up and the drainage of Lakes Erie and Huron changed from a northern to a southern discharge. Carbon isotopic compositions similarly show major changes and have been interpreted as being due to changing densities in aquatic vegetation in response to changing climate and water depth.

FRESHWATER LAKES AND SEDIMENTS

Freshwater lakes are very complicated hydrological and biological systems and the chemical and isotopic compositions of their sediments reflect this. Simple interpretations are usually not possible. The following sections attempt to describe and summarize the significance of these parameters with special emphasize on those on which the description of paleoenvironments depends.

^{13}C in aqueous carbonate and organic matter

In a discussion of the paleoclimatic information stored in the ^{13}C/^{12}C ratios in organic matter from sediments in the Gulf of Mexico, Roger and Koons (1969) suggested that planktonic organisms produced isotopically lighter carbon when surface water temperature was lower since both kinetic and equilibrium isotope effects which occur during photosynthesis are temperature dependent. According to this interpretation, warm climatic periods were characterized by δ^{13}C values between -24 and $-26\text{\textperthousand}$. How-

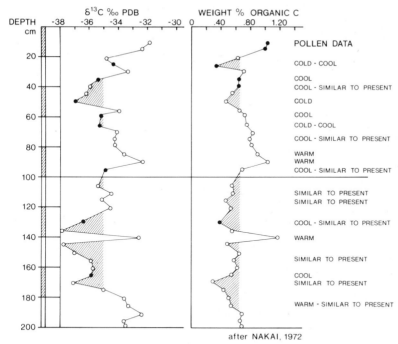

Fig. 12-3. A composition of ^{13}C data with organic carbon contents and pollen data from Lake Biwa (Japan) to document the agreement between paleoclimates and varying δ^{13}C values (after Nakai, 1972).

ever, Sackett and Rankin (1970) felt that in the Gulf of Mexico area the temperature changes were too high to account for these relatively small ^{13}C shifts and Parker et al. (1972) suggested that a more probable explanation for the observed correlation between ^{13}C shifts and climate was related to variable terrestrial input as proposed earlier by Sackett (1964).

Similarly, a comparison of pollen data with ^{13}C contents in the organic matter of sediments from Lake Biwa, Japan, led Nakai (1972) to the conclusion that there the ^{13}C contents reflected the climate which existed during the deposition of these sediments: higher ^{13}C contents were associated with warm climates whereas decreasing temperature resulted in decreasing ^{13}C concentrations. Parallel to these changes the organic productivity appeared to be higher during warm climates than during colder ones resulting in higher organic carbon contents in sediments deposited during the warm intervals (Fig. 12-3).

Two possible explanations were given by Nakai (1972) to account for the variable ^{13}C contents: (1) either at higher temperatures the photosynthetic reaction discriminated less towards the heavier $^{13}CO_2$ molecules and this would result in higher δ^{13}C values in the organic matter, or (2) higher δ^{13}C

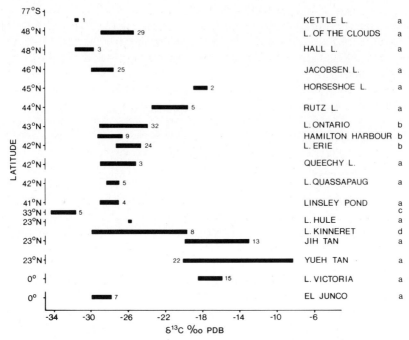

Fig. 12-4. ^{13}C contents in lake sediments as a function of latitude. A decrease with increasing latitudes seems to exist. Data are (a) from Stuiver (1975); (b) from Drimmie (1976); and (c) from Stiller (1976).

values in warmer lake water result from the increased use of isotopically heavier bicarbonate (via molecular CO_2) for photosynthesis. However, as a result of more detailed investigations on samples collected in lakes in North America, Stuiver (1975) emphasizes a much greater complexity of the organic carbon system than that envisaged by Nakai. Yet he points out that "for some lakes, the sedimentary ^{13}C record appears a sensitive indicator of climatic change with other factors causing smaller variations superimposed on the long term record. However, other lakes react little to climatic change and in these instances the ^{13}C record is useless for climate studies" (Stuiver, 1975, p. 261). Although this suggests that each lake reacts to a different degree towards climatic change a general tendency towards lower ^{13}C/^{12}C ratios is observed when the climate is cooler. This is reflected in lake sediments as a general but highly irregular decrease in δ^{13}C values with increasing latitudes (Fig. 12-4).

The overall range in δ^{13}C values for organic carbon in young and old freshwater sediments is close to 30‰ and varies between −7 and −38‰ (see also Chapter 9). Such a range is similar to the one found for terrestrial plants where different photosynthetic cycles are primarily responsible for the differential ^{13}C fixation in plant materials (Smith and Epstein, 1971). For the lakes, however, this large range must also be attributed to the different sources of carbon which contribute to lake systems as well as the biologic carbon isotope fractionation effects which occur during photosynthesis of algae and phytoplankton.

Photosynthesis discriminates against ^{13}C-bearing CO_2 or bicarbonate molecules and the produced organic matter is depleted in ^{13}C. Studies on oceanic algae show that the maximum isotopic difference between aqueous bicarbonate and algae is about $28^0/_{00}$ at $0°C$ and $25.6^0/_{00}$ at $30°C$ (Sackett et al., 1965; Deuser et al., 1968; Stuiver, 1975). Laboratory experiments with blue-green algae confirm the existence of these isotope effects but at the same time they demonstrate that the degree of photosynthetic fractionation of carbon isotopes is dependent in a non-linear fashion on the CO_2 contents of the feed gas. Virtually no isotope fractionation occurred and isotopically heavy organic matter was produced if the partial pressure of CO_2 in the feed gas was close to 2.5×10^{-3} atmosphere whereas at partial pressures above 10^{-2} atmosphere fractionation effects between 25 and $30^0/_{00}$ were obtained (Calder, 1972). Similarly, pH and temperature can influence the isotopic fractionation of carbon isotopes (Deuser et al., 1968) but again no simple relationship between carbon isotopic compositions and environmental parameters is known to exist.

These biologic effects on the $^{13}C/^{12}C$ ratios in autochthonous organic matter can be enhanced or dampened by the isotope contents of the CO_2 or bicarbonate available for photosynthesis. A very comprehensive investigation of the isotopic compositions of dissolved inorganic carbon was presented by Oana and Deevey (1960) who show that the CO_2 produced by biological activities within the lake and its sediments, the isotopic composition of groundwater carbonates entering the lake and isotopic exchange with atmosphere CO_2 contribute to varying degrees to the final ^{13}C contents of the

TABLE 12-1

$\delta^{13}C$ values ($^0/_{00}$ PDB) of dissolved inorganic carbon in Perch Lake. The values are average values for a given date from up to eight sampling stations (from Killey, 1977)

Depth (m)	1975					1976		
	May 11	May 30	June 26	July 20	August 9	January 29	March 31	April 21
0	n.d.	−9.3	−9.7	−11.5	−10.4	ice	−16.7	−15.9
1.1—1.5	n.d.	−9.5	−10.3	−11.4	−10.0	−14.3	−14.6	−16.5
2.4	−5.7	n.d.	n.d.	n.d.	−11.6	−16.9	n.d.	−17.9
3	−7.2	−18.5	−16.6	−15.9	−16.7	−16.5	−18.4	−17.3
3.4	n.d.	n.d.	n.d.	−20.0	n.d.	n.d.	n.d.	n.d.

n.d. = not determined.

inorganic carbon. This often results in a ^{13}C stratification within lakes. Table 12-1 lists average $\delta^{13}C$ values from a small (~0.5 km^2) and shallow (maximum depth 35 m) lake on the Canadian Shield (Killey and Fritz, 1978) and includes samples collected between 11 May 1975 and 21 April 1976. These data clearly demonstrate the seasonal variability one may expect. Two samples collected on 11 May just before the ice left the lake gave $\delta^{13}C$ values of -5.7 and $-7.2^0/_{00}$. This suggests that the lake became partially anaerobic during winter and ^{13}C-depleted methane and ^{13}C-rich CO_2 were simultaneously produced from the gyttia which has a $\delta^{13}C = -30.6\%_{00}$. In similar studies, Oana and Deevey (1960) collected samples of mud gas from methane-producing lake sediments and determined for the CO_2 a $\delta^{13}C = -5.6\%_{00}$, and Fritz and Poplawski (1974) found $\delta^{13}C$ values as high as $+4\%_{00}$ in shallow lakes during and immediately following the winter months.

During the summer, the reducing bottom conditions disappear in this small lake (Table 12-1) and already by the end of May the distribution of ^{13}C had attained a pattern that existed for the rest of the summer: low ^{13}C values in bottom waters and higher values towards the surface which clearly reflects the interplay between atmospheric CO_2 and CO_2 produced from and within lake sediments. It is possible to use these data to determine the CO_2 fluxes across the water-air interface (Killey and Fritz, 1978).

The anaerobic conditions which existed during the winter of 1975 were not repeated in 1976. However, as expected, the stratification disappears under ice cover and sediment-derived isotopically higher carbon dominates.

Inorganic carbonate entering a lake with inflowing groundwater will usually have $\delta^{13}C$ values between -10 and $-5\%_{00}$ (Chapter 9). Such contributions can be important and Oana and Deevey (1960) as well as Stuiver (1975) report that, for example, such contributions to Queechy Lake in New York State result in a present-day ^{14}C deficiency with regard to atmospheric CO_2 of about 30% because the groundwater carbonate is virtually free of ^{14}C. Similarly, the $\delta^{13}C$ values of dissolved inorganic carbon in hard water lakes are much higher and closer to values observed in marine carbonate rocks than in low carbonate lakes.

Depending on the size of the lake systems and its drainage basin inorganic carbon carried by surface water into the lakes could be quantitatively and isotopically very important. There again, one can expect rather large isotope variations depending on the biologic productivity in the streams and rivers discharging into the lakes, exchange with the atmosphere and possibly the origin of this water and the types of carbonate rocks dissolved in it. Hitchon and Krouse (1972) examined the Mackenzie River drainage system in Canada from this point of view and found that bicarbonate of biogenic origin dominated the $\delta^{13}C$ values of the dissolved inorganic carbon (Fig. 12-5). Exchange with atmospheric carbon dioxide and rock carbonate were of minor importance.

The ^{14}C contents of the inorganic carbon in surface waters, however,

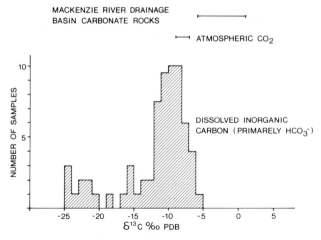

Fig. 12-5. Dissolved inorganic carbon in the Mackenzie River system (after Hitchon and Krouse, 1972).

should stay close to modern atmospheric values, since this biogenic CO_2 is "modern" and therefore has ^{14}C contents close to those found in the atmosphere. This argument is supported by only few measurements, one by Thurber and Broecker (1969) who in 1966 measured ~160% modern ^{14}C contents in a creek in California, analyses by the authors of the inorganic carbonate in the Grand River in southern Ontario which had in 1976 a ^{14}C content of 126% modern ^{14}C and a $\delta^{13}C = -12.6^0/_{00}$, and one 1976 result from the Seine River in Paris (France) with 96.7 ± 1.1% modern ^{14}C and $\delta^{13}C = -11.3^0/_{00}$ (J.Ch. Fontes, personal communication, 1978).

The existence of biological isotope effects and the variability of the isotopic composition of the inorganic carbon utilized for photosynthesis make it impossible to predict the ^{13}C values of the organic or inorganic components of a lake system. However, from the limited data available in the literature it appears that although one lake may be isotopically quite different from another within a given area, these differences are rather consistent and will only change if the environment changes. Prevailing climatic conditions play here certainly a role as documented by the above mentioned studies carried out in Japan and North America.

The climatic conditions have not only significant influence on the isotopic composition of the autochthonous organic matter deposited in a lake but also on its production and destruction. However, the total organic carbon found in lake sediments has also a significant allochthonous component in the form of terrestrial organic carbon carried into the lakes by rivers and shoreline erosion. This allochthonous, terrestrial component can isotopically be very variable. As shown in Chapter 9, the different photosynthetic pathways selected by different plants bring about very large variations of ^{13}C contents of the organic matter produced by land plants. Since the flora of a region is very strongly dependent on climatic conditions it is conceivable that the isotopic composition of the terrestrial component will also depend on and change with the climate. At present, not sufficient detailed studies on modern lakes and their allochthonous and autochthonous organic carbon components have been carried out to discuss this aspect further.

P. Deines (Chapter 9) briefly discusses the effects of diagenesis on the carbon isotopic composition of sedimentary carbon and lists those factors that could cause post-depositional changes: bacterial degradation, preferential decomposition of certain components and decarboxiliation. Both enrichment and depletion in ^{13}C would occur but indications are that shifts of more than $1-2^0/_{00}$ are unlikely. This represents approximately the difference between humic acids and total organic carbon whereby the humic acid has the lower $\delta^{13}C$ values (Degens, 1969; Drimmie, 1976).

Source differences, biologic fractionation effects and possibly diagenesis can all contribute and define the ^{13}C contents of organic matter in freshwater lake sediments. Fortunately most parameters are relatively constant in time and change only if the environmental conditions are altered to a significant degree. This leads to rather stable, but unique isotopic composition in each lake system. There can be little doubt, however, that no unique formula will be found which relates climate and ^{13}C contents in aquatic organic sediments be it in marine or continental systems. But, as Stuiver (1975) points out "in sensitive lakes, the ^{13}C record may provide a useful parameter indicating climate and productivity changes".

Calcium carbonate in lake sediments

Calcium carbonate is an important component of many lake sediments, be it in the form of marls or as finely disseminated mineral matter in organic and other sediments. Many lakes go in their evolution through a marl stage and it is common to find extensive deposits almost immediately following the glacio-lacustrine sediments deposited in ice-contact lakes. Both an allochthonous and autochthonous component can be present although only few and variably successful attempts have been made to differentiate between the two.

Degens and Ross (1971) describe biogenic and inorganic authigenic calcium carbonate deposits which occur together with detrital calcium carbonate in Black Sea sediments. They note that "carbonates present below 1.5 m have $\delta^{18}O$ values between -5 and $-7^0/_{00}$ (on the PDB scale) which is a common range for freshwater carbonates". In contrast, the $\delta^{13}C$ values of these carbonates vary between -2 and $+2^0/_{00}$ which is more in line with a marine origin. This discrepancy can best be explained in assuming that: (1) the oxygen exchange reservoir (interstitial waters) is large while that for carbon is comparatively limited, and (2) the oxygens of a CO_3 group are placed in better exchange positions than the carbon. In this way, the oxygen isotopes in the carbonate phase may completely equilibrate with pore fluids while exchange for carbon isotopes is restricted.

Re-equilibration of fine crystalline, detrital or authigenic carbonates in lake sediment could be an important process with respect to their isotope compositions. It is therefore interesting to compare inorganic and biogenic carbonates from the same horizon as was done, for example by Stuiver (1970): marl, an inorganic calcite precipitate, was always enriched in ^{13}C by about 3—4‰ if compared to associated aragonitic mollusk shells. This enrichment cannot be explained by the isotope fractionation effects which exist between syngenetic calcite and aragonite where calcite is enriched by ~1.8‰ with respect to aragonite, but is probably a result of photosynthetic activities in the lake. Aquatic plants preferentially utilize isotopically light carbon and since the photosynthetic CO_2 removal can result in a local supersaturation with respect to calcite, this mineral will then precipitate and have rather high $\delta^{13}C$ values. Such interpretation could indicate that many marls have preserved their original isotopic composition.

Contrary to the isotopic behaviour of marls, gastropods living in a pond weed environment in which isotopically enriched CO_2 might be available to them, do not show any ^{13}C enrichment but, if anything, are even depleted in ^{13}C if compared to gastropods living in environments with little vegetation. This is thought to be due to the release of isotopically light carbon during the decay of plant tissues under relatively stagnant conditions and emphasizes the very localized controls which determine the isotopic composition of marls.

The marls analyzed by Stuiver (1970) were systematically depleted in ^{18}O if compared to mollusks. Taking into account isotope effects between calcite and aragonite the average depletion was close to 1.3%. Provided these differences are not due to kinetic effects as they may occur during rapid precipitation this difference too has to be explained as being due to variable environmental conditions (e.g. seasonal variability of ^{18}O in lake water and temperature) between the time of calcite precipitation and growth of shells. A preservation of primary isotopic compositions is thus also supported by the ^{18}O data.

Detrital carbonate is an important fraction in older sediments from Lake Geneva whereas more recent ones have an increasingly significant authigenic component. ^{18}O and ^{13}C data indicate also in this case that the isotopic composition of both was closely preserved and it is possible to show that the detrital carbonates have their origin in the Mesozoic carbonate of the surrounding drainage basins whereas the authigenic ones were deposited in or close to equilibrium with the lake water.

A somewhat different lake carbonate has been studied by Fontes and Pouchan (1975) in Lake Abbé at Djibouti. There warm, calcium sulphate waters enter a carbonate-rich lake. The mixing of the two water types results in a calcite saturation and precipitation. The ^{18}O content shows that isotopic equilibrium with the water is achieved and since the lake carbonate is in such equilibrium with the atmosphere this is also the case for the solids. ^{14}C then permits a direct dating of these deposits whose formation started about 6300 years ago.

^{14}C in lake sediments

Paleoclimatic studies of lake cores demand that their sedimentologic history be described on an absolute time scale. ^{14}C dating of sediments is thereby a standard technique.

Two principal parameters will determine the ^{14}C record obtainable from lake sediments: the origin of the carbon-bearing waters, usually organic matter or inorganic and biogenic carbonates, and the depositional history of such material. Diagenetic changes will have little effect on the ^{14}C contents of both organic and inorganic matter since it is generally assumed that the isotope fractionation effects involving ^{14}C are about double those known for ^{13}C ($\sim 20‰$) and thus for ^{14}C should not exceed 4%. Clearly, this is no longer the case once carbon from the outside is added to the sediment, as it may occur if, for example, a lake carbonate recrystallizes in a secondary environment.

The influence of the source materials which contribute directly or indirectly to the ^{14}C budget of the sediments is similar to the one discussed for the $^{13}C/^{12}C$ ratios in the sediments. The terrestrial component consists primarily of the dissolved and particulate organic load brought into the lakes from eroding and leached landscapes, decaying organic matter and the phytoplankton produced within the rivers and streams. As a consequence the ^{14}C contents of this component can be very variable and, because of the presence of old matter, in general will be lower than the ^{14}C content of the atmosphere (Keith and Anderson, 1963).

Furthermore, groundwater can carry significant amounts of low-^{14}C carbonate into lake systems. This carbon can, if not lost through exchange with the atmosphere, be incorporated into the inorganic carbonate deposits found in many lakes or can provide a carbon supply for aquatic vegetation.

Neither for organic matter nor the inorganic carbonates, mainly marls, is it possible to quantitatively assess the amounts of "dead" carbon which have been added. However, Stuiver (1975) observes that in the New England lakes marls show lower ^{14}C contents than does the organic matter, i.e., the "dead carbon effect" is more significant and age differences between organic and inorganic carbon of more than 2000 years are known. Bortolami et al. (1977) note a difference of about 6500 years between ^{14}C ages for organic matter and associated calcareous mud. Degens and Ross (1971) find even larger differences in Black Sea sediments where at a sediment depth of 5 m a ^{14}C age difference of $\approx 15,000$ years exists between the two substances. These differences are principally caused by the mixing of ^{14}C-free ancient carbonates with autigenic Black Sea precipitates.

Unfortunately only indirect methods can be used to develop suitable procedures for the transformation of these relative ages to absolute ages and these methods are usually based on a comparison of ^{14}C data and sedimenta-

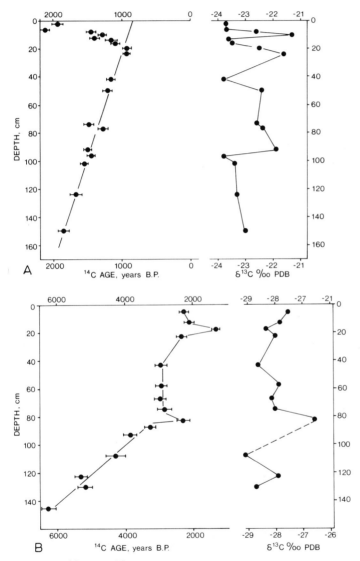

Fig. 12-6. ^{14}C and ^{13}C in lake sediments. A. Samples from a core collected in the Baltic Sea (Erlenkeuser et al., 1974). B. Samples from Lake Ontario. In both cases modern samples have a high ^{14}C age and ^{14}C changes with depth indicate that simple extrapolation towards the surface to obtain the "initial" ^{14}C contents of older sediments has to be done with caution.

tion rates. Attempts by Stiller (1976) to quantify through ^{18}O and ^{13}C analyses the autochthonous and allochthonous components of calcium carbonate in a small lake in Israel were somewhat inconclusive since daily and seasonal variations affected the abundance of both isotopes in the autochthonous calcium carbonate precipitate.

The indirect approach has been discussed by Stuiver (1968, 1975), Erlenkeuser et al. (1974) and others. Data from two cores are shown in Fig. 12-6, one from the Baltic Sea, the other from Lake Ontario. The Baltic Sea core data indicate a continuous and constant sedimentation for the past 2000 years, neglecting the anthropogenic contributions of the past 150 years. An extrapolation to the sediment surface yields a modern sediment age of 850 years. Similar extrapolations were carried out by Stuiver (1975) who obtained "ages" for modern sediments ranging from 200 years in a lake with a very high organic productivity to about 2500 years for a hard water lake. Such extrapolation is not possible for the Lake Ontario core since here evidently major disturbances had occurred. Also, if one were to attempt to extrapolate from the data obtained on samples from below 1 m to the surface one would obtain future ages. This is clearly impossible and may have to be explained by either differential sediment compaction which would distort any such extrapolation towards more modern ages, or increased sedimentation rates during the deposition of the younger overlying sediments.

Modern sediments are in most lake systems contaminated through man's activities. Therefore, they do not give any indication about the dead carbon contributions to the old sediments and no simple correction formula based on other geochemical parameters has yet been found to evaluate these. Isotope data of the very young sediments provide, however, valuable environmental information and indirectly reflect the magnitude of modern pollution. Spiker and Rubin (1975) and Kölle (1970) carried this argument further and analyzed directly the dissolved organic carbon of rivers for their carbon isotopic compositions in order to distinguish between fossil organic carbon from petrochemical effluents and modern organic carbon from domestic wastes and natural decaying organic matter. They could then show that, for example, the organic carbon dissolved in the Potomac River in Washington, D.C., contains up to 30% fossil fuel carbon.

It was mentioned above, that isotope effects involving ^{14}C are twice those of ^{13}C participating in the same reaction. This provides a basis for a minor but sometimes necessary correction based on $\delta^{13}C$ values in organic matter. The total range of $\delta^{13}C$ in organic matter is close to $30^0/_{00}$ and therefore 6% for ^{14}C. The measured age difference between an isotopically heavy and light wood would thus be close to 400 years. To overcome this by convention all ^{14}C dates on organic matter are normalized to a $\delta^{13}C = -25^0/_{00}$, a value which is close to the average value of most terrestrial plants (Broecker and Walton, 1959). For samples with $\delta^{13}C$ values within $5^0/_{00}$ of this value this correction is usually smaller than the analytical error and therefore is often neglected.

DEUTERIUM IN ORGANIC MATTER AS PALEOCLIMATIC INDICATOR

The dependence of the deuterium contents in plant matter on the isotopic composition of the water available to them and its possible use as paleothermometer has first been proposed by Schiegl and Vogel (1970). Smith and Epstein (1970) documented the relationship between water and plant for salt marsh biota, and Schiegl (1972) discussed the deuterium content in peat and the paleoclimatic implications of such analyses whereas Libby and Pandolfi (1974), Schiegl (1974) and Wilson and Grinsted (1975) attempted to correlate deuterium contents in tree rings with historical temperature records.

The analyses by Libby and Pandolfi (1974) included $^{18}O/^{16}O$ and $^{13}C/^{12}C$ determinations. However, as Epstein et al. (1976) point out the $^{18}O/^{16}O$ ratios in land plants are poorly correlated with those of the associated environmental water and, therefore, will not be further discussed.

Similarly inconclusive are carbon isotope data on total wood or cellulose samples from tree rings. For example, Craig (1954) presented data on sequoia tree rings which grew between about 1100 B.C. and 1650 A.D. and found that although the $\delta^{13}C$ values in the oldest part of the tree were lower by about $2^0/_{00}$ most of the fluctuations observed could neither be correlated with environmental nor physiological parameters. This observation was also made by Yapp and Epstein (1977) who determined $^{13}C/^{12}C$ variation on cellulose separated for deuterium analyses.

Most deuterium studies on plant material are based on deuterium determinations in the total plant hydrogen. However, plants are extremely complex chemical systems and, similar to what is known for carbon isotopic compositions (Chapter 9), isotopic differences between different chemical compounds in any given plant could determine the $^2H/^1H$ ratio observed for the total plant. For example, Smith and Epstein (1970) found that the lipid fraction of plants has about $100^0/_{00}$ less deuterium than the total plant hydrogen. Such observations imply that deuterium measurements made on "total organic matter" and difference observed between different plants, for example, might well be as much due to the chemical heterogeneities between different plant species as they are a function of environmental parameters.

To overcome this problem Epstein et al. (1976) and Epstein and Yapp (1976) investigated the usefulness of a single chemical substance, cellulose, which is common to all photosynthetic plants. The same choice was made by Wilson and Grinsted (1975). In these studies it was then found that two types of hydrogen do occur in plants and that carbon-bound hydrogen in cellulose is non-exchangeable whereas the hydroxyl or oxygen-bound hydrogens exchange readily. Preparation techniques have to aim at the total removal of all hydrogen which might have undergone secondary exchange before the non-exchangeable hydrogen can be extracted for deuterium analyses. These techniques are described in detail by Epstein et al. (1976).

TABLE 12-2

δ^2H values of plant extracted cellulose nitrate and associated environmental water (from Epstein et al., 1976)

Sample*	Location	Cellulose nitrate		Environmental water δ^2H (‰)
		δ^2H (‰)	H$_2$ yield (%)	
Turtle grass	open ocean near Miami, Florida	−1(1)**	93	+7
Turtle grass	open ocean off Puerto Rico	−4(3)	100	+7
Red mangrove	open ocean near Miami, Florida	−7(3)	91	+7
Phyllospadix	open ocean near Santa Catalina, California	−40(4)	98	−3
Grass (unidentified)	University Bay, Madison, Wisconsin	−44(2)	93	−32
Box Elder tree	University Bay, Madison, Wisconsin	−44(1)	91	−32
Oak tree	near NASA Space Center, Houston, Texas	−45(2)	102	−24
Silverweed	Crespi Pond, Monterey Peninsula, California	−60(2)	98	−28
Pondweed	Crespi Pond, Monterey Peninsula, California	−66(2)	96	−28
Scots pine	Loch Affric, Scotland	−63(1)	103	−42
Eurasian water milfoil	Lake Mendota, Wisconsin	−71(2)	106	−51
Grass (unidentified)	Wisconsin River near Spring Green	−85(2)	100	−69
Hard maple	Wisconsin River near Spring Green	−84(2)	100	−69
Grass (unidentified)	marsh near Oconto, Wisconsin	−104(2)	100	−65
White birch	marsh near Oconto, Wisconsin	−92(2)	98	−65
Coulter pine	Kalamalka Lake, British Columbia	−103(2)	100	−97
Polygonum natans	Bridgeport Lake, California	−110(2)	97	−107
Spruce tree	Lac Des Roches, British Columbia	−134(2)	96	−113
Sedge	Nymph Lake, Colorado	−140(2)	96	−104
Poplar tree	Nalte Lake, British Columbia	−140(2)	98	−117
Poplar tree	Okanagan Lake, British Columbia	−142(2)	98	−106
Poplar tree	Southwest corner Stuart Lake, British Columbia	−153(2)	97	−139
Poplar tree	Southeast corner Stuart Lake, British Columbia	−154(2)	93	−144
Pine tree	Seeley Lake, British Columbia	−154(2)	100	−136
White spruce	Kluane Lake, Yukon Territory	−181(2)	100	−174

* For trees, the youngest 2 or 3 rings were analyzed.
** Number in parenthesis indicates number of analyses.

Comparisons between the ^{2}H/^{1}H ratios in plants and in the water available to them are shown in Table 12-2 and clearly document that very good correlation between these two does exist. Since, in turn, the δ^2H value of freshwater is strongly dependent on climatic parameters (Chapter 1) the plant data reflect paleoclimatic conditions. To further document this Epstein and Yapp (1976) determined among others the deuterium contents in the growth rings of a Scots pine from Scotland which grew between 1841 and 1970 and bristlecone pines from the White Mountains of California which cover a total period from about 970 to 1974 A.D. Fig. 12-7 shows the bristlecone data and emphasizes the broad, qualitative agreement with other estimates of

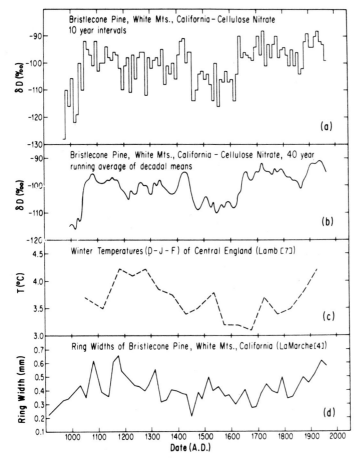

Fig. 12-7. The δ^2H record obtained on the nitrate cellulose of two bristlecone pines from North America is compared with the approximate winter temperatures of Central England (Lamb, 1966) and tree-ring widths of a bristlecone pine (LaMarche, 1974; from Epstein and Yapp, 1976).

climatic variations. Similarly the δ^2H values obtained from the Scots pine correlate well with historic temperature records, and it was concluded that the deuterium records from both trees reflect a sensitivity to large-scale climatic trends.

Another interesting feature of the bristlecone data is the apparent existence of a 22-year periodicity which has also been noted in the occurrence of droughts in the Great Plains of North America. No obvious periodicity of this size has been observed in the Scots pine deuterium record.

The peat data presented by Schiegl (1972) cover the period of about 50,000 years B.P. to present and are characterized by δ^2H values for the total organic matter of -60 to $-70‰$ between $\sim$10,000 years B.P. and present whereas before 10,000 years B.P. values as low as $-95‰$ have been measured.

In a more recent paper Yapp and Epstein (1977) carry their studies one step further and discuss the isotopic composition of meteoric water over North America between 22,000 and 9500 years B.P. Fig. 12-8 is a reproduction of figure 4 in their publication and emphasises the most important

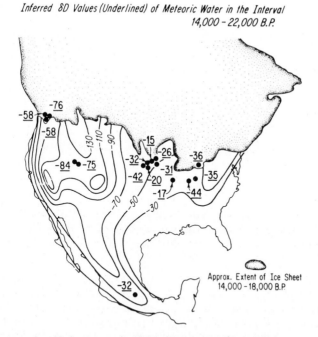

Fig. 12-8. Composition and regional distribution of inferred δ^2H in paleoprecipitations with modern meteoric waters. Note that at a given location paleowaters appear enriched in deuterium with respect to modern waters.

observations made: the δ^2H values of glacial-age meteoric waters in the ice free parts of North America were apparently more positive than corresponding modern values. They state (p. 347): "inferred D/H ratios of meteoric waters suggest that the average climatic temperature of the ice-free areas of North America during the late glacial Wisconsin were similar to that of the present. However, the North American Wisconsin ice sheet must have been maintained by summer temperature lower than that of the present which would reduce rates of ablation and by winter temperatures higher than the present which would allow more snow to reach the ice sheets during the winter months."

"The transition from glacial to interglacial climatic conditions, as indicated by the δD data from ancient North American trees, occurred in a relatively short time interval of 2000—3000 years duration."

These observations are especially interesting if compared to data obtained from ^{18}O, ^{2}H and ^{14}C analyses of groundwaters: although a great number of studies in the literature report ^{14}C ages for groundwaters for the period discussed by Yapp and Epstein (1977), very few report a significant isotope shift for either ^{18}O or deuterium towards lower δ-values. This could lend support to the deductions on isotopically heavy meteoric waters during glaciation made on the basis of plant-deuterium data. However, some exceptions are known: e.g. groundwaters believed to have formed during the last ice age in North Africa and the Sinai Peninsula seem to have had about 20—30‰ lower deuterium contents than modern waters in the area (Gonfiantini et al., 1974; Gat and Issar, 1974) and groundwaters in aquifers underlying postglacial Lake Agassiz in Southern Manitoba (Canada) have up to 80‰ less deuterium than modern precipitations (Fritz et al., 1974). Little is known about the paleohydrology of the North African groundwater systems but intensive studies are currently in progress to elucidate the history of the groundwaters in southern Manitoba. Yapp and Epstein (1977) report δ^2H values of close to —100‰ for pro-glacial Lake Agassiz in southern Manitoba which compares with values between —140 and —150‰ for groundwaters *within* Lake Agassiz sediments and underlying aquifers. Furthermore, ^{14}C dates on these groundwaters indicate ages between about 10,000 and 15,000 years B.P. These groundwater data thus seem to conflict with the conclusions of Yapp and Epstein (1977), and it is hoped that further studies will clarify this point.

The temperature dependence of deuterium contents in plants was discussed by Wilson and Grinsted (1975) and Epstein et al. (1976, p. 249). Although disagreement exists between the two studies further experimental work might be worthwhile. For example, Epstein et al. (1976) observed for aquatic plants a 25‰ difference in deuterium contents for a temperature difference of 11°C. This apparent temperature dependence of the fractionation factor causes the δ^2H of cellulose carbon bound hydrogen to increase with increasing temperature, which is the same direction as the temperature response of the δ^2H of meteoric water. Consequently, the two effects tend to enhance one another. However, the number of data points and the controls on them are insufficient to justify definitive conclusions. This important question can be resolved by laboratory studies under controlled conditions.

REFERENCES

Allen, P. and Keith, M.L., 1965. Carbon isotope ratios and paleosalinities of Purbeck-Wealden carbonates. *Nature*, 208: 1278—1280.

Allen, P., Keith, M.L., Tan, F.C. and Deines, P., 1973. Isotope ratios and Wealden environments. *Paleontology*, 16: 607—621.

Bortolami, G.C., Fontes, J.C., Markgraf, V. and Saliege, J.F., 1977. Land, sea and climate in the northern Adriatic region during Late Pleistocene and Holocene. *Palaeogeogr., Palaeoclimatol., Palaeoecol.*, 21: 139—156.

Broecker, W.S. and Walton, A., 1959. The geochemistry of ^{14}C in freshwater systems. *Geochim. Cosmochim. Acta*, 16: 15—38.

Buchardt, B., 1977. Oxygen isotope composition of shell material from the Danish middle Paleocene (Selandian) deposits and their interpretation as paleotemperatures. *Palaeogeogr., Palaeoclimatol., Palaeoecol.*, 22: 209—230.

Calder, J.A., 1972. Geochemical implications of induced changes in ^{13}C fractionation by blue-green algae. *Geochim. Cosmochim. Acta*, 37: 133—140.

Clayton, R.N. and Degens, E.T., 1959. Use of carbon isotope analyses of carbonates for differentiating freshwater and marine sediments. *Bull. Am. Assoc. Pet. Geol.*, 43: 890—897.

Covich, A. and Stuiver, M., 1974. Changes in oxygen-18 as a measure of long-term fluctuations in tropical lake levels and molluskan populations. *Limnol. Oceanogr.*, 19: 682—691.

Craig, H., 1954. Carbon-13 variations in sequoia rings and the atmosphere. *Science*, 119: 141—143.

Craig, H., 1961. Isotope variations in meteoric waters. *Science*, 133: 1702—1703.

Craig, H., 1965. The measurement of oxygen isotope paleotemperatures. In: E. Tongiorgi (Editor), *Stable Isotopes in Oceanographic Studies and Paleotemperatures, Spoleto 1965*. CNR, Pisa, pp. 161—182.

Dansgaard, W., 1954. The ^{18}O abundance in fresh water. *Geochim. Cosmochim. Acta*, 6: 241—260.

Dansgaard, W., 1964. Stable isotopes in precipitation. *Tellus*, 16: 436—468.

Dansgaard, W., Johnsen, S.J., Clausen, H.B. and Gundestrup, N., 1973. Stable isotope glaciology. *Medd. Gron.*, 197 (2).

Dansgaard, W., Johnsen, S.J., Reeh, N., Gundestrup, N., Clausen, H.B. and Hammer, C.U., 1975. Climatic changes, Norsemen's fate — and modern man's. *Nature*, 225: 24—28.

Degens, E.T., 1969. Biochemistry of stable carbon isotopes. In: C. Eglington and M.T.J. Murphy (Editors), *Organic Geochemistry*. Springer Verlag, New York, N.Y., pp. 304—329.

Degens, E.T. and Ross, D.A., 1971. Chronology of the Black Sea over the last 25,000 years. *Chem. Geol.*, 10: 1—16.

Deines, P., Langmuir, D. and Harmon, R.S., 1974. Stable carbon isotope ratios and the existence of a gas phase in the evolution of carbonate groundwaters. *Geochim. Cosmochim. Acta*, 38: 1147—1154.

Deuser, E.G., Degens, E.T. and Guillard, R.R.L., 1968. Carbon isotope relationships between plankton and seawater. *Geochim. Cosmochim. Acta*, 32: 657—660.

Dodd, J.R. and Stanton, R.J., 1975. Paleosalinities within a Pliocene bay, Rettleman Hills, California: a study of the resolving power of isotope and faunal techniques. *Geol. Soc. Am. Bull.*, 86: 51—64.

Drimmie, R.J., 1976. *Stable Carbon Isotope Composition of Organic Matter in Lake Erie and Lake Ontario.* M.Sc. Thesis, University of Waterloo, Waterloo, Ont. (unpublished).

Durazzi, J.T., 1977. Stable isotopes in the Ostracode shell, a preliminary study. *Geochim. Cosmochim. Acta*, 41: 1168—1170.

Emrich, K., Ehhalt, D.H. and Vogel, J.C., 1970. Carbon isotope fractionation during the precipitation of calcium carbonate. *Earth Planet. Sci. Lett.*, 8: 363—371.

Epstein, S., 1956. Variation of the $^{18}O/^{16}O$ ratio of fresh water and ice. In: *Nuclear Processes in Geologic Settings.* Natl. Acad. Sci. U.S.A., Publ. 400: 20.

Epstein, S. and Yapp, C.J., 1976. Climatic implications of the D/H ratio of hydrogen in C—H groups in tree cellulose. *Earth Planet. Sci. Lett.*, 30: 252—261.

Epstein, S., Buchsbaum, R., Lowenstam, H.A. and Urey, H.C., 1951. Carbonate-water isotopic temperature scale. *Geol. Soc. Am. Bull.*, 62: 417—425.

Epstein, S., Buchsbaum, R., Lowenstam, H.A. and Urey, H.C., 1953. Revised carbonate-water isotopic temperature scale. *Geol. Soc. Am. Bull.*, 64: 1315—1326.

Epstein, S., Yapp, C.J. and Hall, J.H., 1976. The determination of the D/H ratio of non-exchangeable hydrogen in cellulose extracted from aquatic and land plants. *Earth Planet. Sci. Lett.*, 30: 241—251.

Erlenkeuser, H., Suess, E. and Willkomm, H., 1974. Industrialization affects heavy metal and carbon isotope concentrations in recent Baltic Sea sediments. *Geochim. Cosmochim. Acta*, 38: 813—822.

Fontes, J.C. and Pouchan, P., 1975. Les cheminées du Lac Abbé (Djibouti): stations hydroclimatiques de l'Holocène. *C.R. Acad. Sci. Paris*, 280: 383—386.

Fritz, P. and Poplawski, S., 1974. ^{18}O and ^{13}C in the shells of fresh water mollusks and their environments. *Earth Planet. Sci. Lett.*, 24: 91—98.

Fritz, P., Drimmie, R.J. and Render, F., 1974. Stable isotope contents of a major prairie aquifer in Central Manitoba. In: *Isotope Techniques in Groundwater Hydrology.* IAEA, Vienna, pp. 379—397.

Fritz, P., Anderson, T.W. and Lewis, C.F.M., 1975. Late-Quaternary climatic trends and history of Lake Erie from stable isotopic studies. *Science*, 190: 267—269.

Gasse, F., Fontes, J.C. and Rongnon, P., 1974. Variations hydrologiques et extensions des lacs Holocènes du désert Danakil. *Palaeogeogr., Palaeoclimatol., Palaeoecol.*, 15: 109—148.

Gat, J.R. and Issar, A., 1974. Desert hydrology: water sources of the Sinai Desert. *Geochim. Cosmochim. Acta*, 38: 1117—1131.

Gonfiantini, R., Togliatti, V., Tongiorgi, E., Breuck, W.D. and Picciotto, E., 1963. Snow stratigraphy and oxygen isotope variations in the glaciological pit of King Bandon Station, Queen Maud Land, Antarctica. *J. Geophys. Res.*, 68: 2791—2798.

Gonfiantini, R., Panichi, C. and Tongiorgi, E., 1968. Isotopic disequilibrium in travertine deposition. *Earth Planet. Sci. Lett.*, 5: 55—58.

Gonfiantini, R., Dinçer, T. and Derekoy, A.M., 1974. Environmental isotope hydrology in the Hodna region, Algeria. *Isotope Techniques in Groundwater Hydrology*, 1. IAEA, Vienna, pp. 293—316.

Gross, M.G., 1964. Variations in $^{18}O/^{16}O$ and $^{13}C/^{12}C$ ratios of diagenetically altered limestones in the Bermuda Islands. *J. Geol.*, 72: 170—194.

Hendy, C.H., 1971. The isotopic geochemistry of speleothems, I. The calculation of the effects of different modes of formation on the isotopic composition of speleothems and their applicability as paleoclimatic indicators. *Geochim. Cosmochim. Acta*, 35: 801—824.

Hitchon, B. and Krouse, H.R., 1972. Hydrogeochemistry of the surface waters of the

Mackenzie River drainage basin, Canada, III. Stable isotopes of oxygen, carbon and sulfur. *Geochim. Cosmochim. Acta*, 36: 1337—1557.

Keith, M.L. and Anderson, G.M., 1963. Radiocarbon dating: fictitious results with mollusc shells. *Science*, 141: 634—637.

Keith, M.L. and Weber, J.N., 1964. Carbon and oxygen isotopic composition of selected limestones and fossils. *Geochim. Cosmochim. Acta*, 28: 1787—1816.

Killey, R.W.D., 1977. *Carbon Isotopes in the Groundwater and Perch Lake in the Perch Lake Basin, Chalk River, Ontario*. M.Sc. Thesis, University of Waterloo, Waterloo, Ont. (unpublished).

Killey, R.W. and Fritz, P., 1979. The carbon dioxide flux across a lake-atmosphere interface as determined by carbon isotope data. In: *The Application of Nuclear Techniques to the Study of Lake Dynamics*. IAEA, Vienna (in press).

Kölle, W., 1970. Radiocarbon measurement of organic pollutants of the Rhine. In: *Nuclear Techniques in Environmental Pollution, Proceeding of an IAEA Symposium, Salzburg 1970*. IAEA, Vienna, pp. 497—506.

Lamb, H.H., 1966. *The Changing Climate*. Methuen, London.

LaMarche, V.C., Jr., 1974. Paleoclimatic inferences from long tree-ring records. *Science*, 183: 1043.

Libby, L.M. and Pandolfi, L.J., 1974. Temperature dependence of isotopic ratios in tree rings. *Proc. Natl. Acad. Sci. U.S.A.*, 71: 2482.

Longinelli, A., 1966. Ratios of oxygen-18 : oxygen-16 in phosphate and carbonate from living and fossil marine organisms. *Nature*, 211: 923—927.

Longinelli, A. and Nuti, S., 1965. Oxygen isotopic composition of phosphate and carbonate from living and fossil marine organisms. In: E. Tongiorgi (Editor), *Stable Isotopes in Oceanographic Studies and Paleotemperatures, Spoleto 1965*. CNR, Pisa, 183—197.

Longinelli, A. and Sordi, S., 1966. Oxygen isotope composition of phosphate from shells of some living crustaceans. *Nature*, 211: 727.

Longinelli, A. and Nuti, S., 1973. Revised phosphate-water isotopic temperature scale. *Earth Planet. Sci. Lett.*, 19: 373—376.

Lowenstam, H.A. and Epstein, S., 1953. Paleotemperatures of the Post-Aptian Cretaceous as determined by the oxygen isotope method. *J. Geol.*, 62: 207—248.

McCrea, J.M., 1950. On the isotopic chemistry of carbonates and a paleotemperature scale. *J. Chem. Phys.*, 18: 849—857.

Mook, W.C., 1970. Stable carbon and oxygen isotopes of natural waters in the Netherlands. In: *Isotope Hydrology*. IAEA, Vienna, pp. 163—188.

Mook, W.G., 1971. Paleotemperatures and chlorinities from stable carbon and oxygen isotopes in shell carbonate. *Palaeogeogr., Palaeoclimatol., Palaeoecol.*, 9: 245—264.

Mook, W.G. and Eisma, D., 1976. Shell characteristics, isotopic composition and trace element content of some euryhaline molluscs as indicator of salinity. *Palaeogeogr., Palaeoclimatol., Palaeoecol.*, 19: 39—62.

Mook, W.G. and Vogel, J.C., 1968. Isotopic equilibrium between shells and their environment. *Science*, 159: 874—875.

Nakai, N., 1972. Carbon isotopic variation and the paleoclimate of sediment from Lake Biwa. *Proc. Jpn. Acad.*, 48: 516—521.

Oana, S. and Deevey, E.S., 1960. Carbon-13 in lake waters, and its possible bearing on paleolimnology. *Am. J. Sci.*, 258-A: 253—272.

Parker, P.L., Behrens, E.W., Calder, I.A. and Shultz, D., 1972. Stable carbon isotope ratio variations in the organic carbon from Gulf of Mexico sediment. *Contrib. Mar. Sci.*, 16: 139—147.

Rogers, M.A. and Koons, C.B., 1969. Organic carbon ^{13}C values from Quaternary marine sequences in the Gulf of Mexico: A reflection of paleotemperature changes. *Trans. Gulf Coast Assoc., Geol. Soc.*, 19: 529—534.

Rothe, P., Hoefs, J. and Sonne, V., 1974. The isotopic composition of Tertiary carbonates from the Mainz Basin: an example of isotopic fractionations in "closed basins". *Sedimentology*, 21: 373—395.

Rubin, M., Likins, R.C. and Berry, E.G., 1963. On the validity of radiocarbon dates from snail shells. *J. Geol.*, 71: 84.

Rubinson, M. and Clayton, R.N., 1969. Carbon-13 fractionation between aragonite and calcite. *Geochim. Cosmochim. Acta*, 33: 997—1002.

Sackett, W.M., 1964. The depositional history and isotopic organic carbon composition of marine sediments. *Mar. Geol.*, 2: 173—185.

Sackett, W.M. and Rankin, J.G., 1970. Paleotemperatures for the Gulf Coast of Mexico. *J. Geophys. Res.*, 75: 4457—4560.

Sackett, W.H., Eckelmann, W.R., Bender, M.L. and Be, A.W.H., 1965. Temperature dependence of carbon isotopic composition in marine plankton and sediment. *Science*, 148: 235—237.

Schiegl, W.E., 1972. Deuterium content of peat as a paleoclimatic recorder. *Science*, 175: 512—513.

Schiegl, W.E., 1974. Climatic significance of deuterium abundance in growth rings of Picea. *Nature*, 251: 582.

Schiegl, W.E. and Vogel, J.C., 1970. Deuterium content of organic matter. *Earth Planet. Sci. Lett.*, 7: 307.

Schwarcz, H.P., Harmon, R.S., Thompson, P. and Ford, D.C., 1976. Stable isotope studies of fluid inclusions in speleothems and their paleoclimatic significance. *Geochim. Cosmochim. Acta*, 40: 657—665.

Serruya, C. and Vergnaud-Grazzini, C., 1966. Evolution paléoclimatologique des sédiments du Lac Léman. *Arch. Sci. Genève*, 19: 197—210.

Smith, B.N. and Epstein, S., 1970. Biogeochemistry of the stable isotopes of hydrogen and carbon in salt-marsh biota. *Plant Physiol.*, 46: 738.

Smith, B.N. and Epstein, S., 1971. Two categories of $^{13}C/^{12}C$ ratios for higher plants *Plant. Physiol.*, 47: 380—384.

Spiker, E.C. and Rubin, M., 1975. Petroleum pollutants in surface and groundwater as indicated by the carbon-14 activity of dissolved organic carbon. *Science*, 187: 61—64.

Stahl, W. and Jordan, R., 1969. General considerations on isotopic paleotemperature determinations and analyses on Jurassic ammonites. *Earth Planet. Sci. Lett.*, 6: 173—178.

Stiller, M., 1976. Origin of sedimentation component in Lake Kinneret traced by their isotopic composition (manuscript).

Stuiver, M., 1968. Oxygen-18 content of atmospheric precipitation during last 11,000 years in the Great Lakes region. *Science*, 162: 994—997.

Stuiver, M., 1970. Oxygen and carbon isotope ratios of fresh water carbonates as climatic indicators. *J. Geophys. Res.*, 75: 5247—5257.

Stuiver, M., 1975. Climate versus changes in ^{13}C content of the organic component of lake sediments during the late Quarternary. *Quart. Res.*, 5: 251—262.

Tamers, M.A., 1970. Validity of radiocarbon dates on terrestrial snail shells. *Am. Antiquity*, 35: 94—100.

Tan, F.C. and Hudson, J.D., 1971. Isotopic composition of carbonates in a marginal marine formation. *Nature Phys. Sci.*, 232: 87—88.

Tan, F.X. and Hudson, J.D., 1974. Isotopic studies of the paleoecology and diagenesis of the Great Estuarine Series (Jurassic) of Scotland. *Scott. J. Geol.*, 10: 91—128.

Tarutani, T., Clayton, R.N. and Mayeda, T., 1969. The effect of palymorphism and magnesium substitution on oxygen isotope fractionation between calcium carbonate and water. *Geochim. Cosmochim. Acta*, 33: 987—996.

Taylor, H.P., Jr., 1974. The application of oxygen and hydrogen isotope studies to problems of hydrothermal alterations and ore deposition. *Econ. Geol.*, 69: 843.

Urey, H.C., Lowenstam, H.A., Epstein, S. and McKinney, C.R., 1951. Measurement of paleotemperatures and temperatures of the upper Cretaceous of England, Denmark and the Southeastern United States. *Geol. Soc. Am. Bull.*, 62: 399—416.

Veizer, J., 1974. Chemical diagenesis of Belemnite shells and possible consequences for paleotemperature determinations. *Neues Jahrb. Geol. Paläontol., Abh.*, 145: 279—305.

Thurber, D.L. and Broecker, W., 1969. The behaviour of radiocarbon in surface waters of the Great Basin. *Columbia Univ., CU-2493-10 (App. A)*, 40 pp.

Weber, J.N., 1973. Deep sea ahermatypic scleractinian corals: isotopic composition of the skeleton. *Deep-Sea Res.*, 20: 901—909.

Weber, J.N. and La Rocque, A., 1963. Isotope ratios in marine mollusk shells after prolonged contact with flowing fresh water. *Science*, 142: 1666.

Weber, J.N. and Raup, D.M., 1968. Comparison of $^{13}C/^{12}C$ and $^{18}O/^{16}O$ in skeletal calcite of recent and fossil echinoids. *J. Paleontol.*, 42: 37—50.

Weber, J.N. and Woodhead, P.M.J., 1972. Temperature dependence of oxygen-18 concentration in reef coral carbonates. *J. Geophys. Res.*, 77: 463—473.

Wilson, A.J. and Grinsted, M.J., 1975. Paleotemperatures from tree rings and the D/H ratio of cellulose as a biochemical thermometer. *Nature*, 257: 387.

Yapp, C.J. and Epstein, S., 1977. Climatic implications of D/H ratios of meteoric water over North America (9500—22,000 B.P.) as inferred from ancient wood cellulose C—H hydrogen. *Earth Planet. Sci. Lett.*, 34: 333—350.

REFERENCES INDEX

Abelson, P.H. and Hoering, T.C., 335, 337, 341, 343, *393*
Adams, J.A.S., see Kronfeld, J. and Adams, J.A.S.
Aldaz, L. and Deutsch, S., 143, 144, *174;* see also Piccioto, E. et al.
Alekseev, F.A., Gottikh, R.P., Zverev, V.L., Spiridov, A.I. and Grumbkov, A.P., 274, 276, *279*
Allan, J.R. and Metthews, R.K., 307, *321*
Allaway, W.G., Osmond, C.B. and Troughton, J.H., 343, *393*
Allen, E.R., see Kellogg, W.W. et al.
Allen, E.T. and Day, A.L. 180, *219*
Allen, P. and Keitts, M.L., 481, *500*
Allen, P., Keitts, M.L., Tan, F.C. and Deines, P., 481, *500*
Allen, R.C., Gavish, E., Friedman, G.M. and Sanders, J.E., 357, *393*
Allison, G.B. and Hughes, M.W., 88, *134*
Allsup, J.R., see Eckelmann, W.R. et al.
Aly, A.I.M., see Freyer, H.D. and Aly, A.I.M.
Amarger, N., Mariotti, A. and Mariotti, F., 415, *429*
Ambach, W., see Behrens, H. et al.; Deutsch, S. et al.; Moser, H. et al.
Ambach, W. and Dansgaard, W., 159, 160, *174*
Ambach, W., Dansgaard, W., Eisner, H. and Möller, J., 145, 159, *174*
Ambach, W., Eisner, H., Elsässer, M., Löschhorn, V., Moser, H., Rauert, W. and Stichler, W., 159, 160, 167, 168, 169, 172, *174*
Ambach, W., Eisner, H. and Pessl, K., 150, *174*
Ambach, W., Eisner, H. and Url, M., 166, *174*
Ambach, W., Elsässer, M., Moser, H., Rauert, W., Stichler, W. and Trimborn, P., 146, *174*
Andersen, L.J. and Sevel, T., 87, 88, *134*
Anderson, G.M., see Keitts, M.L. and Anderson, G.M.
Anderson, T.F. and Cole, S.A., 306, *321*
Anderson, T.F., see Lawrence, J.R. et al.
Anderson, T.W., see Fritz, P. et al.
Andreae, M.O., 359, 360, *393*
Andrieux, C., 102, *134*
Arana, V. and Panichi, C., 181, *219*
Arnason, B., 141, 147, 157, 158, 163, 169, *175*, 186, 188, 197, 202, *219*
Arnason, B., Buason, T., Martinec, J. and Theodorsson, P., 152, 162, *175*
Arnason, B. and Sigurgeirsson, T., 115, *134*
Arnold, J.R., see Lal, D. et al.
Arnorsson, S., 197, *219*
Atakan, Y., Roether, W., Münnich, K.O. and Matthess, G., 91, *134*
Athavale, R.N., Lal, D. and Rama, S., 24, *44*
Atkinson, T.C., 60, *71*
Aust, H., see Stahl, W. et al.

Baas-Becking, L.G.M., 461, *467*
Bacastow, R. and Keeling, C.D., 386, 387, *393*
Back, W., see Rightmire, C.T. et al.
Baertschi, P., 11, *17*, 21, *44*, 292, *321*, 338, 339, 347, *393*
Baht, S.F. and Krishnaswami, S., 270, 278, *279*
Bailey, I.H., Hulston, J.R., Macklin, W.C. and Stewart, J.R., 39, *44*
Bailey, N.J.L., Jobson, A.M. and Rogers, M.A., 368, 372, *393*

Bailey, N.J.L., see Rogers et al.
Bailey, S.A. and Smith, J.W., 218, *219*
Bainbridge, A.E., Suess, E. and Friedman, I., 390, *394*
Bakalowicz, M., Blavoux, B. and Mangin, A., 120, *134*
Bakalowicz, M. and Olive, P., 117, 118, *134*
Baker, D.R. and Claypool, G.E., 359, 360, *394*
Baker, D.R., see James, A.T. and Baker, D.R.
Baldi, P., Ferrara, G.C. and Panichi, C., 184, *219*
Banwell, C.J., 208, *219*
Baranov, V.L., Surkov, Yr. A. and Vilenskii, V.D., 261, *279*
Bardin, R., Domenach, A.M. and Chalamet, A., *417, 429*
Barker, F. and Friedman, I., 359, 360, *394*
Barker, J.F., see Fritz, P. et al.
Bartholomew, M.V., see Hauck, R.D. et al.
Barnes, I., see White, D.E. et al.
Barrer, R.M. and Denny, A.F., 292, *321*
Bassham, J.A., 329, *394*
Batts, B.D., see Smith, J.W. and Batts, B.D.
Baxter, M.S., see Farmer, J.G. and Baxter, M.S.
Bé, A.W.H., see Sackett, W.M. et al.
Becker, R.H. and Clayton, R.N., 409, 411, *429*
Bedinger, M.S., see Pearson, F.J., Jr. et al.
Bedinger, M.S., Pearson, F.J., Jr., Reed, J.E., Smiegocki, R.J. and Stone, C.G., 60, *71*
Begemann, F., 26, 27, *44*
Begemann, F. and Libby, W.F., 22, *44*, 209, *220*
Begun, G.M. and Fletcher, W.H., 412, *429*
Behrendt, M., see Degens, E.T. et al.
Behrens, E.W. and Frishman, S.A., 346, 356, *394*
Behrens, E.W., see Newman, J.W. et al.; Parker, P.L. et al.
Behrens, H., Bergmann, H., Moser, H., Rauert, W., Stichler, W., Ambach, W., Eisner, H. and Pessl, K., 167, *175*
Behrens, H., Moser, H., Rauert, W.,
Stichler, W., Ambach, W. and Kirchlechner, P., 166, *175*
Behrens, W., Eichmann, R., Plate, A. and Kroepelin, H., 410, *429*
Behrens, W., see Eichmann, R. and Behrens, W.; Wollanke, G. et al.
Bein, A., see Issar, A. et al.
Belau, L., see Gaissler, C. and Belau, L.
Ben Halima, A., 425, 427, 428, *429*; see also Mariotti, A. et al.
Bender, M.L., see Sackett, W.M. et al.
Bender, M.M., 331, 332, *394*
Bender, M.M., Rouhami, I., Vines, H.M. and Black, C.C., 332, *394*
Benedict, C.R., see Smith, B.N. and Benedict, C.R.; see also Whelan, T., et al.
Benson, B.B., see Klots, C.A. and Benson, B.B.
Benson, B.B. and Parker, P.D.M., 414, *429*
Bergmann, H., see Behrens, H. et al.
Berner, R.A., 234, *255*, 459, *467*; see also Clayton, R.N. et al.
Berry, E.G., see Rubin, M. et al.
Berry, F.A.F., see Kharaka, Y.K. et al.
Berry, J.A., see Tregunna, E.B. et al.
Bigeleisen, J., 51, *71*
Bigeleisen, J. and Mayer, M.G., 228, 229, 231, *255*
Binda, P.L., see Fritz, P. et al.
Bir, R., 278, *280*
Black, A.S. and Waring, S.A., 418, *429*
Black, C.C., see Bender, M.M. et al.
Black, R.F., see Bowser, C.J. et al.
Blackmer, A.M. and Bremner, J.M., 417, *429*
Blanc, P. and Dray, M., 99, *134*
Blattner, P., 215, *220*
Blavoux, B., see Bakalowicz, M. et al.; Fontes, J.Ch. et al.; Mariotti, A. et al.
Blavoux, B. and Siwertz, E., 85, *134*
Bleeker, W., Dansgaard, W. and Lablans, W.N., 37, *45*
Boato, G., see Craig, H. et al.
Bodden, M., see Pearson, F.J., Jr. and Bodden, M.
Bogdanov, Yu. V., Golubchina, M.N., Prilutsky, R.E., Joksubayev, A.I. and Geoktistov, V.P., 436, *467*
Bogdanow, W.M., see Lebedev, W.C. et al.
Boigk, H., Hagemann, H.W., Stahl, W. and

Wollanke, G., 374, 378, 383, *394*
Boigk, H. and Stahl, W., 383, *394*
Boigk, H., Stahl, W., Teichmüller, M. and Teichmüller, R., 382, *394*
Boigk, H., see Stahl, W. et al.
Bokhoven, C. and Theuwen, H.J., 410, *429*
Bolin, B., 26, 27, 29, 38, *45*
Bommerson, J.C., see Mook, G.W. et al.
Bond, J.G., see Koons, C.B. et al.
Booker, D.V., 29, *45*
Boone, L.V., see Meints, W.V. et al.
Bortolami, G.C., see Fontes, J.Ch. et al.
Bortolami, G.C., Fontes, J.Ch., Markgraf, V. and Saliege, J.F., 492, *500*
Bortolami, G.C., Fontes, J.Ch., and Panichi, C., 84, 99, 131, *134*
Bortolami, G.C., Fontes, J.Ch. and Zuppi, G.M., 116, *134*
Bosch, B., Guégan, A., Marcé, A. and Siméon, C., 99, *134*
Bosch, B., see Garcia-Loygorri, A. et al.
Bottinga, Y., 8, *17*, 188, 195, 198, 199, 201, 203, 205, *220*, 357, 362, 376, 386, *394*; see also Richet, P. et al.
Bottinga, Y. and Craig, H., 7, *17*, 188, 203, 205, *220*
Bowen, P.A., see Nehring, N.L. et al.
Bowser, C.J., Rafter, T.A. and Black, R.F., 309, 310, *321*
Bradley, E., Brown, R.M., Gonfiantini, R., Payne, B.R., Przewlocki, K., Sauzay, G., Yen, C.K. and Yurtsever, Y., 180, *220*
Bredenkamp, D.B., Schutte, J.M. and Du Toit, G.J., 89, *135*
Bredenkamp, D.B. and Vogel, J.C., 67, *71*
Bremner, J.M., see Blackmer, A.M. and Bremner, J.M.; Cheng, H.H. et al.; Hauck, R.D. et al.
Bremner, J.M., Cheng, H.H. and Edwards, A.P., 407, 408, *429*
Bremner, J.M. and Keeney, D.R., 408, *429*
Bremner, J.M. and Tabatabai, M.A., 408, 417, 418, 420, 421, 425, *429*
Brenck, W.D., see Gonfiantini, R. et al.
Brickwedde, I.G., see Urey, H. et al.
Briel, L.I., 267, 268, 269, 279, *280*
Brigode, N., see Riga, A. et al.
Brinkman, R., Eïchler, R., Ehhalt, D. and Münnich, K.O., 85, *135*
Brinkman, R., Münnich, K.O. and Vogel, J.C., 55, *71*, 127, *135*
Broadbent, F.E., see Hauck, R.D. et al.
Broecker, W., see Thurber, D.L. and Broecker, W.
Broecker, W.S., see Eckelmann, W.R. et al.; Hamza, M.S. and Broecker, W.S.; Lawrence, J.R. et al.
Broecker, W.S. and Olson, E.A., 348, *394*
Broecker, W.S. and Walton, A., 68, *71*, 494, *500*
Brooks, J.M., Gormly, J.R. and Sackett, W.M., 374, *394*
Brown, A.E., see Mook, G.W. et al.
Brown, H.M., see Hitchon, B. et al.
Brown, L.J. and Taylor, C.B., 109, *135*
Brown, L.L. and Drury, J.S., 412, *429*, *430*
Brown, R.M., 159, *175*; see also Bradley, E. et al.; Fritz, P. et al.
Brown, R.M. and Grummitt, 22, *45*
Brown, W.V., see Smith, B.N. and Brown, W.V.
Brown, W.V. and Smith, B.N., 332, *394*
Browne, P.R.L., Roedder, E. and Wodzicki, A., 216, *220*
Brownell, L.E., see Isaacson, R.E. et al.
Bruckert, S., see Mariotti, A. et al.
Bryant, R., see Laishley, E.J. et al.
Buason, T., 152, *175*; see also Arnason, B. et al.
Buchardt, B., 479, 480, *500*
Bucher, P., see Oeschger, H. et al.
Bucher, P., Möll, M., Oeschger, H., Stauffer, B. and Patersen, W.S.B., 170, *175*
Buchsbaum, R., see Epstein, S. et al.
Buckley, H.A., see Schackleton, N.J. et al.
Buddemeier, R.W., see Hufen, T.H. et al.
Burdon, D.J., Eriksson, E., Papadimitropoulos, T., Papakis, N. and Payne, B.R., 117, *135*
Burger, A., Mathey, B., Marcé, A. and Olive, P., 120, *135*
Burns, M.S., see Chambers, L.A. et al.

Cain, W.F. and Suess, H.E., 388, *394*
Calder, J.A., 487, *500*; see also Johnson, R.W. and Calder, J.A.; Kellogg, W.W. et al.; Parker, P.L. et al.
Calder, J.A., Harvath, G.J., Shultz, D.J. and Newman, J.W., 351, *394*
Calder, J.A. and Parker, P.L., 337, 342,

343, 353, *394*
Cameron, R., see Picciotto, E. et al.
Campbell, F.A., see Ryznar, G. et al.
Campbell, F.A., Evans, T.L. and Krouse, H.R., 437, *467*
Campbell, J.B., see Krouse, H.R. et al.
Card, K.A., see Troughton, J.H. et al.
Carey, B.D., see Stahl, W. and Carey, B.D.
Carmi, I., see Gat, J.R. and Carmi, I.; Mazor, E. et al.
Carmi, I. and Gat, J.R., 29, *45*
Carter, J.A., see Thompson, G.M. et al.
Castleman, A.W., Munkelwitz, H.R. and Manowitz, B., 466, *467*
Celati, R., see Panichi, C. et al.
Celati, R., Noto, P., Panichi, C. and Squarci, P. and Taffi, L., 209, *220*
Chaigneau, M., see Roblot, M. et al.
Chalamet, A., Bardin, R. et al.; see Domenach, A.M. and Chalamet, A.
Chalov, P.I., 261, *280*
Chalov, P.I. and Merkulova, K.I., 262, *280*
Chambers, L.A. Trudinger, P.A., Smith, J.W. and Burns, M.S., 457, *467*
Chassaing, B., see Mariotti, A. et al.
Chawla, V.K., see Weiler, R.R. and Chawla, V.K.
Chebotarev, Ye, N., see Matrosov, A.G. et al.
Chemerys, J.C., see Hobba, W.A. et al.
Cheminée, J., Létolle, R. and Olive, P., 385, *394*
Cheney, E.S., see Vredenburgh, L.D. and Cheney, E.S.
Cheng, H.H., see Bremner, J.M. et al.; Hauck, R.D. et al.
Cheng, H.H., Bremner, J.M. and Edwards, A.P., 417, 418, 420, 421, *430*
Cherdyntsev, V.V., 261, *280*
Cherry, J.A., see Fritz, P. et al.
Choitz, H.C., see Gaebler, O.H. et al.
Choquette, P.M., 308, *321*
Christ, C.L., see Garrels, M. and Christ, C.L.
Chukhrov, F.V., Churikov, V.S., Yermilova, L.P. and Nosik, L.P., 239, *255*
Church, T., 304, 305, *321*
Churikov, V.S., see Chukhrov, F.V. et al.
Clausen, H.B., 170, *175*; see also Dansgaard, W. et al.; Hammer, G.U. et al.; Johnsen, S.J. et al.; Reeh, N. et al.

Clausen, H.B. and Dansgaard, W., 160, 163, *175*
Claypool, G.E., see Baker, D.R. and Claypool, G.E.
Claypool, G.E., Holser, W.T., Kaplan, I.R., Sakai, H. and Zak, I., 308, 309, *321*
Claypool, G.E., Presley, B.J. and Kaplan, I.R., 356, *394*
Clayton, R.N. and Degens, E.T., 481, *500*
Clayton, R.N. and Epstein, S., 307, *321*
Clayton, R.N., Friedman, I., Graf, D.L., Mayeda, T.K., Meents, W.F. and Shimp, N.F., 182, 183, *220*, 320, *321*
Clayton, R.N., Goldsmitts, J.R., Kavel, K.J., Mayeda, T.K. and Newton, R.C., 229, *255*
Clayton, R.N., Jones, B.F. and Berner, R., 307, *321*
Clayton, R.N. and Mayeda, T.K., 3, *17*
Clayton, R.N., Muffler, L.J.P. and White, D.E., 215, *220*
Clayton, R.N., O'Neil, J.R. and Mayeda, T.K., 285, 287, *321*
Clayton, R.N., Rex, R.W., Syers, J.K. and Jackson, M.L., 285, 302, 312, *321*
Clayton, R.N., Skinner, H.C.W., Berner, R.A. and Rubinson, M., 307, *321*
Clayton, R.N. and Steiner, A., 183, 215, *220*
Clayton, R.N., see Becker, R.H. and Clayton, R.N.; Engel, A.E.J. et al.; Northrop, D.A. and Clayton, R.N.; O'Neil, J.R. and Clayton, R.N.; O'Neil, J.R. et al.; Robinson, M. and Clayton, R.N.; Rex, R.W. et al.; Sharma, T. and Clayton, R.N.; Tarutani, T. et al.
Cline, J.D. and Kaplan, I.R., 416, *430*
Clynne, M.A., see Potter, R.W. and Clynne, M.A.; Potter, R.W. et al.
Coantic, M., see Merlivat, L. and Coantic, M.
Cole, J.A. and Wilkinson, W.B., 84, *135*
Cole, S.A., see Anderson, T.F. and Cole, S.A.
Colombo, U., see Teichmüller, R., et al.
Colombo, U., Gazzarrini, F., Gonfiantini, R., Kneuper, G., Teichmüller, M. and Teichmüller, R., 361, 362, 363, 364, *395*
Colombo, U., Gazzarrini, F., Sironi, G., Gonfiantini, R. and Tongiorgi, E., 374,

377, 378, 381, *394*, 413, *430*
Commoner, B., see Feigin, A.D. et al.; Kohl, D.A. et al.; Shearer, G. et al.
Compston, W., 361, *395*; see also Jeffrey, P.M. et al.
Conrad, G., 121, *135*; see also Gonfiantini, R. et al.
Conrad, G. and Fontes, J.Ch., 108, 122, 127, *135*
Conrad, G., Fontes, J.Ch., Létolle, R. and Roche, M.A., 108, 122, *135*
Conrad, G., Marcé, A. and Olive, P., 83, 84, *135*
Cook, A.C., 362, *395*
Cook, F.D., see Krouse, H.R. et al.; Šmejkal, V. et al.; Wellman, R.P. et al.
Cook, F.D., Wellman, R.P. and Krouse, H.R., 416, *430*
Cook, G., see Sackett, W.M. and Cook, G.
Coplen, T.B., 215, *220*
Coplen, T.B. and Hanshaw, B.B., 320, *321*
Corlis, J.B., see Dymond, J. et al.
Cortecci, G., 197, 204, *220*, 238, *255*; see also Longinelli, A. and Cortecci, G.; Schwarcz, H.P. and Cortecci, G.; Tenu, A. et al.
Cortecci, G. and Longinelli, A., 237, *255*, 304, 305, *322*, 449, 450, *467*
Cotecchia, V., Tazioli, G.S. and Magri, G., 125, *135*
Covich, A. and Stuiver, M., 482, *500*
Cowart, J.B. and Osmond, J.K., 272, 275, 277, *280*
Cowart, J.B., Kaufman, M.I. and Osmond, J.K., 273, *280*
Cowart, J.B., see Osmond, J.K. and Cowart, J.B.; Osmond, J.K. et al.
Craig, C.B., see Thode, H.G. et al.
Craig, H., 3, 11, 12, 13, 14, 15, *17*, *18*, 21, 32, *45*, 51, 57, *71*, 78, *135*, 179, 180, 181, 182, 183, 192, 193, 197, 198, 199, 200, 210, 211, 212, *220*, 294, *322*, 331, 332, 334, 338, 347, 350, 351, 353, 359, 360, 361, 362, 365, 376, 382, 385, 386, 387, 389, *395*, 413, *430*, 473, 475, *500*
Craig, H., Boato, G. and White, D.E., 179, 182, *220*
Craig, H. and Gordon, L., 28, 30, 33, 34, 36, 40, 41, *45*
Craig, H., Gordon, L. and Horibe, Y., 35, *45*

Craig, H. and Hom, B., 141, *175*
Craig, H. and Horibe, Y., 33, *45*
Craig, H. and Keeling, C., 15, *18*, 387, *395*
Craig, H. and Lupton, J.E., 214, 219, *220*
Craig, H., see Bottinga, Y. and Craig, H.; Kroopnick, P.M. and Craig, H.; Longinelli, A. and Craig, H.; Pineau, F. et al.
Crienko, V.A., see Malinin, S.D. et al.
Crouzet, E., Hubert, P., Olive, P., Siwertz, E. and Marce, A., 103, *135*
Crozaz, C., see Picciotto, E. et al.
Culp, J.H., see Presley, B.J. et al.
Cusicangui, H., Mahon, W.A.J. and Ellis, A.J., 186, 193, *221*

Dal Ollio, A., see Salati, E. et al.
Dalrymple, D., see Sackett, W.M. et al.
D'Amore, F., 214, 219, *221*
Dance, L.R., see Sondheimer, E.W.A. et al.
Dansgaard, W., 30, 32, 34, 37, 43, 44, *45*, *46*, 77, 78, *135*, *137*, 143, 154, *175*, 475, 482, *500*; see also Ambach, W. and Dansgaard, W.; Bleeker, W. et al.; Clausen, H.B. and Dansgaard, W.; Gat, J.R. and Dansgaard, W.; Hammer, C.V. et al.; Johnsen, S.J. et al.; Reeh, N. et al.
Dansgaard, W. and Johnsen, S.J., 170, 173, 174, *175*
Dansgaard, W., Johnsen, S.J., Clausen, H.B. and Gundestrup, N., 143, 156, 157, 172, 174, *175*, 482, *500*
Dansgaard, W., Johnsen, S.J., Moeller, J. and Langway, C.C., Jr., 130, *135*, 173, *175*
Dansgaard, W., Johnsen, S.J., Reeh, N., Gundestrup, N., Clausen, H.B. and Hammer, G.U., 173, *175*, 482, *500*
Dasch, E.J., 298, *322*
Dasch, E.J., see Dymond, J. et al.
David, G., see Tataru, S. et al.
Davies, G.R. and Krouse, H.R., 437, *467*
Davis, R., Jr., see Harned, H.S. and Davis, R., Jr.
Day, E.T., see Allen, E.T. and Day, A.L.
Deak, J., 109, *135*, 110, *135*
Dean, L.A., see McAuliffe, C.D. et al.
Deevey, E.S. and Nakai, N., 240, *255*
Deevey, E.S., Nakai, N. and Stuiver, M., 240, *255*

Deevey, E.S. and Stuiver, M., 332, 333, *395*
Deevey, E.S., see Stuiver, M. and Deevey, E.S.; Oana, S. and Deevey, E.S.
Degens, E.T., 129, *136*, 346, 354, *395*, 490, *500*; see also Clayton, R.N. and Degens, E.T.; Deuser, W.G. et al.; Hattswey, J.C. and Degens, E.T.; Knetsch, C. et al.; Taylor, H.P. et al.
Degens, E.T., Behrendt, M., Gotthardt, B. and Reppmann, E., 332, 333, 336, 337, *395*
Degens, E.T. and Epstein, S., 307, *322*
Degens, E.T., Guillard, R.R.L., Sacket, W.M. and Hellebust, J.A., 333, 336, 339, 342, *395*
Degens, E.T. and Ross, D.A., 490, 492, *500*
Deines, P. and Gold, D.P., 386, *395*
Deines, P., Langmuir, D. and Harmon, R.S., 58, 61, 63, *71*, 477, *500*
Deines, P., see Allen, P. et al.
De Laeter, J., see Jeffrey, P.M. et al.
Delmas, R., see Raynaud, D. and Delmas, R.
Delwiche, C.C. and Steyn, P.L., 415, 416, 417, *430*
De Maere, X., see Picciotto, E. et al.
Dement'yev, V.S., 279, *280*
Dement'yev, V.S. and Syromayatnikov, N.G., 259, *280*
De Niro, M.J. and Epstein, S., 353, *395*
Denny, A.F., see Barrer, R.M. and Denny, A.F.
Dequasie, H.L. and Grey, D.C., 389, *396*
De Quervain, M., 142, 148, *175*
De Quervain, M.R., see Martinec, J. et al.
Derekoy, A.M., see Gonfiantini, R. et al.
Deuser, W.G., 351, 357, *396*
Deuser, W.G., Degens, E.T. and Guillard, R.R.L., 333, 339, 343, *396*, 487, *500*
Deutsch, S., Ambach, W., and Eisner, H., 156, *175*,
Deutsch, S., see Aldaz, L. and Deutsch, S.; Picciotto, E. et al.
Dickson, F.W., see Sakai, H. and Dickson, F.W.
Dimitroulas, C., see Leontiatis, J. and Dimitroulas, C.
Din, G.A., see McCready, R.G.L. et al.
Dinçer, T., Noory, M., Javed, A.R.K., Nuti, S. and Tongiorgi, E., 113, *136*

Dinçer, T. and Payne, B.R., 117, 119, *136*
Dinçer, T., Payne, B.R., Florkowski, T., Martinec, J. and Tongiorgi, E., 103, *136*, 164, 169, *175*
Dinçer, T., see Gonfiantini, R. et al.
Dodd, J.R. and Stanton, R.J., 481, *501*
Doe, B.R., Hedge, C.E. and White, D.E., 212, *221*
Doe, B.R., see Rosholt, J.N. et al.
Domenach, A.M. and Chalamet, A., 417, *430*
Domenach, A.M., see Bardin, R. et al.
Dostal, K., see May, F. et al.
Douglas, R.G. and Savin, S.M., 306, *322*
Douglas, R.G., see Savin, S.M. and Douglas, R.G.
Downing, R.A., Smith, D.B., Pearson, F.J., Jr., Monkhouse, R.A. and Otlet, R.L., 130, 132, *136*
Downing, R.A., see Smith, D.B. et al.
Downton, W.J.S., see Tregunna, E.B. et al.
Dray, M., see Blanc, P. and Dray, M.
Drimmie, R.J., 486, 490, *501*; see also Fritz, P. et al.; Jackson, T.A. et al.
Drury, J.S., see Brown, L.L. and Drury, J.S.
Drubrova, N.V. and Nesmelova, Z.N., 363, 374, *396*
Dudley, W.C., 306, *322*
Dudley, W.C., see Margolis, S.V. et al.
Duffy, J., see Shearer, G. et al.
Dunford, H.B., see Thode, H.G. et al.
Dunning, H.N., see Park, R. and Dunning, H.N.
Durazzi, J.T., 476, *501*
Durazzi, J.T., see Lindrotts, K.J. et al.
Du Toit, G.J., see Bredenkamp, D.B. et al.
Dyck, W., see Lowdon, J.A. and Dyck, W.
Dymond, J., Corliss, J.B., Heatts, G.R., Field, C.W., Dasch, E.J. and Veeh, H.H., 289, 301, *322*
Dymond, J.R., see Garlick, G.D. and Dymond, J.R.

Eadie, B.J., 332, 333, *396*; see also Sackett, W.M. et al.
Eckelmann, W.R., Broecker, W.S., Whitlock, D.W. and Allsup, J.R., 350, 351, *396*

511

Eckelmann, W.R., see Sackett, W.M. et al.
Edwards, A.P., 425, *430*
Edwards, A.P., see Bremner, J.M. et al.; Cheng, H.H. et al.; Hauck, R.D. et al.
Edwards, K.W., 279, *280*
Ehhalt, D., Knott, K., Nagel, J.F. and Vogel, J.C., 43, *45*
Ehhalt, D. and Ostlund, G., 38, *45*
Ehhalt, D.H., 25, 26, 29, 38, 39, *45*, 143, *175*, 390, *396*; see also Brinkman, R. et al., Emrich, K. et al.; Vogel, J.C. and Ehhalt, D.; Zimmermann, O. et al.
Eichler, R., 85, *136*
Eichler, R., see Brinkman, R. et al.
Eichmann, R., Plate, A., Behrens, W. and Kroepelin, H., 410, *430*
Eichmann, R. and Schidlowski, M., 359, *396*; Eichmann, R., see Behrens, W. et al.; Schidlowski, M. et al.
Eisma, D., see Mook, G.W. and Eisma, D.
Eisner, H., see Ambach, W. et al.; Behrens, H. et al.; Deutsch, S. et al.
Ekdahl, C.A. and Keeling, C.D., 387, *396*
Ellis, A.J. and Mahon, W.A.J., 179, 211, *221*
Ellis, A.J., Mahon, W.A.J. and Ritchie, J.A., 217, *221*
Ellis, A.J., see Cusicangui, H. et al.
Elsässer, M., see Ambach, W. et al.
Emery, K.O., 351, *396*; see also Kaplan, I.R. et al.
Emiliani, C., 306, *322*
Emrich, K., Ehhalt, D.H. and Vogel, J.C., 63, *71*, 477, *501*
Engel, A.E.J., Clayton, R.N. and Epstein, S., 307, *322*
Engelkemeir, A., see Stevens, C.M. et al.
Engelkeimeir, A.G., see Holt, B.D. et al.
Epstein, S., 482, *501*; see also Clayton, R.N. and Epstein, S.; Degens, E.T. and Epstein, S.; De Niro, M.J. and Epstein, S.; Engel, A.E.J. et al.; Gow, A.J. and Epstein, S.; Jacobson, B.S. et al.; Knauth, L.P. and Epstein, S.;Kolodny, Y. and Epstein, S.; Land, L.S. and Epstein, S.; Lowenstam, H.A. and Epstein, S.; O'Neil, J.R. and Epstein, S.; Park, R. and Epstein, S.; Savin, S.M. and Epstein, S.; Silverman, S.R. and Epstein, S.; Smitts, B.N. and Epstein, S.; Suzuoki, T. and Epstein, S.; Taylor, H.P. and Epstein, S.; Urey, H.C. et al.; Yapp, C.J. and Epstein, S.;

Yeh, H.W. and Epstein, S.
Epstein, S., Buchsbaum, R., Lowenstam, H.A. and Urey, H.C., 286, *322*, 475, 476, *501*
Epstein, S. and Mayeda, T.K., 3, *18*
Epstein, S. and Sharp, R.P., 170, *175*
Epstein, S., Sharp, R. and Gow, A.J., 42, *45*, 146, 161, *176*
Epstein, S., Sharp, R.P. and Goddard, I., 146, 147, 154, *176*
Epstein, S., Yapp, C.J. and Hall, J.H., 3, *18*, 495, 496, 497, 499, *501*
Eriksson, E., 24, 27, 28, *45*, 70, *71*, 125, 126, *136*, 438, 449, *468;* see also Bourdon, D.J. et al.
Erlenkeuser, H., Suess, E. and Willkomm, H., 351, 354, *396*, 493, 494, *501*
Erlenkeuser, H., see Wilkomm, H. and Erlenkeuser, H.
Eslinger, E.V. and Savin, S.M., 215, *221*, 285, 287, 299, *322*
Eslinger, E.V., see Hower, J. et al.
Evans, T.L., see Campbell, F.A. et al.
Evin, J., see Margrita, R. et al.
Evin, J. and Vuilaume, Y., 57, *71*
Exner, M.E., see Sackett, W.M. et al.

Facy, L., Merlivat, L., Nief, G. and Roth, E., 39, *45*
Fairbairn, A.W., see Hurley, P.M. et al.
Falting, V. and Harteck, P.Z., 22, *45*
Farlekas, G.M., see Winograd, I.J. and Farlekas, G.M.
Farmer, J.G. and Baxter, M.S., 389, 390, *396*
Farvolden, R.N., see Sklash, M.G. et al.
Fautts, H., see Matthes et al.; Wendt, I. et al.
Feigin, A.D., Koll, D.H., Shearer, G.B. and Commoner, B., 416, 421, *430*
Feigin, A.D., Shearer, G.B., Kohl, D.H. Commoner, B., 420, *430*
Fergusson, G.J., 387, *396*
Fergusson, G.J. and Knox, F.B., 210, *221*
Ferrara, G., Gonfiantini, R. and Pistoia, P., 213, *221*
Ferrara, G., see Ferrara, G.C. et al.
Ferrara, G.C., Ferrara, G. and Gonfiantini, R., 198, 211, *221*, 385, 386, *396*
Ferrara, G.C., see Baldi, P. et al; Ferrara, G. et al.; Panichi, C. et al.
Fiebiger, W., see Schidlowski, M. et al.
Field, C.W., see Dymond, J. et al.

Filip, G., see Tataru, S. et al.
Fischer, W., Maass, I. and Sontag, E., 361, *396*
Fischer, W., Maass, I., Sontag, E. and Süss, M., 347, 361, 362, *396*
Fisher, D.W., 237, 238, *255*
Fisher, D.W., see Hobba, W.A. et al., Pearson, F.J., Jr. and Fisher, D.W.
Flamm, E.J., see Nir, A.
Flandrin, J., see Margrita, R. et al.
Fletcher, W.H., see Begun, G.M. and Fletcher, W.H.
Florenski, K.P., see Volynets, V.P. et al.
Florkowski, T. see Dinçer, T. et al.
Focht, D.D., 416, 421, *430*
Folinsbee, R.E., see Fritz, P. et al.
Fontes, J.Ch., 3, 101, 119, 120, *136*, 233, *255*, 291, 292, 308, 309, *322*; see also Bortolami et al.; Conrad, G. and Fontes, J.Ch.; Conrad, G. et al.; Gasse, F. et al.; Gonfiantini, R. and Fontes, J.Ch.; Gonfiantini, R. et al.; Zuppi, G.M. et al.
Fontes, J.Ch., Bortolami, G.C. and Zuppi, G.M., 116, *136*
Fontes, J.Ch. and Garnier, J.M., 64, *71*, 128, 130, 132, *136*
Fontes, J.Ch. and Gonfiantini, R., 33, 41, *45*, 121, 122, *136*, 291, 292, 308, *322*
Fontes, J.Ch., Gonfiantini, R. and Roche, M.A., 101, *136*
Fontes, J.Ch., Létolle, R. and Nesteroff, W.D., 309, *322*
Fontes, J.Ch., Létolle, R., Olive, P. and Blavoux, B., 92, *136*
Fontes, J.Ch. and Nielsen, H., 309, *322*
Fontes, J.Ch. and Olivry, J.C., 77, *136*
Fontes, J.Ch. and Pouchan, P., 16, *18*, 477, 491, *501*
Fontes, J.Ch. and Zuppi, G.M., 95, 96, *136*, 241, 244, *255*
Ford, D.C., see Schwarcz, H.P. et al.; Thompson, P. et al.
Ford, H.T., see Hoering, T.C. and Ford, H.T.
Forrest, J. and Newman, L., 443, *468*
Forrest, J., see Newman, L. et al.
Fournier, R.O. and Truesdell, A.H., 204, *221*
Fournier, R.O., White, D.E. and Truesdell, A.H., 200, *211*
Fournier, R.O., see Mazor, E. and Fournier, R.O.; Truesdell, A.H. and Fournier, R.O.; White, D.E. et al.
Frank, D.J., Gromly, J.R. and Sackett, W.M., 383, *396*, *397*
Frechen, J., see Taylor, H.P. et al.
Freund, W., see May, F. et al.
Freyer, H.D. and Aly, A.I.M., 414, 416, 420, 424, 425, *430*
Freyer, H.D. and Kasten, M., 417, 418, *430*
Freyer, H.D. and Wiesberg, L., 388, 389, 390, *396*, *397*
Friedman, G.M., see Allen, R.C. et al.
Friedman, I., 32, *45*; see also Bainbridge, A.E. et al., Barker, F. and Friedman, I.; Clayton, R.N. et al.; Graf, D.L. et al.; Hitchon, B. and Frieman, I.; Judy, C. et al.; Kharaka, Y.K. et al.; Matsuo, S. et al.; Meiman, J. et al.; Murata, K.J. et al.; Pearson, F.J., Jr. and Frieman, I.; Picciotto, E. et al.
Friedman, I. and Hardcastle, K., 3, *18*, 313, 319, 320, *322*, 323
Friedman, I. and Irsa, A.P., 387, *397*
Friedman, I., Machta, L. and Soller, R., 38, *46*
Friedman, I. and O'Neil, J.R., 6, 12, 13, 14, *18*, 197, *221*, 232, 233, *255*, 286, 287, *323*
Friedman, I., Redfield, A.C., Schoen, B. and Harris, J., 154, *176*
Friedman, I. and Smith, G., 143, 144, 146, *176*
Friedrich, H.U. and Jüntgen, H., 364, 365, 382, *397*
Friend, J.P., 438, *468*
Frishman, S.A., see Behrens, E.W. and Frishman, S.A.
Fritz, P., Anderson, T.W. and Lewis, C.F.M., 476, 479, 482, 483, 484, *501*
Fritz, P., Binda, P.L., Folinsbee, R.E. and Krouse, H.R., 357, *397*
Fritz, P., Cherry, J.A., Weyer, K.U. and Sklash, M., 106, 107, *136*
Fritz, P., Drimmie, R.J. and Render, F., 499, *501*
Fritz, P. and Poplawski, S., 475, 476, 478, 479, 481, 488, *501*
Fritz, P., Reardon, E.J., Barker, J., Brown, R.M., Cherry, J.A., Killey, R.W.D. and McNaughton, D., 127, *136*
Fritz, P., Silva, C., Suzuki, O. and Salati,

E., 99, *136*
Fritz, P. and Smith, D.G.W., 285, *323*
Fritz, P., see Jackson, T.A. et al.; Killey, R.W. and Fritz, P.; Michel, F.A. and Fritz, P.; Reardon, E.J. and Fritz, P.; Sklash, M.G. et al.

Gaebler, O.H., Choitz, H.C., Vitti, J.C. and Vumirovich, R., 417, *430*
Gaines, A.G., see Orr, W.L. and Gaines, A.G.
Galimov, E.M., 60, *71*, 340, 341, 348, 351, 374, 376, 379, 381, 382, 389, *397*
Galimov, E.M., Mamchur, G.P. and Kuznetsova, N.G., 359, *397*
Galimov, E.M., see Vinogradov, A.P. and Galimov, E.M.
Gallo, G., see Gonfiantini, R. et al.
Games, L.M. and Hayes, J.M., 357, 358, *397*
Garcia, E., see Payne, B.R. and Garcia, E.
Garcia-Loygorri, A., Bosch, B. and Marce, A., 361, *397*
Garlick, G.D., 227, 255, 288, *323*
Garlick, G.D. and Dymond, J.R., 300, *323*
Garlick, G.D., see Mopper, K. and Garlick, G.D.
Garnier, J.M., see Fontes, J.Ch. and Garnier, J.M.
Garrels, M. and Christ, C.L., 58, *71*, 234, 255
Garrels, R.M., see Hostetler, P.B.
Gasse, F., Fontes, J.Ch. and Rognon, P., 481, *501*
Gat, J.R., 37, *46*; see also Carmi, I. and Gat, J.R.; Mazor, R. et al.; Salati, E., et al.; Sofer, Z. and Gat, J.
Gat, J.R. and Carmi, I., 41, *46*, 78, 129, *137*
Gat, J.R. and Dansgaard, W., 35, 44, *46*
Gat, J.R. and Issar, A., 83, 129, *137*, 499, *501*
Gat, J.R., Karfunkel, V. and Nir, A., 40, *46*
Gat, J.R. and Levy, Y., 36, *46*
Gat, J.R. and Tzur, Y., 82, *137*
Gavelin, S., 359, 360, *397*
Gavish, E., see Allen, R.C. et al.
Gazzarrini, F., see Colombo, U. et al.
Gehlen, K. and Nielsen, H., 436, *468*

Geissler, C. and Belau, L., 362, *397*
Geoktistov, V.P., see Bogdanov, Yu.V. et al.
Geyh, M.A., 56, 62, 70, *71*; see also Matthess, G. et al.; Wendt, I. et al.
Geyh, M.A. and Mairhofer, J., 125, 126, *137*
Geyh, M.A., Merkt, J. and Muller, H., 70, *71*
Geyh, M.A. and Wendt, I., *71*
Giaugne, W.F. and Johnston, H.L., 21, *46*
Gieskes, J.M., see Lawrence, J.R. et al.
Giggenbach, W.F., 181, 186, 187, 192, 193, 194, 205, 207, 212, *221*
Giletti, B.J., Semet, M.P. and Yund, R.A., 215, *219*
Gleuson, J.D., see Gunter, B.D. and Gleuson, J.D.; Murata, K.J. et al.
Godl, L., see Korkisch, J. and God, L.
Goddard, I., see Epstein, S. et al.
Godfrey, J.D., 3, *18*
Gold, D.P., see Deines, P. and Gold, D.P.
Goldberg, E.D., see Rex, R.W. and Goldberg, E.D.; Koide, M. and Goldberg, E.D.
Goldhaber, M.B., 234, 235, *256*
Goldhaber, M.B. and Kaplan, I.R., 232, 234, 236, 240, 250, *256*, 458, 459, *468*
Goldsmith, J.R., see Clayton, R.N. et al.
Golubchina, M.N., see Bogdonov, Yu.V. et al.
Gonfiantini, R., 7, 9, 11, 13, 14, *18*, 144, *176*, 180, *221*; see also Bradley, E. et al.; Colombo, U. et al.; Ferrara, G.C. et al.; Fontes, J.Ch. and Gonfiantini, R.; Fentes, J.Ch. et al.; Panichi, C. and Gonfiantini, R.; Panichi, C. et al.
Gonfiantini, R., Conrad, G., Fontes, J.Ch., Sauzay, G. and Payne, B.R., 57, 72, 84, 108, *137*
Gonfiantini, R., Dinçer, T. and Derekoy, A.M., 60, 72, 109, 123, 124, 129, *137*, 499, *501*
Gonfiantini, R., Gallo, G., Payne, B.R. and Taylor, C.B., 92, 93, 115, *137*
Gonfiantini, R., Gratziu, S. and Tongiorgi, E., 42, *46*
Gonfiantini, R. and Fontes, J.Ch., 309, *323*
Gonfiantini, R., Panichi, C. and Tongiorgi, E., 474, 476, *501*

Gonfiantini, R., Togliatti, V. and Tongiorgi, E., 99, *137*
Gonfiantini, R., Togliatti, V., Tongiorgi, E., Breuck, W.D. and Picciotto, E., 482, *501*
Goodney, D.E., see Margolis, S.V. et al.
Gordon, L., see Craig, H. and Gordon, L.; Craig, H. et al.
Gordon, L.I., see Williams, P.M. and Gordon, L.I.
Gormly, J.R., see Brooks, J.M. et al.
Gotthardt, B., see Degens, E.T. et al.
Gottikh, R.P., see Alekseev, F.A. et al.
Götz, D., see Heinzinger, K. and Götz, D.
Goudie, A., see Salomons, W. et al.
Gow, A.J. and Epstein, S., 170, *176*
Gow, A.J., see Epstein, S. et al.
Gradztan, E., see Kronefeld, J. et al.; Yaron, F. et al.
Graf, D.L., Friedman, I. and Meents, W.F., 320, *323*
Graf, D.L., see Clayton, R.N. et al.
Granat, L., Hallberg, R.O. and Rodhe, H., 438, *468*
Granger, H.C. and Warren, C.G., 234, *255*
Gratziu, S., see Gonfiantini, R. et al.
Gray, D.C., see Dequasie, H.L. and Gray, D.C.
Greenhalgh, D., see Jeffrey, P.M. et al.
Grey, D.C. and Jensen, M.L., 465, *468*
Grinenko, L.N., Kononova, V.A. and Grinenko, V.A., 436, *468*
Grinenko, V.A., see Grinenko, L.N. et al.; Vinogradov, A.P. et al.
Grindsted, M.J., see Wilson, A.J. and Grinsted, M.J.
Groeneveld, D.J., 23, 40, *46*; see also Mook, G.W. et al.
Gromly, J.R., see Frank, D.J. et al.
Grootes, P.M., see Vogel, J.C. et al.
Gross, M.G., 306, *323*, 480, *501*
Gross, W.H., see Shima, M. et al.
Grosse, A., Johnston, W.M., Wolfgang, R.L. and Libby, W.F., 22, *46*
Grumbkov, A.P., see Alekseev, F.A. et al.
Grummitt, see Brown, R.M. and Grummit
Guegan, A., see Bosch, B. et al.
Gugelmann, A., see Oeschger, H. et al.
Guillard, R.R.L., see Degens, E. et al.; Deuser, E.G. et al.
Gulbransen, E.A., see Nier, A.O. and Gulbransen, E.A.

Gundestrup, N., see Dansgaard, W. et al.; Hammer, G.U. et al.; Reeh, N. et al.
Gunter, B.D. and Gleuson, J.D., 381, *397*
Gunter, B.D. and Musgrave, B.C., 198, 200, 202, 203, 212, *221*, 385, 386, *397*
Gupta, M.L., Saxena, V.K. and Sukhija, B.S., 187, *222*
Gutsalo, L.K., 214, *222*

Hageman, R., see Lorius, C. et al.
Hageman, R., Nief, G. and Roth, E., 13, 18, 21, *46*
Hagemann, H.W., see Boigk, H. et al.
Hahn-Weinheimer, P., 359, 360, *397*, *398*
Hahn-Weinheimer, P., Markl, G. and Raschka, H., 359, 360, *398*
Halevy, E., 293, *323*
Hall, E.S., see Young, C.P. et al.
Hall, J.H., see Epstein, S. et al.
Hall, R.L., 331, *398*
Hall, W., see McAuliffe, C.D. et al.
Hallberg, R.O., see Granat, L. et al.
Hammer, G.U., Clausen, H.B., Dansgaard, W., Gundestrup, N., Johnsen, S.J. and Reeh, N., 163, 172, *176*
Hammer, G.U., see Dansgaard, W. et al.
Hamza, M.S. and Broecker, W.S., 287, *323*
Hagemann, R., see Lorius, C. et al.
Hanshaw, B.B., see Coplen, T.P. and Hanshaw, B.B.; Pearson, F.J., Jr. and Hanshaw, B.B.; Rightmire, C.T. and Hanshaw, B.B.; Rightmire, C.T. et al.
Hardcastle, K., see Friedman, I. and Hardcastle, K.; Meiman, J. et al.
Harmon, R.S., see Deines, P. et al.; Schwarcz, H.P. et al.
Harned, H.S. and Davis, R., Jr., 64, *72*
Harned, H.S. and Scholes, S.R., 64, *72*
Harris, J., see Friedman, I. et al.
Harrison, A.G. and Thode, H.G., 232, 256, 457, *468*
Harteck, P.Z., see Falting, V. and Harteck, P.Z.
Harvath, G.J., see Calder, J.A. et al.
Hashimoto, T., see Sakanone, M. and Hashimoto, T.
Hattersley-Smith, G., Krouse, H.R. and West, K.E., 160, *176*
Hattori, A., see Wada, E. and Hattori, A.
Hattsway, J.C. and Degens, E.T., 357, *398*

Hauck, R.D., Bartholomew, W.V., Bremmer, J.M., Broadbent, F.E., Cheng, H.H., Edwards, A.P., Keeney, D.R., Legg, J.O., Olsen, S.R. and Porter, L.K., 425, *431*
Hawkins, J.W., see Pineau, F. et al.
Hay, R.L., see O'Neil, J.R. and Hay, R.L.
Hayes, J.M., see Games, L.M. and Hayes, J.M.
Healy, J., see Mazor, E. et al.
Heatts, G.R., see Dymond, J. et al.
Hedge, C.E., see Doe, B.R. et al.
Hedges, J., 351, 353, 354, *398*
Heezen, B.C., see Hurley, P.M. et al.
Heinzinger, K. and Götz, D., 292, *323*
Hellebust, J.A., see Degens, E.T. et al.
Hendricks, S.B., see McAuliffe, C.D. et al.
Hendy, C.H., 62, 72, 348, *398*, 474, *501*; see also Troughton, J.H. et al.
Herrmann, A. and Stichler, W., 150, 153, 162, *176*
Hitchcock, D.R., 466, *468*
Hitchon, B., Brown, H.M. and Krouse, H.R., 437, *468*
Hitchon, B. and Friedman, I., 320, *323*
Hitchon, B. and Krouse, H.R., 239, *256*, 437, 451, *468*, 488, 489, *501*
Hobba, W.A., Jr., Fisher, D.W., Pearson, F.J., Jr. and Chemerys, J.C., 209, *222*
Hodgson, W.A., 357, *398*
Hoefs, J., 227, 228, *256*, 407, *431*; see also Rothe, P. et al.; Welte, D.H. et al.
Hoefs, J. and Schildlowski, M., 3, *18*, 359, *398*
Hoering, T.C., 359, *398*, 417, 424, *431*; see also Abelson, P.H. and Hoering, T.C.
Hoering, T.C. and Ford, H.T., 415, *431*
Hoering, T.C. and Moore, H.E., 382, *398*
Holser, W.T., 235, 236, 237, *256*; see also Claypool, G.E. et al.
Holser, W.T. and Kaplan, I.R., 235, 236, 238, *256*, 437, *468*
Holt, B., 3, *18*, 435, 439, *468*
Holt, B.D., Engelkeimeir, A.G. and Venters, A., 449, 451, *468*
Hom, B., see Craig, H. and Hom, B.
Horgan, T., see Wollanke, G. et al.
Horibe, Y., see Craig, H. and Horibe, Y.; Craig, H. et al.
Horodyski, R.J., see Schopf, J.W. et al.
Hostetler, P.B. and Garrels, R.M., 259, *280*

Houtermans, J., Suess, H.E. and Munk, W., 387, *398*
Hower, J., Eslinger, E.V., Hower, M. and Perry, E.A., 315, *323*
Hower, J., see Perry, E.A. and Hower, J.
Hower, M., see Hower, J. et al.
Hsu, J.G. and Smith, B.N., 338, *398*
Hsü, K.J., see Lloyd, R.M. and Hsü, K.J.
Hubert, P., Marce, A., Olive, P. and Siwertz, E., 126, *137*
Hubert, P., Marin, E., Meybeck, M., Olive, P. and Siwertz, E., 103, *137*
Hubert, P., see Crouzet, E. et al.; Meybeck, M. et al.
Hudson, B.E., see Smith, P.V. and Hudson, B.E.
Hudson, J.D., see Tan, F.C. and Hudson, J.D.
Hufen, T.H., Lan, L.S. and Buddemeier, R.W., 54, 72, 127, *137*
Hughes, W.W., see Allison, G.B. and Hughes, W.W.
Hulston, J.R., 195, 198, 200, 201, 203, *222*; see also Bailey, I.H. et al.; Kaplan, I.R. et al.; Lyon, G.L. and Hulston, J.R.; Thode, H.G. et al.
Hulston, J.R. and McCabe, W.J., 198, 199, 211, 212, 213, *222*, 385, 386, *398*
Hunt, J.M., 350, 351, 354, *398*
Hurley, P.M., Heezen, B.C., Pinson, W.H. and Fairbairn, H.W., 298, *323*
Husain, S.A., see Krouse, H.R. et al.
Hutton, L.G., see Mazor, E. et al.

I.A.E.A., 81, *137*
Igumnov, S.A., 453, *468*
Ingerson, E. and Pearson, F.J., 56, 59, *73*, 127, 130, *137*
Irsa, A.P., see Friedman, I. and Irsa, A.P.
Isaacson, R.E., Brownell, L.E., Nelson, R.W. and Roetman, E.L., 89, *137*
Issar, A., Bein, A. and Michaeli, A., 129, *138*
Issar, A., see Gat, J.R. and Issar, A.
Ivonov, M.V., see Matrosov, A.G. et al.

Jackson, M.L., see Clayton, R.N. et al.; Rex, R.W. et al.
Jackson, T.A., Fritz, P. and Drimmie, R., 359, *398*
Jacobsen, R.L. and Langmuir, D., 64, 72
Jacobson, B.S., Laties, G.G., Smith, N.N.,

Epstein, S. and Laties, B., 337, *399*
Jacobson, B.S., Smith, B.N., Epstein, S. and Laties, G.G., 338, *399*
James, A.T. and Baker, D.R., 288, 293, *323*
James, R., 193, *222*
Jansen, H.S., 389, 390, *399*
Javed, A.R.K., see Dinçer, T. et al.
Javoy, M., see Pineau, F. et al.; Richet, P. et al.
Jeffrey, L.M., 353, *399*
Jeffrey, P.M., Canpston, W., Greenhalgh, D. and De Laeter, J., 359, 361, 362, *399*
Jensen, M.L. and Nakai, N., 237, *256*, 449, *468*
Jensen, M.L., see Grey, D.C. and Jensen, M.L.; Nakai, N. and Jensen, M.L.
Johns, L.E., see Rennie, D.A. et al.
Johnsen, S.J., 157, 158, 160, 161, *176*; see also Dansgaard, W. and Johnsen, S.J.; Dansgaard, W. et al.; Hammer, G.U. et al.; Reeh, N. et al.
Johnsen, S.J., Dansgaard, W., Clausen, H.B. and Langway, C.C., Jr., 157, 163, 170, 174, *176*
Johnson, R.W. and Calder, J.A., 346, 351, 356, *399*
Johnson, W.M., see Grosse, A. et al.
Johnston, H.L., see Giaugne, W.F. and Johnston, H.L.
Joksubayev, A.I., see Bogdanov, Yu.V. et al.
Jones, B.F., see Clayton, R.N. et al.; Pearson, F.J., Jr. et al.; Plummer, L.N. et al.; Truesdell, A.H. and Jones, B.F.
Jones, D.C., see Kreitler, C.W. and Jones, D.C.
Jones, D.F., see Pinder, G.F. and Jones, D.F.
Jones, G.E. and Starkey, R.L., 457, *468*
Jordan, R., see Stahl, W. and Jordan, R.
Jouzel, J., Merlivat, L. and Pourchet, M., 160, *176*
Jouzel, J., Merlivat, L. and Roth, E., 39, *46*
Jouzel, J., see Merlivat, L. et al.
Judy, C., Meiman, J.R. and Friedman, I., 151, *176*
Junge, C.E., 438, *468*
Junge, C.E. and Ryan, T., 233, *256*
Junghans, H.G., see Roether, W. and Junghans, H.G.

Jüntgen, H., see Friedrich, H.U. and Jüntgen, H.

Kadonaga, T., see Wada, E. et al.
Kahout, F.A., 275, 276, *280*
Kalkreuth, W., see Welte, D.H. et al.
Kanonov, V.I. and Polak, B.G., 179, 214, *222*
Kaplan, I.R., 438, *468*; see also Cline, J.D. and Kaplan, I.R.; Claypool, G.E. et al.; Goldhaber, M.B. and Kaplan, I.R.; Holser, W.T. and Kaplan, I.R.; Kolodny, Y. and Kaplan, I.R.; McCready, R.G.L. et al.; Nissenbaum, A. and Kaplan, I.R.; Nissenbaum, A. et al.; Presley, B.J. and Kaplan, I.R. Presley, B.J. et al.; Smith, J.W. et al.; Sweeney, R.E. et al.
Kaplan, I.R., Emery, K.O. and Rittenberg, S.C., 459, *468*
Kaplan, I.R., Rafter, T.A. and Hulston, J.R., 458, 461, *468*
Kaplan, I.R. and Rittenberg, S.C., 232, 239, 250, *256*, 346, *399*, 457, 458, *469*
Karfunkel, V., see Gat, J.R. et al.
Kasten, M., see Freyer, H.D. and Kasten, M.
Katz, A., see Yaron, F. et al.
Kauder, L.N., see Stern, M.J. et al.
Kaufman, A., see Mazor, E. et al.
Kaufman, M.I., 270, *280*; see also Osmond, J.K. et al.
Kaufman, M.I., Rydell, H.S. and Osmond, J.K., 266, *280*
Kaufman, S. and Libby, W.F., 22, *46*
Kavel, K.J., see Clayton, R.N. et al.
Keeling, C.D., 52, *72*, 338, 387, 388, 391, *399*; see also Bacastow, R. and Keeling, C.D.; Craig, H. and Keeling, C.D.; Ekdahl, C.A. and Keeling, C.D.
Keeney, D.R., see Bremner, J.M. and Keeney, D.R.; Hauck, R.D. et al.
Keith, M.L., see Allen, P. and Keith, M.L.; Allen, P. et al.
Keith, M.L. and Anderson, G.M., 478, 492, *502*
Keith, M.L. and Weber, J.N., 360, 386, *399*, 478, 481, *502*
Kellogg, W.W., Cadle, R.D., Allen, E.R., Lagrus, A.L. and Martell, E.A., 438, *469*
Kemp, A.L.W. and Thode, H.G., 424,

431, 457, *469*
Kendall, C., 214, *222*
Kharaka, Y.F., see O'Neil, J.R. and Kharaka, Y.F.
Kharaka, Y.K., Berry, F.A.F. and Friedman, I., 183, *222*, 320, *323*
Kigoshi, K., 262, 267, 274, 276, 279, *280*
Killey, R.W.D., 487, *502*; see also Fritz, P. et al.
Killey, R.W. and Fritz, P., 488, *502*
Kirchlechner, P., see Behrens H. et al.
Kirschenbaum, I., 38, *46*
Kiyosu, Y., see Nakai, N. et al.
Kleerekoper, H., see Thode, H.G. et al.
Klots, C.E. and Benson, B.B., 414, *431*
Klyen, L.E., 217, *222*
Knauss, K.G., see Ku, T.L. et al.
Knauth, L.P., 290, 291, 303, 311, 312, *323*
Knauth, L.P. and Epstein, S., 288, 291, 311, 312, 313, 314, *323*
Knetsch, C., Shata, A., Degens, E., Münnich, K.O., Vogel, J.C. and Shazly, M.M., 129, *138*
Kneuper, G., see Colombo, U. et al.; Teichmuller, R. et al.
Knott, K., see Ehhalt, D. et al.
Knox, F.B., see Ferguson, G.J. and Knox, F.B.
Koch, J., see Stahl, W. and Koch, J.
Koene, B.K.S., see Mook, G.W. and Koene, B.K.S.
Kohl, D.A., Shearer, G.B. and Commoner, B., 407, 425, *431*
Kohl, D.H., see Feigin, A.D. et al.; Meints, V.W. et al.; Shearer, G. et al.
Koide, M. and Goldberg, E.D., 264, *280*
Kölle, W., 494, *502*
Kolodny, Y., see Yaron, F. et al.
Kolodny, Y. and Epstein, S., 311, 313, *323*
Kolodny, Y. and Kaplan, I.R., 262, 271, *280*
Kondrat'eva, E.N., see Mekhtieva, V.L. and Kondrat'eva, E.N.
Kononova, V.A., see Grinenko, L.N. et al.
Koons, C.B., Bond, J.G. and Pierce, F.C., 367, 368, 371, *399*
Koons, C.B., see Rogers, M.A. and Koons, C.B.
Korkisch, J. and Godl, L., 279, *281*
Kreitler, C.W., 3, *18*, 415, 427, *431*

Kreitler, C.W. and Jones, D.C., 427, *431*
Krejci-Graf, K. and Wickman, F.E., 351, 359, 367, *399*
Kreutz, W., see Zimmermann, U. et al.
Krishnaswami, S., see Baht, S.F. and Krishnaswami, S.
Kroepelin, H., see Behrens, W. et al.; Eichmann, R. et al.
Kronfeld, J., see Yaron, F. et al.
Kronfeld, J. and Adams, J.A.S., 272, 276, 279, *281*
Kronfeld, J., Gradztan, E., Muller, H.W., Radin, J., Yaniv, A. and Zach, R., 272, 274, 276, *281*
Kroopnick, P., 387, *399*; see also Margolis, S.V. et al.
Kroopnick, P.M. and Craig, H., 236, *256*
Kropotova, O.I., see Malinin, S.D. et al.
Krouse, H.R., 437, 438, 459, 463, 464, *469*; see also Campbell, F.A. et al.; Cook, F.D. et al.; Davies, G.R. and Krouse, H.R.; Fritz, P. et al.; Hattersley-Smith, G. et al.; Hitchon, B. and Krouse, H.R.; Hitchon, B. et al.; Laishley, E.J. et al.; Lowe, L.F. et al.; McCready, R.G.L. et al.; Mitchell, R.H. and Krouse, H.R.; Ryznar, G. et al.; Sakai, H. and Krouse, H.R.; Sasaki, A. and Krouse, H.R.; Smejkal, V. et al.; Van Everdingen, R.O. and Krouse, H.R.; Wellman, R.P. et al.; West, K.E. and Krouse, H.R.
Krouse, H.R., Cook, F.D., Sasaki, A. and Smejkal, V., 457, 458, *469*
Krouse, H.R. and Mackay, J.R., 452, *369*
Krouse, H.R., McCready, R.G.L., Husain, S.A. and Campbell, J.B., 457, *469*
Krouse, H.R. and Sasaki, A., 457, *469*
Krouse, H.R. and Smith, J.L., 152, *176*
Krout, L., see Stevens, C.M. et al.
Kruger, P., see Stoker, A.K. and Kruger, P.
Kruger, P., Stoker, A.K. and Umana, A., 213, *222*
Krugger, S.J., see Nir, A. et al.
Ku, T.L., 261, 270, 276, *281*
Ku, T.L., Knauss, K.G. and Mathieu, G.G., 270, *281*
Kudryavtseva, A.J., see Matrosov, A.G. et al.
Kuhn, W. and Thürkauf, M., 141, *176*
Kullerud, G., see Puchelt, H. and

Kullerud, G.
Kunkler, J.L., 348, *399*
Kurtz, L.T., see Meints, W.V. et al.
Kusachi, I., see Matsubaya, I. et al.
Kusakabe, M., 204, 207, 212, 218, *222*
Kuznetsova, N.G., see Galimov, E.M. et al.
Kvenvolden, K. and Squires, R., 351, 367, *399*
Kvenvonden, K.A., see Oehler, D.Z. et al.; Schopf, J.W. et al.

Labeyrie, L., 286, 302, *323*
Lablans, W.N., see Bleeker, W. et al.
Lagrus, A.L., see Kellogg, W.W. et al.
Laishley, E.J., McCready, R.G.L., Bryant, R. and Krouse, H.R., 415, *431*, 457, *469*
Laishley, E.J., see McCready, R.G.L. et al.
Lal, D., Arnold, J.R. and Somayeijulu, B.L.K., 279, *281*
Lal, D., Nijampurkar, V.N. and Rama, S., 127, *138*
Lal, D. and Peters, B., 24, *46*
Lal, D., see Athavale, R.N. et al.
Lalou, C., see Van, N.H. and Lalou, C.
La Marche, V.C., Jr., 497, *502*
Lamb, H.H., 497, *502*
Lambert, S.J., 216, *222*
Lan, L.S., see Hufen, T.H. et al.
Land, L.S., 307, *324*
Land, L.S. and Epstein, S., 307, *324*
Landergren, S., 359, 360, *399*
Landergren, S. and Parwel, A., 351, *399*
Lang, W.B., 211, *222*
Lange, H., see Rösler, J.H. and Länge, H.
Langmuir, D., see Deines, P. et al.; Jacobsen, R.L. and Langmuir, D.
Langway, C.C., Jr., see Dansgaard, W. et al.; Johnsen, S.J. et al.
La Rogue, A., see Weber, J.N. and La Rogue, A.
Laties, B., see Jacobson, B.S. et al.
Laties, G.G., see Jacobson, B.S. et al.
Lawrence, J.R., 293, 294, 295, 296, 319, *324*
Lawrence, J.R., Gieskes, J. and Anderson, T.F., 319, *324*
Lawrence, J.R., Gieskes, J.M. and Broecker, W.S., 319, 320, *324*
Lawrence, J.R. and Taylor, H.P., Jr., 289, 294, 295, 296, 297, 298, *324*

Lebedev, V.S., 351, 363, *400*
Lebedev, W.C., Owsjannikow, W.M., Mogilewskij, G.A. and Bogdanow, W.M., 348, 363, *400*
Ledoux, R.L. and White, J.L., 293, *324*
Legg, J.O., see Hauck, R.D. et al.
Leontiadis, J. and Dimitroulas, C., 117, *138*
Lerman, J.C., 60, 72, 348, *400;* see also Vogel, J.C. et al.
Lerman, J.C., Mook, W.G. and Vogel, J.C., 50, *72*, 332, 387, *400*
Lerman, J.C. and Queiroz, O., 343, *400*
Létolle, R., 407, *431*; see also Cheminée, J. et al.; Conrad, G. et al.; Fontes, J.Ch. et al.; Mariotti, A. and Létolle, R.; Mariotti, A. et al.; Zuppi, G.M. et al.
Létolle, R. and Mariotti, A., 421, *431*
Levitte, D., see Mazor, E. et al.
Levy, H., 391, *400*
Levy, Y., see Gat, J.R. and Levy, Y.
Lewis, C.F.M., see Fritz, P. et al.
Leythaeuser, D., 359, *400*
Libby, L.M. and Pandolfi, L.J., 388, *400*, 495, *502*
Libby, W.F., 22, *46*, 49, *72*; see also Begemann, F. and Libby, W.F.; Grosse, A. et al.; Kaufman, S. and Libby, W.F.
Libby, W.F., see Von Buttlar, H. and Libby, W.F.
Lindrots, K.J., Miller, L.G., Durazzi, J.T., McIntyre, A. and Van Donk, 306, *324*
Lingenfelder, R.E., see Nir, A. et al.
Linkins, R.C., see Rubin, M. et al.
Liu, K.K., see Sweeney, R.E. et al.
Lloyd, R.M., 203, 204, *222*, 230, 232, 233, 236, 243, *256*, 287, 303, 304, 308, *324*
Lloyd, R.M. and Hsü, K.J., 308, 309, *324*
Loehr, R.C., see Wong Chong, G.M. and Loehr, R.C.
Loewe, F., 42, *46*
Loijens, H.S., see Prantl, F.A. and Loijens, H.S.
London, M., 412, 415, *431*
Longinelli, A., 305, *324*, 476, *502*; see also Cortecci, G. and Longinelli, A.
Longinelli, A. and Cortecci, G., 239, *256*, 451, 467, *469*
Longinelli, A. and Craig, H., 3, *18*, 236, *256*, 303, *324*

Longinelli, A. and Nuti, S., 3, *18*, 286, 287, 305, *324*, *325*, 476, *502*
Longinelli, A. and Sordi, S., 476, *502*
Loosli, H., see Oeschger, H. et al.
Lorius, C., 143, *176;* see also Merlivat, L. et al.
Lorius, C. and Merlivat, L., 143, 145, *176*
Lorius, C., Merlivat, L. and Hageman, R., 143, 144, 147, *176*
Lorius, R., see Raynaud, D. and Lorius, R.
Löschhorn, U., see Ambach, W. et al.
Lowdon, J.A. and Dyck, W., 332, 334, 335, *400*
Lowe, L.F., Sasaki, A. and Krouse, H.R., 238, *256*, 462, *469*
Lowenstam, H.A., see Epstein, S. et al.; Urey, H.C. et al.
Lowenstam, H.A. and Epstein, S., 479, 502
Lumsden, D.N., see Thompson, G.M. et al.
Lupton, J.E., see Craig, H. and Lupton, J.E.
Lyon, G.L., 197, 199, 211, 212, *222*, 356, *400*
Lyon, G.L. and Hulston, J.R., 197, 199, 202, *223*

Maass, I., see Fischer, W. et al.
Macdonald, W.P.J., 199, *223*
Machta, L., 387, *400*; see also Friedman, I. et al.
Mackay, J.R., see Krouse, H.R. and Mackay, J.R.
Macklin, W.C., Merlivat, L. and Stevenson, C.M., 39, *46*
Macklin, W.C., see Bailey, I.H. et al.
Macnamara, J., see Thode, H.G. et al.
Macnamara, J. and Thode, H.G., 461, *469*
Madsen, B.M., see Murata, K.J. et al.
Magaritz, M. and Taylor, H.P., 318, *325*
Magri, G., see Cotecchia, V. et al.
Mahenc, J., 412, *431*
Mahon, W.A.J., see Cusicanqui, H. et al; Ellis, A.J. and Mahon, W.A.J.; Ellis, et al.
Mahoney, M.E., see Margolis, S.V. et al.
Mairhofer, J., see Geyh, M.A. and Mairhofer, J.
Majoube, M., 7, *18*, 34, *46*; see also Roblot, M. et al.

Malinin, S.D., Kropotova, O.I. and Crienko, V.A., 195, 200, 201, *223*
Mamchur, G.P., see Galimov, E.M. et al.
Mamryn, B.A., 214, *223*
Mangin, A., see Bakalowicz, M. et al.
Manowitz, B., see Castleman, A.W. et al.; Newman, L. et al.
Marce, A., see Bosch, B. et al.; Burger, A. et al.; Conrad, G. et al.; Crouzet, E. et al.; Garcia-Loygorri, A. et al.; Hubert et al.
Margolis, S.V., Kroopnick, P.M., Goodney, D.E., Dudley, W.C. and Mahoney, M.E., 306, *325*
Margrita, R., Evin, J., Flandrin, J. and Paloc, H., 121, *138*
Marin, E., see Hubert, P. et al.
Mariner, R.H. and Willey, L.M., 186, *223*
Mariotti, A., see Amarger, N. et al.; Létolle, R. and Mariotti, A.
Mariotti, A., Berger, G. and Ben Halima, A., 425, *432*
Mariotti, A. and Létolle, R., 420, 425, *431*
Mariotti, A., Létolle, R., Blavoux, B. and Chassaing, B., 425, 426, 428, *431*
Mariotti, A., Pierre, D., Vedy, J.C. and Bruckert, S., 416, 418, 420, *432*
Mariotti, F., see Amarger, N. et al.
Markgraf, V., see Bortolami, G.C. et al.
Markl, G., see Hahn-Weinheimer, P. et al.
Martell, E.A., 24, *46*; see also Kellogg, W.W. et al.
Martin, E.A., see Ross, P.J. and Martin, E.A.
Martin, J.M., Nijampurkar, V. and Salvadori, F., 271, *281*
Martin, J.M., see Meybeck, M. et al.
Martinec, J., 163, 164, *176*, *177*; see also Arnason, B. et al.; Dinçer, T. et al.
Martinec, J., Moser, H.H., De Quervain, M.R., Rauert, W. and Stichler, W., 149, 150, 151, 162, 163, *177*
Martinec, J., Siegenthaler, U., Oescheger, H. and Tongiorgi, E., 103, 105, *138*, 165, 168, 169, *177*
Mathey, B., see Burger, A. et al.
Mathieu, G.G., see Ku, T.L. et al.
Mathieu, M., see Van Donk, J. and Mathieu, M.
Matrosov, A.G., Chebotarev, Ye.N., Kudryavtseva, A.J., Zyukun, A.M. and

Ivonov, M.V., 240, *256*
Matsubaya, O., Sakai, H., Kusachi, I. and Setake, H., 184, *223*
Matsubaya, O., see Sakai, H. and Matsubaya, O.
Matsui, E., see Salati, E. et al.
Matsuo, S., Friedman, I. and Smith, G.I., 291, 292, *325*
Matsuo, S., see Wada, E. et al.
Matthess, G., see Atakan, Y. et al.
Matthess, G., Fautts, H., Geyh, M.A. and Wendt, I., 55, *72*
Matthess, G., Münnich, K.O. and Sonntag, C., 84, *138*
Mattick, L.R., see Sondheimer, E.W.A. et al.
May, F., Freund, W., Müller, E.P. and Dostal, K.P., 381, 382, *400*
Mayeda, T.K., see Clayton, R.N. and Mayeda, T.K.; Clayton, R.N. et al.; Epstein, S. and Mayeda, T.K.; O'Neil, J.R. et al.; Tarutani, T. et al.
Mayer, M.G., see Bigeleisen, J. and Mayer, M.G.
Mayne, K.I., 410, *432*
Mazor, E., 213, 219, *223*
Mazor, E. and Fournier, R.O., 213, *223*
Mazor, E., Kaufman, A. and Carmi, I., 187, 210, *223*
Mazor, E., see Potter, R.W. et al.; Verhagen, B.T. et al.; Wasserburg, G.J. et al.
Mazor, E., Levitte, D., Truesdell, A.H., Healy, J., Gat, J. and Nissenbaum, A., 209, *223*
Mazor, E., Verhagen, B.T., Sellschop, J.P.F., Robins, N.S. and Hutton, L.G., 60, 70, *72*
Mazor, E. and Wasserburg, G.J., 213, *223*
McAlary, J.D., see Rogers, M.A. et al.
McAuliffe, C.D., Hall, W.S., Dean, L.A. and Hendricks, S.B., 293, *325*
McCabe, W.J., see Hulston, J.R. and McCabe, W.J.
McConnel, J.C., see Wofsy, S. et al.
McCready, R.G.L., see Krouse, H.R. et al.; Laishley, E.J. et al.
McCready, R.G.L., Kaplan, I.R. and Din, G.A., 457, *469*
McCready, R.G.L., Laishley, E.J. and Krouse, H.R., 457, *469*
McCreer, J.M., 3, *18*, 292, *325*

McElcheran, D., see Thode, H.G. et al.
McElroy, M.B., Yung, Y.L. and Nier, A.O., 414, *432*
McElroy, M.B., see Wofsy, S. et al.
McIntyne, A., see Lindrotts, K.J. et al.
McKenzie, W.F. and Truesdell, A.H., 197, 200, 204, 205, *223*
McKinney, C.R., see Urey, H.C. et al.
McKirdy, D.M. and Powell, T.G., 359, 360, *400*
McNaughton, D., see Fritz, P. et al.
Meents, W.F., see Clayton, R.N. et al.; Graf, D.L. et al.
Meiman, J.R., see Judy, C. et al.
Meiman, J.R., Friedman, I. and Hardcastle, K., 165, *177*
Meints, V.W., Shearer, G., Kohl, D.H. and Kurtz, L.T., 416, *432*
Meints, W.V., Boone, L.V. and Kurtz, L.T., 420, *432*
Mekhtieva, V.L., 457, *469*; see also Yeremenko, N.A. and Mekhtieva, V.L.
Mekhtieva, V.L. and Kondrat'eva, E.N., 458, *469*
Mendes Compos, M., see Salati, E. et al.
Menendez, R., see Sackett, W.M. and Menendez, R.
Meneres Lear, J., see Salati, E. et al.
Merkt, J., see Geyh, M.A. et al.
Merkulova, K.I., see Chalov, P.I. and Merkulova, K.I.
Merlivat, L., 37, *46*; see also Facy, L. et al.; Jouzel, J. et al.; Lorius, C. and Merlivat, L.; Lorius, C. et al.; Macklin, W.C. et al.
Merlivat, L., Jouzel, J., Robert, J. and Lorius, C., 145, 160, *177*
Merlivat, L. and Nief, G., 7, *18*, 141, *177*
Merlivat, L., Ravoire, J., Vergnaud, J.P. and Lorius, C., 145, *177*
Metthews, R.K., see Allan, J.R. and Metthews, R.K.
Meybeck, M., Hubert, P., Martin, J.M. and Olive, P., 103, *138*
Meybeck, M., see Hubert, P. et al.
Michaeli, A., see Issar, A. et al.
Micheicek, M., see Sonejkal, V. et al.
Michel, F.A. and Fritz, P., 84, *138*
Mikhailov, B.A., see Starik, I.E. et al.
Miller, L.G., see Lindrotts, K.J. et al.
Mitchell, R.F., 279, *281*
Mitchell, R.H. and Krouse, H.R., 436, *469*

Miyake, Y., Sugimura, Y. and Uchida, T., 279, *281*
Mizutani, Y., 203, 204, 218, *223*; see also Rafter, T.A. and Mizutani, Y.
Mizutani, Y. and Rafter, T.A., 203, 204, *223*, 236, 237, *256*, 303, 304, *325*, 449, 450, 451, 467, *469*
Mo, T., see Sackett, W.M. et al.
Moeller, J., see Ambach et al.; Dansgaard, W. et al.
Mogilewskij, G.A., see Lebedev, W.C. et al.
Möll, M., see Bucher, P. et al.; Oeschger, H. et al.
Monkhouse, R.A., see Downing, R.A. et al.; Smith, D.B. et al.
Monster, J., 351, 368, *400*
Monster, J., see Thode, H.G. and Monster, J.; Thode, H.G. et al.
Mook, G.W., Bommerson, J.C. and Staverman, W.H., 63, *73*, 195, 200, 201, *223*
Mook, G.W. and Eisma, D., 481, *502*
Mook, G.W., Groenveld, D.J., Brown, A.E. and Van Granswijk, A.J., 106, *138*
Mook, G.W. and Koene, B.K.S., 64, *73*
Mook, W.G. and Vogel, J.C., 478, 481, *502*
Mook, W.G., 55, 62, 65, 68, *72*, 475, 478, *502*; see also Lerman, J.C. et al.; Solomons, W. and Mook, W.G.; Solomons, W. et al.; Vogel, J.C. et al.
Moore, H.E., see Hoering, T.C. and Moore, H.E.
Moore, W.S., 279, *281*
Mopper, K. and Garlick, G.D., 302, *325*
Morgan, J.J. see Stumm, W. and Morgan, J.J.
Morgan, T.D., see Scalan, R.S. and Morgan, T.D.
Morgan, V.I., 170, *177*
Mörner, N.A., 128, *138*
Moser, H., see Ambach et al.; Behrens, H. et al.; Martinec, J. et al.
Moser, H., Rauert, W., Stichler, W., Ambach, W. and Eisner, H., 167, *177*
Moser, H., Silva, C., Stichler, W. and Stowhas, L., 150, 153, *177*
Moser, H. and Stichler, W., 44, *46*, 142, 144, 145, 146, 148, 149, 150, 154, 155, 156, *177*

Muehlenbachs, K., 300, *325*
Muffler, L.J.P., see Clayton, R.N. et al.; White, D.E. et al.
Müller, E.P., see May, F. et al.
Muller, H., see Geyh, M.A. et al.
Muller, H.W., see Kronfeld, J. et al.
Müller, P. and Wienholz, R., 374, *400*
Munk, W., see Houtermans, J. et al.
Munkelwitz, H.R., see Castleman, A.W. et al.
Münnich, K.O., 56, *73*, 127, *138*; see also Atakan, Y. et al.; Brinkman, R. et al.; Knetsch, C. et al.; Matthess, G. et al.; Thilo, L. and Münnich, K.O.; Zimmermann, U. et al.
Münnich, K.O. and Roether, W., 56, *73*, 126, *138*
Münnich, K.O., Roether, W. and Thilo, L., 54, 56, *73*
Münnich, K.O. and Vogel, J.C., 127, *135*, 129, *138*
Murata, K.J., Friedman, I. and Gleason, J.D., 311, 312, *325*
Murata, K.J., Friedman, I. and Madsen, B.M., 357, *400*
Murphey, B.F. and Nier, A.O., 331, *400*
Murphy, G.M., see Urey, H. et al.
Musgrave, B.C., see Gunter, B.D. and Musgrave, B.C.
Myiake, Y. and Wada, E., 412, 414, 416, *432*

Nagel, J.F., see Ehhalt, D. et al.
Nagy, B., 351, *400*
Nakai, N., 347, 349, 374, *400*, 485, 486, *502*; see also Deevey, E.S. and Nakai, N.; Deevey, E.S. et al.; Jensen, M.L. and Nakai, N.
Nakai, N. and Jensen, M.L., 237, *257*, 449, 455, *469*
Nakai, N., Wada, H., Kiyosu, Y. and Takimoto, M., 459, *470*
Nakaparksin, S., see Sackett, W.M. et al.
Nathenson, M., see Truesdell, A.H. et al.
Naughton, J.J. and Terada, K., 385, *400*
Nehring, N.L., Bowen, P.A. and Truesdell, A.H., 218, *223*
Nehring, N.L. and Truesdell, A.H., 217, *223*
Nehring, N.L., see Truesdell, A.H. and Nehring, N.L.
Nelson, R.W., see Isaacson, R.E. et al.

Nesmelova, A.N., see Dubrova, N.V. and Nesmelova, Z.N.
Nesteroff, W.D., see Fontes, J.Ch. et al.
Newell, R.S., 24, *47*
Newman, J.W., see Calder, J.A. et al.
Newman, J.W., Parker, P.L. and Behrens, E.W., 351, 352, 353, 355, *401*
Newman, L., see Forrest, J. and Newman, L.
Newman, L., Forrest, J. and Manowitz, B., 455, *470*
Newton, R.C., see Clayton, R.N. et al.
Nicki, H., see Weinstock, B. and Nicki, H.
Nief, G., see Facy, L. et al.; Hagemann, R. et al.; Merlivat, L. and Nief, G.
Nielsen, H., 235, 236, *256*, 439, *470*; see also Fontes, J.Ch. and Nielsen, H.; Nielsen, H. and Gehlen, K.
Nielsen, H. and Ricke, W., 437, *470*
Nielson, R.L., see Sheppard, S.M.E. et al.
Nier, A.O., see MeElroy, M.B. et al.; Murphey, B.F. and Nier, A.O.
Nier, A.O. and Gulbransen, E.A., 331, *401*
Nijampurkar, V., see Martin, J.M. et al.
Nijampurkar, V.N., see Lal, D. et al.
Nir, A., Krugger, S.J., Lingenfelder, R.E. and Flamm, E.J., 22, *46*
Nir, A., see Gat, J.R. et al.
Nissenbaum, A., 346, 347, 353, *401*
Nissenbaum, A. and Kaplan, I.R., 347, 353, *401*, 459, *470*
Nissenbaum, A., Presley, B.J. and Kaplan, I.R., 356, 357, *401*, 459, *470*
Nissenbaum, A. and Refter, R.A., 232, 239, *257*
Nissenbaum, A. and Schallinger, K.M., 346, 347, *401*
Nissenbaum, A., see Mazor, E. et al.
Noble, D.C., see Rosholt, J.N. and Noble, D.C.
Noory, M., see Dinçer, T. et al.
Northrop, D.A. and Clayton, R.N., 285, 307, *325*
Nosik, L.P., see Chukhrov, F.V. et al.
Noto, P., see Celati, R. et al.; Panichi, C. et al.; Tenu, A. et al.
Nriagu, J.O., 455, 466, *470*; see also Pang, P.C. and Nriagu, J.O.
Nuti, S., see Dinçer, T. et al.; Longinelli, A. and Nuti, S.; Tenu, A. et al.

Oana, S. and Deevey, E.S., 332, 347, 358, 363, *401*, 475, 488, *502*
Oakes, D.B., see Young, C.P. et al.
Oehler, D.Z., Schopf, J.W. and Kvenvonden, K.A., 359, *401*
Oehler, D.Z., see Schopf, J.W. et al.
Oescher, H., see Bucker, P. et al.; Martinec, J. et al.; Siegenthaler, U. et al.
Oeschger, H., Gugelmann, A., Loosli, H., Schotterer, U., Siegenthaler, U. and Wiest, W., 210, *224*
Oeschger, H., Gugelmann, A., Loosli, H., Schotterer, U. and Wiest, W., 127, *138*
Oeschger, H., Stauffer, B., Bucher, P. and Loosli, H., 173, *177*
Oeschger, H., Stauffer, B., Bucher, P. and Moell, M., 171, 172, *177*
Ohmoto, H., 3, *18*, 235, 240, *257*
Ohmoto, H. and Rye, R.O., 182, *224*, 231, 235, 248, *257*
Ohmoto, H., see Rye, R.O. and Ohmoto, H.
Olive, P., see Bakalowicz, M. and Olive, P.; Burger, A. et al.; Cheminée, J. et al.; Conrad, G. et al.; Crouzet, E. et al.; Fontes, J.Ch. et al.; Hubert, P. et al.; Meybeck, M. et al.
Olivry, J.C., see Fontes, J.Ch. and Olivry, J.C.
Olsen, D.R., see Puphal, K. and Olsen, D.R.
Olsen, S.R., see Hauck, R.D. et al.
Olson, E.A., see Broecker, W.S. and Olson, E.A.
Olsson, I.U., 14, *18*
O'Neil, J.R., 7, *18*, 141, *177;* see also Clayton, R.N. et al.; Friedman, I. and O'Neil, J.; White, D.E. et al.
O'Neil, J.R. and Clayton, R.N., 289, *325*
O'Neil, J.R., Clayton, R.N. and Mayeda, T.K., 285, 287, *325*
O'Neil, J.R. and Epstein, S., 285, 307, *325*
O'Neil, J.R. and Hay, R.L., 310, *325*
O'Neil, J.R. and Kharaka, Y.F., 294, 296, 297, *325*
O'Neil, J.R. and Taylor, H.P., 285, 293, *325*
Orr, W.L.., 351, 367, 371, *401*, 437, *470*
Orr, W.L. and Gaines, A.G., 459, *470*
Osmond, C.B., 331, *401*
Osmond, C.B., Ziegler, H., Stichler, W.

and Trimborn, P., 332, 343, *401*
Osmond, C.B. and Allaway, W.G. et al.
Osmond, J.K., see Kaufman, M.I. et al.
Osmond, J.K. and Cowart, J.B., 260, 261, 262, 265, 266, 267, 269, 270, 272, 273, 274, 275, 276, 277, *281*
Osmond, J.K., Kaufman, M.I. and Cowart, J.B., 267, *281*
Osmond, J.K., Rydell, H.S. and Kaufman, M.I., 266, *281*
Östlund, G., 237, *257*
Östlund, G., see Ehhalt, D. and Östlund, G.
Otlet, R.L., see Downing, R.A. et al.
Otlet, R.L., see Smith, D.B. et al.
Owsjannikow, W.M., see Lebedev, W.C. et al.

Paloc, H., see Margrita, R. et al.
Pandolfi, L.J., see Libby, L.M. and Pandolfi, L.J.
Pang, P.C. and Nriagu, J., 425, *432*
Panichi, C., see Arana, V. and Panichi, C.; Baldi, P. et al.; Bortolami, G.C. et al.; Celati, R. et al.; Gonfiantini, R. et al.
Panichi, C., Celati, R., Noto, P., Squarci, P., Taffi, L. and Tongiorgi, E., 181, 191, *224*
Panichi, C., Ferrara, G.C. and Gonfiantini, R., 197, 198, 199, 205, 211, 218, *224*
Panichi, C. and Gonfiantini, R., 180, *224*
Panichi, C. and Tongiorgi, E., 211, *224*
Papadimitropoulos, T., see Burdon, D.J. et al.
Papakis, N., see Burdon, D.J. et al.
Park, R. and Dunning, H.N., 351, 368, *401*
Park, R. and Epstein, S., 334, 337, 338, 343, 347, *401*
Parker, P.D.M., see Benson, B.B. et al.
Parker, P.L., 332, 334, 337, 350, 351, 352, *401*; see also Calder, J.A. and Parker, P.L.; Newman, J.W. et al.
Parker, P.L., Behrens, E.W., Calder, I.A. and Shultz, D., 485, *502*
Parwel, A., see Landergren, S. and Parwel, A.
Paterson, W.S.B., see Bucher, P. et al.
Paul, E.A., see Rennie, D.A. and Paul, E.A.; Rennie, D.A. et al.
Payne, B.R., 99, 122, 124, *138;* see also Bradley, E. et al.; Burdon, D.J. et al.; Dinçer, T. and Payne, B.R.; Dinçer, T. et al.; Gonfiantini, R. et al.
Payne, B.R. and Yurtsever, Y., 93, *138*
Pearson, F.J., Jr., 59, 73; see also Bedinger, M.S. et al.; Downing, R.A. et al.; Fisher, D.W. et al.; Hobba, W.A. et al.; Ingerson, E. and Pearson, F.J., Jr.; Rightmire, C.T. et al.; Smith, D.B. et al.; Wigley, T.M.L. et al.
Pearson, F.J., Jr., Bedinger, M.S. and Jones, B.F., 56, *73*, 210, *224*
Pearson, F.J., Jr. and Bodden, M., 217, *224*
Pearson, F.J., Jr. and Fisher, D.W., 237, 238, *257*
Pearson, F.J., Jr. and Friedman, I., 54, *73*, 127, *138*
Pearson, F.J., Jr. and Hanshaw, B.B., 55, 59, 61, 67, *73*, 347
Pearson, F.J., Jr. and Rettman, P.L., 250, *257*
Pearson, F.J., Jr. and Swarzenki, W.V., 56, *73*
Pearson, F.J., Jr. and Truesdell, A.H., 209, *224*
Pearson, F.J., Jr. and White, D.E., 131, *138*
Perry, E.A., see Hower, J. et al.
Perry, E.A. and Hower, J., 315, *325*
Perry, E.C., Jr., 314, *326*
Pessl, K., see Ambach, W. et al.; Behrens, H. et al.
Peters, B., see Lal, D. and Peters, B.
Petrowski, C., see Presley, B.J. et al.
Picciotto, E., see Gonfiantini, R. et al.
Picciotto, E., Cameron, R., Crozaz, C., Deutsch, S. and Wilgain, S., 161, 163, 170, 171, 172, *177*
Picciotto, E., De Maere, X. and Friedman, I., 143, *177*
Picciotto, E., Deutsch, S. and Aldaz, L., 146, 163, *177*
Pierce, F.C., see Koons, C.B. et al.
Pierre, D., see Mariotti, A. et al.
Pilot, J., 410, *432*
Pinder, G.F. and Jones, D.F., 102, *139*
Pineau, F., Javoy, M., Hawkins, J.W. and Craig, H., 300, *326*
Pinson, W.H., see Hurley, P.M. et al.
Plate, A., see Behrens, W. et al.; Eichmann, R. et al.

Plummer, L.N., 62, *73*; see also Wigley, T.M.L. et al.
Plummer, L.N., Jones, B.F. and Truesdell, A.H., 64, 66, *73*
Polach, H.A., see Taylor, C.B. et al.
Polak, B.G., see Kanonov, V.I. and Polak, B.G.
Poplawski, S., see Fritz, P. and Poplawski, S.
Porter, L.K., see Hauck, R.D. et al.
Potter, R.W.I.I. and Clynne, M.A., 213, *224*
Potter, R.W.I.I., Mazor, E. and Clynne, M.A., 213, *224*
Pouchan, P., see Fontes, J.Ch. and Pouchan, P.
Pourchet, M., see Jouzel, J. et al.
Powell, T.G., see McKirdy, D.M. and Powell, T.G.
Prantl, F.A. and Loijens, H.S., 168, *178*
Presley, B.J., see Claypool, G.E. et al.; Nissenbaum, A. et al.
Presley, B.J., Culp, J.H., Petrowski, C. and Kaplan, I.R., 356, *402*
Presley, B.J. and Kaplan, I.R., 356, *401, 402*, 459, *470*
Presley, B.J., Petrowski, C. and Kaplan, I.R., 356, *402*
Prilutsky, R.E., see Bogdanov, Yu.V. et al.
Przewlocki, K., see Bradley, E. et al.; Rozkowski, A. and Przewlocki, K.
Puchelt, H. and Kullerud, G., 3, 16, *18*
Puphal, K. and Olsen, D.R., 279, *281*
Putintseva, V.S., see Veselovsky, V.N. et al.

Queiroz, O., see Lerman, J.C. and Queiroz, O.

Rabinovitch, A.L., 459, *470*; see also Veselovsky, V.N. et al.
Radin, J., see Kronfeld, J. et al.; Yaron, F. et al.
Rafter, T.A., 3, *19*, 218, *224*; see also Bowser, C.J. et al.; Kaplan, I.R. et al.; Mizutami, Y. and Rafter, T.A.; Nissenbaum, A. and Rafter, T.A.; Steiner, A. and Rafter, T.A.; Taylor, C.B. et al.
Rafter, T.A. and Mizutani, Y., 305, *326*
Rafter, T.A. and Wilson, S.H., 207, 212, 218, *224*, 436, *470*

Rama, S., see Athavale, R.N. et al.; Lal, D. et al.
Rankin, J.G., see Sackett, W.M. and Rankin, J.G.
Raschka, H., see Hahn-Weinheimer, P. et al.
Rauert, W., see Ambach, W. et al.; Behrens, H. et al.; Moser, H. et al.
Rauert, W. and Stichler, W., 116, *139*
Raup, D.M., see Weber, N. and Raup, D.M.
Ravoire, J., see Merlivat, L. et al.
Raynaud, D. and Delmas, R., 171, *178*
Raynaud, D. and Lorius, C., 174, *178*
Reardon, E.J. and Fritz, P., 64, *73*
Reardon, E.J., see Fritz, P. et al.
Redfield, A.C., see Friedman, I. et al.
Reed, J.E., see Bedinger, M.S. et al.
Reeh, N., see Dansgaard, W. et al.; Hammer, C.U. et al.
Reeh, N., Clausen, H.B., Dansgaard, W., Gundestrup, N., Hammer, C.U., Johnsen, S.J., 160, *178*
Reeh, N., Clausen, H.B., Gundestrop, N. and Johnsen, S.J., 160, 161, *178*
Rees, C.E., 236, *257*, 421, 424, *432*, 437, 457, *470*
Rees, C.E., see Thode, H.G. et al.
Reid, H.C., 170, *178*
Renaud, A., 144, *177*
Render, F., see Fritz, P. et al.
Rennie, D.A. and Paul, E.A., 417, 418, 420, *432*
Rennie, D.A., Paul, E.A. and Johns, L.E., 421, *432*
Reppmann, E., see Degens, E.T. et al.
Rettman, P.L., see Pearson, F.J., Jr. and Rettman, P.L.
Rex, R.W., see Clayton, R.N. et al.
Rex, R.W. and Goldberg, E.D., 302, *326*
Rex, R.W., Syers, J.K., Jackson, M.L. and Clayton, R.N., 302, *326*
Reynolds, J.H., see Zartman, R.E. et al.
Richards, H.J., see Smith, D.B. et al.
Richet, P., 412, 413, *432*
Richet, P., Bottinga, Y. and Javoy, Y. and Javoy, M., 201, 203, *224*, 228, 229, 231, *257*, 413, *432*
Ricke, W., see Nielsen, H. and Ricke, W.
Riga, A., Van Praag, H.T. and Brigode, N., 416, 417, 418, *432*
Rightmire, C.T., 60, *73*

Rightmire, C.T. and Hanshaw, B.B., 348, *402*
Rightmire, C.T., Pearson, F.J., Jr., Back, W., Rye, R.O. and Hanshaw, B.B., 241, 245, 246, 247, 250, 251, 252, *257*
Rightmire, C.T., Young, H.W. and White, R.L., 186, *224*
Ritchie, J.A., see Ellis, A.J. et al.
Rittenberg, S.C., see Kaplan, I.R. and Rittenberg, S.C.; Kaplan, I.R. et al.
Robbins, R.C., see Robinson, E. and Robbins, R.C.
Robello, A. and Wagener, K., 389, *402*
Robert, J., see Merlivat, L. et al.
Roberts, F.B., see Vogel, J.C. et al.
Roberts, H.H. and Whelan, T., 357, *402*
Robinovitch, A.L., see Veslovsky, V.N. et al.
Robins, N.S., see Mazor, E. et al.
Robinson, B.W., 206, 207, *224*, 229, 230, 231, *257*, 453, *470*
Robinson, E. and Robbins, R.C., 438, *470*
Roblot, M., Chaigneau, M. and Majoube, M., 359, *402*
Roche, M.A., 101, *139*; see also Conrad, G. et al.; Fontes, J.Ch. et al.
Rodhe, H., see Granat, L. et al.
Roedder, E., see Browne, P.R.L. et al.
Roether, W., see Atakan, Y. et al., Münnich, K.O. and Roether, W.; Münnich, K.O. et al.; Zimmermann, U. et al.
Roether, W. and Junghans, H.G., 33, *47*
Roetman, E.L., see Isaacson, R.E. et al.
Rogers, M.A. and Koons, C.B., 351, 354, *402*, 484, *502*
Rogers, M.A., McAlary, J.D. and Bailey, N.J.L., 351, 368, 371, 372, *402*
Rogers, M.A., Van Hinte, J.R. and Sugden, J.G., 351, 352, 355, *402*
Rognon, P., see Gasse, F. et al.
Rognon, P. and Williams, M.A.J., 128, *139*
Rosenfeld, W.D. and Silverman, S.R., 346, 357, 358, *402*
Rosholt, J.N., Doe, B.R. and Tatsumoto, M., 264, 265, *281*
Rosholt, J.N. and Noble, D.C., 278, *281*
Rösler, H.J. and Länge, H., 409, *432*
Ross, D.A., see Degens, E.T. and Ross, D.A.

Ross, L.E., see Stevens, C.M. et al.
Ross, P.J. and Martin, E.A., 408, *432*
Roth, E., see Facy, L. et al; Hageman, R. et al.; Jouzel, J. et al.
Rothe, P., Hoefs, J. and Sonne, V., 481, *503*
Rouhami, I., see Bender, M.M. et al.
Rowe, P.C., see Smith, D.B. et al.
Roy, D.M. and Roy, R., 293, *326*
Roy, R., see Roy, D.M. and Roy, R.
Rozkowski, A. and Przewlocki, K., 124, 125, *139*
Rubin, M. Likins, R.C. and Berry, E.G., 478, *503*
Rubin, M., see Spiker, E.C. and Rubin, M.
Rubinson, M., see Clayton, R.N. et al.
Rubinson, M. and Clayton, R.N., 63, *73*, 477, *503*
Ryan, T., see Junge, C.E. and Ryan, T.
Rydell, H.S., see Kaufman, M.I. et al.; Osmond, J.K. et al.
Rye, R.O., see Ohmoto, H. and Rye, R.O.; Rightmire, C.T. et al.; Schoen, R. and Rye, R.O.; Truesdell, A.H. et al.
Rye, R.O. and Ohmoto, H., 249, *257*
Ryznar, G., Campbell, F.A. and Krouse, H.R., 436, *471*

Sackett, W.M., 346, 351, 352, 353, 354, 365, 374, 383, *402*, 485, *503*; see also Brooks, J.M. et al.; Degens, E.T. et al.; Frank, D.J. et al.; Spalding, R.F. and Sackett, W.M.; Whelan, T. et al.
Sackett, W.M. and Cook, G., 264, 270, *281*
Sackett, W.M., Eadie, B.J. and Exner, M.E., 332, 333, 337, 351, 352, 354, 355, *403*
Sackett, W.M., Eckelmann, W.R., Bender, M.L. and Bé, A.W.H., 332, 333, *403*, 487, *503*
Sackett, W.M. and Menendez, R., 359, *403*
Sackett, W.M., MO, T., Spalding, R.F. and Exner, M.E., 270, *281*
Sackett, W.M., Nakaparksin, S. and Dalrymple, D., 364, 371, *403*
Sackett, W.M. and Rankin, J.G., 355, *403*, 485, *503*
Sackett, W.M. and Thompson, R.R., 346, 350, 352, 353, *403*
Sakai, H., 197, 204, 206, *225*, 231, 248,

257, 436, 453, *470, 471;* see also Claypool, G.E. et al.; Matsubaya, I. et al.
Sakai, H. and Dickson, F.W., 206, 207, *225*
Sakai, H. and Krouse, H.R., 3, *19,* 218, *225*
Sakai, H. and Matsubaya, O., 183, 184, 187, 193, *225*
Sakanone, M. and Hashimoto, T., 261, *281*
Salati, E., Dal Ollio, A., Matsui, E. and Gat, J.R., 42, *47*
Salati, E., Meneres Lear, J., Mendes Compos, M., 57, *73*
Salati, E., see Fritz, P. et al.
Saliege, J.F., see Bortolami, G.C. et al.
Salomons, W., Goudie, A. and Mook, G.W., 67, *73*
Salomons, W. and Mook, W.G., 56, *73*
Salvadori, F., see Martin, J.M. et al.
Sanders, J.E., see Allen, R.C. et al.
Sangster, D.F., 437, *471*
Sasaki, A., see Krouse, H.R. and Sasaki, A.; Krouse, H.R. et al.; Lowe, L.F. et al.
Sasaki, A. and Krouse, H.R., 437, *471*
Sauzay, G., see Bradley, E. et al.; Gonfiantini, R. et al.
Savin, S.M., 288, 293, *326;* see also Douglas, R.G. and Savin, S.M.; Eslinger, E.V. and Savin, S.M.; Yeh, H.W. and Savin, S.M.
Savin, S.M. and Douglas, R.G., 306, *326*
Savin, S.M. and Epstein, S., 288, 289, 290, 294, 295, 296, 299, *326*
Saxena, V.K., see Gupta, M.L. et al.
Scalan, R.S. and Morgan, T.D., 350, 351, *403*
Schallinger, K.M., see Nissenbaum, A. and Schallinger, K.M.
Scharpenseel, H.W., see Tamers, M.A. and Scharpenseel, H.W.
Schaubach, K., see Zimmermann, U. et al.
Schidlowski, M., see Eichmann, R. and Schidlowski, M.; Hoefs, J. and Schidlowski, M.
Schidlowski, M., Eichmann, R. and Fiebiger, W., 359, *403*
Schiegl, W.E., 495, 498, *503*
Schiegl, W.E. and Vogel, J., 3, *19,* 495, *503*

Schneider, J.L., 101, *139*
Schoen, B., see Friedman, I. et al.
Schoen, R. and Rye, R.O., 244, *257,* 458, 459, *471*
Scholes, S.R., see Harned, H.S. and Scholes, S.R.
Schopf, J.W., see Oehler, D.Z. et al.; Smith, J.W. et al.
Schopf, J.W., Oehler, D.Z., Horodyski, R.J. and Kvenvolden, K., 359, *403*
Schotterer, U., see Oeschger, H. et al.; Siegenthaler, U. and Schotterer, U.
Schutte, J.M., see Bredenkamp, D.B. et al.
Schwarcz, H.P., 436, *471;* see also Thompson, P. et al.
Schwarcz, H.P. and Cortecci, G., 234, *257*
Schwarcz, H.P., Harmon, R.S., Thompson, P. and Ford, D.C., 474, *503*
Scott, M.R., 264, 279, *281*
Sears, S.O., 348, *403*
Seiler, W. and Zankl, H., 390, *403*
Sellschop, J.P.F., see Mazor, E. et al.; Verhagen, B.T. et al.
Semet, M.P., see Giletti, B.J. et al.
Setake, H., see Matsubaya, I. et al.
Sevel, T., see Andersen, L.J. and Sevel, T.
Shackleton, N.J., see Vincent, E. and Shackleton, N.J.
Shackleton, N.J., Wiseman, J.D.H. and Buckley, H.A., 306, *326*
Sharma, T. and Clayton, R.N., 13, *19*
Sharp, R.P., see Epstein, S. and Sharp, R.P.; Epstein, S. et al.
Shata, A., see Knetsch, C. et al.
Shazly, M.M., see Knetsch, C. et al.
Shearer, G., Duffy, J., Kohl, D.H. and Commoner, B., 416, 421, *432*
Shearer, G.B., Kohl, D.H. and Commoner, B., 417, 418, 420, *432*
Shearer, G.B., see Feigin, A.D. et al.; Kohl, D.A. et al.; Meints, V.W. et al.
Sheppard, S.M.E., Nielson, R.L. and Taylor, H.P., Jr., 290, *326*
Shima, M., Gross, W.H. and Thode, H.G., 436, *471*
Shima, M., see Thode, H.G. et al.
Shimp, N.F., see Clayton, R.N. et al.
Shultz, D.J., see Calder, J.A. et al.; Parker, P.L. et al.
Siegel, D., see Zimmermann, U. et al.

Siegenthaler, U., 169, *178;* see also Martinec, J. et al.; Oescher, H. et al.
Siegenthaler, U., Oeschger, H. and Tongiorgi, E., 92, *139*
Siegenthaler, U. and Shotterer, U., 99, *139*
Sigurgeirsson, T., see Arnason, B. and Sigurgeirsson, T.
Silva, C., see Fritz, P. et al.; Moser, H. et al.
Silverman, S.R., 360, 363, 368, 369, 370, 371, 372, *403;* see also Rosenfeld, W.D. and Silverman, S.R.; Sondheimer, E.W.A. et al.
Silverman, S.R. and Epstein, S., 337, 351, 359, 365, 367, 368, 374, *403*
Simeon, C., see Bosch, B. et al.
Sironi, G., see Colombo, U. et al.
Siwertz, E., see Blavoux, B. and Siwertz, E.; Crouzet, E. et al.; Hubert, P. et al.
Skinner, H.C.W., see Clayton, R.N. et al.
Sklash, M., see Fritz, P. et al.
Sklash, M.G., Farvolden, R.N. and Fritz, P., 106, *139*
Smejkal, V., see Krouse, H.R. et al.
Smejkal, V., Cook, F.D. and Krouse, H.R., 457, 458, *471*
Smejkal, V., Michelicek, M. and Krouse, H.R., 459, *471*
Smiegocki, R.T., see Bedinger, M.S. et al.
Smith, B.N., 338, 347, *403*; see also Brown, W.V. and Smith, B.N.; Hsu, J.G. and Smith, B.N.; Jacobson, B.S. et al.; Tregunna, E.B. et al.
Smith, B.N. and Benedict, C.R., 334, 337, *404*
Smith, B.N. and Brown, W.V., 332, *404*
Smith, B.N. and Epstein, S., 331, 332, 337, 352, *404*, 487, 495, *503*
Smith, D.B., see Downing, R.A. et al.
Smith, D.B., Downing, R.A., Monkhouse, R.A., Otlet, R.L. and Pearson, F.J., Jr., 60, *73*, 84, 130, 132, *139*
Smith, D.B., Wearn, P.L., Richards, H.J. and Rowe, P.C., 89, 90, *139*
Smith, D.G.W., see Fritz, P. and Smith, D.G.W.
Smith, G., see Friedman, I. and Smith, G.
Smith, G.I., see Matsuo, S. et al.
Smith, J.L., see Krouse, H.R. and Smith, J.L.
Smith, J.W., 439, *471*; see also Bailey, S.A. and Smith, J.W.; Chambers, L.A. et al.
Smith, J.W. and Batts, B.D., 438, *471*
Smith, J.W., Schopf, J.W. and Kaplan, I.R., 359, *404*
Smith, N.N., see Jacobson, B.S. et al.
Smith, P.V. and Hudson, B.E., 410, *432*
Sofer, Z. and Gat, J., 3, *19*
Soller, R., Friedman, I., Machta, L. and Soller, R.
Somayajulu, B.L.K., see Lal, D. et al.
Sondheimer, E.W.A., Dance, L.R., Mattick, L.R. and Silverman, S.R., 352, *404*
Sonne, V., see Rothe, P. et al.
Sonntag, C., see Matthess, G. et al.
Sontag, E., see Fischer, W. et al.
Sordi, S., see Longinelli, A. and Sordi, S.
Spalding, R.F., see Sackett, W.M. et al.
Spalding, R.F. and Sackett, W.M., 270, *281*
Spiker, E.C. and Rubin, M., 494, *503*
Spindel, W., 412, *433;* see also Stern, M.J. et al.
Spiridonov, A.I. and Tyminskiy, V.G., 276, *281*
Spiridov, A.I., see Alekseev, F.A. et al.
Squarci, P., see Celati, R. et al.; Panichi, C. et al.
Squires, R., see Kvenvolden, K. and Squires, R.
Stahl, W., 199, *225*, 332, 347, 348, 361, 362, 363, 374, 377, 378, 383, 385, *404*; see also Boigk, H. and Stahl, W.; Boigk, H. et al.; Wendt, J. et al.
Stahl, W., Aust, H. and Dounas, A., 93, *139*
Stahl, W. and Carey, B.D., 374, 383, *404*
Stahl, W. and Jordan, R., 479, *503*
Stahl, W. and Koch, J., 383, *404*
Stahl, W. and Tang, C.H., 374, 383, *404*
Stahl, W., Wollanke, G. and Boigk, H., 410, 413, *433*
Stanton, R.J., see Dodd, J.R. and Stanton, R.J.
Starik, F.E., see Starik, I.E. et al.
Starik, I.E., Starik, F.E. and Mikhailov, B.A., 261, *281*
Starinsky, A., see Yaron, F. et al.
Starkey, R.L., see Jones, G.E. and Starkey, R.L.
Stauffer, B., see Bucher, P. et al.;

Oescher, H. et al.
Staverman, W.H., see Mook, G.W. et al.
Steiner, A., see Clayton, R.N. and Steiner, A.
Steiner, A. and Rafter, T.A., 212, *225*
Stern, M.J., Kauder, L.N. and Spindel, W., 412, *433*
Stevenson, C.M., see Maklin, W.C. et al.
Stevens, C.M., Krout, L., Walling, D., Venters, A., Engelkemeir, A. and Ross, L.E., 391, 392, 393, *404*
Stewart, J.R., see Bailey, I.H. et al.
Stewart, M.K., 37, *47*, 181, 183, 215, *225*, 291, 292, 309, *326*
Steyn, P.L., see Delwiche, C.C. and Steyn, P.L.
Stichler, W., see Ambach, W. et al.; Behrens, H. et al.; Herrmann, A. and Stichler, W.; Martinec, J. et al.; Moser, H. and Stichler, W.; Moser, H. et al.; Osmond, C.B. et al.; Rauert, W. and Stichler, W.
Stichler, W. and Herrmann, A., 157, *178*
Stiller, M., 486, 494, *503*
Stoker, A.K., see Kruger, P. et al.
Stoker, A.K. and Kruger, P., 213, 219, *225*,
Stone, C.G., see Bedinger, M.S. et al.
Stowhas, L., see Moser, H. et al.
Strakhov, N.M., 459, *471*
Stuiver, M., 347, 349, *404*, 482, 484, 486, 487, 488, 490, 491, 492, 494, *503*; see also Covich, A. and Stuiver, M.; Deevey, E.S. and Stuiver, M.; Deevey, E.S. et al.
Stuiver, M. and Deevey, E.S., 331, *404*
Stumm, W. and Morgan, J.J., 234, *257*
Suess, E., see Bainbridge, A.E. et al.; Cain, W.F. and Suess, H.E.; Erlenkeuser, H. et al.; Hautermans, J. et al.
Suess, H.E., 387, *404*
Sugden, J.G., see Rogers, M.A. et al.
Sugimura, Y., see Miyake, Y. et al.
Sukhija, B.S. et al.
Surkov, Yr.A., see Baranov, V.L. et al.
Süss, M., see Fischer, W. et al.
Suszczynski, E.F., 113, *139*
Suzuki, O., see Fritz, P. et al.
Suzuoki, T. and Epstein, S., 290, 291, 320, *326*
Swarzenki, W.V., see Pearson, F.J., Jr. and Swarzenki, W.V.

Sweeney, R.E., Liu, K.K. and Kaplan, I.R., 407, *433*
Syers, J.K., see Clayton, R.N. et al.; Rex, R.W. et al.
Syromayatnikov, N.G., see Dement'yev, V.S. and Syromayatnikov, N.G.
Syromyatnikov, N.F., 261, *282*

Tabatabai, M.A., see Bremner, J.M. and Tabatabai, M.A.
Taffi, L., see Celati, R. et al.; Panichi, C. et al.
Takimoto, M., see Nakai, N. et al.
Tamers, M.A., 59, 73, 74, 127, *139*, 478, *503*
Tamers, M.A. and Scharpenseel, H.W., 59, 74
Tan, F.C., see Allen, P. et al.
Tan, F.C. and Hudson, J.D., 479, 481, *503*
Tang, C.H., see Stahl, W. and Tang, C.H.
Tarutani, T., Clayton, R.N. and Mayeda, T., 477, *503*
Tataru, S., David, G. and Filip, G., 279, *282*
Tatsumoto, M., see Rosholt, J.N. et al.
Tauber, H., 331, *405*
Taylor, C.B., 24, 33, 42, 44, *47*; see also Brown, L.J. and Taylor, C.B.; Gonfiantini, R. et al.
Taylor, C.B., Polach, H.A. and Rafter, T.A., 208, *225*
Taylor, H.P., Jr., 181, 182, 183, *225*, 498, *503*; see also Lawrence, J.R. and Taylor, H.P., Jr.; Margaritz, M. and Taylor, H.P., Jr.; O'Neil, J.R. and Taylor, H.P., Jr.; Sheppard, S.M.E. et al.; Wenner, D.B. and Taylor, H.P., Jr.
Taylor, H.P., Jr. and Epstein, S., 3, *19*, 294, *326*
Taylor, H.P., Jr., Frechen, J. and Degens, E.T., 385, *405*
Tazioli, G.S., see Cotecchie, V. et al.
Teichmüller, M., see Boigk, H. et al.; Colombo, U. et al.; Teichmüller, R. et al.
Teichmüller, R., see Boighk, H. et al.; Colombo, U. et al.
Teichmüller, R., Teichmüller, M., Colombo, U., Gazzarrini, F., Gonfiantini, R. and Kneuper, G., 361, 362, 363, 364, 382, *405*

Teis, R.V., 232, 236, *257*
Tenu, A., Noto, P., Cortecci, G. and Nuti, S., 130, 132, *139*
Terada, K., see Naughton, J.J. and Terada, K.
Theodorsson, P., 171, 172, *178;* see also Arnason, B. et al.
Theuwen, H.J., see Bokhoven, C. and Theuwen, H.J.
Thilo, L., see Münnich, K.O. et al.
Thilo, L. and Münnich, K.O., 56, *74*
Thode, H.G., see Harrison, A.G. and Thode, H.G.; Kemp, A.L.W. and Thode, H.G.; Macnamara, J. and Thode, H.G.; Tudge, A.P. and Thode, H.G.; Shima, M. et al.
Thode, H.G., Craig, C.B., Hulston, J.R. and Rees, C.E., 3, *19*, 206, *225*, 231, *257*
Thode, H.G., Dunford, H.B. and Shima, M., 436, *471*
Thode, H.G., Kleerekoper, H. and McElcheran, D., 457, *471*
Thode, H.G., Macnamara, J. and Collins, C.B., 435, *471*
Thode, H.G. and Monster, J., 437, *471*
Thode, H.G., Monster, J. and Dunford, H.B., 436, *471*
Thode, H.G., Wanless, R.K. and Wallouch, R., 461, *471*
Thompson, G.M., Lumsden, D.N., Walker, R.L. and Carter, J.A., 276, *282*
Thompson, P., see Schwarcz, H.P. et al.
Thompson, P., Ford, D.C. and Schwartz, H.P., 276, 278, *282*
Thompson, R.R., see Sackett, W.M. and Thompson, R.R.
Thorstenson, D.C., 234, *257*
Thurber, D.L., 261, 264, 276, *282*
Thurber, D.L. and Broecker, W., 489, *503*
Thürkauf, M., see Kuhn, W. and Thürkauf, M.
Togliatti, V., see Gonfiantini, R. et al.
Tongiorgi, E., see Colombo, U. et al.; Dinçer, T. et al.; Gonfiantini, R. et al.; Martinec, J. et al.; Panichi, C. and Tongiorgi, E.; Panichi, C. et al.; Siegenthaler, U., et al.
Tregunna, E.B., Smith, B.N., Berry, J.A. and Downton, W.J.S., 331, *405*
Trimborn, P., see Ambach, W. et al.; Osmond, C.B. et al.

Troughton, J.H., 60, *74*, 332, 334, 342, 343, *405*
Troughton, J.H., Hendy, C.H. and Card, K.A., 332, *405*
Trudinger, P.A., see Chambers, L.A. et al.
Truesdell, A.H., 196, 197, *225*, 243, *257*; see also Fournier, R.O. and Truesdell, A.H.; Fournier, R.O. et al.; Mazor, E. et al.; McKenzie, W.F. and Truesdell, A.H.; Nehring, N.L. and Truesdell, A.H.; Nehring, N.L. et al.; Pearson, F.J., Jr. and Truesdell, A.H.; Plummer, L.N. et al.; White, D.E. et al.
Truesdell, A.H. and Fournier, R.O., 197, 200, 205, 209, *225*
Truesdell, A.H. and Jones, B.F., 64, *74*
Truesdell, A.H., Nathenson, M. and Rye, R.O., 186, 188, 189, 190, 191, 197, 205, 209, *225*
Truesdell, A.H. and Nehring, N.L., 192, 212, *226*
Truesdell, A.H. and White, D.E., 191, *225*
Tudge, A.P., 3, *19*, 305, *326*
Tudge, A.P. and Thode, H.G., 231, *258*, 437, 453, *471*
Tyminskiy, V.G., see Spiridonov, A.I. and Tyminskiy, V.G.
Tzur, Y., 38, *47*
Tzur, Y., see Gat, J.R. and Tzur, Y.

Uchida, T., see Miyake, Y. et al.
Umana, A., see Kruger, P. et al
Umemoto, S., 264, *282*
Usitalo, E., 292, *326*
Ustinov, V.I., see Vinogradov, A.P. et al.
Urey, H.C., 228, 229, 231, 232, *258*, 412, *433*; see also Epstein, S. et al.
Urey, H.C., Brickwedde, I.G. and Murphy, G.M., 21, *47*
Urey, H.C., Lowenstam, H.A., Epstein, S. and McKinney, C.R., 56, *74*, 475, 476, *504*
Url, M., see Ambach, W. et al.

Van, N.H. and Lalou, C., 276, 278, 279, *282*
Van Donk, J., see Lindrotts, K.J. et al.
Van Donk, J. and Mathieu, M., 356, *405*
Van Everdingen, R.O. and Krouse, H.R., 461, *471*
Van Granwijk, A.J., see Mook, G.W. et al.
Van Hinte, J.E., see Rogers, M.A. et al.
Van Praag, H.T., see Riga, A. et al.

Van Urk, H., see Vogel, J.C. and Van Urk, H.
Vedy, J.C., see Mariotti, A. et al.
Veeh, H.H., 270, 276, *282*; see also Dymond, J. et al.
Veizer, J., 479, *504*
Venters, A., see Holt, B.D. et al.; Stevens, C.M. et al.
Vergnaud, J.P., see Merlivat, L. et al.
Verhagen, B.T., see Mazor, E. et al.
Verhagen, B.T., Mazor, E. and Sellschop, J.P.F., 83, *139*
Veselsky, J., 279, *282*
Veselovsky, V.N., Rabinovitch, A.L. and Putintseva, V.S., 436, 451, *471*
Vilenskii, V.D., see Baranov, V.L. et al.
Vincent, E. and Shackleton, N.J., 306, *326*
Vines, H.M., see Bender, M.M. et al.
Vinogradov, A.P. and Galimov, A.M., 351, 363, 368, 368, *405*
Vinogradov, A.P., Grinenko, V.A. and Ustinov, V.I., 459, 461, 471
Vitti, T.C., see Gaebler, O.H. et al.
Voge, H.H., 229, 230, *258*
Vogel, J.C., 57, 61, 67, *74*, 127, 130, *139*; see also Bredenkamp, D.B. and Vogel, J.C.; Brinkman, R. et al.; Ehhalt, D. et al.; Emrich, K. et al.; Knetsch, C. et al.; Lerman, J.C. et al.; Mook G.W. and Vogel, J.C.; Münnich, K.O. and Vogel, J.C.; Schiegl, W.C. and Vogel, J.C.
Vogel, J.C. and Ehhalt, D., 55, 61, *74*
Vogel, J.C., Grootes, P.M. and Mook, G.W., 63, *74*
Vogel, J.C., Lerman, J.C. and Mook, W.G., 99, *139*
Vogel, J.C., Lerman, J.C., Mook, W.G. and Roberts, F.B., 130, *139*
Vogel, J.C. and Van Urk, H., 131, *139*
Volynets, V.P., Zadorozhny, I.K. and Florenski, K.P., 410, *433*
Von Buttlar, H. and Libby, W.F., 22, *47*
Vredenburgh, L.D. and Cheney, E.S., 351, 367, 371, *405*, 437, *471*
Vumirovich, R., see Gaebler, O.H. et al.

Wada, E., see Myiake, Y. and Wada, E.
Wada, E. and Hattori, A., 425, *433*
Wada, E., Kadonaga, T. and Matsuo, S., 415, 416, 424, *433*
Wada, H., see Nakai, N. et al.

Wagener, K., see Rebello, A. and Wagener, K.
Wakshal, E. and Yaron, F., 274, *282*
Walker, R.L., see Thompson, G.M. et al.
Wallick, E.I., 60, *74*
Walling, D., see Stevens, C.M. et al.
Wallouch, R., see Thode, H.G. et al.
Walton, A., see Broecker, W.S. and Walton, A.
Wanless, R.K., see Thode, H.G. et al.
Waring, S.A., see Black, A.S. and Waring, S.A.
Warren, C.G., see Granger, H.C. and Warren, C.G.
Wasserburg, G.J., see Mazor, E. and Wasserburg, G.J.; Zartman, R.E. et al.
Wasserburg, G.J., Mazor, E. and Zartman, R.E., 199, *226*, 363, 374, 385, *405*
Wearn, P.L., see Smith, D.B. et al.
Weber, J.J., see Keith, M.L. and Weber, J.J.
Weber, J.N., 314, *326*, 360, *405*, 476, *504*; see also Keith, M.L. and Weber, J.N.
Weber, J.N. and La Rogue, A., 479, 480, *504*
Weber, J.N. and Raup, D.M., 476, *504*
Weiler, R.R. and Chawla, V.K., 270, *282*
Weinstock, B. and Nicki, H., 391, *405*
Wellman, R.P., see Cock, F.D. et al.
Wellman, R.P., Cook, F.D. and Krouse, H.R., 416, *433*
Welte, D.H., 360, 368, 369, 370, 372, *405*
Welte, D.H., Kalkreuth, W. and Hoefs, J., 360, *405*
Wendt, I., see Gey, M.A. and Wendt, I.
Wendt, I., Stahl, W., Geyh, M.A. and Fauth, F., 57, 61, *74*, 127, *140*
Wenner, D.B. and Taylor, H.P., Jr., 285, 318, *326*, 327
West, K.E., see Hattersley-Smith, G. et al.
West, K.E., and Krouse, H.R., 170, *178*
West, S.S., 365, *405*
Weston, R.E., Jr., 141, *178*
Weyer, K.U., see Fritz, P. et al.
Whelan, T., see Roberts, H.H. and Whelan, T.
Whelan, T., Sackett, W.M. and Benedict, C.R., 337, 339, *405*
White, D.E., 181, 183, 197, 215, *226*, 385, *405*; see also Craig, H. et al.; Doe, B.R. et al.; Fournier, R.O. et al.;

Pearson, F.J., Jr. and White, D.E.; Truesdell, A.H. and White, D.E.
White, D.E., Barnes, I. and O'Neil, J.R., 184, *226*
White, D.E., Fournier, R.O., Muffler, L.T.P. and Truesdell, A.H., 187, *226*
White, D.E., Muffler, L.P.J. and Truesdell, A.H., 187, 193, *226*
White, J.L., see Ledroux, R.L. and White, J.L.
White, R.L., see Rightmire, C.T. et al.
Whitlock, D.W., see Eckelmann, W.R. et al.
Wickman, F.E., 331, 332, 342, 347, 351, 361, 362, 365, *405*; see also Krejci-Graf, K. and Wickman, F.E.
Wienholz, R., see Müller, P. and Wienholz, R.
Wiesberg, L., see Freyer, H.D. and Wiesberg, L.
Wiest, W., see Oeschger, H. et al.
Wigley, T.M.L., Plummer, L.N. and Pearson, F.J., Jr., 65, 66, 68, *74*
Wilgain, S., see Picciotto, E. et al.
Wilkinson, W.B., see Cole, J.A. and Wilkinson, W.B.
Willey, L.M., see Mariner, R.H. and Willey, L.M.
Williams, D.F., 306, *327*
Williams, J.A., 351, 359, 367, *405*
Williams, M.A.J., see Rognon, P. and Williams, M.A.J.
Williams, P.M., 353, *405*
Williams, P.M. and Gordon, L.I., 332, *405*
Willkomm, H., see Erlenkeuser, H. et al.
Willkomm, H. and Erlenkeuser, H., 59, *74*
Wilson, A.J. and Grinsted, M.J., 495, 499, *504*
Wilson, S.H., 207, 209, 212, *226*
Wilson, S.H., see Rafter, T.A. and Wilson, S.H.
Winograd, I.J. and Farlekas, G.M., 55, 59, *74*
Wiseman, J.D.H., see Shackleton, N.J. et al.
Wlotzka, F., 407, *433*
Wodzicki, A., see Browne, P.R.L. et al.
Wolfgang, R.L., see Grosse, A. et al.
Wofsy, S., McConnell, J.C. and McElroy, M.B., 391, *405*
Wollanke, G., Behrens, W. and Horgan, T., 382, *405*, 410, *433*
Wollanke, G., see Boigk, H. et al.; Stahl, W. et al.

Wollenberg, H.A., 275, *282*
Wong, Chong, G.M.. and Loehr, R.C., 423, *433*

Yaniv, A., see Kronfeld, J. et al.; Yaron, F., et al.
Yapp, C.J., see Epstein, S. et al.
Yapp, C.J. and Epstein, S., 495, 498, 499, *504*
Yaron, F., see Wakshal, E. and Yaron, F.
Yaron, F., Kronfeld, J., Gradsztan, E., Radin, J., Yaniv, A., Zach, R., Starinsky, A., Kolodny, Y. and Katz, A., 262, *282*
Yeh, H.W. and Epstein, S., 296, 298, *327*
Yeh, H.W. and Savin, S.M., 285, 287, 296, 298, 299, 315, 316, 317, *327*
Yen, C.K., see Bradley, E. et al.
Yeremenko, N.A. and Mekhtieva, V.L., 457, *471*
Yermilova, L.P., see Chukhrov, F.V. et al.
Young, C.P., Hall, E.S. and Oakes, D.B., 84, *140*
Young, H.W., see Rightmire, C.T. et al.
Yund, R.A., see Giletti, B.J. et al.
Yung, Y.L., see McElroy, M.B. et al.
Yurtsever, Y., see Bradley, E. et al.; Payne, B.R. and Yurtsever, Y.
Yurtsever, Y., 30, 31, 33, *47*

Zach, R., see Kronfeld, J. et al.; Yaron, F. et al.
Zadorozhny, I.K., see Volynets, V.P. et al.
Zak, I., see Claypool, G.E. et al.
Zankl, H., see Seiler, W. and Zankl, H.
Zartman, R.E., see Wasserburg, G.J. et al.
Zartman, R.E., Wasserburg, G.J. and Reynolds, J.H., 374, *405*
Ziegler, H., see Osmond, C.B. et al.
Zimmermann, U., Ehhalt, D. and Münnich, K.O., 41, *47*
Zimmermann, U., Münnich, K.O. and Roether, W., 85, 86, 87, *140*
Zimmermann, U., Münnich, K.O., Roether, W., Kreutz, W., Schaubach, K. and Siegel, D., 85, *140*
Zuppi, G.M., see Bortolami, G.C. et al.; Fontes, J.Ch. and Zuppi, G.M.
Zuppi, G.M., Fontes, J.Ch. and Létolle, R., 93, 95, 96, *140*, 241, 242, 243, 244, *258*
Zverev, V.L., see Alekseev, F.A. et al.
Zyukun, A.M., see Matrosov, A.G. et al.

SUBJECT INDEX

Ablation, 154, 165
Activity ratios, 17, 259—279
Adsorption of sulphate, 455
Age dating
    geothermal waters, 207—210
    groundwater, 57—66, 70, 125—132
    sediments, 492
    snow and ice, 160, 169—172
Alabama, 350
Alanin, 341
Alberta, Canada, 443, 460, 462
Algae, 332, 343
Algeria, 108, 129
Alpha recoil, 261
Alps, 116, 144, 153—159
Altitude effect
    precipitation (^{18}O, ^2H and ^3H), 30, 40, 77, 92—97, 144, 172
    ^{13}C in plants, 342
    pseudo-altitude effect, 44
Altitude of recharge, 92—97
Amazon basin, 42
Amino acids, 330, 335—340
Ammonium fertilizers, 414
Amount effect, 30, 38, 40—44, 78, 146
Analytical techniques, 3
    gases in geothermal systems, 217
    geothermal waters, 217
    nitrogen isotopes, 408
    ^{234}U and ^{238}U analysis, 278
Ancient seawater, 183
Anhydrite, 234, 287, 308
Animal matter
    carbon isotopes, 51, 353
    nitrogen-15, 417
Annual layering of snow and ice, 156, 159, 160
Antarctica, 42, 143—147, 156, 168, 171—173, 309, 352, 482

Anthracite, 361
Anthropogenic sulphur compounds, 465
Apatite, 305
Aquatic vegetation, 332, 491
Aqueous carbonate, 51, 54, 57, 61, 197, 340, 475, 497, 484
Aqueous sulphur compounds (^{18}O and ^{34}S), 227, 235—238, 436, 449, 463
Aquifer interactions, 272
Aragonite, 476, 480
Arctic, 333
Argentina, 69, 130
Argon, in geothermal systems, 213
Argon-39, 127
    concentration in ice, 170
    geothermal systems, 210
    ice dating, 170
Arima, Japan, 187
Arkansas, 210
Artesian water, 109
Aspartic acid, 330
Asphaltenes, 368
Atlantic Ocean, 354, 473
Atlas Plateau, Algeria, 129
Atmosphere-ocean exchange, 28
Atmospheric carbon compounds, 384
    carbon dioxide, 15, 51, 52, 54, 69, 387
    carbon monoxide, 391—392
    carbon tetrachloride, 384
    methane, 390—392
Atmospheric moisture, 26, 33, 36
Atmospheric oxygen, 236
Atmospheric sulphur compounds, 443—449
Australia, 32, 69, 88
Austria, 149, 159, 165, 167

Bacterial carbon isotope fractionation,

346, 358, 374, 379
Bacterial sulphate reduction, 232
Bacterial sulphur isotope fractionation, 232, 456—461
Bahamas, 307, 350
Baltic Sea, 354, 493, 494
Barbados, 308
Barites, 304, 436
Basalt
 carbon, 212
 submarine, 300
Base flow, 168
Bavaria, 163
Bay of Quinte, Ontario, 455
Beggiatoa, S isotopes, 456, 458, 461
Belt Chert, 314
Bentonite, 289
Bermuda, 306
Beta activity in precipitation and snow, 158—159
Bicarbonate
 groundwater, 54, 57, 61, 125
 oceans, 51, 340
 surface waters, 488—494
 see also Isotope fractionation
Biogenic
 calcium phosphates, 286, 476
 carbon dioxide, 15—54, 60, 154, 335—338, 347, 356
 silica, 286
 sulphur deposits, 461
Biologic sulphur isotope fractionation, 456—461
Biopolymers, 344
Biotite, 290
Bitumen, 368—369
Black Sea, 41, 490, 492
Blue-green algae, 335, 487
 ^{13}C fractionation, 342—343
Biodegradation, ^{13}C effects, 372
Bomb tritium, 22—26, 79, 81, 143, 171, 461
Borax, 291
Brachiopods, 476
Brackish environment, 481
Brine, 125, 183
British Honduras, 350
Broadlands, New Zealand, 182, 187, 188, 198, 212, 215
Bruneau-Grandview, Idaho, 186, 204
Burial depth
 silica diagenesis, 314
 ^{13}C in marine sediments, 352

Burial diagenesis, 354
Butane, 369

Calcite, *see* Carbonate; Isotope fractionation
Calcite precipitation in aquifers, 68
Calcium phosphates, biogenic, 285, 476
Calcrete, 67
California, 183, 186, 188, 191, 193, 205, 208, 215, 307, 311, 318, 459, 481, 489
Calvin-Benson cycle, 60, 329
CAM cycle, 331
Cambrian
 evaporates, 236, 308, 438
 petroleum, 438
Camp Century, Greenland, 156, 171
Campi Flegrei, Italy, 185
Canada, 166, 169, 239, 461
Canadian Shield, 488
Canary Islands, 93, 115
Canon Diablo meteorite, 16, 436
Carbohydrate formation, 330
Carbon-14, 2, 14, 49—71
 abundance, 50
 correction, 51, 494
 decay, 52
 general, 14, 15
 geothermal systems, 209—210
 geothermal water, 187
 groundwater dating, 53—70, 125
 ice dating, 169, 170
 production, 49
 reference standard, 14, 51
 specific activity, 15, 50, 52
 surfacewaters, 488, 492, 489, 494
 terrestrial plants, 69
Carbon cycle, 50, 57
Carbon dioxide
 atmosphere, 52, 54, 69, 384, 385, 387
 coalification, 361, 362
 decay of organic matters, 356
 fossil fuel, 388
 geothermal systems, 211, 385
 natural gas, 376
 oxygen, 12, 13
 respiration, 154, 335, 337, 338
 soil gas, 15, 54, 60, 347, 348
 volcanic-magmatic, 56, 199, 385
Carbon isotopes, 2, 477
 reference standards, 14, 51
Carbon isotope effects
 biodegradation of petroleu, 372

carbon dioxide and plant components, 337, 340, 341
coalification, 361, 362, 364, 365
deasphalting of petroleum, 372
diffusion of methane, 381, 382
influence of water on CAM plants, 343
natural gas, 375, 376
oil maturation, 370, 371
petroleum maturation, model for, 371, 372
photosynthesis, 338, 339, 341, 342, 343, 344
— carbon dioxide availability, 341, 343
— environmental conditions, 341
— light level, 341, 342, 343
— oxygen availability, 342, 343
— pH, 341, 342
— temperature, 341, 342
— water availability, 342, 343, 344
pyrolysis, 364
thermal alteration of petroleum, 372
thermal cracking, 371, 383
total plant and plant components, 336, 337, 341
water washing of petroleum, 372
see also Isotope fractionation
Carbon isotope geothermometers, 198
Carbon monoxide, 390—392
Carbon reservoirs, 50
Carbon tetrachloride, 384—385
C_3 cycle, 60, 329
plants, 332
C_4 cycle, 329
plants, 332
Carbonate
carbonatite, 212, 386, 436
diogenetic, 306, 357
freshwater, 386, 474—477, 490—492
geothermal, 215
marine, 54, 306—308
precipitate in aquifers, 68
shells, 474—477, 480
soil, 67
see also Isotope fractionation
Carbonate aquifers, 56
Carbonate dissolution, 54, 114
Carbonatities, 212, 386
Carboniferous
coal, 364
coal gas, 384
natural gas, 199
natural gas-^{15}N, 414

Caribbean, 374
Carrizo aquifer, Texas, 277
Cellulose
^{13}C, 335, 341, 388, 495—496
^{2}H, 495
Central Africa, 32, 99
Central Europe, 85
Central Italy, 99
Chad Basin, 99
Charcoal, 361
Chert, 290, 310—314
Chile, 153, 186, 188, 192, 193
Chlorine-36
concentration in ice, 170
ice dating, 170
Chlorite, 315
Chromatium, 450, 456, 458
Clays
detrital vs. authigenic, 298, 302, 316
mineral-water isotope exchange, 285—302
Climatic variations, 168, 172, 474
Closed system, ^{14}C groundwater dating, 58
Clostridium, S isotopes, 456, 457, 458
Cloud models, 37
Coal
C isotopes, 360—361
S isotopes, 438
Coal gas, 362, 378
Coccoliths, 306
Coccothophores, 335
Concretions
barite, 305, 436
phosphorite, 305
pyrite, 436
Condensation
isotope effects, 33
Connecticut, 240
Continental air, 41
Continental effects (^{2}H and ^{18}O), 28, 79, 81, 93
Continental rain, 81
Continental Intercalaire, Algeria, 108
Coorong Lagoon, Australia, 307
Cosmic radiation, 22
Crassulacean Acid Metabolism (CAM) cycle, 60, 331
Cretaceous
barite, 305
evaporites, 438
marine, 315
petroleum, 438

Crude oils, 365
Crustaceans, 476
Crystal lattice damage, 262
Cyprus, 318
Cyrenaican Lakes (Libya), 461
Czechoslovakia, 103, 164, 459

d-parameter, 32, 40—44, 129
Dead carbon effect, 491
Dead Sea, 459
Dead Sea Rift, Israel, 209
Death Valley chert, 314
Deep Sea Drilling Project, 299, 309, 311, 319, 459
Deep-seated carbon, 212
Deep Springs Lake, California, 307
Delta definition, 5
Denitrification, 416
Denmark, 87
Detrital carbonates, 492
Detrital clay, 316, 298—302
Deuterium, 6
    isotope effects, see Isotope fractionation
    magmatic water, 181
    plant matter, 495
    precipitations, 29—33
    reference standard, 11
    thermal waters, 181
    troposphere, 33
Deuterium excess, 32, 40, 129
Devon Island, Canada, 169
Devonian
    evaporites, 308, 438
    petroleum, 438
Deweylite, 289
Diagenesis, 298, 307, 314, 344, 346, 479
    burial, 314
    carbonate, 306, 357
    cherts, 310—314
    organic matter in soils (^{15}N), 418
    shales, 314—317
    silica minerals, 311
Diamonds, 386
Diatoms, 303, 335
Diffusion
    ^{14}C, 128
    H_2S, 253
    in ice, 156—158
    natural gas, 381, 382
    — methane, ^{13}C effects, 381
    nitrogen, 382, 413
    precipitations, 37

Dihydroxyacetone phosphate (DHAP), 330
Dilution-factor, ^{14}C groundwater dating, 59
Dinoflagelates, 335
Diphosphoglyceric acid, 330
Direct run off, 102, 164—167
Discharge hydrograph, 102—107, 164—167
Dispersion, 82, 125
Dissolution-exchange model (^{14}C), 58, 60—68
Dissolved inorganic carbon, 15, 57—71
Djibouti, 83, 491
Dolomite formation, 307
    see also Isotope fractionation
Dryopteris, ^{13}C latitude dependence, 342

East Africa, 310
East Pacific Rise, 301
Echinoderms, 476
Edwards aquifer, Texas, 250—254
Egypt, 129
El Tatio, Chile, 182, 186—188, 192, 193
Endogenous carbon, 345, 352
England, 89, 130
Enrichment factor, 5, 165
Eocene
    evaporites, 246
    marine sediments, 320
Equatorial Pacific, 320
Equilibrium fractionation factors, see Isotope fractionation
Estuaries, 354, 481
Ethane, 356, 368—373
Ethiopian Plateau, 99
Europe, 28, 69, 117, 192
Evaporation, 3, 28, 33—37, 41, 76, 78, 82, 85, 98, 308
    closed basins, 121, 308
    geothermal systems, 185, 192
    groundwater, 108—113
    in fractured rocks, 117
    lakes, 98—102
    oceans, 41
    rain, 33—37, 43, 93
    Sebkha, 121
    snow, 148, 149
    soils, 85
    tritium in precipitation, 28
Evaporite minerals, 235, 236, 246, 308, 436—438

Evapotranspiration, 26, 41, 83, 89, 91
Exchange reaction, see Isotope exchange; Isotope fractionation
Exogenous carbon, 345
Exponential model, 125

Fatty acids, 336
Fertilizers, 414, 420
Firn
   dating, 171
   profiles, 156, 158
Flood hydrograph, 102, 119—120
Florida, 245, 265—268, 271—275, 350
Florida Bay, 307
Floridan aquifer, 245—250, 276
Flow rates based on ^{14}C, 131
Fluid inclusions, 181
Food effect, 478
Foraminifera, 306
Formation waters, 182, 320
Fossil brine in mines, 125
Fossil fuel, 52, 360—384
Fractionation effects, see Isotope fractionation
Fractionation factor, 5
   see also Isotope fractionation
Fractured rock hydrology, 113
France, 83, 89, 116, 132, 144, 425
Franciscan Formation, California, 318
Freon, 384
Freshwater carbonates, 386, 474—477, 490—492
Freshwater environments, 480
Freshwater mollusks, 478—480
Fructose-6-phosphate, 330
Fructose diphosphate, 330
Fulvic acids, 344, 346
Fumarole, 181, 192

Gas migration, 381
Gelbstoff, 353
Geothermal systems, 57, 179—219
   brines, 183
   calcite, 215
   ^{14}C, 209
   carbon isotopes, 211
   fluids, 115, 179—219
   tritium, 208
   methane, 385
   rare gases, 213
   uranium, 275
   water-rock ratio, 215

Geothermometer, 3, 6, 195—207, 312, 317
   calcite-water, 475
   carbon dioxide, 198, 205
   carbon dioxide-dissolved bicarbonate, 200
   hydrogen isotope, 201
   hydrogen-water, 202
   methane-hydrogen, 202
   oxygen isotope, 203
   quartz-clay, 317
   serpentine, 318
   serpentine-magnetite, 318
   sulphate-water, 204
   sulphur isotope, 206
   water-steam, 205
Germany, 87, 91, 199, 410
Geysers, 182, 187, 191, 193, 205, 208—212, 215
Gibbsite, 289, 294, 295
Glacial drift gas, 199
Glacier, 145, 165—172
   age determination, 169—172
   discharge, 165
   enrichment effects, 165
   homogenization, 165
   ice formation, 154
   residence time of water, 167
   run-off, 166
Glauconite, 288
Glutamic acid, 340
Glycine, 341
Grain size and isotopic fractionation, 287
Grand Erg Occidental, Algeria, 108
Graphite, 359
Great Lakes, 269, 482—494
Greece, 93
Greenland, 130, 144, 156, 160, 168, 171
   Camp Century, 156, 171
   ice sheets, 171, 482
Groenedel IAEA-WMO station, 30
Groningen, The Netherlands, 69
Ground air carbon dioxide, 61
Groundwater, 24
   age dating
     — age limit, 70
     — ^{14}C, 49—71, 125—132
     — tritium, 125—132
     — uranium, 276
   ^{15}N, 425
   recharge, 58, 82—98
   residence times, 126, 167

storage, 167
sulphur isotopes, 234, 240—254, 458—461
Growler, California, 192
Gulf of Mexico, 315, 352, 354, 484, 485
Gunflint Chert, 314
Gyttia, 488
Gypsum, 233, 308—310

Hail, 29, 33—39
Half-life, 2
   ^{3}H, 2, 170
   ^{14}C, 2, 53, 170
   ^{32}Si, 170
   ^{35}S, 227
   ^{36}Cl, 170
   ^{39}Ar, 170, 210
   ^{210}Pb, 170
   ^{222}Rn, 213
   ^{226}Ra, 263
   ^{230}Th, 263
   ^{234}Pa, 263
   ^{234}Th, 262, 263
   ^{234}U, 2, 261, 263
   ^{235}U, 2
   ^{238}U, 261, 263
Halloysite, 293
Halmyrolysis, 298
Hammat Gader, Israel, 187, 210
Hardwater effect, 59, 488, 492
Hatch-Slack cycle, 60, 329
^{3}He/^{4}He ratios, 214
Helium in geothermal systems, 213
Hemi-cellulose, 335
Heptane, 371, 373
Hexane, 373
Hoarfrost, 148
Hodna Plain, Algeria, 109
Holocene
   climate, 128, 484
   lake sediments, 484
   marine sediments, 307, 350
Homogenization of snow and ice profiles, 157, 165
Hornblende, 290
Hot springs, 95, 181, 209
Humic acid, 55, 344, 346
Humic substances in marine sediments, 353
Humidification, 346
Hungary, 109
Hurricane, 38

Hydration of gypsum, 233
Hydrocarbons in natural gas, 375
Hydrogen isotope fractionation, 197, 290
   see also Isotope fractionation
Hydrogen isotope shift, 182
Hydrogen isotopes, 2
   reference standard, 11, 21
Hydrogen sulphide, 228, 235, 253, 437, 439, 453, 456, 462
Hydrothermal calcite, 210, 215
Hydroxides, 297, 301
Hypsithermal interval, 484

IAEA-WMO network, 22, 23 30
Ice and snow, 6, 141—173
Ice lenses, 151, 157
Ice sheets, 168, 172
Iceland, 115, 151, 182, 186, 202
Idaho, 182, 186, 204
Illite, 182, 186, 204
In-storm variations (^{18}O), 146
India, 187
Indian Ocean, 41
Industrial sulphur compounds, 449, 465
Initial ^{14}C (groundwater dating), 57
Ion exchange, 415
Iron oxides/manganese oxides, 301
Isoleucine, 341
Isotope effects
   during soil formation, 294
   during weathering, 294
   see also Carbon isotope effects; Isotope fractionation
Isotope exchange, 3, 5, 15, 56, 60, 150, 156, 490
   calcite-aqueous carbon, 56, 60, 182, 490
   clay-pore fluids, 315
   clay-water, 293—298
   CO_2-H_2O, 200
   CO_2-HCO_3, 60, 63, 200
   CO_2 (aq)-CO_2 (gas), 60, 200
   detrital clay-seawater (^{18}O), 298
   ice-vapour, 150
   mixed layer illite-smectite, 293
   rock-water, 185
   SO_4-H_2O, 204
   sulphate-sulphide, 229, 453
   sulphate-water, 230, 232
   sulphite-water, 233
   water-ice, 151
   see also Isotope fractionation

Isotope exchange reaction, 3, 5
  see also Isotope fractionation
Isotope fractionation, 3—6
  carbon-13 and carbon-14
  — CO_2-CH_4, 195, 197
  — CO_2 (aq)-CO_2 (gas), 477
  — CO_3^{2-}-CO_2 (gas), 477
  — HCO_3^--CO_2 (gas), 197, 477
  — SO_3^{2-}-H_2O, 304
  — see also Carbon isotope effects
  deuterium
  — biotite-water, 290
  — borax-water, 291
  — $CaSO_4$-water, 233
  — CH_4-H_2, 197, 201
  — chert-water, 290
  — deweylite-water, 289
  — gibbsite-water, 289
  — glauconite-water, 288
  — hornblende-water, 290
  — H_2O_l-CH_4, 201
  — H_2O_l-H_2, 197, 201
  — H_2O_l-H_2O_s, 7, 150
  — H_2O_v-CH_4, 201
  — H_2O_v-H_2, 201
  — H_2O_v-H_2O_l, 7, 34, 188, 189, 203
  — manganese nodule-water, 289
  — muscovite-water, 290
  — smectite-water, 289
  — trona-water, 291
  nitrogen-15, 412
  oxygen-18
  — anhydrite water, 287
  — biogenic silica water, 286
  — $CaSO_4$-water, 233
  — calcite-aragonite, 476, 491
  — calcite-phosphoric acid, 14
  — — dependence on grain size, 287
  — calcite-water, 215, 285, 287, 475
  — CO_2-H_2O_l, 12, 197, 203
  — deweylite-water, 289
  — dolomite-water, 285
  — H_2O_l-H_2O_s, 7, 150
  — H_2O_v-H_2O_l, 7, 34, 188, 189, 194, 203
  — HSO_4-H_2O_l, 203
  — illite-water, 285, 287, 288
  — kaolinite water, 289
  — magnetite-water, 289
  — manganese nodule-water, 289
  — mirabilite-water, 291
  — mixed-layer illite/smectite-water, 285
  — muscovite-water, 285
  — phillipsite-water, 288
  — phosphate-water, 287
  — proto-dolomite water, 285
  — quartz-water, 215, 285, 287, 288
  — sepolite-water, 289
  — serpentine-magnetite, 318
  — serpentine-water, 285
  — shell-phosphate water, 286
  — smectite-water, 285, 287, 289
  — SO_4^{2-}-water, 287
  — SO_4^{2-}-H_2O_l, 197, 203
  sulphur-34
  — H_2SO_4-H_2S, 206, 231
  — HS^--H_2S, 206, 231
  — HSO_4^--H_2S, 231
  — S^{2-}-H_2S, 231
  — SO_2-H_2S, 206
  — SO_4^{2-}-H_2S, 197, 206
  — SO_4^{2-}-H_2S, 231
  — see also Sulphur isotope fractionation
  uranium-234, 261
Isotopic composition
  carbon, 2
  hydrogen, 2
  nitrogen, 2
  oxygen, 2, 4
  sulphur, 2
  strontium, 2, 16
  uranium, 2
Isotopic equilibrium, 2
  see also Isotope fractionation
Israel, 187, 209, 274, 494
Italy, 93, 95, 131, 185, 198, 207, 208, 237—241

Jamaica, 307
Japan, 183—193, 204, 485
Jurassic
  evaporites, 438
  petroleum, 438
Juvenile carbon, 212

Kalahari desert, 83
Kaolinite, 289, 294, 295
Karsts, 114, 117—121, 241, 245, 267
Katagenesis, 345
Kenya, 101
Kerogen, 357
Kimberlite carbonate, 386
Kinetic isotope effects, 4, 6, 37, 194, 292, 340, 413, 453—456

Kranz photosynthetic cycle, 329
Krypton in geothermal systems, 213

Lacustrine plants, 332
Lake Abbe, 477, 491
Lake Biwa, Japan, 485
Lake Chad, 83, 101, 480
Lake Erie, 269, 482, 484
Lake Geneva, 41, 85, 103, 491
Lake Huron, 269, 484
Lake Michigan, 269
Lake Ontario 269, 493, 494
Lake sediments, 349, 486
   calcium carbonate, 490, 492
   ^{13}C, 347—349, 484, 487, 493
   ^{14}C, 488, 492, 493
   sulphur isotopes, 240
Lake Superior, 269
Laken Vanda, Antarctica, 459
Lakes, 28, 310, 484
   suphur isotopes, 239, 459
Lanzarote, California, 182
Larderello, Italy, 182—212
Lassen Volcanic Park, 192—214
Latitude effect
   precipitation, 24, 30, 40, 43, 143
   ^{13}C in plants, 342
Lead-210
   concentration in ice and snow, 170, 171
   ice dating, 170, 171
Lead-zinc deposits, S isotopes, 437
Leakage from
   lakes, 99—102, 119
   rivers, 98—99
Leucine, 341
Lichens, S isotopes, 463
Lignin, 335, 354, 388
Lignite, 361
Lipid fraction
   ^{13}C, 335, 341, 370
   ^{2}H, 495
Lithification and ^{13}C in sediments, 359
London Basin, U.K., 89, 130
Long Valley, California, 186
Los Angeles, California, 69
Louisiana, 461
Lysimeter, 85
Lysine, 341

Madras, India, 461
Magadite, 310

Magmatic carbon dioxide, 56
Magmatic water, 180
Magnetite, 289
Malate, 330
Malic acid, 330
Maketu, New Zealand, 199
Manganese nodules, 289, 301, 304
Manikaran, India, 187
Manitoba, 106
Marine
   atmosphere, 40
   bicarbonate, 51, 340
   carbonates, 54, 306
   organisms, 237
   phosphates, 305
   plants, 332, 333
   rain, 30, 81
   sediments, 307, 350—353
   sedimentation, 298
   sulphates, 235—237, 246, 303—309, 436, 450, 459
Marls, 490—492
Mars (^{15}N), 414
McKenzie River, Canada, 239, 488
Mean residence time of groundwater, 125
Mediterranean Sea, 28, 29, 41, 78, 241
Melarchez experimental basin, France, 425
Melt water, 151—154
Melting of ice and snow, 142
Membrane filtration, 183
Mesozoic
   chert nodules, 312
   evaporite, 308
Meta-sediments, 359
Metamorphic carbon dioxide, 210
Metamorphic waters, 183
Metamorphism, 334, 345
Meteoric water line, 32, 40, 43, 77
Meteorites, S isotopes, 436
Methane
   atmospheric, 390
   ^{14}C groundwater dating, 56
   coalification, 362
   decay of organic matter, 488
   diffusion, 381—382
   fossil fuel, 363, 369, 377, 378
   geothermal drift gas, 199, 363
   natural gas, 374
   sediments, 356, 363
   soil gas, 363

volcanic, 385
Methane-forming bacteria, 357
Mexico, 124, 482
Middle East, 32, 117
Migration of natural gas, 413
Mine water, 125
Mineral-water interactions, 283—321
Miocene
    diatomite, 303
    evaporites, 308, 309
Mirabilite, 291, 309, 310
Mississipi, 350
Mississipi River, 270
Mississippian
    evaporites, 308, 438
    petroleum, 438
Mixing between aquifers, 108
Modern ^{14}C, 51
Molecular diffusion, 36, 85, 89, 157
Molecular exchange, 28, 38
Mont Blanc, 116, 144, 147
Monterey Formation, 311
Morocco, 108
Mosses, S isotopes, 463
Mount Kilimanjaro, 144
Muscovite, 285

Natural gass, 363, 373
    ^{13}C, 363—373
    ^{15}N, 408, 410
    ^{34}S, 437, 443
NBS-1 standard, 11—13, 21
NBS-1A standard, 12, 13
NBS-14 standard, 16
NBS-20 standard, 12, 14
NBS-21 standard, 14
NBS-120 standard, 16
NBS-987 standard, 17
Neon in geothermal systems, 213
Nevada, 200
New York State, 488
New Zealand, 69, 109, 183, 198—215, 237
Nicaragua, 93
Nitrates
    aqueous, 420
    minerals, 410
    soils, 420
Nitrogen-15, 2, 407—433
    abundance, 2, 407—429
    analytical techniques, 408
    biochemical process, 415
    geothermal, 213
    isotopes, 2, 407
    magmatic/metamorphic rocks, 410
    natural gas, 408, 410
    oxidation of nitrogen compounds, 415
    precipitations, 424
    reference standard, 16, 407
    surface water, 425
Nitrogen isotope fractionation, 412
    aqueous and gaseous compounds, 412
    gas migration, 382
    see also Isotope fractionation
Nord IAEA-WMO station, 30
North African plateau, 109
North America, 30, 32, 499
North Pacific, 299, 302
Northern Africa, 117, 499
Northern Atlantic, 32, 352, 482
Northern Hemisphere, 23, 24, 50, 69, 80
Nubian sandstones, 108
Nuclear bomb fallout
    ^{14}C, 52
    tritium, 22, 24, 79, 143, 171, 461

Ocean
    evaporation, 40, 41
    vapour exchange, 33
    see also Marine
Oceanic serpentines, 318
Octane, 371
Ohnuma, Japan, 204
Oil maturation, 371
Oman, 83
Ontario, 106, 489
Opaline silica, 302
Open system ^{14}C groundwater dating, 58, 60
Ophiolite, 318
Ordovician
    evaporites, 438
    petroleum, 438
Organic matter, 54, 329—393
    decay, 347, 358
    diagenesis, 344, 356
    dissolved organic carbon
    — seawater, 353
    — freshwater, 484
    H/C ratio, 345
    ^{2}H and ^{18}O, 495
    metamorphism, 345
    metamorphosis, 360
    ^{15}N, 417

peat, 348
 sediments, 344, 350, 484
 soils, 348
Ostracods, 476
Otake, Japan, 188, 204
Ottawa, Canada, 22, 89
Oxalacetic acid, 330
Oxalic acid standard, 14, 51
Oxidation
 inorganic sulphur compounds, 232, 244, 303, 455, 456—461
 nitrogne compounds, 415
 organic matter, 55
 organic sulphur, 239
 uranium compounds, 262
Oxygen isotopes
 abundances, 2—4
 atmospheric oxygen, 236
 isotope effect, see Isotope fractionation; Isotope exchange
 precipitations, 29—33
 reference standard, 11
Oxygen isotope shift, 97, 182, 183, 187, 192

Paleoclimate, 79, 128, 474
 ^{13}C in lake sediments, 484—485, 349
 ^{13}C in marine sediments, 354, 355
 freshwater mollusks, 482
 ^{2}H in organic matter, 495
 ice cores, 172
 ^{18}O in organic matter, 495
Paleohydrology, 128
Paleotemperatures, 306, 349, 473, 475
Paleotemperature equation, 475
Paleowaters, 129, 499
Paleozoic
 chert nodules, 312
 see also Cambrian; Devonian; Mississippian; Ordovician; Pennsylvanian; Permian; Silurian
Pampa del Tamarugal, Chile, 99
Paraffin, 369
Paris Basin, France, 83, 89
PDB standard, 11—14
Peat, 55, 348, 462, 495
Pectin, 335, 336
Pedosphere, S isotopes, 461—465
Pennsylvanian
 evaporites, 438
 petroleum, 438
Pentane, 373

Perch Lake, Ontario, 487
Permafrost, 84
Permian
 evaporites, 308, 438
 natural gas, 199
 natural gas-^{15}N, 414
 petroleum, 438
Persian Gulf, 41
Petroleum
 ^{13}C, 351, 365—372
 ^{34}S, 436—438
Petroleum gases, 378
Phanerozoic chert, 313
Phenylalanine, 341
Phillipsite, 288, 301
Phosphate, 286, 476
Phosphate nodules, 262, 305
Phosphoenolpyruvate carboxylase, 331
Phosphoenolpyruvic acid, 330
Phosphoglyceraldehyde, 330
Phosphoglyceric acid, 330
Photo-respiration of plants, 337
Photosynthesis, 329, 486—489
 see also Carbon isotope effects
Phytoplankton, 492
Piston flow, 88, 89
Plankton, 333, 339
 ^{13}C dependence on environment, 342, 355
 lacustrine, 333
 marine, 332—333
Plants
 ^{13}C and altitude, 342
 ^{13}C and environmental conditions, 338—343
 carbon isotopic composition
 — lacustrine, 332
 — marine, 332
 — terrestrial, 15, 60, 69, 331
 sulphur isotopes, 463
 total plant versus components, 336—341
Pleistocene, 168, 306, 307
 climate, 302
 limestones, 306
 marine sediments, 320, 355
 organic matters in marine sediments, 353
Poland, 125
Polar diagrammes, δ vs. wind direction, 447—449
Polar glacier, 154—157

Pollen, 334
Polymerization, 371
Pore water, 319—320, 356—357
Porphyrins, 368
Potomac River, U.S.A., 494
Power plant emissions, S isotopes, 439, 455
Precambrian
  cherts, 314
  evaporites, 438
  graphite, 359, 360
  kerogen, 359
  organic matter, 360
Precipitations
  ^{2}H, ^{3}H and ^{18}O isotope contents, 21—44
  — altitude effect, 30, 40, 77, 92—97, 144, 172
  — amount effect, 30, 38, 40, 78, 146
  — latitude effect, 24, 30, 40, 43, 143
  — seasonal variations, 23, 29, 30, 77, 156, 167
  — topographic effect, 93
  vapour exchange, 29
  nitrogen compounds, 408, 424
  sulphur compounds, 237, 244, 450, 455
Preservation of isotope content in mollusk shells, 480
Pretoria, South Africa, 69
Propane, 369—374
Proterozoic
  evaporites, 308
Proto-dolomite, 285
Pseudo-altitude effect, 44
Pyrenees, 120
Pyrite
  oxidation, 234
  S isotopes, 436
Pyruvate, 330

Qatar, 83
Quarternary
  climates, 128
  glaciation, 128
  marine sediments, 302, 354
  soils, 296, 297
Quartz
  detrital vs. authigenic, 302
  geothermal systems, 215
  see also Isotope fractionation
Quartz-clay isotopic temperatures, 317
Queechy Lake, New York State, 488

Radiolarian tests, 302
Radium-226 in calcite, 214
Raft river, Idaho, 204
Rain, see Precipitation
Rare gases, 213
Rate of isotope exchange, 229, 293, 303
Rayleigh fractionation, 6—8, 10, 34, 38, 40, 68
Reaction half-life, 230
Reaction rates, 229
Recharge rate, 85—81, 125
Recoil transfer, 261—264, 272
Recrystallization of shell carbonate, 479
Red Sea, 320
Reducing barriers (uranium isotopes), 272, 274
Reduction
  sulphate, 55, 247, 249, 304
  sulphur compounds, 55, 232, 303, 456
Reduction of snow cover, 153
Reef carbonates, 307
Relative groundwater ages, 66
Relative humidity, 28, 79
Reservoir characteristics, 120
Residence time
  groundwater in aquifers, 126, 167
  sulphate in seawater, 303
  uranium in seawater, 270
Respiration carbon dioxide, 335
Rhodopseudomonas, 456, 458
Ribulose, 330—331
River infiltration, 98
Rivers, 98
  ^{13}C, 488—489, 494
  ^{14}C, 489, 494
  ^{15}N, 427
  sediments (^{13}C), 350
  sulphur isotopes, 239, 449—453
Rocky Mountains, 166
Roosevelt hot springs, Utah, 188, 205
Root respiration, 54
Ross Sea, Antarctica, 354
Roumania, 132
Rubidium, 17
Run-off
  hydrograph separation, 102—107, 119—120
  nitrogen, 427
  snowmelt and glaciers, 164—167
  tritium balance, 164
  uranium, 265, 266

Saanich Inlet, British Columbia, 357, 459

Sacchoromyces cerevisiae, S isotopes, 456, 458
Sahara, 83, 108, 121, 127, 129
Saline aquifer, 275
Salinization of groundwater, 121—125
Salmonella, 456, 458
Salton Sea, California, 182, 202, 210—215
Saskatchewan, 460
Saudi-Arabia, 83, 109
Schiermonnikoog, The Netherlands, 127
Sea level, 128
Seawater-freshwater interface, 307
Seawater intrusions, 125, 275
   see also Marine
Sebkha el Melah, Algeria, 121
Sediments
   calcium carbonate, 490
   ^{14}C dating, 488, 492
   lakes, 276, 349, 486, 490—492
   marine, 276, 307, 350—353
   organic matter in, 344, 484
   river, 350
   siliceous, 302
   uranium, 276
Seeds, 334
Seine River (France), 489
Semi-arid regions, 60
Separation
   multistage, 191
   single-stage steam, 202
   steam, 189, 191, 193
Sepiolite, 288, 289, 318, 285
Serpentinization, 317
Sewage oxidation, 460
Shales, 314—317
Shell carbonate 474—477, 480
Sierra Nevada, California, 144, 186
Siliceous sediments, 302
Silicon-32, 127, 169
   ice dating, 169
Silurian
   evaporites, 438
   petroleum, 438
Sinai Peninsula, 83, 499
SLAP standard, 11—14
Smectite, 289, 296, 301, 315
SMOW standard, 11—13, 21
Snow, 29, 39, 141—173
   age determinations, 160, 169—172
   accumulation rate, 156, 157, 160
   cover, 142
   cover isotopes mass balance, 161—163
   evaporation, 149, 165
   lysimeter, 152, 163
   melt, 102, 148, 164
   pack, 151
Solid state diffusion, 56
Soil
   air, 54
   carbon dioxide, 15, 54, 60, 347, 348
   carbonates, 54, 56, 61—65
   formation, 294
   moisture, 28, 84—91, 272
   ^{15}N, 417, 418—424
   organic matter, 346—347, 418—424
   sulphur compounds, 461
   sulphur isotopes, 238
   water, 85, 272
   zones, 54, 57
Sour gas, S isotopes, 443—450
South America, 69
South Atlantic, 320
South Australia, 307
South Pacific, 302
Southern Hemisphere, 23, 24, 50, 69, 80
Steamboat Springs, California, 182, 200, 201, 210—212
Stratosphere, 24—26, 79, 446
Strontium isotopes, 2, 16, 17
Sublimation of snow, 142
Submarine volcanic rock, 300
Sudan, 99
Suess effect, 52
Sugar, 335
Sulphur bacteria, 456—461
Sulphate
   adsorption on sediments, 455
   reduction, 55, 232, 303, 456
   surface waters, 239, 241—245, 449—453, 459
   see also Evaporite minerals; Groundwater; Marine; Precipitation; Sediments
Sulphate oxygen (^{18}O), 236, 308
Sulphide
   aqueous, 240—242, 247, 461
   minerals, 236
Sulphur-35, 227
Sulphur compounds
   antrhopogenic, 443—449, 465
   atmospheric, 443
   volcanic, 244
   see also Sulphate; Sulphur isotopes

Sulphur cycle, 439—442
Sulphur geochemistry, 234, 440
Sulphur isotopes
   abundance, 2, 227, 435
   evaporites, see Evaporite minerals
   fractionation
   — biologic processes, 456—461
   — industrial processes, 455
   — hydrosphere, 458—461
   — kinetic effects, 453—455
   — see also Isotope fractionation
   geothermal systems, 206—213
   groundwater, 235, 238
   oceans, 235, 436
   petroleum, 438
   plants, 463
   precipitations, 237, 244, 449
   reference standard, 16, 436
   rivers, 238, 449—452
   soils, 238, 461—463
   springs, 241—245, 459
Sulphuretem, 461
Summer ablation, 152
Summer rains, 30, 79
Switzerland, 92, 99, 120, 148, 150, 164

Tank experiments
   mollusks, 478
Tanzania, 90
Tar, 368—369
Temperate glacier, 145, 154—159
Terrestrial humic substances, 353
Terrestrial sediments, 345—348
Tertiary
   natural gas, 199
   phosphate nodules, 305
   soils, 296, 297
   volcanic glass, 300
Texas, 131, 250, 277, 350, 461
The Netherlands, 106, 481
Thermal convection, 183
Thermal cracking, 371, 410
Thermal water, 180
Thermonuclear tests, 22, 24, 79
Thule IAEA-WMO station, 30
Transvall, 89
Tree rings
   ^{13}C, 50, 53, 388, 495
   ^{2}H, 495—499
   ^{34}S, 464
Triassic
   evaporites, 236—238, 438
   petroleum, 438

Tritium, 14, 22, 25, 26
   activity, 14, 22
   atmosphere, 26, 27, 29
   concentration in ice, 170
   continental effects, 28
   geothermal water, 187
   groundwater, 24, 75—134
   groundwater dating, 125
   hail, 29
   hot springs, 209
   ice dating, 170
   latitude dependence, 24
   ocean, 24, 26, 27
   precipitation, 22—29, 69, 79—81
   production, 22, 79
   relative humidity, 28
   run-off, 164
   shallow groundwater, 84
   snow, 29, 145
   snow cover, 164
   soil moisture, 28, 84—91
   standard, 14
   stratosphere, 24, 26
   thermo-nuclear, 22—26, 79, 81, 143, 171, 461
   troposphere, 24
Tritium unit, 14, 22
Trolite, 16
Trondheim, Norway, 69
Troodos ophiolite complex, Cyprus, 318
Tropical environments, 480
Tropical island stations, 33
Tropical stations, 23, 24
Troposphere, 24, 25, 33, 49, 79
Turkey, 117
Tuscany, 239
Tyrosine, 341

Ultramafic rocks, serpentinization, 317
Unconfined aquifer, 127
Unsaturated zone, 54—56, 60—61, 82—85
Uranium
   activity ratio, 17, 259—279
   continental water, 260—269
   decay series, 263
   disequilibrium, 259—279
   geochemical cycle, 266
   geochemistry, 259
   in recharging water, 266, 272
   isotopes, 2, 17
   leaching, 263, 272
   mixing diagrams, 267
   oceans, 276, 271

orebodies, 272
oxidation, 262
phosphate, 271
precipitation, 272
Uranium-234 excess, 264—271
$^{234}U/^{230}Th$ activity ratio, 265
$^{238}U/^{230}Th$ activity ratio, 265
$^{234}U/^{238}U$ activity ratios, 17, 259—279
Uranium-thorium fractionation, 264
Utah, U.S.A., 188, 205

V-SMOW standard, 11—14, 21
Vegetation, see Plants
Vertebrates, 476
Vienna, Austria, 30
Virginia, U.S.A., 209
Vital effects in mollusks 293, 476, 478
Volcanic
    carbon dioxide, 56, 199, 385
    glass, 300
    methane, 385
    sulphur compounds, 244, 436

Wairakei, New Zealand, 182, 198, 204—215

Washington, 494
Washington State, 89
Water of crystallization
    borax, 291
    gaylussite, 291
    gypsum, 233, 291, 309
    mirabilite, 291
    trona, 291
Water/rock ratio in geothermal systems, 215
Wax, 369
Weathering, 266, 294
Wieser, Idaho, 186
Winter precipitations, 79
Wisconsin ice sheet, 499
Wyoming, U.S.A., 186, 189, 190, 193, 200, 209

Xenon in geothermal systems, 213
Xylite, 361

Yellowstone, Wyoming, 182, 186—190, 200—204, 209, 274
Yukatan Peninsula, Mexico, 482